Progress in Nonlinear Differential Equations and Their Applications

Volume 2

Partial Differential Equations and the Calculus of Variations

Essays in Honor of Ennio De Giorgi

Volume II

Edited by
F. Colombini
A. Marino
L. Modica
S. Spagnolo

1989

Birkhäuser
Boston · Basel · Berlin

Editors
Ferruccio Colombini, Antonio Marino,
Luciano Modica, Sergio Spagnolo
Dipartimento di Matematica
Universita di Pisa
Via F. Buonarroti 2
56100 Pisa
Italy

Library of Congress Cataloging–in–Publication Data
Partial differential equations and the calculus of variations : essays
in honor of Ennio De Giorgi / Ferruccio Colombini . . . [et al.].
p. cm. — (Progress in nonlinear differential equations and
their applications)
ISBN 0-8176-3424-X (v. 1 : alk. paper). — ISBN 0-8176-3425-8 (v.
2 : alk. paper). — ISBN 0-8176-3426-6 (set : alk. paper)
1. Differential equations, Partial. 2. Calculus of variations.
3. Giorgi, Ennio De. I. Giorgi, Ennio De. II. Colombini, F.
(Ferruccio) III. Series.
QA377.P297 1989
515′.353—dc20 89-9746

Printed on acid-free paper.

Volume I
ISBN 0-8176-3424-X
ISBN 3-7643-3424-X

Volume II
ISBN 0-8176-3425-8
ISBN 3-7643-3425-8

2-Volume set
ISBN 0-8176-3426-6
ISBN 3-7643-3426-6

Camera-ready copy prepared by the editors using TEX.
Printed and bound by Edwards Brothers, Ann Arbor, Michigan.
Printed in the U.S.A.

9 8 7 6 5 4 3 2 1

A Ennio De Giorgi
i suoi allievi ed amici
con affetto

Courtesy of Foto Frassi, Pisa.

Contents

Volume I

Volume II

ON A CLASS OF NONLINEAR DIAGONAL ELLIPTIC SYSTEMS WITH CRITICAL GROWTH AND C^α-REGULARITY

JENS FREHSE

Dedicated to Ennio De Giorgi on his sixtieth birthday

1. Introduction and statement of theorem 1. In this paper we consider diagonal uniformly elliptic systems

$$\text{div}\,(A\nabla u_\nu) = a_\nu(., u, \nabla u) \quad , \quad \nu = 1, \ldots, N, \tag{1.1}$$

$u = (u_1, \ldots, u_N)$ in a bounded open subset of $\mathbf{R}^n, n \geq 2, N \geq 2$. The right hand side of (1.1) is allowed to have *quadratic growth* in ∇u, namely

$$|a(x, u, \nabla u)| \leq K_C |\nabla u|^2 + f_C(x) \ , \ x \in \Omega, \ |u| \leq C \tag{1.2}$$

with some constant K_C, and a function f_C which is contained in the Morrey space $L^{1,q}$ for some $q > n - 2$. By the latter we mean that

$$\int_{B_R} |f_C| dx \leq K R^q \text{ for all balls } B_R \subset \Omega \text{ of radius } R.$$

Concerning the regularity hypotheses we shall assume

$$a_{ik} \in L^\infty(\Omega) \ , \ A = (a_{ik})_{i,k=1}^n, \text{ and} \tag{1.3}$$

(1.4) $\quad a_\nu(x,u,\eta)$ satisfies the Carathéodory conditions

i.e. a_ν is measurable in x and continuous in $(u,\eta) \in \mathbf{R}^n \times \mathbf{R}^{Nn}$. Furthermore we require the ellipticity condition

$$\xi^T A\xi \geq \lambda|\xi|^2 \quad , \quad \xi \in \mathbf{R}^n, \tag{1.5}$$

with some constant $\lambda > 0$.

By a bounded local weak solution u of (1.1) we mean a function

$$u \in L^\infty \cap H^{1,2}_{\text{loc}}(\Omega;\mathbf{R}^N)$$

such that

$$-\int \nabla\varphi^T A\nabla u_\nu dx = \int a_\nu(.,u,\nabla u)\varphi dx \ , \ \nu = 1,...,N,$$

for all $\varphi \in C_o^\infty(\Omega)$. Here and in the following $\int ...dx$ denotes integration over Ω. The usual Sobolev space $H^{1,2}(\Omega;\mathbf{R}^N)$ consists of all (equivalence classes of) functions $u : \Omega \to \mathbf{R}^N$ with $\nabla u \in L^2$. We emphasize that the matrix function A does not depend on the index ν in (1.1). ∇u_ν is a column vector function.

There are several methods to obtain bounded weak solutions of (1.1), say, for the Dirichlet problem. For this and the numerous result concerning C^α-regularity of bounded weak solutions to (1.1) see the surveys [Hi1],[Fr2],[Gia2]. We confine the discussion to an interesting case which has been unsolved up to now.

Let $a = (a_1,...,a_N)$ satisfy a *"one sided condition"*

$$a(x,u,\eta)\cdot u \geq -\lambda^*|\eta|^2 - f_o, \tag{1.6}$$

for all $x \in \Omega$, $u \in \mathbf{R}^N$, $\eta \in \mathbf{R}^{nN}$, where

$$f_o \in L^{1,q} \text{ with some } q > n-2. \tag{1.7}$$

Condition (1.6) guarentees that the nonlinear form

$$Q(u,u) = \sum_{\nu=1}^N \int \nabla u_\nu^T A\nabla u_\nu \ dx + \int a(.,u,\nabla u)u \ dx$$

is coercive on $H^{1,2}_o \cap \{u \in L^\infty | \quad \|u\|_\infty \le C\}$.

We emphasize that no quantitative restriction on the growth constant K_C in (1.2) has been imposed. Due to this critical growth behaviour of the right hand side $a(x, u, \nabla u)$, however, one cannot apply the usual theory of Visik-Leray-Lions, c.f. Morrey's book [Mo1,§5.12] to obtain the existence of weak solutions. Up to now it is not known whether the conditions (1.2) up to (1.7) are sufficient for the existence problem for $n \ge 3$. For $n = 2$ and zero Dirichlet boundary conditions it was solved in [Fr 1], where also non diagonal elliptic principal part was treated. For the case $n = 2$ and *non homogeneous* Dirichlet boundary condition of §4 of this paper.

The proof for the case $n = 2$ is based on C^α-a-priori bounds which are sufficiently strong to apply the ideas of Leray-Lions.

Interestingly for $n \ge 3$ it was shown by counterexamples [Iv1], [Str1] that C^α-a-priori estimates and regularity results are not possible under the above conditions. (By C^α we denote the space of Hölder continuous functions with Hölder exponent α.)

Important progress has recently made by Landes [La1]. He replaces the one sided condition (1.6) by an *angle condition.* In view of C^α-regularity a mild reformulation of his condition in our setting is

$$(1.8) \qquad a(x, u, \eta).(u - v) \ge -\lambda^* |\eta|^2 - \bar{f}_o \ , \ \bar{f}_o \ge 0,$$

for all

$$x \in \Omega, \ u, v \in \mathbf{R}^N, \ |v| \le |u| \le C.$$

Here $\lambda^* > 0$ has to be sufficiently small and $\bar{f}_o$ satisfies (1.7). The angle condition includes the case that a has the form

$$a(x, u, \eta) = uh(x, u, \eta) \ , \ \text{e.g.} \ = u|\eta|^2$$

where h is a nonnegative scalar function. With his additional condition Landes was able to prove the existence of weak solutions in the case of zero boundary conditions. He was also able to treat more general principal parts and the L^p-case. His technique does not use C^α-estimates or reverse Holder inequality; it is based on the truncation of *vector* functions.

In the setting presented here Landes' technique gives the existence of weak solutions $u \in H_o^{1,\infty} \cap L^\infty(\Omega, \mathbf{R}^N)$. The additional L^∞-estimate for u follows from a uniform condition similar to (1.8) and can be obtained with a simple addition to Landes' proof. We further remark that also the *inhomogeneous* case can be treated since an $L^\infty \cap H^{1,2}$-estimate is possible (cfr. §4).

The purpose of this paper is to present the following theorem concerning the Holder continuity of local weak solutions $u \in H^1_{\mathrm{loc}} \cap L^\infty(\Omega; \mathbf{R}^N)$.

Theorem 1.1. *Let $u \in L^\infty \cap H^1_{\mathrm{loc}}(\Omega; \mathbf{R}^N)$ be a weak solution of equation (1.1) in a bounded domain of $\mathbf{R}^n$, $n \geq 3, N \geq 2$, and assume that the Carathéodory conditions (1.3), (1.4), the growth assumption (1.2), the ellipticity condition (1.5) and the angle condition (1.8) are satisified with $C = \|u\|_\infty$ and $\lambda^* > 0$ sufficiently small. Then u is Hölder continuous with some exponent α on interior domains $\Omega_o \subset\subset \Omega$.*

The interior C^α-norms of u are uniform as long as the data

$$\|u\|_\infty,\ \|a_{ik}\|_\infty,\ \lambda^{-1},\ \|f_C\|_{1,q},\ \|f_o\|_{1,q},\ \lambda^* < \lambda_o^*,\ (q-2)^{-1}$$

vary on bounded sets of $\mathbf{R}$. The smallness of λ^* which is necessary for our proof depends on $\lambda, \|u\|_\infty$ and the growth constant K_C in (1.2).

We shall prove this theorem in Section 3 using the "hole filling method". An alternative and quite similar proof is possible by establishing a reverse Hölder inequality which implies an $L^{2+\epsilon}$-estimate for ∇u, and then proceed as in Giaquinta-Giusti's paper [Gia 2]. However, on the whole, we found the methods presented here to be the simpler one.

It is not hard to modify our proof in order to obtain C^α-regularity and a priori estimates up to the boundary for solutions u of the Dirichlet problem. For this, one has to assume the usual Wiener condition for $\partial\Omega$. Furthemore it is possible to obtain an H^1-estimate for the nonhomogeneous Dirichlet problem; in this setting this is non-trivial. We state the corresponding

Theorem 1.2. *Let $g \in H^{1,2}(\Omega; \mathbf{R}^N) \cap L^\infty$ represent the boundary condition and let the assumptions (1.2) up to (1.8) with $\lambda^* > 0$*

small enough and arbitrary $C \in \mathbf{R}$ in (1.8) be satisfied. Then there exists a weak solution $u \in g + L^\infty \cap H_o^{1,2}(\Omega; \mathbf{R}^N) \cap C^\alpha_{\text{loc}}$ of (1.1). If $\partial\Omega$ satisfies a uniform Wiener condition and $\nabla g \in L^{2,q}$ with $q > n-2$ then $u \in C^\alpha$ up to the boundary with some $\alpha \in]0,1[$. Under similar conditions to those of Theorem 1.1., uniform control of the C^α-norm of u is also obtained.

2. Preliminary inequalities. We shall deal with *truncated* vector functions. This idea was successfully applied by Meier [Me1] and Landes [La1]. For $\xi \in \mathbf{R}^N, \ell \in \mathbf{R}_+$ we define $|\xi|_\ell = \max\{|\xi|, \ell\}$ and

$$[\xi]_\ell = \ell \xi |\xi|_\ell^{-1}.$$

We have $[\xi]_\ell = \xi$ for $|\xi| \leq \ell$ and $[\xi]_\ell = \ell\xi|\xi|^{-1}$ for $|\xi| \geq \ell$. $[\xi]_\ell$ is a bounded Lipschitz function with respect to ξ and hence

$$[u]_\ell \in L^\infty \cap H^1(\Omega; \mathbf{R}^N), \quad \text{provided that} \quad u \in H^1(\Omega; \mathbf{R}^N).$$

We have $\|[u]_\ell\|_\infty \leq \ell$ and a.e.

$$\begin{aligned} D_i[u]_\ell &= D_i u && \text{on } (|u| < \ell) \\ D_i[u]_\ell &= \ell D_i u |u|_\ell^{-1} - \ell (D_i u . u)|u|^{-3} && \text{on } (|u| \geq \ell). \end{aligned}$$

Note that $D_i u \cdot u = D_i(u^2) = 0$ a.e. on $(|u| = \ell)$.

Lemma 2.1 *Let $u \in H^1_{\text{loc}}(\Omega; \mathbf{R}^N)$ and $\ell > 0$. Let $A = (a_{ik})$ satisfy (1.3) and (1.5). Then*

$$\sum_{\nu=1}^{N} \nabla u_\nu^T A \nabla [u]_{\ell,\nu} \geq \lambda |\nabla u|^2 \chi_\ell \ \text{ a.e. in } \Omega. \tag{2.1}$$

Here

$$\chi_\ell = 1 \ \text{ on } \ (|u| \leq \ell) = \{x \in \Omega \,|\, |u(x)| \leq \ell\} \ \text{ and } \ \chi_\ell = 0 \quad \text{otherwise.}$$

The ν-the components of $[u]_\ell$ is $[u]_{\ell,\nu}$. This lemma is given in Landes' paper in a more general setting.

PROOF. The nontrivial part of inequality 2.1 is for values of x in $(|u| > \ell)$. We have to prove that

$$\sum_{\nu=1}^{N} \nabla u_\nu^T A(\nabla u_\nu - u_\nu |u|^{-2} \sum_{\mu=1}^{N} u_\mu \nabla u_\mu) \geq 0$$

or

$$|u|^2 \sum_{\nu=1}^{N} \nabla u_\nu^T A \nabla u_\nu \geq (\sum_{\nu=1}^{N} u_\nu \nabla u_\nu^T) A (\sum_{\mu=1}^{N} u_\mu \nabla u_\mu). \tag{2.2}$$

For all $x \in \Omega, v, w \in \mathbf{R}^N$ we define the positive semidefinite quadratic form in v, w

$$q(v,w) = \sum_{\nu=1}^{N} v_\mu \nabla u_\nu^T A \nabla u_\nu w_\nu .$$

Via Hölder's inequality one proves by induction that

$$q(u,u) \leq |u|^2 \sum_{\mu=1}^{N} q(e_\mu, e_\mu) \ , \ e_\mu = \mu - \text{th unit vector}. \tag{2.3}$$

This only uses the fact that q is semidefinite, symmetric and bilinear. Then this is (2.2) as required.

For the convenience of the reader we present the induction argument here. For $i = 1, \ldots, N$ and all $u \in span \ < e_1, ..., e_i >$ we shall prove

$$q(u,u) \leq |u|^2 \sum_{\mu=1}^{i} q(e_\mu, e_\mu). \tag{2.4}$$

Inequality (2.4) is obviously true for $i = 1$. From (2.4) we shall conclude the analogous inequality for $i+1$. Set $u = v + \alpha e_{i+1}$, $v \in span < e_1, \ldots, e_i >$. Then

$$q(u,u) = q(v,v) + \alpha^2 q(e_{i+1}, e_{i+1}) + 2\alpha q(v, e_{i+1}).$$

Estimating $2\alpha q(v, e_{i+1})$ by

$$\alpha^2 |v|^{-2} q(v,v) + |v|^2 q(e_{i+1}, e_{i+1})$$

we obtain, using (2.4) for v,

$$q(u,u) \leq (1+\alpha^2|v|^{-2})q(v,v) + (\alpha^2+|v|^2)q(q_{i+1},e_{i+1})$$
$$\leq (|v|^2+\alpha^2)\sum_{\mu=1}^{i} q(e_\mu,e_\mu) + (\alpha^2+|v|^2)q(e_{i+1},e_{i+1}).$$

But this just (2.4) with i replaced by $i+1$.

For $u \in L^\infty \cap H^{1,2}_{\text{loc}}(\Omega;\mathbf{R}^N)$ we define the mean value

$$c_\ell = c_\ell(x_o) = |B_{3R}-B_{2R}|^{-1}\int_{B_{3R}-B_{2R}} [u]_\ell dx \in \mathbf{R}^N$$

on the truncated vector function $[u]_\ell$. Here $|V|$ is the n-dimensional volume of $V \subset \mathbf{R}^N$, and B_{2R}, B_{3R} are concentric with center x_o. Since $|[u]_\ell| \leq \ell$ we have

$$|c_\ell| \leq \ell. \tag{2.5}$$

We shall deal with terms

$$B = \sum_{\mu=1}^{N}\{[u]_{\ell,\mu} - c_{\ell\mu}\}D_k u_\mu$$

where $[u]_{\ell,\mu}$ and $c_{\ell\mu}$ denote the μ-th component of $[u]_\ell$ and c_ℓ. We have a.e. on $(|u|<\ell)$

$$B = \frac{1}{2}D_k|u-c_\ell|^2$$

and on $(|u|\geq\ell)$

$$B = \sum_{\mu=1}^{N}(\ell|u|^{-1}u_\mu - c_{\ell\mu})D_k u_\mu = \sum_{\mu=1}^{N} D_k(\ell|u| - \frac{1}{2}\ell^2 - c_\ell u + \frac{1}{2}|c_\ell|^2).$$

Consider the function $\psi : \mathbf{R}^N \to \mathbf{R}$ by

$$\psi(\xi) = \tfrac{1}{2}(\xi - c_\ell)^2 \qquad \text{for } |\xi|<\ell$$

$$\psi(\xi) = \ell|\xi| - \tfrac{1}{2}\ell^2 - c_\ell\xi + \tfrac{1}{2}|c_\ell|^2 \quad \text{for } |\xi|\geq\ell.$$

Obviously

$$B = D_k(\psi(u)) \tag{2.6}$$

and we have

Lemma 2.2. *The function ψ defined above is Lipschitz, non-negative and satisfies*

$$\psi(\xi) \le \frac{1}{2}|\xi - c_\ell|^2. \tag{2.7}$$

PROOF. Obviously

$$\psi(\xi) \to \frac{1}{2}\ell^2 + \frac{1}{2}c_\ell^2 - c_\ell \xi^* \quad \text{for} \quad \xi \to \xi^*,$$

$|\xi| < \ell$ as well as $|\xi| > \ell$. Since further $\nabla_\xi \psi$ exists on $(|\xi| \ne \ell)$ and is locally bounded we conclude that ψ is Lipschitz.
On $(|\xi| < \ell)$ obviously $\psi \ge 0$. On $(|\xi| \ge \ell)$ we estimate

$$\psi(\xi) \ge \ell|\xi| - |c_\ell||\xi| + \frac{1}{2}|c_\ell|^2 - \frac{1}{2}\ell^2 = |\xi|(\ell - |c_\ell|) = \frac{1}{2}|c_\ell|^2 - \frac{1}{2}\ell^2$$
$$\ge \ell(\ell - |c_\ell|) + \frac{1}{2}|c_\ell|^2 - \frac{1}{2}\ell^2 = \frac{1}{2}\ell^2 - \ell|c_\ell| + \frac{1}{2}|c_\ell|^2 \ge 0.$$

Thus $\psi \ge 0$ on all $\mathbf{R}^N$. Finally we prove that (2.7) holds also on $(|\xi| \ge \ell)$. (On $(|\xi| \le \ell)$, (2.7) holds by the definition of ψ.) Define

$$\varphi(\xi, h) = h|\xi| - c_\ell \xi + \frac{1}{2}c_\ell^2 - \frac{1}{2}h^2.$$

With respect to the argument h the function φ is monotone increasing, provided that $|\xi| \ge h \ge \ell$. In fact,

$$\frac{\partial}{\partial h}\varphi(\xi, h) = |\xi| - h \ge 0.$$

Thus

$$\psi(\xi) = \varphi(\xi, \ell) \le \varphi(\xi, |\xi|) = \frac{1}{2}|\xi|^2 - c_\ell \xi + \frac{1}{2}c_\ell^2$$

which proves (2.7).

3. Proof of Theorem 1.1. The proof of Theorem 1.1 uses the "hole filling" technique, cf. for example [Hi1]. For this, one proves for the solution u the inequality

(3.1)
$$\int_{B_R} |\nabla u|^2 |x-x_o|^{2-n} dx \leq K \int_{B_{4R}-B_R} |\nabla u|^2 |x-x_o|^{2-n} dx + KR^\beta$$

for all $x_o \in \Omega$ and concentric balls

$$B_R = B_R(x_o) \subset B_{4R}(x_o) = B_{4R} \subset\subset \Omega.$$

Here K and $\beta > 0$ are constants not depending on x_o and R. Inequality (3.1) implies interior Hölder continuity for u with some Hölder exponent α which depends only on K and β, the interior $C^\alpha(\Omega_o)$-norms depend only on K, β and $\Omega_o \subset\subset \Omega$.

Inequality (3.1) is equivalent to

$$\int_{B_R} |\nabla u|^2 G_z dx \leq KR^{2-n} \int_{B_{4R}-B_R} |\nabla u|^2 dx + KR^\beta. \tag{3.2}$$

Here and in the following, we use the same symbol K for constants not depending on the relevant parameters. G_z, $z = x_o$, denotes the Green function of the operator $L = -\mathrm{div}(A\nabla)$ under Dirichlet boundary conditions with respect to some ball $B \supset\supset \Omega$, with singularity at $z \in \Omega$. This means that

$$\int_B \nabla G_z^T A \nabla \varphi_y dx = \varphi(z) \text{ for all } \varphi \in C_c^\infty(B).$$

It is known, cf. [Sta1],[Grü1] that such a function

$$G_z \in H_o^{1,q}(B),\ 1 \leq q < n/(n-1)$$

exists, and that $G_z \geq 0$ and that there exists a constant K with

$$K^{-1}|x-z|^{2-n} \leq G_z(x) \leq K|x-z|^{2-n} \text{ for all } x \in \Omega. \tag{3.3}$$

Note that inequality (3.3) is a consequence of De Giorgi's famous theorem [De G 1], cf. [Sta1], [Grü 1].

We also shall deal with the approximate Green functions

$$G_z^\rho \in L^\infty \cap H_o^{1,2}(B)$$

which are defined by

$$\int_B (\nabla G_z^\rho)^T A \nabla \varphi dx = |B_\rho|^{-1} \int_{B_\rho(z)} \varphi dx.$$

It is well known that, for $\rho \to 0$,

$$G_z^\rho \to G_z \quad \text{weakly in} \quad H^{1,q} \quad \text{for all} \quad q \in [1, n/(n-1)[$$

and strongly in $L^\infty(\Omega - B_s(z))$ and $H^{1,2}(\Omega - B_s(z))$ for all $s > 0$.

Furthermore, it is known (cf. [Grü1], proposition 1.4) that for $\rho < \operatorname{dist}(z, \partial Q)$

$$G_z^\rho(x) \leq K|z - x|^{2-n}. \tag{3.4}$$

Let $z \in \Omega$ and $R > 0$. For $g, h \in L^1(\Omega; \mathbf{R})$ we write

$$h \subset_{R,z} g$$

iff there exist constants K and $\beta > 0$ not depending on z and R such that for $B_R = B_R(z) \subset B_{4R}(z) \subset \Omega$

$$\int \chi(B_R) h dx \leq K R^{2-n} \int \chi(B_{4R} - B_R) |\nabla u|^2 dx + K R^\beta + \\ + \int \chi(B_R) g dx.$$

Here u is the solution of the differential equation in discussion. The integration over M is denoted by $\int \chi(M) ... dx$. In order to prove the hole filling condition (3.1) we have to show that

$$|\nabla u|^2 G_z \subset_{R,z} 0. \tag{3.5}$$

For establishing (3.5) we prove that there exists a number δ depending on λ, $\lambda^* > 0$, the growth constant K_c in (1.2) and on $\|u\|_\infty$ such that for $k = 0, 1, \ldots, M$, $M = \min\{m \in \mathbf{N} | m\delta \geq \|u\|_\infty\}$

$$\begin{aligned} &\tau^2 |\nabla u|^2 G_z \chi(|u| \leq k\delta) \subset_{R,z} \\ &\subset_{R,z} K_o \tau^2 |\nabla u|^2 G_z \{\chi(|u| \leq (k-1)\delta) + \lambda^* \chi(|u| > M\delta)\}. \end{aligned} \tag{3.6}$$

Here τ is the usual localization function i.e.

$$\tau \in H_o^{1,\infty}(B_{3R}(z)) \;,\; |\nabla\tau| \leq KR^{-1},\; \tau = 1 \text{ on } B_{2R} = B_{2R}(z).$$

By $\chi(|u| \leq C)$ we mean the characteristic function of the set

$$(|u| \leq c) = \{x \in \Omega \mid |u(x)| \leq c\},$$

i.e. $\chi = 1$ on this set and 0 else.
In fact, for $k = 0$ the first summation in the right hand side of (3.6) is 0 and by induction we arrive after M steps at the inequality

$$\begin{aligned} &\tau^2|\nabla u|^2 G_z\chi(|u| \leq M\delta) \subset_{R,z} \\ (3.7) \qquad &\subset_{R,z} \Big(\sum_{i=0}^{M+1} K_o^i\Big)\lambda^*\tau^2|\nabla u|^2 G_z\chi(|u| \leq M\delta). \end{aligned}$$

Assume λ^* is small enough, (3.5) follows. Clearly, the constant K in the inequality defining the relation $\subset_{R,z}$ may alter at each step. Thus (3.1) follows.

This induction argument together with inequality (3.1) is the key of the proof and is analogous to Landes' compactness proof [La1].

Inequality (3.7) is deduced from the equation for the weak solution u by choosing the following test function

$$\varphi = ([u]_\ell - c_\ell)G_z^\rho\tau^2 \;,\; \ell = k\delta$$

where $[u]_\ell$ is the truncated vector function defined in §2, c_ℓ is the mean value of $[u]_\ell$ taken over $B_{3R}(z)$ and τ, G_z^ρ are defined as above. The constant δ is chosen later. Note that $|c_\ell| \leq \ell$. We obtain

$$(3.8) \quad A_o := \sum_{\nu=1}^{N} \int \nabla\varphi_\nu^T \; A\nabla u_\nu dx = -\int a_o(.,u,\nabla u).\varphi dx =: B_o$$

and estimate A_o and B_o. We have by Lemma 2.1 that

$$\sum_{\nu=1}^{N} \nabla[u]_{\ell,\nu}^T A\nabla u_\nu \geq \lambda\chi(|u| < \ell)|\nabla u|^2 \text{ and hence}$$

$$A_o \geq \lambda \int \chi(|u| < \ell)|\nabla u|^2 G_z^\rho dx +$$

$$+2\sum_{\nu=1}^{N}\int \tau\nabla\tau^T A\nabla u_\nu([u]_{\ell,\nu}-c_{\ell,\nu})G_z^\rho dx$$

$$+2\sum_{\nu=1}^{N}\int \tau^2(\nabla G_z^\rho)^T A\nabla u_\nu([u]_{\ell,\nu}-c_{\ell,\nu})dx =: A_1+A_2+A_3.$$

The second term A_2 in the last sum can be estimated from below by an integral $\int h dx$ where the function h satisfies

$$h \subset_{R,z} R^{-n}|[u]_\ell - c_\ell|^2\chi(B_{3R}-B_{2R}) \subset_{R,z} 0. \tag{3.9}$$

The first relation $\subset_{R,z}$ in (3.9) is due to the fact that support $\nabla\tau \subset B_{3R}-B_{2R}$ and $|\nabla\tau| \leq KR^{-1}$, and that G_z^ϱ satisfies inequality (3.4) for ρ small enough. Also Hölder's inequality for $\nabla\tau^T A\nabla u_\nu$ and the estimate (3.3) are used. The second relation $\subset_{R,z}$ in (3.9) follows via Poincaré's inequality for $[u]_\ell - c_\ell$ on $B_{3R}-B_{2R}$, using the estimate $|\nabla[u]_\ell| \leq K|\nabla u|$. Note that $\ell = k\delta, k=1,\ldots,M$, is bounded from below by δ.

The third term is rewritten using the function ψ in Lemma 2.2, namely

$$A_3 = 2\int \tau^2(\nabla G_z^\rho)^T A\nabla\psi_\ell(u)dx$$
$$= 2\int (\nabla G_z^\rho)^T A\nabla(\tau^2\psi_\ell(u))dx - c_o \geq -c_o$$

where

$$c_o = 4\int (\nabla G_z^\rho)^T A\nabla\tau\tau\psi_\ell(u)dx.$$

The last inequality follows from the definition of G_z^ϱ and the fact that $\psi_\ell \geq 0$ according to Lemma 2.2.
By Lemma 2.2 we estimate

$$\psi_\ell(u) \leq \frac{1}{2}|[u_\ell]-c_\ell|^2 \quad \text{on } (|u|\leq \ell).$$

On $|u| \geq \ell$ we claim that

$$\psi_\ell(u) \leq |u_\ell - c_\ell|^2 + |u-c|^2$$

where c is the mean value of u taken over $B_{3R} - B_{2R}$. In fact, we have

$$\psi_\ell(u) = \frac{1}{2}\left\{|u - c_\ell|^2 - (|u| - \ell)^2\right\}.$$

Since

$$\begin{aligned}|c - c_\ell|^2 &= \left|\fint \chi(u \geq \ell)(u - \ell u|u|^{-1})\, dx\right|^2 \\ &\leq \fint \chi(u \geq \ell)(|u| - \ell)^2\, dx\end{aligned}$$

we conclude that

$$\begin{aligned}&\int\limits_{B_{3R}\setminus B_R} \chi(u \geq \ell)\psi_\ell(u)\, dx \\ &\qquad\leq \frac{1}{2}\int\limits_{B_{3R}\setminus B_R} \left\{\chi(u \geq \ell)|u - c|^2 - |c - c_\ell|^2\right\}\, dx \\ &\qquad\leq \frac{1}{2}\int\limits_{B_{3R}\setminus B_R} \left\{|u - c_\ell|^2 - |c - c_\ell|^2\right\}\, dx \\ &\qquad\leq \frac{1}{2}\int\limits_{B_{3R}\setminus B_R} |u - c|^2\, dx.\end{aligned}$$

Thus we have

$$\int\limits_{B_{3R}\setminus B_R} \psi_\ell(u)\, dx \leq \int\limits_{B_{3R}\setminus B_R} \left\{|[u_\ell] - c_\ell|^2 + |u - c|^2\right\}\, dx.$$

Hence

$$\begin{aligned}A_3 &\geq -R^{-1}K\int \chi(B_{3R} - B_{2R})|\nabla G_z^\rho|(|[u_\ell] - c_\ell|^2 + |u - c|^2)dx \\ &\geq -KR^{-n}\int \chi(B_{3R} - B_{2R})(|[u_\ell] - c_\ell|^2 + |u - c|^2)dx + \\ &\quad - KR^{n-2}\int \chi(B_{3R} - B_{2R})(|[u_\ell] - c_\ell|^2 + |u - c|^2)|\nabla G_z^\rho|^2 dx \\ &:= -C_1 - C_2.\end{aligned}$$

Again, the term C_1 is of the form $\int hdx$ where $h \subset_{R,z} 0$. This follows as before via Poincaré's inequality. The same is true for the term C_2. For proving this we use the defining equation for G_z^ϱ and take the function $\varphi = G_z^\varrho[[u_\ell] - c_\ell]\zeta^2$ as test function where ζ is Lipschitz, $|\nabla\zeta| \leq KR^{-1}$ and $\zeta = 1$ on $B_{3R} - B_{2R}$, $\zeta = 0$ on B_R and $\mathbf{R}^n - B_{4R}$. Simple applications of Hölder's and Poincaré's inequality and the estimate 3.4 for G_z^ϱ then lead to the relation $C_2 = \int hdx, h \subset_{R,z} 0$. Collecting our results we arrive at the inequality

$$(3.10)\quad \lambda \int \chi(|u| < \ell)|\nabla u|^2 G_z^\varrho \tau^2 dx \leq A_o - A_2 + C_1 + C_2 \leq A_o + C_3$$

where

$$C_3 = -A_2 + C_1 + C_2 \quad \text{and} \quad C_3 = \int hdx$$

where $h \subset_{R,z} 0$. Let us analyze the term B_o in (3.8). We set $\ell = k\delta$ and decompose

$$\begin{aligned} a_1 &= a_o(.,u,\nabla u)([u]_\ell - c_\ell)G_z^\varrho\tau^2 = E_1 + E_2 + E_3 \\ &= a_1\chi(|u| \geq k\delta) + a_1\chi((k-1)\delta \leq |u| < k\delta) \\ &\quad + a_1\chi(|u| < (k-1)\delta). \end{aligned}$$

On the set $(|u| \geq \ell)$ we have $|[u]_\ell| = \ell \geq |c_\ell|$ and $[u]_\ell = \ell|u|^{-1}$. Hence the angle condition yields

$$\begin{aligned} a_o(.,u,\nabla u)([u]_\ell - c_\ell) &= \ell|u|^{-1}a_o(.,u,\nabla u)(u - \ell^{-1}|u|c_\ell) \\ &\geq -\ell|u|^{-1}\lambda^*|\nabla u|^2 - \ell|u|^{-1}\bar{f}_o \geq -\lambda^*|\nabla u|^2 - \bar{f}_o \end{aligned}$$

on $(|u| \geq \ell)$. We have used the fact that $\left|\ell^{-1}|u|c_\ell\right| \leq |u|$ and $\ell|u|^{-1} \leq 1$. Thus we have

$$(3.11)\qquad E_1 \geq -(\lambda^*|\nabla u|^2 + \bar{f}_o)\chi(|u| \geq k\delta).$$

On the set $(|u| \leq (k-1)\tau\delta)$ we simply use the growth condition and the L^∞-bound for u to obtain

$$(3.12)\qquad -E_2 \leq K(|\nabla u|^2 + f_o)\tau^2 G_z^\varrho \chi(|u| \leq (k-1)\delta).$$

Finally, on the set $((k-1)\delta \leq |u| \leq k\delta)$ we have $[u]_{k\delta} = u$ and

$$(1-k)^{-1}c_{k\delta} \leq (1-k^{-1})k\delta \leq (k-1)\delta \leq |u|.$$

We decompose

$$a_o(.,u,\nabla u)([u]-c_{k\delta}) = k^{-1}c_{k\delta}a_o(.,u,\nabla u)+ \\ + a_o(.,u,\nabla u)(u-(1-k^{-1})c_{k\delta}).$$

The first term is estimated from below by the growth condition, the second by the angle condition. This yields

$$a_o(.,u,\nabla u)(u_{k\delta}-c_{k\delta}) \geq \delta K_c|\nabla u|^2 - Kf_o + \lambda^*|\nabla u|^2 - K\bar{f}_o$$

and

$$E_2 \geq -\delta K_c|\nabla u|^2 G_z^\rho \tau^2 - K\tilde{f}_o G_z^\rho \tau^2 - \lambda^*|\nabla u|^2 G_z^\rho \tau^2$$

where $\tilde{f}_o$ satisfies a Morrey condition analogously to the one for $\bar{f}_o$ and f_o. Note that $\int K\tilde{f}_o G_z^\rho \tau^2 dx = \int h dx$ where $h \subset_{R,z} 0$. We collect our results and arrive at the inequality

$$\begin{aligned} B_o = &- \int (E_1+E_2+E_3)dx + \int h dx \leq \\ &\leq \lambda^* \int \chi(|u| \geq (k-1)\delta)|\nabla u|^2 G_z^\rho \tau^2 dx \\ &+ \delta K_c \int \chi((k-1))\delta \leq |u| \leq k\delta)|\nabla u|^2 G_z^\rho \tau^2 dx \\ &+ K_c \int \chi(|u| \leq (k-1)\delta)|\nabla u|^2 G_z^\rho \tau^2 dx + \int h dx \end{aligned}$$

where $h \subset_{R,z} 0$. We combine this inequality (3.10) and obtain, by choosing $\delta K_c < \frac{1}{2}\lambda$,

$$\begin{aligned} &\frac{1}{2}\lambda \int \chi(|u| < \ell)|\nabla u|^2 G_z^\rho \tau^2 dx \leq \\ &\leq K \int \chi(|u| \leq (k-1)\delta)|\nabla u|^2 G_z^\rho dx \\ &+ \lambda^* \int \chi(|u| \geq (k-1)\delta)|\nabla u|^2 G_z^\rho \tau^2 dx + \int h dx \end{aligned} \tag{3.13}$$

where $h \subset_{R,z} 0$. Recall $\ell = k\delta$. In inequality (3.13) we may pass to the limit $\rho \to 0$ since $G_z^\rho \to G_z$ in measure, $0 \leq G^\rho \leq KG$ and $\int |\nabla u|^2 G_\rho dx \leq K$ uniformly as $\rho \to 0$. The latter inequality follows from (3.13), however can also be deduced directly from the differential equation. So we obtain inequality (3.13) with G_z^ρ replaced by G_z and this is just (3.6), This concludes the proof of Theorem 1.1.

4. An H^1-estimate and further results. For the proof of Theorem 1.2 where the *inhomogeneous* boundary condition

$$u \in g + H_o^1(\Omega, \mathbf{R}^N)$$

is considered, we approximate the system (1.1) by replacing the right hand side a_ν by

$$a_\nu^m = a_\nu/(1 + \frac{1}{m}|a_\nu|) \ , \ m = 1, 2, \dots \ .$$

Then there are solutions $u^m \in L^\infty \cap H^1$ of the approximate system

$$\operatorname{div}(A\nabla u_\nu^m) = a_\nu^m(., u, \nabla u) \ , \ \nu = 1, \dots, N$$
$$u^m \in g + H_o^1(\Omega, \mathbf{R}^N).$$

For the proof of Theorem 1.2 it remains to establish a uniform bound for the u^m in $L^\infty \cap H^1$. Once this is done Theorem 1.2 implies a uniform local C^α-estimate which in turn implies local H^1-compacteness via the monotonicity argument of Visik-Leray-Lions, cf. [Mo1, §5.12]. In order to simplify the notation we drop the index m of u^m and prove that $L^\infty \cap H^1$-solutions of (1.1) have bounds in L^∞ and H^1 which depend only on the data

$$K_C, \quad f_c, \|a_{ik}\|_\infty, \quad \lambda^{-1}, \quad {\lambda^*}^{-1}, f_o, \quad \text{vol}\,\Omega,$$

in the conditions (1.2) up to (1.8) and further on $\|g\|_\infty + \|g\|_{H^1}$. Note that (1.8) has to hold uniformly for all $C \in \mathbf{R}_+$ while this is not required in (1.2).

The L^∞-bound is well known; we have

Lemma 4.3. *Let $u \in g + L^\infty \cap H_o^1(\Omega; \mathbf{R}^N)$ be a weak solution of (1.1) and assume that (1.2) up to (1.8) with $\lambda^* < \lambda$ hold. Then* *there exists a constant K depending only on* $\|a_{ij}\|_\infty$, λ^{-1}, $\|f_o\|_{L^{1,q}}$, q, n, *vol* Ω *and* $\|g\|_\infty$ *such that*

$$\|u\|_\infty \leq K.$$

For the proof, one can choose the function

$$\varphi = u_\nu(|u|^2 - \|g\|^2_\infty)G^\rho_z$$

as test functions. Here $(\zeta)_+$ denotes the positive part of ζ and $G^\rho_z \in H^1_o(Q), Q \supset\supset \Omega$, is defined as in the proof of Theorem 1.1. Note that $\varphi = 0$ on $\partial\Omega$. With this procedure, one obtains an L^∞-bound for $(|u|^2 - \|g\|^2_\infty)^2_+$.

Theorem 4.1. *Let* $u \in g + L^\infty \cap H^1_o(\Omega; \mathbf{R}^N)$ *be a weak solution of (1.1) and assume that (1.2) up to (1.8) hold where* $\lambda^* > 0$ *is sufficiently small. Then there exists a constant K depending only on*

$$\|u\|_\infty, \quad \|g\|_{H^1}, \quad \|f_o\|_{L^1}, \|f_C\|_{L^1}, \quad \lambda^{-1}, \|a_{ik}\|_\infty, \quad K_C$$

with $C = \|u\|_\infty$, *and* $\lambda^* < \lambda^*_o$ *such that*

$$\|u\|_{H^1} \leq K.$$

The smallness of λ^*_o depends on $\|u\|_\infty, K_C$, and λ^{-1}.

Proof. Again we use the truncated vector functions $[u]_\ell$ introduced in §2 and set $\ell = k\delta$ where δ will be choosen small enough and $k = 0, 1, \ldots, M$, $\|u\|_\infty = M\delta$. We choose the vector φ of the test functions

$$\varphi = [u]_\ell - [g]_\ell$$

which is contained in $L^\infty \cap H^1_o$. This yields

$$\begin{aligned}(4.1) \qquad \sum_{\nu=1}^N \int \nabla([u]_{\ell_\nu} - [g]_{\ell_\nu})^T A \nabla u_\nu dx + \\ + \int a_\nu(., u, \nabla u) \cdot ([u]_\ell - [g]_\ell) dx = 0.\end{aligned}$$

As in the proof of Theorem 1.1 we estimate

$$\sum_{\nu=1}^{N} \nabla[u]_{\ell\nu}^{T} A \nabla u_\nu \geq \lambda\chi(|u| \leq \ell)|\nabla u|^2 \tag{4.2}$$

and

$$\int \nabla[g]_{\ell\nu}^{T} A \nabla u_\nu dx \leq K_\delta \|\nabla u\|_{L^2}. \tag{4.3}$$

The second summand in (4.1) is split into the terms $E_1 + E_2 + E_3$ where

$$E_1 = \int \chi(|u| \leq (k-1)\delta) a \cdot ([u]_\ell - [g]_\ell) dx$$

$$E_2 = \int \chi((k-1)\delta < |u| \leq k\delta) a \cdot ([u]_\ell - [g]_\ell) dx$$

$$E_3 = \int \chi(|u| > k\delta) a \cdot ([u]_\ell - [g]_\ell) dx \ , \ \ell = k\delta.$$

We estimate using the growth condition for a

$$|E_1| \leq K_C \int \chi(|u| \leq (k-1)\delta)|\nabla u|^2 dx + K. \tag{4.4}$$

where $K_C = 2\|u\|_\infty$ times the constant K_C in (1.2).

The term E_2 is rewritten

$$E_2 = \int \chi((k-1)\delta < |u| \leq k\delta) a_o \cdot ([u]_\ell - (1-k^{-1})[g]_\ell + k^{-1}[g]_\ell) dx.$$

The part

$$a_o(., u, \nabla u)([u]_\ell - (1-k^{-1})[g]_\ell)$$

can be estimated from below by Landes' angle condition (1.8) since

$$\|(1-k^{-1})[g]_\ell\|_\infty \leq (1-k)\delta \ \text{ and } \ (1-k)\delta \leq |u|$$

on $((k-1)\delta < |u| \leq k\delta)$. This yields

$$E_2 \geq -\lambda^* \int \chi((k-1)\delta < |u| \leq k\delta)|\nabla u|^2 dx - K +$$

$$+ \int \chi((k-1)\delta < |u| \leq k\delta) a_o k^{-1} [g]_\ell dx.$$

Note that $\|k^{-1}[g]_\ell\| \leq \delta$. We choose δ so small such that $\delta K_C \|u\|_\infty < \frac{\lambda}{2}$ and hence

$$|a_o k^{-1}[g]_\ell| \leq \frac{\lambda}{2}|\nabla u|^2 + f_C.$$

Thus we obtain

$$E_2 \geq -(\lambda^* + \frac{\lambda}{2}) \int \chi((k-1)\delta < |u| \leq k\delta)|\nabla u|^2 dx - K. \tag{4.5}$$

Finally we use the angle condition (1.8) to estimate the term E_3 from below. Note that by (1.8)

$$a(x,u,\nabla u).(u - \ell^{-1}|u|[g]_\ell) \geq -\lambda^*|\nabla u|^2 - \bar{f}_o$$

since $|\ell^{-1}|u|[g]_\ell| \leq |u|$. On $(|u| \geq \ell)$ we have $\ell/|u| \leq 1$ and this yields

$$a(x,u,\nabla u)\dot{(}[u]_\ell - [g]_\ell) \geq -\lambda^*|\nabla u|^2 - \bar{f}_o$$

on $(|u| \geq \ell)$. Thus we may estimate

$$E_3 \geq -\lambda^* \int \chi(|u| \geq \ell)|\nabla u|^2 dx - K.$$

Thus we arrive at the inequality

$$\begin{aligned}\lambda \int \chi(|u| \leq k\delta)|\nabla u|^2 dx \leq K + K_C \int \chi(|u| \leq (k-1)\delta)|\nabla u|^2 dx + \\ + (\lambda^* + \frac{\lambda}{2}) \int \chi((k-1)\delta < |u| \leq k\delta)|\nabla u|^2 dx \\ + \lambda^* \int \chi(|u| > \ell)|\nabla u|^2 + K\|\nabla u\|_2^2\end{aligned}$$

and finally

$$\begin{aligned}&\frac{\lambda}{2}\int \chi(|u| \leq k\delta)|\nabla u|^2 dx \leq \\ &\leq K + \frac{3}{2}K_C \int \chi(|u| \leq (k-1)\delta)|\nabla u|^2 dx + \\ &+ \lambda^* \int \chi(|u| \geq M\delta)|\nabla u|^2 dx.\end{aligned} \tag{4.6}$$

Again we apply an induction argument to estimate

$$\int \chi(|u| \le M\delta)|\nabla u|^2 dx \le$$
$$\le K + \lambda^* \sum_{i=0}^{M} (\frac{3K_C}{\lambda})^i \int \chi(|u| \ge M\delta)|\nabla u|^2 dx.$$

The H^1-estimate follows if $\lambda^* < [\sum_{i=0}^{M}(3K_C/\lambda)^i]^{-1}$.

Further results and open problems. The condition for λ^* in Theorem 1.1 and 1.2 is very restrictive. One should analyze if it sufficies to assume $\lambda^* < \lambda$. One can combine other regularity techniques with the one presented in Theorem 1.1 to obtain more general structure conditions for $a(x, u, \nabla u)$. An example is the following case
Let $b = (b_1, \ldots, b_N)$ be a vector function such that

$$b_\nu(x, u, \nabla u) = a_\nu(x, u, \nabla u) + Q(x, u, \nabla u).\nabla u_\nu$$

where $a = (a_1, \ldots, a_N)$ satisfies the assumptions of Theorem 1.1 and the vector function $Q = (Q_1, \ldots, Q_N)$ satisfies the Carathéodory conditions and the growth condition

$$|Q(x, u, \nabla u)| \le K|\nabla u| + K. \tag{4.7}$$

Furthemore let $A = (a_{ik})$ satisfy (1.3).

Then every weak solution $u \in L^\infty \cap H^1(\Omega; \mathbf{R}^N)$ of the system

$$\text{div } A(\nabla u) = b(x, u, \nabla u)$$

is Hölder continuous on interior domains of Ω.

For the proof one has to choose test functions of the type

$$\varphi = ([u]_\ell - c_\ell)\exp\{\alpha\psi(u)\}\tau^2 G_z^\varrho$$

where C_ℓ and ψ are defined in Section 2.

The parameter α is chosen large. It is possible to prove the hole filling condition (3.1) by a similar induction argument as in Theorem 1.1.

Finally, we remark that Theorem 1.1 implies the Hölder continuity of the *extremals* (not necessarily minimizers) of variational integrals

$$\int a(|u|^2)|\nabla u|^2 dx$$

where $a \in C^1$ and $a' \geq 0$, $a \geq \alpha_0 > 0$.

I wish to thank Michael Meier pointing out an error (which has been corrected).

References

[De G 1] E.De Giorgi, *Sulla differenziabilità e l'analiticità delle estremali degli integrali multipli regolari.* Mem. Accad. Sci. Torino cl. Sci. Fis. Mat. Nat. **3** (1957), 25-43.

[Fr 1] J.Frehse, *On two-dimensional quasilinear elliptic systems.* Manuscripta math. **28** (1959), 21-50.

[Fr 2] J.Frehse, *Existence and perturbation theorems of nonlinear elliptic systems.* From: Nonlinear partial diff. eq. and their applications, Collège de France Seminar IV. Ed. H.Brezis, J.L. Lions, Pitman 1983.

[Gia 1] M.Giaquinta, *Multiple integrals in the calculus of variations and nonlinear elliptic systems.* Vorlesungsreihe SFB 72, No.6; Univ. Bonn.

[Gia 2] M.Giaquinta, E.Giusti, *Nonlinear elliptic systems with quadratic growth.* Manuscripta math. **24** (1978), 323-349.

[Grü 1] M.Grüter, K.O.Widmann *The Green function for uniformly elliptic equations.* Manuscripta math. **37** (1982), 303-342.

[Hi 1] S.Hildebrandt, *Nonlinear elliptic systems and harmonic mappings.* Proceedings of Beijing Symp. on Diff. Geom. and Diff. Equ. Sci. 1980 Science Press, Beijing, China.

[I 1] P.A.Ivert, *On quasilinear systems in diagonal form.* Math. Z. **170** (1980), 283-286.

[L 1] R.Landes, *On the existence of weak solutions of perturbed systems with critical growth.* To appear in Crelle's Journal.

[Mei 1] M. Meier, *Reguläre und singulare Lösungen quasilinearer elliptischer Gleichungen und Systeme.* Diss. Bonn 1978.

[Mo 1] C.B.Morrey jr., *Multiple integrals in the calculus of variations.* Springer-Verlag, Berlin-Heidelberg -New York, (1968).

[Sta 1] G.Stampacchia, *Le problème de Dirichlet pour les équations elliptiques du second ordre à coefficients discontinus,* Ann. Inst. Fourier **15** (1965), 189-258.

[Str 1] M.Struwe, *A counterexample in elliptic regularity theory,* Manuscripta math. **34** (1981), 85-92.

Institut für Angewandte Mathematik
Beringstrasse 4-6
D-5300 BONN

HIGHER INTEGRABILITY FROM REVERSE JENSEN INEQUALITIES WITH DIFFERENT SUPPORTS

NICOLA FUSCO CARLO SBORDONE

Dedicated to Ennio De Giorgi on his sixtieth birthday

In many recent papers dealing with higher integrability of the gradient of solutions of elliptic p.d.e. or of minima of variational integrals, a crucial role is played by a reverse Hölder inequality (see [4], [7]).

The first result of this kind is contained in a paper by F.W. Gehring ([3]), where he proves that, if g is in $L^q_{\text{loc}}(\Omega)$, $q > 1$, and verifies

$$\fint_Q g^q dx \leq b\left(\fint_Q g dx\right)^q$$

for any cube $Q \subset \Omega$ with sides parallel to the axes; Ω bounded open set in $\mathbf{R}^n$, q, $b > 1$; then $g \in L^p_{\text{loc}}(\Omega)$ for some $p > q$.

Successively, in [5] this result has been extended to the case in which, on the left hand side, the mean value of g is taken over the "half" cube of Q, i.e. the cube with the same center of Q and half side.

The reason for this extension is that this second type of reverse inequality is verified mostly by the gradient of the solution of variational problems.

In this paper we prove a higher integrability result from a reverse Jensen inequality with different supports of the type

$$A^{-1}\left(\fint_{Q/2} A(f)dx\right) \leq b \fint_Q f dx + T$$

for any cube $Q \subset \Omega$ with sides parallel to the axes, where A is a convex function verifying suitable conditions.

Precisely, we prove that there exists $p > 1$ such that, setting $A_p(t) = A^p(t)/t^{p-1}$, the function $A_p(f)$ is in $L^1_{\text{loc}}(\Omega)$.

The proof we give here follows the line of a paper of Bojarski-Iwaniec [2], and, in the case $A(t) = t^q$, $q > 1$, it is different from the one given by Giaquinta-Modica ([5]).

1. Preliminaries. In the following we shall indicate with A a function verifying the following assumptions:

i) $A : [0, +\infty[\to [0, +\infty[$ is an N-function.

ii) There exists $k > 2$ such that

$$A(2t) \leq k\ A(t) \quad \forall t \geq 0,$$

i.e. A verifies the so called Δ_2-condition.

iii) $\int_o^1 A(t)/t^2 dt < \infty$.

iv) There exist $k_1, k_2, h_1, h_2 > 0$ such that

$$k_1 A(s)A(t) - h_1 \leq A(st) \leq k_2 A(s)A(t) + h_2$$

for any $s \geq 1$ and $0 \leq t \leq 1$.

For the properties of N-functions see [1] pag. 228 and also [6].

Remark 1.1. From i), ii) it follows easily that

$$\frac{A(t)}{t} \leq A'(t) \leq (k-1)\frac{A(t)}{t}.$$

From the second inequality, one can deduce

$$A(\lambda t) \leq \lambda^{k-1} A(t) \qquad \forall \lambda \geq 1, \quad \forall t \geq 0.$$

Remark 1.2. If $A(t)$ verifies i), ii), iv) then there exist $k_1', k_2', h_1', h_2' > 0$ such that, for any $s \geq 1$ and $0 \leq t \leq 1$

$$k_1' A^{-1}(s) A^{-1}(t) - h_1' \leq A^{-1}(st) \leq k_2' A^{-1}(s) A^{-1}(t) + h_2'.$$

Moreover $A^{-1}(t)/t$ is decreasing.

For $p > 1$, we shall also consider the function

$$A_p(t) = \frac{[A(t)]^p}{t^{p-1}}.$$

Remark 1.3. $A_p(t)$ is an increasing function verifying a.e.

$$\frac{A_p(t)}{t} \leq A_p'(t) \leq [1 + p(k-2)] \frac{A_p(t)}{t}$$

hence

$$A_p(t) \geq A^p(1)\, t \qquad \forall t \geq 1.$$

Let us now recall the definition of the complementary function $\tilde{A}$ of A. Let $a(t) = A'(t)$ and $\tilde{a}(s) = \sup\{t : a(t) \leq s\}$. Then we set

$$\tilde{A}(s) = \int_o^s \tilde{a}(\sigma) d\sigma.$$

The following Young inequality

$$st \leq A(t) + \tilde{A}(s), \qquad \forall s, t \geq 0$$

is well known.

Let us quote also the inequality

$$\tilde{A}\left(\frac{A(t)}{t}\right) \leq A(t) \quad \forall t > 0. \tag{1.1}$$

Let Ω be a bounded open set in $\mathbf{R}^n$. If A verifies i), ii), the vector space of all measurable functions u defined on Ω and satisfying

$$\int_\Omega A(|u(x)|) dx < \infty$$

is denoted by $L_A(\Omega)$ and is a Banach space under the norm

$$\|u\|_{L_A(\Omega)} = \inf\{\lambda > 0 : \fint_\Omega A(\frac{|u(x)|}{\lambda})dx \leq 1\}.$$

Remark 1.4. From condition ii) it follows that, if $\|u\|_{L_A} \neq 0$, then

$$\fint_\Omega A\left(\frac{|u(x)|}{\|u\|_{L_A}}\right) dx = 1.$$

In fact, if it were

$$\delta = \fint_\Omega A\left(\frac{|u(x)|}{\|u\|_{L_A}}\right) dx < 1$$

then, if $0 < \epsilon < \|u\|_{L_A}$, using Remark 1.1,

$$\begin{aligned}\fint_\Omega A(\frac{u(x)}{\|u\|_{L_A} - \epsilon}) &= \fint_\Omega A(\frac{u(x)}{\|u\|_{L_A}} \frac{\|u\|_{L_A}}{\|u\|_{L_A} - \epsilon})dx \\ &\leq \delta(\frac{\|u\|_{L_A}}{\|u\|_{L_A} - \epsilon})^{k-1}\end{aligned}$$

and the last term can be less than 1 for some $\epsilon > 0$, which is absurd.

We note that, under the above assumptions on A, one can control the $L_A(\Omega)$-norm of u in terms of $A^{-1}(\fint_\Omega A(|u(x)|)dx)$ as it is shown in the following

Proposition 1.1. *If A verifies i), ii), iv), and*

$$\inf_{0<t<1} A(t)A(\frac{1}{t}) \geq m > 0,$$

then there exist two constants $k_3, k_4 > 0$, such that for any $u \in L^1(\Omega)$

$$A^{-1}\left(\fint_\Omega A(|u|)dx\right) \leq k_3\left(1 + \|u\|_{L_A(\Omega)}\right)$$

$$\|u\|_{L_A(\Omega)} \leq k_4\left[1 + A^{-1}\left(\fint_\Omega A(|u|)dx\right)\right].$$

(For a result of similar type see [8]).

PROOF. We may always suppose $A(1)=1$ and u bounded. Let us prove the first inequality. If $\|u\|_{L_A}\le 1$, since $⨍_\Omega A(\frac{|u|}{\|u\|_{L_A}})dx=1$, then

$$⨍_\Omega A(|u|)dx \le \|u\|_{L_A} ⨍_\Omega A(\frac{|u|}{\|u\|_{L_A}})dx = \|u\|_{L_A} \le 1$$

and also

$$A^{-1}\Big(⨍_\Omega A(|u|)dx\Big) \le 1.$$

If $\|u\|_{L_A}>1$, then, by iv):

$$\begin{aligned} 1 = ⨍_\Omega A(\frac{|u|}{\|u\|_{L_A}})dx &\ge \frac{1}{|\Omega|}\int_{\Omega\cap\{|u|\ge 1\}} A(\frac{|u|}{\|u\|_{L_A}})dx \\ &\ge k_1 A(\frac{1}{\|u\|_{L_A}})\cdot\frac{1}{|\Omega|}\int_{\Omega\cap\{|u|\ge 1\}} A(|u|)dx - h_1 \\ &\ge \frac{k_1 m}{A(\|u\|_{L_A})}\cdot\frac{1}{|\Omega|}\int_{\Omega\cap\{|u|\ge 1\}} A(|u|)dx - h_1, \end{aligned}$$

from which it follows

$$\frac{1}{|\Omega|}\int_\Omega A(|u|)dx \le c[A(\|u\|_{L_A})+1]$$

and also the assertion.

Let us now prove the second inequality.

That inequality is obvious if $⨍_\Omega A(|u|)dx\le 1$, because, in this case, $\|u\|_{L_A}\le 1$. Otherwise, set

$$\xi = A^{-1}\Big(⨍_\Omega A(|u|)dx\Big)$$

and so $\xi\ge 1$.

For any $\lambda>1$, we have:

$$\fint_{\Omega} A\left(\frac{|u|}{\lambda\xi}\right)dx \leq$$
$$\frac{1}{\lambda}A\left(\frac{1}{\xi}\right)\frac{1}{|\Omega|}\int_{\Omega\cap\{|u|\leq 1\}} dx + \frac{1}{\lambda}\cdot\frac{1}{|\Omega|}\int_{\Omega\cap\{|u|\geq 1\}} A\left(\frac{|u|}{\xi}\right)dx.$$

Using iv), we have

$$\fint_{\Omega} A\left(\frac{|u|}{\lambda\xi}\right)dx \leq$$
$$\frac{1}{\lambda}A\left(\frac{1}{\xi}\right) + \frac{1}{\lambda}A\left(\frac{1}{\xi}\right)\frac{k_2}{|\Omega|}\int_{\Omega\cap\{|u|\geq 1\}} A(|u|)dx + \frac{h_2}{\lambda}.$$

Using once more iv), we have

$$A(\frac{1}{\xi}) \leq \frac{1+h_1}{k_1}\frac{1}{\fint_{\Omega} A(|u|)dx}$$

so, there exists $c > 0$, such that

$$\fint_{\Omega} A(\frac{|u|}{\lambda\xi})dx \leq \frac{c}{\lambda}.$$

Choosing λ such that $c/\lambda \leq 1$, we have

$$\|u\|_{L_A} \leq \lambda\xi = \lambda A^{-1}\left(\fint_{\Omega} A(|u|)dx\right)$$

i.e. the result.

If Ω is a bounded open set of $\mathbf{R}^n$ and $f \in L^1(\Omega)$ a non negative function, we shall denote by $(Mf)(x)$ the cubic maximal function of f in the sense of Hardy-Littlewood, i.e.

$$(Mf)(x) = \sup_{\substack{x\in Q\\ Q\subset\Omega}} \fint_{Q} f(y)dy$$

where Q is a cube with sides parallel to the axes and $\fint_Q f(y)dy = \frac{1}{|Q|}\int_Q f(y)dy$.

We shall also use the weak maximal function

$$(WMf)(x) = \sup\{ \fint_{Q/2} f(y)dy : x \in Q/2, \quad Q \subset \Omega\}$$

where $Q/2$ denotes the cube with the same center of Q and half side.

Let us recall the following

Proposition 1.2. *If $Q_o \subset \mathbf{R}^n$ is a cube and $f \in L^1(Q_o)$, $f \geq 0$, for any $t \geq \fint_{Q_o} f\, dx$:*

$$|\{x \in Q_o : Mf(x) > t\}| \geq \frac{1}{2^n t} \int_{\{f>t\}} f\, dx.$$

For the proof see e.g. [2]. In the same paper, one can find also the proof of the next

Proposition 1.3. *If Q_o is a cube of $\mathbf{R}^n$; $f \in L^1(Q_o)$, $f \geq 0$, then, for any $t > 0$*

$$|\{x \in Q_o : Mf(x) > t\}| \leq \frac{10^n}{t} \int_{\{f>\frac{t}{2}\}} f\, dx.$$

Let us now prove the following lemmas.

Lemma 1.1. *If f is a nonnegative, bounded measurable function in the cube Q_o, and A is a function verifying i),..., iv), then there exist $c_1, c_2 > 0$ such that*

$$\fint_{Q_o} A_p(Mf)dx \leq c_1 + c_2 \fint_{Q_o} A_p(f)dx,$$

when $p > 1$ and c_i are bounded if p is bounded.

Proof. Let us denote by $\lambda(t)$ the distribution function of Mf, i.e.

$$\lambda(t) = |\{x \in Q_o : Mf(x) > t\}|.$$

Then, using proposition 1.3, we have

$$\begin{aligned}
\fint_{Q_o} A_p(Mf)dx &= \frac{1}{|Q_o|}\int_o^{+\infty} A_p'(t)\lambda(t)dt \\
&\leq \frac{c}{|Q_o|}\int_o^{+\infty} \frac{A_p'(t)}{t}\int_{\{f>\frac{t}{2}\}} f dx \\
&\leq \frac{c}{|Q_o|}\int_o^{+\infty} \frac{A(t)}{t^2}\int_{\{f>t\}} f\, dx
\end{aligned}$$

where we used remark 1.3 and Δ_2 condition for A_p. So, using Fubini's theorem, we get

$$\begin{aligned}
&\fint_{Q_o} A_p(Mf)dx \leq c\fint_{Q_o} f(x)\, dx \int_o^{f(x)} \frac{A_p(t)}{t^2}dt \\
&\leq c\int_o^1 \frac{A_p(t)}{t^2}dt + \frac{c}{|Q_o|}\int_{Q_o\cap\{f\geq 1\}} f(x)dx \int_o^1 \frac{A_p(t)}{t^2}dt \\
&\qquad + \frac{c}{|Q_o|}\int_{Q_o\cap\{f\geq 1\}} f(x)dx \int_1^{f(x)} \frac{A_p(t)}{t^2}dt \qquad (1.2)\\
&\leq c\int_o^1 \frac{A_p(t)}{t^2}dt[1 + \frac{1}{|Q_o|}\int_{Q_o\cap\{f\geq 1\}} A_p(f(x))dx] \\
&\qquad + \frac{c}{|Q_o|}\int_{Q_o\cap\{f\geq 1\}} dx \int_{\frac{1}{f(x)}}^1 \frac{A_p(sf(x))}{s^2}ds.
\end{aligned}$$

Now, using iv), the last integral is controlled by

$$\begin{aligned}
&\frac{c}{|Q_o|}\int_{Q_o\cap\{f\geq 1\}} dx \int_{\frac{1}{f(x)}}^1 \frac{(k_2A(s)A(f(x)) + h_2)^p}{s^{p+1}(f(x))^{p-1}}ds \\
&\quad \leq \frac{c}{|Q_o|}\int_{Q_o\cap\{f\geq 1\}} dx\ A_p(f(x))\cdot\int_{\frac{1}{f(x)}}^1 \frac{A_p(s)}{s^2}ds \\
&\qquad + \frac{c}{|Q_o|}\int_{Q_o\cap\{f\geq 1\}} dx \int_{\frac{1}{f(x)}}^1 \frac{1}{s^{p+1}(f(x))^{p-1}}ds \leq
\end{aligned}$$

$$\leq \frac{c}{|Q_o|} \int_o^1 \frac{A_p(t)}{t^2} dt \int_{Q_o \cap \{f \geq 1\}} A_p(f(x)) dx + \frac{c}{|Q_o|} \int_{Q_o \cap \{f \geq 1\}} f(x) dx.$$

So, from (1.2) we get

$$\fint_{Q_o} A_p(Mf) dx \leq c \int_o^1 \frac{A_p(t)}{t^2} dt \Big[1 + \frac{1}{|Q_o|} \int_{Q_o} A_p(f) dx\Big] + \frac{c}{|Q_o|} \int_{Q_o} A_p(f) dx.$$

Finally

$$\int_o^1 \frac{A_p(s)}{s^2} ds = \int_o^1 \Big(\frac{A(s)}{s}\Big)^{p-1} \frac{A(s)}{s^2} ds \leq A^{p-1}(1) \int_o^1 \frac{A(s)}{s^2} ds \ \leq \ c$$

and this gives the result.

Lemma 1.3. *If f is a nonnegative, bounded measurable function in the cube Q_o and A is a function verifying i) and ii), then*

$$\fint_{Q_o} A_p(f) dx \leq 2A_p\Big(A^{-1}\Big(\fint_{Q_o} A(f) dx\Big)\Big) + c(p-1) \fint_{Q_o} A_p(A^{-1}(M(Af))).$$

Proof. Let $t_o = \fint_{Q_o} A(f) dx$. Then

$$\fint_{Q_o} A_p(f) dx = \frac{1}{|Q_o|} \int_{\{A(f) \leq t_o\}} A(f) \Big(\frac{A(f)}{f}\Big)^{p-1} dx + \frac{1}{|Q_o|} \int_{\{A(f) \geq t_o\}} \frac{A^p(f)}{f^{p-1}} dx = I + II.$$

Since $A(t)/t$ is increasing, then:

$$
(1.4)\qquad
\begin{aligned}
I &= \frac{1}{|Q_o|}\int_{\{f\le A^{-1}(t_o)\}} A(f)\Big(\frac{A(f)}{f}\Big)^{p-1}dx\\
&\le \frac{t_o^{p-1}}{(A^{-1}(t_o))^{p-1}}\fint_{Q_o} A(f)dx = A_p\Big(A^{-1}\Big(\fint_{Q_o} A(f)dx\Big)\Big).
\end{aligned}
$$

If we set $h(t) = \int_{\{f\ge t\}} f dx$, then

$$
(1.5)\qquad
\begin{aligned}
II &= \frac{1}{|Q_o|}\int_{\{f\ge A^{-1}(t_o)\}} \frac{A^p(f)}{f^{p-1}}dx = -\frac{1}{|Q_o|}\int_{A^{-1}(t_o)}^{+\infty}\frac{A^p(s)}{s^p}dh(s)\\
&= -\frac{1}{|Q_o|}\int_{A^{-1}(t_o)}^{+\infty}\Big(\frac{A(t)}{t}\Big)^{p-1}\Big(\int_t^{+\infty}\frac{A(s)}{s}dh(s)\Big)'\\
&= II_1 + II_2,
\end{aligned}
$$

where

$$
(1.6)\qquad
\begin{aligned}
II_1 &= -\frac{1}{|Q_o|}\Big(\frac{A(A^{-1}(t_o))}{A^{-1}(t_o)}\Big)^{p-1}\int_{A^{-1}(t_o)}^{+\infty}\frac{A(s)}{s}dh(s)\\
&= \Big(\frac{t_o}{A^{-1}(t_o)}\Big)^{p-1}\cdot\frac{1}{|Q_o|}\int_{\{f\ge A^{-1}(t_o)\}} A(f)dx\\
&\le \frac{t_o^p}{(A^{-1}(t_o))^{p-1}} = A_p\Big(A^{-1}\Big(\fint_{Q_o} A(f)dx\Big)\Big)
\end{aligned}
$$

and

$$
\begin{aligned}
II_2 &= \frac{(p-1)}{|Q_o|}\int_{A^{-1}(t_o)}^{+\infty}\Big(\frac{A(t)}{t}\Big)^{p-2}\frac{A'(t)t-A(t)}{t^2}dt\int_t^{+\infty}\frac{A(s)}{s}dh(s)\\
&\le \frac{(k-2)(p-1)}{|Q_o|}\int_{A^{-1}(t_o)}^{+\infty}\frac{A(t)^{p-1}}{t^p}dt\int_{\{f\ge t\}}A(f)dx\\
&= \frac{c(p-1)}{|Q_o|}\int_{A^{-1}(t_o)}^{+\infty}\frac{A(t)^{p-1}}{t^p}dt\int_{\{A(f)\ge A(t)\}}A(f)dx.
\end{aligned}
$$

Then, using Proposition 1.2:

$$II_2 \le \frac{c(p-1)}{|Q_o|} \int_{A^{-1}(t_o)}^{+\infty} \frac{A_p(t)}{t} \big|\{x : M(A(f))(x) > A(t)\}\big| dt$$
$$= \frac{c(p-1)}{|Q_o|} \int_{A_p(A^{-1}(t_o))} DA_p^{-1}(s) \frac{s}{A_p^{-1}(s)} \cdot$$
$$\cdot \big|\{x : M(A(f))(x) > A(A_p^{-1}(s))\}\big| ds$$

Now, from Remark 1.3 it follows that

$$D\ A_p^{-1}(s) \frac{s}{A_p^{-1}(s)} = \frac{1}{A_p'(A_p^{-1}(s))} \cdot \frac{s}{A_p^{-1}(s)} \le 1,$$

and so:

(1.7)
$$II_2 \le \frac{c(p-1)}{|Q_o|} \int_{A_p\left(A^{-1}(t_o)\right)}^{+\infty} \big|\{x : A_p(A^{-1}(M(A(f))(x))) > s\}\big| ds$$
$$\le c(p-1) \fint_{Q_o} A_p\big(A^{-1}(M(A(f))(x))\big) dx$$

Then, putting together (1.3)-(1.7), we have the assertion.

Combining the two previous lemmas, one immediately gets:

Proposition 1.4. *Under the same assumptions of lemma 1.1 we have*

$$\fint_{Q_o} A_p(Mf) dx \le c_1 + c_2\ A_p\Big(A^{-1}\big(\fint_{Q_o} A(f) dx\big)\Big) +$$
$$+ c_3(p-1) \fint_{Q_o} A_p\Big(A^{-1}(M(A(f))(x))\Big) dx$$

2. Proof of the main result. In this section we shall prove the following

Theorem 2.1. *If Ω is a bounded open set of $\mathbf{R}^n$, $f \geq 0$ is measurable, A is a function verifying i),..., iv) and $A(f) \in L^1(\Omega)$, and for any cube $Q \subset \Omega$*

$$A^{-1}\Big(\fint_{Q/2} A(f)dx\Big) \leq b \fint_Q f\ dx + T, \tag{2.1}$$

then, there exists $p > 1$; $b_1, b_2 > 0$, all depending only on b, T, n, k, k_1, k_2, h_1, h_2, such that $A_p(f) \in L^1_{\text{loc}}(\Omega)$ and, for any cube $Q \subset \Omega$

$$A_p^{-1}\Big(\fint_{Q/2} A_p(f)dx\Big) \leq b_1 + b_2 \fint_Q f\ dx. \tag{2.2}$$

From now on, Q_o will denote a fixed cube such that $Q_o \subset\subset \Omega$. In order to prove that (2.2) holds for $Q = Q_o$, we would like to reduce ourselves to the case that f is bounded. This can be done by mean of the following

Proposition 2.1. *If f satisfies (2.1) and $Q_o \subset\subset \Omega$, then there exists a sequence of bounded non negative functions f_n in Q_o such that $\forall Q \subset Q_o$*

$$A^{-1}(\fint_{Q/2} A(f_n)dx) \leq \tilde{b} \fint_Q f_n dx + \tilde{T}$$

where $\tilde{b}, \tilde{T}$ do not depend on n and $f_n \to f$ in $L^1(Q_o)$.

The proof of Proposition 2.1 is based on the following Minkowski-type inequality.

Lemma 2.1. *If $X \subset \mathbf{R}^n$, $Y \subset \mathbf{R}^m$ are measurable sets of positive measure; if A verifies i), ii), iv) and $A(f(x,y)) \in L^1(X \times Y)$, where f is non negative, then*

$$\begin{aligned} &A^{-1}\Big(\fint_Y A\Big(\fint_X f(x,y)dx\Big)dy\Big) \\ &\qquad\qquad \leq c_1 + c_2 \fint_X A^{-1}\Big(\fint_Y A(f(x,y))dy\Big)dx. \end{aligned}$$

PROOF. Of course we may suppose $A(1) = 1$. Let us split $f = f_1 + f_2$ where $f_1(x,y) = f(x,y)$ if $f \geq 1$, otherwise $f_1 = 0$. We also set, for any y in Y

$$\varphi(y) = \fint_X f(x,y)dx$$

and

$$Y^+ = \{y \in Y : \varphi(y) \geq 2\}.$$

Then we can write

$$\begin{aligned}\fint_Y A\Big(\fint_X f(x,y)dx\Big)dy &= \fint_Y A(\varphi(y))dy \\ &\leq \frac{|Y^+|}{|Y|}\fint_{Y+} A(\varphi(y))dy + A(2).\end{aligned} \tag{2.3}$$

Then we have, using Fubini's theorem

$$\begin{aligned}&\fint_{Y+} A(\varphi(y))dy = \fint_{Y+} \frac{A(\varphi(y))}{\varphi(y)}\Big(\fint_X f(x,y)dx\Big)dy \\ &= \fint_X \cdot\Big(\fint_{Y+} \frac{A(\varphi(y))}{\varphi(y)} f(x,y)dy\Big)dx \\ &\leq \fint_X \Big(\fint_{Y+} \frac{A(\varphi(y))}{\varphi(y)} f_1(x,y)dy\Big)dx + \fint_{Y+} \frac{A(\varphi(y))}{\varphi(y)}dy.\end{aligned} \tag{2.4}$$

Set now, for x fixed

$$\alpha = \Big(\fint_{Y+} A(\varphi(y))dy\Big)A^{-1}\big(\frac{1}{\fint_{Y+} A(\varphi(y))dy}\big)$$

$$\bar{\alpha} = A^{-1}\Big(\frac{1}{\fint_{Y+} A(\varphi)dy}\Big)$$

$$\beta = A^{-1}\Big(\fint_{Y+} A(f_1(x,y))dy\Big)$$

and note that

$$\begin{aligned}&\bar{\alpha} \leq A^{-1}\big(\frac{1}{A(2)}\big) \quad , \quad \beta \geq 1 \\ &\alpha \geq A(2)A^{-1}\big(\frac{1}{A(2)}\big).\end{aligned} \tag{2.5}$$

Then, using Young inequality, (1.1) and iv)

$$
\begin{aligned}
&\fint_{Y+} \frac{A(\varphi(y))}{\alpha\varphi(y)} \frac{f_1(x,y)}{\beta} dy \leq \\
&\leq \fint_{Y+} A\left(\frac{\lambda f_1(x,y)}{\beta}\right) dy + \fint_{Y+} \tilde{A}\left(\frac{A(\varphi(y))}{\lambda\alpha\varphi(y)}\right) dy
\end{aligned}
$$

with $\lambda > 1$ to be chosen.
Now we estimate the second integral

$$
\begin{aligned}
\fint_{Y+} \tilde{A}\left(\frac{A(\varphi(y))}{\lambda\alpha\varphi(y)}\right) dy &= \fint_{Y+} \tilde{A}\left(A(\varphi(y)) \frac{A(\bar{\alpha})}{\lambda\bar{\alpha}\varphi(y)}\right) dy \\
&\leq \frac{1}{|Y+|} \int_{Y+\cap\{\bar{\alpha}\varphi\leq 1\}} \tilde{A}\left(\frac{1}{\lambda} A(\frac{1}{\bar{\alpha}}) A(\bar{\alpha})\right) dy \\
&\quad + \frac{1}{|Y+|} \int_{Y+\cap\{\bar{\alpha}\varphi\geq 1\}} \tilde{A}\left(\frac{A(\bar{\alpha}\varphi) + h_1}{k_1\lambda\bar{\alpha}\varphi}\right) dy \\
&\leq \tilde{A}\left(\frac{A(1) + h_1}{k_1\lambda}\right) \\
&\quad + \frac{1}{|Y+|} \int_{Y+\cap\{\bar{\alpha}\varphi\geq 1\}} \tilde{A}\left(\left(1 + \frac{h_1}{A(1)}\right) \frac{1}{\lambda k_1} \frac{A(\bar{\alpha}\varphi)}{\bar{\alpha}\varphi}\right) dy \quad .
\end{aligned}
$$

Now we choose λ such that $\left(1 + \frac{h_1}{A(1)}\right) \frac{1}{\lambda k_1} = 1$, so

$$
\begin{aligned}
\fint_{Y+} \tilde{A}(\frac{A(\varphi(y))}{\lambda\alpha\varphi(y)}) dy &\leq c + \fint_{Y+} \tilde{A}(\frac{A(\bar{\alpha}\varphi(y))}{\bar{\alpha}\varphi(y)}) dy \\
&\leq c + c' \fint_{Y+} A(\bar{\alpha}\varphi(y)) dy \leq c''.
\end{aligned}
$$

Hence

$$
\begin{aligned}
\fint_{Y+} \frac{A(\varphi(y))}{\alpha\varphi(y)} \frac{f_1(x,y)}{\beta} dy &\leq \fint_{Y+} A(\frac{\lambda f_1(x,y)}{\beta}) dy + c'' \\
&\leq c \fint_{Y+} A(f_1(x,y)) A(\frac{1}{\beta}) dy + c''' \leq
\end{aligned}
$$

$$\leq c \fint_{Y+} \frac{A(f_1(x,y))}{A(\beta)} dy + c''' \leq c.$$

Hence

$$\begin{aligned}
\fint_{Y+} \frac{A(\varphi(y))}{\varphi(y)} f_1(x,y)dy &\leq c\alpha\beta \\
&= c\Big(\fint_{Y+} A(\varphi(y))dy\Big) \cdot A^{-1}\Big(\frac{1}{\fint_{Y+} A(\varphi(y))dy}\Big) \cdot \\
&\quad \cdot A^{-1}\Big(\fint_{Y+} A(f_1(x,y))dy\Big) \\
&\leq c\Big(\fint_{Y+} A(\varphi)dy\Big) \frac{1}{A^{-1}\Big(\fint_{Y+} A(\varphi)dy\Big)} \cdot \\
&\quad \cdot A^{-1}\Big(\fint_{Y+} A(f_1(x,y))dy\Big) .
\end{aligned}$$

If we insert this inequality in (2.4) and multiply both sides by

$$A^{-1}\Big(\fint_{Y+} A(\varphi)dy\Big) \Big/ \fint_{Y+} A(\varphi)dy$$

we get

$$\begin{aligned}
A^{-1}&\Big(\fint_{Y+} A(\varphi)dy\Big) \\
&\leq c\fint_X A^{-1}\Big(\fint_{Y+} A(f_1(x,y))dy\Big)dx + \\
&\quad + \fint_{Y+} \frac{A(\varphi)}{\varphi} dy \cdot \frac{A^{-1}\Big(\fint_{Y+} A(\varphi)dy\Big)}{\fint_{Y+} A(\varphi)dy} \\
&\leq c\fint_X A^{-1}\Big(\fint_{Y+} A(f_1(x,y))dy\Big)dx + \\
&\quad + \frac{1}{2} A^{-1}\Big(\fint_{Y+} A(\varphi)dy\Big)
\end{aligned}$$

since on Y^+ we have $\varphi(y) \geq 2$. So we have proved

$$\fint_{Y+} A(\varphi)dy \leq cA\Big(\fint_X A^{-1}\Big(\fint_{Y+} A(f_1(x,y))dy\Big)dx\Big).$$

From this inequality and (2.3) we deduce

$$\begin{aligned}&\fint_Y A\Big(\fint_X f(x,y)dx\Big)dy \leq \\ &\qquad \leq \frac{|Y^+|}{|Y|} cA\Big(\fint_X A^{-1}\Big(\fint_{Y+} A(f_1)dy\Big)dx\Big) + A(2),\end{aligned}$$

and, using twice iv) and noting that

$$\fint_{Y+} A(f_1(x,y))dy \geq 1 \quad \text{for a.e.} \quad x$$

and so also $\fint_X A^{-1}\big(\fint_{Y+} A(f_1)dy\big)dx \geq 1$, while $|Y^+|/|Y| \leq 1$, then we obtain

$$\begin{aligned}&\fint_Y A\Big(\fint_X f(x,y)dx\Big)dy \\ &\qquad \leq c\, A\Big(\fint_X A^{-1}\Big(\frac{|Y^+|}{|Y|}\fint_{Y+} A(f_1(x,y))dy\Big)dx\Big) + c \\ &\qquad \leq c\, A\Big(\fint_X A^{-1}\Big(\fint_Y A(f_1(x,y)\Big)dy\Big)dx + c\end{aligned}$$

from which we get the result.

Now we can give the

Proof of Proposition 2.1. If f satisfies (2.1) and $Q_o \subset\subset \Omega$, we may define for $0 < \epsilon < \operatorname{dist}(Q_o, \partial\Omega)$:

$$f_\epsilon(x) = \fint_{B_1} \eta(y) f(x - \epsilon y)dy$$

where $\eta \in C^1_o(B_1)$, $\eta \geq 0$, $\fint_{B_1} \eta(y)dy = 1$, $\eta(y) \leq c(n)$. Then, if $Q \subset Q_o$, using lemma 2.1

$$\begin{aligned}
&A^{-1}\Big(\fint_{Q/2} A(f_\epsilon(x))dx\Big)\\
&\quad = A^{-1}\Big(\fint_{Q/2} A\Big(\fint_{B_1} \eta(y)f(x-\epsilon y)dy\Big)dx\Big)\\
&\quad \leq c_2 \fint_{B_1} A^{-1}\Big(\fint_{Q/2} A(\eta(y)f(x-\epsilon y))dx\Big)dy + c_1\\
&\quad \leq c_2' \fint_{B_1} A^{-1}\Big(A(\eta(y))\fint_{Q/2} A(f(x-\epsilon y))dx\Big)dy + c_1'\\
&\quad \leq c_2'' \fint_{B_1} A^{-1}\Big(A(\eta(y))A\Big(\fint_{Q} f(x-\epsilon y)dx\Big)\Big)dy + c_2''\\
&\quad \leq \tilde{b} \fint_{B_1} \eta(y)\Big(\fint_{Q} f(x-\epsilon y)dx\Big)dy + \tilde{T}\\
&\quad = \tilde{b} \int_{Q} f_\epsilon(x)dx + \tilde{T}.
\end{aligned}$$

In the following, se suppose that $Q_o = [0,1]^n$. Then, if we set $\|x\| = \max_{1\leq i\leq n} |x_i|$, $\forall x \in \mathbf{R}^n$ and

$$\rho(x) = (1 - \|x\|)^n \quad \forall x \in Q_o,$$

then the following estimates hold:

Proposition 2.2. *If is bounded, $f \geq 0$ in Q_o, and A satisfies i), ii), iv), then, for a.e. $x \in Q_o$:*

$$\begin{aligned}
&A_p(\rho(x)Mf(x))\\
&\qquad \leq c_1\, A_p(M(\rho f)(x)) + c_2\, A_p\Big(A^{-1}\Big(\fint_{Q_o} A(f)dy\Big)\Big)
\end{aligned} \tag{2.6}$$

$$\begin{aligned}
&A_p(A^{-1}(M(A(\rho f))(x)))\\
&\qquad \leq c_1 + c_2 A_p(A^{-1}(A(\rho(x)))WMA(f)(x))\\
&\qquad\quad + c_3\, A_p\Big(A^{-1}\Big(\fint_{Q_o} A(f)dy\Big)\Big)
\end{aligned} \tag{2.7}$$

where $p > 1$.

PROOF. Let us fix $x \in Q_o$ and let Q be a cube such that $x \in Q \subset Q_o$. We split the proof in two cases:

Case 1: $\rho(x) \leq 2^n|Q|$.

In this case we have obviously $\sup_Q \rho(y) \leq 3^n|Q|$. Then

$$
\begin{aligned}
A^{-1}\Big(\fint_Q A(\rho(y)f(y))dy\Big) &\leq A^{-1}\Big(\fint_Q A(3^n|Q|f(y))dy\Big) \\
&\leq c\, A^{-1}\Big(|Q|\fint_Q A(3^n f(y))dy\Big) \qquad (2.8)\\
&\leq c'\, A^{-1}\Big(\fint_{Q_o} A(f(y))dy\Big)
\end{aligned}
$$

$$
\rho(x)\fint_Q f dy \leq 3^n|Q|\fint_Q f dy \leq 3^n A^{-1}\Big(\fint_{Q_o} A(f)dy\Big) \qquad (2.9)
$$

Case 2: $\rho(x) > 2^n|Q|$.

In this case we have easily that, for any $y \in Q$

$$
\big(\frac{2}{3}\big)^n \leq \frac{\rho(x)}{\rho(y)} \leq 2^n.
$$

Then:

$$
\rho(x)\fint_Q f dy \leq 2^n \fint_Q \rho(y)f(y)dy \leq 2^n M(\rho f)(x) \qquad (2.10)
$$

$$
\begin{aligned}
&A^{-1}\Big(\fint_Q A(\rho f)dy\Big)\\
&= A^{-1}\Big(\frac{1}{|Q|}\int_{Q\cap\{f\le 1\}} A(\rho f)dy + \frac{1}{|Q|}\int_{Q\cap\{f>1\}} A(\rho f)dy\Big)\\
&\le A^{-1}\Big(A(1) + \frac{k_2}{|Q|}\int_{Q\cap\{f\ge 1\}} A(\rho)A(f)dy + h_2\Big)\\
&\le A^{-1}\Big(c + k_1 A\big((\tfrac{3}{2})^n\rho(x)\big)\fint_Q A(f)dy\Big)\\
&\le A^{-1}(c + c\,A(\rho(x))WMAf(x))\\
&\le c + c\,A^{-1}(A(\rho(x))WMAf(x)).
\end{aligned}
\tag{2.11}
$$

Then, from (2.9) and (2.10) it follows

$$A_p\Big(\rho(x)\fint_Q f dy\Big) \le cA_p\Big(A^{-1}\Big(\fint_{Q_0} A(f)dy\Big)\Big) + cA_p(M(\rho f)(x))$$

from which we get (2.6). On the other side, from (2.8) and (2.11) we have

$$
\begin{aligned}
A_p\Big(A^{-1}\Big(\fint_Q A(\rho(y)f(y))dy\Big)\Big)&\\
\le cA_p\Big(A^{-1}\Big(\fint_{Q_o} A(f)dy\Big)\Big) + c&\\
+ A_p\Big(A^{-1}(A(\rho(x))WMAf(x))\Big)&
\end{aligned}
$$

which implies (2.7).

We can finally give the

Proof of Theorem 2.1. By Proposition 2.1 it is clear that we can limit ourselves to prove the theorem in the case f is bounded. If $Q_o = [0,1]^n$, integrating both sides of (2.6), we obtain

$$
\begin{aligned}
\fint_{Q_o} A_p(\rho Mf)dx \le c_1 \fint_{Q_o} A_p(M(\rho f))dx +&\\
+ c_2\, A_p\Big(A^{-1}\Big(\fint_{Q_o} A(f)dx\Big)\Big).&
\end{aligned}
$$

Then, applying proposition 1.4 to ρf, we get:

$$\begin{aligned}\fint_{Q_o} A_p(\rho Mf)dx \le c_1' + c_2' A_p\Big(A^{-1}\Big(\fint_{Q_o} A(\rho f)dx\Big)\Big)\\ + c_3'(p-1)\fint_{Q_o} A_p(A^{-1}(M(A(\rho f))))dx\\ + c_2 A_p\Big(A^{-1}\Big(\fint_{Q_o} A(f)dx\Big)\Big).\end{aligned}$$

Hence, using (2.7)

$$\begin{aligned}\fint_{Q_o} A_p(\rho Mf)dx \le c_1'' + c_2'' A_p\Big(A^{-1}\Big(\fint_{Q_o} A(f)dx\Big)\Big)\\ + c_3''(p-1)\fint_{Q_o} A_p(A^{-1}(A(\rho)WMA(f)))dx\end{aligned}$$

Now we remark that assumption (2.1) implies that, for any $x \in Q_o$, a.e.

$$WMA(f)(x) \le A(bMf(x) + T).$$

Then, from the above estimate, we obtain

$$\begin{aligned}\fint_{Q_o} A_p(\rho Mf)dx &\le c_4 + c_2'' A_p\Big(A^{-1}\Big(\fint_{Q_o} A(f)dx\Big)\Big)\\ &\quad + c_5(p-1)\int_{Q_o} A_p(A^{-1}(A(\rho)A(Mf(x))))dx\\ &\le c_6 + c_2'' A_p\Big(A^{-1}\Big(\fint_{Q_o} A(f)dx\Big)\Big)\\ &\quad + c_7(p-1)\fint_{Q_o} A_p(\rho Mf)dx\end{aligned}$$

where we have applied assumption iv).

Then there exists a suitable $p_o > 1$ such that for $1 < p < p_o$

$$\fint_{Q_o} A_p(\rho(x)Mf)dx \le c_8 + c_9 A_p\Big(A^{-1}\Big(\fint_{Q_o} A(f)dx\Big)\Big)$$

which immediately gives

$$\fint_{Q_o/2} A_p(f)dx \leq c_8' + c_9' A_p\left(A^{-1}\left(\fint_{Q_o} A(f)dx\right)\right)$$
$$\leq c_8'' + c_9''\ A_p\left(\fint_{2Q_o} f dx\right)$$

if $2Q_o \subset \Omega$. Then, by a change of variable one gets that for any cube $Q \subset \Omega$

$$A_p^{-1}\left(\fint_{Q/4} A_p(f)dx\right) \leq b_1 + b_2 \fint_Q f\ dx$$

which, by a simple argument, implies (2.2).

Remark 2.1. If A satisfies the assumptions of theorem 2.1 and there exists $\delta > 0$ such that

$$A(t) \geq c\ t^{1+\delta} \qquad \forall t \geq 1$$

with $c > 0$, then there exists $q > 1$ such that

$$A^q(f) \in L^1_{\text{loc}}(\Omega).$$

In fact, in this case, it is enough to note that, for $t \geq 1$

$$A_p(t) = \left[\frac{A(t)}{t}\right]^{p-1} A(t) \geq ct^{\delta(p-1)} A(t)$$
$$\geq c\left[\frac{A(t)}{A(1)}\right]^{\frac{\delta(p-1)}{k-1}} = c' A^q(t)$$

(see Remark 1.1).

References

[1] R.A.Adams, *Sobolev Spaces*, Academic Press, New York, (1975).
[2] B.V.Bojarskii, T. Iwaniec, *Analitical foundations of the theory of quasi-conformal mappings in* $\mathbf{R}^n$, Annales Ac. Sci. Fen. **8** (1983), 257-324.

[3] F.W.Gehring, *The L^p-integrability of the partial derivatives of a quasi conformal mapping*, Acta Math. **130** (1973), 265-277.
[4] M.Giaquinta, *Multiple integrals in the calculus of variations and nonlinear elliptic systems*, Princeton Univ. Press (1983).
[5] M.Giaquinta, G.Modica, *Regularity results for some classes of higher order non linear elliptic systems.* J. Reine Angew. Math. **311/312** (1979), 145-169.
[6] M.A.Krasnosel'skii, Y.B.Ruticki, *Convex functions and Orlicz Spaces*, Gordon and Breach, New York (1961).
[7] C.Sbordone, *On some integral inequalities and their applications to the calculus of variations*, Boll. UMI, Analisi Funzionale e Appl., Ser. VI, Vol. V, C, **1** (1986), 73-94.
[8] B.Stroffolini, *Medie di tipo Jensen in spazi di Orlicz*, Acc. Sci. Fis. Mat. Napoli (1987), to appear.

Istituto di Matematica
Facoltà di Scienze
Università di Salerno
I-84100 SALERNO

Dipartimento di Matematica e Applicazioni "R.Caccioppoli"
Università di Napoli
Via Mezzocannone 8
I-80134 NAPOLI

PARTIAL REGULARITY OF CARTESIAN CURRENTS WHICH MINIMIZE CERTAIN VARIATIONAL INTEGRALS

MARIANO GIAQUINTA   GIUSEPPE MODICA   JIŘI SOUČEK

Dedicated to Ennio de Giorgi on his sixtieth birthday

1. Introduction. This paper deals with the local regularity of minimizers of functionals of the type

$$\mathcal{F}(u;\Omega) = \int_\Omega F(x, u(x), M(Du(x)))dx \tag{1.1}$$

where Ω is an open set in $\mathbb{R}^n, u : \Omega \to \mathbb{R}^N$, $n \geq 2$, $N \geq 2$, $M(Du(x))$ stands for all minors of the Jacobian matrix Du of u and $F(x,u,M) : \Omega \times \mathbb{R}^\ell \to \mathbb{R}^+$ $\ell := \binom{n+N}{n} - 1$ is a smooth function which is convex in M.

More precisely, we shall indicate by $I(p,n)$ the family of *increasing* multiindeces α of lenght $|\alpha| = p$

$$I(p,n) := \{\alpha = (\alpha_1, .., \alpha_p) : \quad 1 \leq \alpha_1 < \alpha_2 < ... < \alpha_p \leq n\}$$

setting $I(0,n) := \{0\}$. If $\alpha \in I(p,n)$, $\bar{\alpha}$ will denote the multiindex which is *complementary* to α, i.e. whose components together with the components of α give $\{1,...,n\}$ moreover we shall denote by σ_α the permutation which reorders in an increasing way $(\alpha, \bar{\alpha})$. In particular we have $\bar{0} = (1,2,...,n)$, If $G \in M_{N,n}$ is a matrix with N row and n column and $\alpha, \beta \in I(p, \bar{n})$, $\bar{n} := \min(n, N)$, we denote by $M_{\beta\alpha}(G)$ the determinant of the minor of G with β rows and α columns, and we set $M_{oo}(G) = 0$. Finally, $M(G)$ will roughly denote the set of all possible minors $M_{\beta\alpha}(G)$, and we denote its norm by

$$|M(G)|^2 = \Big(\sum_{\substack{|\alpha|+|\beta|=n \\ |\beta|>0}} M_{\beta\bar{\alpha}}(G)^2 \Big)^{1/2}. \tag{1.2}$$

Notice that $(1 + |M(G)|^2)^{1/2}$ is the area element of the graph of the map u.

In [9], assuming that

$$F(x, u, M) \geq |M|^m \quad m > 1 \tag{1.3}$$

we have proved existence of minimizers of (1.1) in the class of the so-called cartesian currents $\text{Cart}^m(\Omega, \mathbb{R}^N)$ under various boundary conditions. Let us recall some definitions and the main steps, referring to [9] for more information and proofs.

To every function $u \in C^1(\Omega, \mathbb{R}^N)$ we associate the n-rectifiable current T_u in $\mathbb{R}^{n+N}$ integration of n-forms over the graph of u and we introduce the "norm"

(1.4)
$$\|T_u\|_{\text{Cart}^m(\Omega,\mathbb{R}^N)} := \Big(\int_\Omega |u|^m dx \Big)^{1/m} + \Big(\int_\Omega |M(Du)|^m dx \Big)^{1/m}.$$

$\text{Cart}^m(\Omega, \mathbb{R}^N)$ is then defined as the sequential closure of smooth graphs with respect to the weak convergence of currents in U with equibounded $\text{Cart}^m(\Omega, \mathbb{R}^N)$-norms. $\text{Cart}^m(\Omega, \mathbb{R}^N)$ is neither a linear space nor a convex set. $\text{Cart}^m(\Omega, \mathbb{R}^N)$ can be obtained by transfinite induction starting from smooth graphs and taking successively weak limits. One sees that if $\{u_n\}$ is an equibounded sequence in $\text{Cart}^m(\Omega, \mathbb{R}^N)$ converging in the sense of currents to a function $u \in \text{Cart}^m(\Omega, \mathbb{R}^N)$, then

$$M(Du_k) \rightharpoonup M(Du) \qquad \text{weakly in } L^m, \tag{1.5}$$

see [9].

The coercitivity condition (1.3) together with the convexity of the integrand F with respect to M and (1.5) which holds for a minimizing sequence (or subsequence) yield at once the existence of minimizers under various boundary conditions.

The next result which is reasonable to expect and one would like to prove is a partial regularity theorem for minimizers u in $\mathrm{Cart}^m(\Omega, \mathbb{R}^N)$, saying that, under certain smoothness and growth conditions on the integrand F a minimizer is actually a C^1 map, except for a closed singular set of measure zero. It is the aim of this paper to prove such a theorem, but unfortunately only for a class of integrands which satisfy a sharp growth condition.

We approach the partial regularity question relying on ideas introduced by De Giorgi [5] and developed by many authors, see e.g. [1], [2], [4], [6] and [10] in the context of the study of minimal surfaces or more generally in the theory of parametric elliptic functionals. Our method of proof is especially close to the ones of [10], [3]. The paper [10] deals with the regularity of rectifiable currents which minimize parametric elliptic functionals, while [3] deals with the regularity of minimizers of non parametric integrals with integrands which depend only on x, u and on the minors of the first order of Du, i.e. on the gradient of u in a convex way. In order to handle the extra difficulties due to the appearance of the minors of order larger than or equal to 2, when dealing with the functional (1.1), one of the key points seems to be the so-called *height bound*, which fives a control of the local oscillation of a minimizer u, provided u is sufficiently flat in the mean. Because of the nonparametric character of our integrals, and especially of the growth $m > 1$ we are not able to prove (or disprove) the validity of such an estimate. We evercome this point by assuming that our integrand is moreover coercive with respect to $|Du|^r$ for some r larger than the dimension n

$$F(x, u, M(Du)) \geq |M(Du)|^m + |Du|^r. \tag{1.6}$$

Of course we can and shall assume that $r \geq m$. Then Sobolev theorem yields the height bound we need, and following in principle [3], [10] we are able to prove a partial regularity theorem. More precisely we shall prove

Theorem 1.1. *Let $F(M) : \mathbb{R}^\ell \to \mathbb{R}$ be a convex function such that*

$$(1.7) \quad |M(Du)|^m + |Du|^r \le F(M(Du)) \le c_1(1 + |M(Du)|^m + |Du|^r)$$

where $m > 1$ and $r > n$, and assume that $u \in \mathrm{Cart}^m(\Omega, \mathbb{R}^N) \cap H^{1,r}(\Omega, \mathbb{R}^N)$ is a local minimum for the functional

$$\mathcal{F}(u;\Omega) = \int_\Omega F(M(Du))dx.$$

If for some $x_o \in \Omega$, $p_o \in \mathbb{R}^{n^N}$, one has

$$(1.8) \qquad \lim_{R\to 0} \fint_{B_R(x_o)} |Du - p_o|^r dx = 0$$

$$(1.9) \qquad F \in C^2(A) \quad \text{for some neighbourhood} \quad A \quad \text{of} \quad M(p_o)$$

$$(1.10) \quad F_{MM}(M(p_o))\xi\xi \ge \lambda|\xi|^2 \quad \text{for all} \quad \xi \in \mathbb{R}^\ell\ , \ \text{for some} \quad \lambda > 0,$$

then $u \in C^{1,\alpha}$ for all $\alpha \in (0,1)$ in a neighbourhood of x_o.

A simple consequence of the above theorem is the following

Corollary 1.1. *Let $F(M)$ be convex and satisfy (1.7). Assume also that F is of class C^2 and that*

$$F_{MM}(M)\xi\xi \ge \lambda(M)|\xi|^2$$

for all $M, \xi \in \mathbb{R}^\ell$, for some function $\lambda(M) > 0$. Then if $u \in \mathrm{Cart}^m(\Omega, \mathbb{R}^N) \cap H^{1,r}(\Omega, \mathbb{R}^N)$ is a local minimum for $\mathcal{F}$ in Ω, there is an open set $\Omega_o \subset \Omega$ such that $u \in C^{1,\alpha}_{\mathrm{loc}}(\Omega_o, \mathbb{R}^N)$ and meas $(\Omega - \Omega_o) = 0$.

Minor variants in the proof will also allow to prove partial Hölder continuity of the first derivatives of minimizers u of integrals in (1.1), assuming moreover Hölder continuity of the integrand F with respect to x and u; but, as this is quite standard by now, compare e.g. [7], [8], we shall not insist on this point.

Finally we notice that our proof works and yields similar results also when the integrand has different growths $m_{\alpha\beta} > 1$ with respect to each minor $M_{\alpha\beta}$

$$\sum |M_{\alpha\beta}|^{m_{\alpha\beta}} + |Du|^r \leq$$

$$\leq F(y, u, M(Du)) \leq c_1(1 + |Du|^r + \sum |M_{\alpha\beta}(Du)|^{m_{\alpha\beta}}).$$

However, in order to avoid a mess in writing and extra technicalities, we shall not carry it out in detail, restricting ourselves to prove the theorem and corollary stated above.

2. Some simple estimates.

We denote by $e_1, ..., e_n$ and $\hat{e}_1, ..., \hat{e}_N$ respectively two basis on $\mathbb{R}^n_x$ and $\mathbb{R}^N_y$. Given a matrix G, the minors of G are naturally associated to the n-plane graph of the map $x \in \mathbb{R}^n \to Gx \in \mathbb{R}^N$. In fact the n-simple vector in $\Lambda_n \mathbb{R}^{n+N}$ associated to this plane is given by

$$e_1 \wedge e_2 \wedge ... \wedge e_n + M(G)$$

where

$$M(G) = \sum_{\substack{|\alpha|+|\beta|=n \\ |\beta|>o}} \text{sign} \quad \sigma_\alpha M_{\beta\bar{\alpha}}(G) e_\alpha \wedge \hat{e}_\beta. \tag{2.1}$$

We shall think of the integrand F in theorem 1.1 as defined on the hyperplane

$$\Lambda_+ := \{\xi \in \Lambda_n \mathbb{R}^{n+N} :< \xi, e_1 \wedge ... \wedge e_n >= 0\}. \tag{2.2}$$

We collect now a few simple estimates on $M(G) : M_{n\times N} \to \Lambda_+$ which will be freely used in the following. Obviously

$$|G| \leq |M(G)| \leq c(|G| + |G|^{\bar{n}}) \quad \bar{n} = \min(n, N). \tag{2.3}$$

Let $\alpha \in I(p,n), \alpha' \in I(k,n) \quad \alpha' \leq \alpha$; we denote by α'' the complementarity multiindex to α' in α and by $\sigma_{\alpha'\alpha''}$ the permutation which reorders (α', α''). Then, for $G, H \in M_{n\times N}$ we have

(2.4)

$$M_{\beta\alpha}(G + H) = \sum_{\substack{\alpha' \leq \alpha \\ \beta' \leq \beta}} \text{sign } \sigma_{\alpha'\alpha''} \text{ sign } \sigma_{\beta'\beta''} M_{\beta'\alpha'}(G) M_{\beta''\alpha''}(H).$$

This yields at once

$$
(2.5)\qquad
\begin{aligned}
&|M(G+H)-M(H)| \le c(1+|M(H)|)|M(G)| \\
&|M(G)-M(H)| \le c(1+|M(H)|)|M(G-H)|
\end{aligned}
$$

moreover writing (2.4) for $G-H$ and $H-H$ and substracting we get

$$
(2.6)\qquad |M(G-H)| \le c(1+|M(H)|)|M(G)-M(H)|.
$$

In the sequel we need to consider also the "linear part" of $M(G)$, defined by

$$
L(G) := \sum_{\substack{|\alpha|+|\beta|=n \\ |\beta|=1}} \operatorname{sign} \sigma_\alpha M_{\beta\bar{\alpha}}(G) e_\alpha \wedge \hat{e}_\beta =
$$

$$
= \sum_{i=1}^{n} \sum_{j=1}^{N} (-1)^{n-i} G_{ij} e_i \wedge \hat{e}_j
$$

and the remainder

$$
R(G) := M(G) - L(G).
$$

Clearly $R(G)$ and $L(G)$ are orthogonal and

$$
(2.7)\qquad
\begin{aligned}
&|L(G)| = |G| \\
&|M(G)|^2 = |L(G)|^2 + |R(G)|^2 \\
&|R(G)| \le |M(G)|.
\end{aligned}
$$

Moreover, since the components of $R(G)$ are the minors of order ≥ 2, we have

$$
(2.8)\qquad |R(G)| \le c(|G|^2 + |G|^{\bar{n}}).
$$

Finally, using Laplace formula, one easily gets

$$
|R(G)| \le c|G||M(G)|
$$

and thus

$$|R(G)| \leq c \min(|M(G)|, |M(G)|^2).$$

In order to simplify our computations, it is also convenient to introduce the following function

$$\gamma_s(t) := (1+t^2)^{s/2} - 1 \qquad t > 0, \quad s > 1. \tag{2.9}$$

The following simple properties are easily checked: γ_s is positive and convex;

$$\min(t^s, t^2) \leq \gamma_s(t) \leq c \min(t^s, t^2);$$

for $\lambda > 1$, $\gamma_s(\lambda t) \leq \lambda^s \gamma_s(t)$;

$$\gamma_s(t+\tau) \leq c\big[\gamma_s(t) + \gamma_s(\tau)\big] \leq 2c\gamma_s(t+\tau);$$

for $t \leq k$, there is a positive vonstant $c(k)$ so that

$$c^{-1}(k)t^2 \leq \gamma_s(t) \leq c(k)t^2.$$

Finally we set

$$d(G) := \gamma_m(|M(G)|). \tag{2.10}$$

Notice that

$$|R(G)| \leq c\, d(G) \tag{2.11}$$

$$\begin{aligned} d(G+H) &\leq c[d(H) + \gamma_m((1+|M(H)|)|M(G)|) \leq \\ &\leq c[d(H) + (1+|M(H)|^m)d(G)] \end{aligned} \tag{2.12}$$

$$d(G) \leq c(k)|G|^2 \quad \text{for} \quad |G| \leq k. \tag{2.13}$$

3. Preliminary results. Let $u \in H^{1,r}(\Omega, \mathbb{R}^N)$ and $B_R(x_o) \subset\subset \Omega$. We denote by $p_R = p_{x_o,R}$ the average of Du on $B_R(x_o)$

$$p_R := \fint_{B_R(x_o)} Du\, dx \tag{3.1}$$

and we introduce the "linear excess"

$$\mathcal{E}(u, B_R) := R^{-n} E(u, B_R)$$

(3.2)
$$E(u, B_R) := \int_{B_R(x_o)} \gamma_r(|Du - p_R|)dx.$$

The, we consider the mollified function

$$u_\epsilon(x) := \fint_{B_\epsilon(x)} u(y)dy$$

and

$$u_{\epsilon,\delta}(x) := u_\epsilon * \eta_\delta(x)$$

where η is a standard radial symmetric mollifying kernel. We have, compare [3]

Proposition 3.1. *Let* $u \in H^{1,r}(\Omega, \mathbb{R}^N)$, $x_o \in \Omega$, $R < \frac{1}{2}\operatorname{dist}(x_o, \partial\Omega)$, $\alpha \in (0,1)$. *If*

$$\mathcal{E} = \mathcal{E}(u, B_R(x_o)) < 1$$

and we choose

(3.4)
$$\epsilon = \delta = \frac{1}{16} R\, \mathcal{E}^\gamma \quad \textit{with} \quad 0 < \gamma < \frac{1}{n + 2\alpha},$$

then

$$\sup_{B_{7R/8}(x_o)} |Du_{\epsilon,\delta} - p_R| + R^\alpha [Du_{\epsilon,\delta}]_{0,\alpha,\frac{7R}{8}} \leq c\mathcal{E}^\beta$$

(3.5)
$$\int_{B_{7R/8}(x_o)} |Du_{\epsilon,\delta} - p_R|^2 dx \leq c\, \mathcal{E}(u, B_R),$$

where β is a positive number with β and c depending only on n, r, α, γ and $[\ ]_{0,\alpha,\frac{7R}{8}}$ denotes the α Hölder seminorm on $B_{\frac{7R}{8}}(x_o)$.

Let F be the integrang in theorem 1.1. We now define a new integrand $f : M_{n\times N} \to \mathbb{R}$ as

(3.6)
$$f(G) := F(L(G - p_R) + M(p_R))$$

where L is the "linear part" of $M(G)$ defined in section 2. It is easily seen that f is convex and of class C^2 in a ball $B_\sigma(p_o)$ provided $p_R \in B_\sigma(p_o)$. Moreover we have

$$|f_{pp}(p) - f_{pp}(\bar{p})| \leq \omega(|p-\bar{p}|) \quad \forall p,\bar{p} \in B_\sigma(p_o) \tag{3.7}$$

where ω is a bounded concave and continuous function with $\omega(0) = 0$, and also

$$\begin{gathered} \lambda|\xi|^2 \leq f_{pp}(p)\xi\xi \leq \Lambda|\xi|^2 \\ \forall \xi \in \mathbb{R}^{nN} \quad \forall p \in B_\sigma(p_o) \quad ; \quad \lambda > 0. \end{gathered} \tag{3.8}$$

Therefore the integrand f is of the type considered in [3] and in particular we have

Proposition 3.2. *Let $v := u_{\epsilon,\delta}$ where the mollifying parameters are determined by (3.4). Then, for all $\tau \in (0, \frac{1}{2})$ there is $a(\tau)$ such that if*

$$p_R \in B_\sigma(p_o), \quad \mathcal{E}(u, B_R(x_o)) < a(\tau) \tag{3.9}$$

then

$$\begin{aligned} &\int_{B_{\tau R}} \gamma_m(|Dv - (Dv)_{\tau R}|)dx \leq \\ &\leq c\,\tau^{n+2} \int_{B_R} \gamma_m(|Du - p_R|)dx + c\,\psi(v, \frac{R}{2}), \end{aligned} \tag{3.10}$$

where

$$\psi(v, \frac{R}{2}) := \int_{B_{R/2}} f(Dv)dx - \int_{B_{R/2}} f(DH)dx \tag{3.11}$$

and H is the solution of the boundary value problem

$$\begin{cases} f_{p^i_\alpha p^j_\beta}(p_R) D_\alpha D_\beta H^i = 0 & in \quad B_{R/2} \\ H = v & on \quad \partial B_{R/2}. \end{cases} \tag{3.12}$$

As in [3] estimate (3.10) will be the starting point for the proof of theorem 1.1, which, modulus an iteration argument, will be reduced to suitably estimating the left-hand-side of (3.11) from below and the error term $\psi(v, R/2)$ from above.

4. Proof of the theorem. Let $u \in \mathrm{Cart}^m(\Omega, \mathbb{R}^N) \cap H^{1,r}(\Omega, \mathbb{R}^N)$ be a minimizer of the functional $\mathcal{F}$ in theorem 1.1. We define for $x_o \in \Omega$, $R < \mathrm{dist}(x_o, \partial\Omega)$

$$\mathcal{D}(u, B_R(x_o)) := R^{-n} D(u, B_R(x_o))$$

(4.1)
$$D(u, B_R(x_o)) := \int_{B_R(x_o)} d(Du - p_R)dx$$

where d is the function defined in (2.10).

Theorem 1.1 follows by an iteration argument similar to the one of [3] sec.2 from the decay estimate for the "excess"

$$\mathcal{E}(u, B_R) + \mathcal{D}(u, B_R)$$

contained in the following

Main Proposition. *Suppose that the assumptions of theorem 1.1 hold. Then for all $\tau \in (0, 1/4)$ there exists $a_1(\tau)$ $0 < a_1(\tau) < 1$ such that, if u is a minimizer, $B_R(x_o) \subset\subset \Omega$ and*

$$\mathcal{E}(u, B_R) < a_1(\tau)$$

then

$$E(u, B_{\tau R}(x_o)) + D(u, B_{\tau R}(x_o)) \le c\, \tau^{n+2}[E(u, B_R(x_o) + D(u, B_R(x_o))].$$

We shall prove that, provided $\mathcal{E}(u, B_R)$ is sufficiently small, we have

(4.2)
$$U(u, B_{\tau R}) + D(u, B_{\tau R}) \le$$

$$\le c \int_{B_{2\tau R}} \gamma_m(|Dv - p_{2\tau R}|)dx + c\, \tau^{n+2}[E(u, B_R) + D(u, B_R)]$$

$$\psi(v, R/2) \le c\,\tau^{n+2}[E(u, B_R) + D(u, B_R)], \tag{4.3}$$

where $v = u_{\epsilon,\delta}$ is the function defined in proposition 3.2. This of course will prove the main proposition taking into account proposition 3.2. In order to do that we have to establish a few preparatory facts.

For each $p^* \in B_\sigma(p_o)$ we set $M^* = M(p^*)$ and we associate to F the new integrand $\bar{F}$ defined as

$$\bar{F}(M) := F(M) - F(M^*) - F_M(M^*)(M - M^*). \tag{4.4}$$

Clearly $\bar{F}$ is convex and vanishes together with its derivatives at M^*. Also one easily sees that there are possitive constants $\underline{c}$, $\bar{c}$ depending on the data and in particular on σ and on M^* so that

$$\begin{aligned} &\underline{c}[\gamma_r(|Du - p^*|) + \gamma_m(|M - M^*|)] \le \\ &\qquad \le \bar{F}(M) \le \bar{c}[\gamma_r(|Du - p^*|) + \gamma_m(|M - M^*|)] \end{aligned} \tag{4.5}$$

and, actually $\underline{c}^{-1}$ and $\bar{c}$ are increasing with respect to $|M^*|+ |F(M^*)| + |F_M(M^*)|$. Moreover for $|M - M^*| \le k$ we have

$$|\bar{F}_M(M)| \le c(k)|M - M^*|. \tag{4.6}$$

Let $u \in \mathrm{Cart}^m(\Omega, \mathbb{R}^N) \cap H^{1,r}(\Omega, \mathbb{R}^N)$, $v \in C^1(\Omega, \mathbb{R}^N)$ and let $B_R \subset\subset \Omega$; consider the function

$$w(x) = \eta(x)u(x) + (1 - \eta(x))v(x) \tag{4.7}$$

where $\eta \in C_o^\infty(\Omega)$ with $\eta \equiv 0$ in $\Omega \backslash B_R$. Then we claim that

$$\begin{aligned} &\int_\Omega [\bar{F}(M(Du)) - \bar{F}(M(Dw))]\,dx = \\ &\qquad = \int_\Omega [F(M(Du)) - F(M(Dw))]\,dx. \end{aligned} \tag{4.8}$$

This is consequence of the following lemma

Lemma 4.1. *Let* $u \in \mathrm{Cart}^m(\Omega, \mathbb{R}^N)$, $v \in C^1(\Omega, \mathbb{R}^N)$ *and* w *be given by (4.7). Then* $w \in \mathrm{Cart}^m(\Omega, \mathbb{R}^N)$ *and*

$$\int_{B_R} M_{\alpha\beta}(Du)dx = \int_{B_R} M_{\alpha\beta}(Dw)dx \quad \forall \alpha, \beta. \tag{4.9}$$

PROOF. If $u \in C^1(\Omega, \mathbb{R}^N)$, the result is well known. Otherwise, let $u_h \in C^1(\Omega, \mathbb{R}^N)$ be a sequence weakly converging to u with $\|u_k\|_{\mathrm{Cart}^m(\Omega,\mathbb{R}^N)} \leq c$. Then $w_k := \eta u_k + (1-\eta)v$ weakly converge to w and are equibounded in Cart^m, that is $w \in \mathrm{Cart}^m(\Omega, \mathbb{R}^N)$. On the other hand (4.9) holds for u and w replaced respectively by u_k and w_k, so the result follows, as the minors also converge weakly. The general case is then proved by transfinite induction.

Finally let us recall the following lemmata which are proved in [3].

Lemma 4.2. *Let* ϵ, δ *be given positive numbers. Then for any two numbers* τ_1, τ_2 *such that*

$$0 < \tau_1 < \tau_2 < R - (\epsilon + \delta)$$

there is a number $\tau \in (\tau_1, \tau_2)$ *such that*

$$\int_{B_\tau} [\bar{F}(M(Du)_{\epsilon,\delta}) - \bar{F}(M(Du))]\, dx \leq \tag{4.10}$$

$$\leq 2\frac{\epsilon+\delta}{\tau_2 - \tau_1} \int_{B_R} \bar{F}(M(Du))dx.$$

Lemma 4.3. *Under the same assumptions of lemma 4.2, suppose that* ρ, τ_1, τ_2 *are given with* $0 < \rho < R/4$ *and* $R/2 \leq \tau_1 < R - (\epsilon + \delta)$. *Then there exist* $\bar{\rho} \in (\rho, 2\rho)$ *and* $\tau \in (\tau_1, \tau_2)$ *such that*

$$\int_{B_\tau \setminus B_\rho} [\bar{F}(M(Du)_{\epsilon,\delta}) - \bar{F}(M(Du))]\, dx \leq \tag{4.11}$$

$$\leq 2\Big(\frac{\epsilon+\delta}{\tau_2 - \tau_1} + \frac{\epsilon+\delta}{\rho}\Big) \int_{B_R} \bar{F}(M(Du))dx.$$

Lemma 4.4. *Let $u \in H^{1,s}(B_R, \mathbb{R}^N)$ and let ϵ, τ, t, k be positive numbers so that $R/2 < t < s < R$ and $\tau + \epsilon + \delta < R$. Then there exists a constant c depending only on s such that*

$$\begin{aligned} &\int_{B_\tau \setminus B_t} \gamma_s(k|u - u_{\epsilon,\epsilon}|)dx \leq \\ &\leq c \, \max((\epsilon k)^2, (\epsilon k)^s) \int_{(B_\tau \setminus B_t)_{2\epsilon}} \gamma_s(|Du|)dx, \end{aligned} \tag{4.12}$$

where

$$(B_\tau \setminus B_t)_{2\epsilon} := \{x \in B_R : \operatorname{dist}(x, B_s \setminus B_t) < 2\epsilon\}.$$

We are now ready to prove (4.2). For any $\tau \in (0, \frac{1}{2})$ set

$$p^* := \int_{B_{\tau R}(x_o)} Dv \, dx \quad M^* = M(p^*).$$

By (3.8) in proposition 3.1 we have

$$|p^* - p_R| \leq \fint_{B_{\tau R}} |Dv - p_k| dx \leq c \, \mathcal{E}^\beta \tag{4.13}$$

and we choose $a_2 \in (0, \frac{1}{2})$ so that $\mathcal{E} < a_2$ implies

$$|p^* - p_R| < \min\left(\frac{\sigma}{2}, 1\right), \quad |M(p^*) - M(p_R)| \leq \frac{\sigma}{2}.$$

Therefore we have

$$\lambda|\xi|^2 \leq \bar{F}_{MM}(M^*)\xi\xi \leq \Lambda|\xi|^2$$

$$|\bar{F}_{MM}(M) - \bar{F}_{MM}(M^*)| \leq \omega(|M - M^*|)$$

for all $M \in B_{\delta/2}(M^*)$ and the estimates (4.5), (4.6) hold with constants depending on $|p_R| + |F(M(p_R))| + |F_M(M(p_R)|$ that from now on will not be specified.

1. The main step consist in proving that for any $\tau \in \left(0, \frac{1}{4}\right)$

$$\begin{aligned}
&\int_{B_{\tau R}} \bar{F}(M(Du))dx \leq \\
(4.14) \qquad &\leq \int_{B_{2\tau R}} \bar{F}(M(Dv))dx + c\, \mathcal{E}^{\gamma_o} \int_{B_R} \bar{F}(M(Du))dx
\end{aligned}$$

where $\mathcal{E} = \mathcal{E}(u, B_R(x_o))$ and $\gamma_o > 0$.
Let s, t be so that $\frac{R}{2} < t < s < \frac{7R}{8}$ and let $\eta \in C_o^\infty(B_s)$, $\eta \equiv 1$ in B_t, $|D\eta| \leq \frac{1}{s-t}$. Finally, let $w = (1-\eta)u + \eta v$. Clearly $w = u$ on $\Omega \backslash B_R$ thus

$$\int_{B_R} F(M(Du))dx \leq \int_{B_R} F(M(Dw))dx$$

and, as $v = u_{\epsilon,\delta}$ is regular, by (4.8) we conclude

$$\int_{B_R} \bar{F}(M(Du))dx \leq \int_{B_R} \bar{F}(M(Dw))dx.$$

Now we have for any $\tau \in \left(0, \frac{1}{2}\right)$
(4.15)

$$\begin{aligned}
\int_{B_{\tau R}} \bar{F}(M(Du))dx &\leq \int_{B_{\tau R}} \bar{F}(M(Dv))dx + \int_{B_t \backslash B_{\tau R}} \left[\bar{F}(M(Du)_{\epsilon,\delta}) + \right. \\
&\left. - \bar{F}(M(Du))\right] dx + \int_{B_t \backslash B_{\tau R}} \left[\bar{F}(M(Dv)) - \bar{F}(M(Du)_{\epsilon,\delta})\right] dx \\
&+ \int_{B_s \backslash B_t} \bar{F}(M(D[(1-\eta)u + \eta v]))dx - \int_{B_s \backslash B_t} \bar{F}(M(Du))dx.
\end{aligned}$$

1.a. We shall now prove that

$$\begin{aligned}
&\int_{B_t \backslash B_{\tau R}} \left[\bar{F}(M(Dv)) - \bar{F}(M(Du)_{\epsilon,\delta})\right] dx \leq \\
(4.16) \qquad &\leq c\, \mathcal{E}^\beta \int_{B_R} \bar{F}(M(Du))dx
\end{aligned}$$

provided $\mathcal{E}$ is sufficiently small. Since $\bar{F}$ is convex we have

$$\bar{F}(M(Dv)) - \bar{F}(M(Du)_{\epsilon,\delta}) \leq$$

$$\leq -\bar{F}_M(M(Dv))\big[M(Du)_{\epsilon,\delta} - M(Dv)\big] \tag{4.17}$$

$$\leq |F_M(M(Dv))||M(Du)_{\epsilon,\delta} - M(Dv)|$$

as $t < \frac{7}{8}R$ by (3.8) $\|Dv\|_{\infty,\frac{7}{8}R} \leq \text{const}$, so by (4.6)

$$|\bar{F}_M(M(Dv))| \leq |M(Dv) - M(p^*)| \leq c|Dv - p^*| \leq c\mathcal{E}^\beta. \tag{4.18}$$

On the other hand, $L(Du)_{\epsilon,\delta} = L(Dv)$, thus

$$|M(Du)_{\epsilon,\delta} - M(Dv)| = |R(Du)_{\epsilon,\delta} - R(Dv)| \tag{4.19}$$

and taking into account the relations

$$M_{\beta\alpha}(Du)_{\epsilon,\delta} = \sum_{\substack{\alpha' \leq \alpha \\ \beta' \leq \beta}} \text{sign } \sigma_{\alpha'\alpha''} \text{sign } \sigma_{\beta'\beta''} M_{\alpha'\beta'}(p^*) \cdot \cdot \Big[M_{\alpha''\beta''}(Du - p^*)\Big]_{\epsilon,\delta}$$

$$M_{\beta\alpha}(Dv) = \sum_{\substack{\alpha' \leq \alpha \\ \beta' \leq \beta}} \text{sign } \sigma_{\alpha'\alpha''} \text{sign } \sigma_{\beta'\beta''} \cdot \cdot M_{\alpha'\beta'}(p^*) M_{\alpha''\beta''}(Dv - p^*)$$

we get

$$|R(Du)_{\epsilon,\delta} - R(Dv)| \leq c|R(Du - p^*)_{\epsilon,\delta} - R(Dv - p^*)|. \tag{4.20}$$

Now, by (2.11) and (4.5)

$$\int_{B_{\frac{7}{8}R}} |R(Du - p^*)_{\epsilon,\delta}| dx \leq \int_{B_{\frac{7}{8}R+\epsilon+\delta}} |R(Du - p^*)| dx \leq$$

$$\leq c \int_{B_R} d(Du - p^*) dx \leq c \int_{B_R} \bar{F}(M(Du)) dx, \tag{4.21}$$

since $\|Dv - p^*\|_{\infty,\frac{7}{8}R} \leq c\,\mathcal{E}^\beta < 1$, we have

$$|R(Dv - p^*)| \leq c\min(|Dv - p^*|, |Dv - p^*|^2) \leq c\,\gamma_m(|Dv - p^*|),$$

hence

$$\int_{B_{\frac{7}{8}R}} |R(Dv - p^*)|dx \leq \int_{B_{\frac{7}{8}R}} \gamma_m(|(Du - p^*)_{\epsilon,\delta}|) \leq$$

$$\leq \int_{B_{\frac{7}{8}R}} \gamma_m(|Du - p^*|)_{\epsilon,\delta} dx \leq \int_{B_{\frac{7}{8}R+\epsilon+\delta}} \gamma_m(|Du - p^*|) \tag{4.22}$$

$$\leq \int_{B_R} d(Du - p^*)dx \leq c \int_{B_R} \bar{F}(M(Du))dx$$

and (4.16) follows at once from (4.17)...(4.22), provided $c\,\mathcal{E}^\beta < 1$.

1.b. We now estimate the fourth term on the right-hand-side of (4.15). We set

$$\hat{u} := u - p^* \cdot (x - x_o) \qquad \hat{v} := v - p^* \cdot (x - x_o)$$

and notice that by (4.5)

$$\bar{F}(M(Dw)) \leq d(Dw - p^*) + \gamma_r(|Dw - p^*|) =$$

$$= d(D[(1-\eta)\hat{u} + \eta\hat{v}]) + \gamma_r(|D[(1-\eta)\hat{u} + \eta\hat{v}]).$$

Now, as $\|D\hat{v}\|_{\infty,\frac{7}{8}R} \leq \|Dv - p^*\|_{\infty,\frac{7}{8}R} \leq c\,\mathcal{E}^\beta < 1$ by (3.8), and as $D\eta(\hat{u} - \hat{v})$ has rank 1, we have

$$d(D[(1-\eta)\hat{u} + \eta\hat{v}]) = d((1-\eta)D\hat{u} + \eta D\hat{v} + D\eta(\hat{v} - \hat{u})) \leq$$

$$\leq c\big\{d(D\hat{v}) + (1 + |M(D\hat{v})|)^m d((1-\eta)D\hat{u} + D\eta(\hat{v} - \hat{u}))\big\}$$

$$\leq c\big\{d(D\hat{v}) + (1 + |D\eta||\hat{u} - \hat{v}|)^m d(D\hat{u}) + \gamma_m(|D\eta||\hat{u} - \hat{v}|)\big\}.$$

As in (4.22)

$$\int_{B_s\setminus B_t} d(D\hat{v})dx \le c \int_{(B_t\setminus\ B_{s_{\epsilon+\delta}})} \bar{F}(M(Du))dx \ ;$$

since osc $_{B_{\frac{7}{8}R}} v \le$ osc $_{B_{\frac{7}{8}R+\epsilon+\delta}} u$, we have

$$\int_{B_s\setminus B_t} (1+|D\eta||\hat{u}-\hat{v}|)^m d(D\hat{u})dx \le$$

$$\le \left(1+\frac{\text{osc } u}{s-t}\right)^m \int_{B_s\setminus B_t} \bar{F}(M(Du));$$

finally by lemma 4.4

$$\int_{B_s\setminus B_t} \gamma_m(|D\eta||\hat{u}-\hat{v}|)dx \le$$

$$\le c\max\left(\left(\frac{\epsilon}{s-t}\right)^2, \left(\frac{\epsilon}{s-t}\right)^m\right) \int_{(B_s\setminus B_t)_{2\epsilon}} \bar{F}(M(Du))dx.$$

Analogously one easily sees that

$$\int_{B_s\setminus B_t} \gamma_r(|Dw-p^*|)dx \le$$

$$\le c\Big[1+\max\left((\frac{\epsilon}{s-t})^2, (\frac{\epsilon}{s-t})^r\right)\Big] \int_{(B_s\setminus B_t)_{2\epsilon}} \bar{F}(M(Du))dx,$$

thus we conclude that

$$\int_{B_s\setminus B_t} \bar{F}(M(Dw))dx \le$$

$$\le c\Big[1+\max\left((\frac{\epsilon}{s-t})^2, (\frac{\epsilon}{s-t})^m, (\frac{\epsilon}{s-t})^r\right)+ \tag{4.23}$$

$$+\left(\frac{\text{osc } u}{s-t}\right)^m\Big] \int_{(B_s\setminus B_t)_{2\epsilon}} \bar{F}(M(Du))dx.$$

On the other hand by Sobolev theorem we have

$$\operatorname{osc}_{B_R} u \leq c\, R\Big(\fint_{B_R} |Du - p_R|^r dx \Big)^{1/r} \leq c\, R\, R^{1/r}(u, B_R),$$

therefore the previous estimates (4.15), (4.16), (4.23) can be summarized as

$$\begin{aligned}
\int_{B_{\tau R}} \bar F(M(Du))dx \leq \int_{B_{\tau R}} \bar F(M(Dv))dx + \\
+ c\, \mathcal{E}^{\beta} \int_{B_R} \bar F(M(Du))dx + \int_{B_t \setminus B_{\tau R}} [\bar F(M(Du))_{\epsilon,\delta}) \\
- \bar F(M(Du))]\, dx + c \int_{(B_s \setminus B_t)_{2\epsilon}} \bar F(M(Du))dx \\
+ c \max\Big((\frac{R\mathcal{E}^{\beta}}{s-t})^2, (\frac{R\mathcal{E}^{\beta}}{s-t})^m, (\frac{R\mathcal{E}^{\beta}}{s-t})^r, (\frac{R\mathcal{E}^{1/r}}{s-t})^m \Big) \cdot \\
\cdot \int_{(B_s \setminus B_t)_{2\epsilon}} \bar F(M(Du))dx.
\end{aligned} \tag{4.24}$$

for all $\tau < 1/2$.

1.c. We shall now see, compare [3], that for a suitable choice of s, t, τ (4.24) implies (4.14). Let $\bar\gamma = \min(\gamma, 1/r)$ and consider the numbers

$$\rho_i := \frac{5}{8}\, R + i\, \frac{R}{400}\, \mathcal{E}^{\bar\gamma/2} \quad \rho_i \in [\frac{5}{8}\, R, \frac{7}{8}\, R].$$

By lemma 4.3, if $\tau < 1/4$, in each interval (ρ_{8j-1}, ρ_{8j}) $j = 1, \dots, k$ $k :=$ integer part of $25(2\mathcal{E}^{\bar\gamma/2})^{-1}$, there is a number t_j and consequently $\tau_j R \in (\tau R, 2\tau R)$ such that

$$\int_{B_{t_j}\setminus B_{\tau_j R}} [\bar F(M(Du)_{\epsilon,\delta}) - \bar F(M(Du))]dx \leq$$

$$\leq 4\epsilon\Big(\frac{1}{\rho_{8j}-\rho_{8j-1}} + \frac{1}{\tau R}\Big)\int_{B_R} \bar F(M(Du))dx \tag{4.25}$$

$$\leq c\big(1+\frac{1}{\tau}\big)\mathcal{E}^{\bar\gamma/2}\int_{B_R} \bar F(M(Du))dx.$$

For each j set

$$s_j = t_j + \frac{R}{400}\,\mathcal{E}^{\bar\gamma/2}$$

and notice that the sets

$$C_j := B_{s_j+2\epsilon}\setminus B_{t_j-2\epsilon} \quad j=1,...,k$$

are pairwise disjoint and contained in B_R. Thus

$$\sum_{j=1}^{k}\int_{C_j} \bar F(M(Du)) \leq \int_{B_R} \bar F(M(Du))dx$$

and there exists at least an integer $j_o \in \{1,...,k\}$ such that

$$\int_{C_j} \bar F(M(Du))dx \leq \frac{1}{10}\,\mathcal{E}^{\bar\gamma/2}\int_{B_R} \bar F(M(Du))dx. \tag{4.26}$$

So writing (4.24) for $\tau = \tau_{j_o}$, $s = s_{j_o}$, $t = t_{j_o}$, using (4.25), (4.26) we obtain

$$\int_{B_{\tau_{j_o}}R} \bar F(M(Du))dx$$

$$\leq \int_{B_{\tau_{j_o}}R} \bar F(M(Dv))dx + c\mathcal{E}^{\gamma_o}\int_{B_R} \bar F(M(Du))dx$$

with $\gamma_o = \min(\beta, \bar\gamma/2)$ which gives (4.16) since $\bar F \geq 0$.

2. We now prove that

2.a.

$$D(u, B_{\tau R}) + E(u, B_{\tau R} \leq c \int_{B_{\tau R}} \bar{F}(M(Du))dx \tag{4.27}$$

provided $\mathcal{E}$ is small enough. Using Jensen inequality, compare (5.5) of [3], we have

$$\begin{aligned} E(u, B_{\tau R}) &= \int_{B_{\tau R}} \gamma_r(|Du - p_{\tau R}|)dx \leq \\ &\leq c \int_{B_{\tau R}} \gamma_r(|Du - p^*|)dx. \end{aligned} \tag{4.28}$$

Also, by (2.12) we get

$$d(Du - p_{\tau R}) \leq c\{d(p^* - p_{\tau R}) + (1 + |M(p^* - p_{\tau R})|)^m d(Du - p^*)\}$$

and, as

$$|p^* - p_{\tau R}| \leq \fint_{B_{\tau R}} |Du - p_R|dx + |p_R - p^*| \leq c\tau^{-n/2}\mathcal{E}^{1/2} + \mathcal{E}^{\beta},$$

we deduce

$$d(Du - p_{\tau R}) \leq c\{d(Du - p^*) + d(p^* - p_{\tau R})\} \tag{4.29}$$

provided $\mathcal{E}$ is sufficiently small. On the other hand, using Jensen inequality

$$\begin{aligned} &d(p^* - p_{\tau R}) \leq c|p^* - p_{\tau R}|^2 \leq c\gamma_m(|p^* - p_{\tau R}|) \leq \\ &\leq c\gamma_m\Big(\fint_{B_{\tau R}} |Du - p^*|dx\Big) \leq c\gamma_m\Big(\fint_{B_{\tau R}} |M(Du - p^*)|dx\Big) \\ &\leq \fint_{B_{\tau R}} d(Du - p^*)dx, \end{aligned}$$

so (4.29) yields

$$D(u, B_{\tau R}) \leq \int_{B_{\tau R}} d(Du - p_{\tau R})dx \leq \int_{B_{\tau R}} d(Du - p^*)dx.$$

This inequality together with (4.28) gives at once (4.27) taking into account (4.5).

2.b. We have, by definition of p^* and (3.8)

$$\begin{aligned}\int_{B_{\tau R}} \bar{F}(M(Dv))dx &\leq c\{E(v, B_{\tau R}) + D(v, B_{\tau R})\} \leq \\ &\leq c \int_{B_{\tau R}} \gamma_m(|Dv - p_{\tau R}|)dx\end{aligned} \tag{4.30}$$

2.c. Finally, we shall now prove that

$$\int_{B_R} \bar{F}(M(Du))dx \leq c\big(1 + \frac{1}{\tau^n}\big)\big[E(u, B_R) + D(u, B_R)\big]. \tag{4.31}$$

Since $|p^* - p_R| \leq c\,\mathcal{E}^\beta < 1$ we have

$$d(Du - p^*) \leq c\{\gamma_n(|p_R - p^*|) + d(Du - p_o)\}$$

and by Jensen inequality

$$\begin{aligned}\gamma_m(|p_R - p^*|) &= \gamma_m\big(\fint_{B_{\tau R}} (p_R - Dv)dx\big) \leq \\ &\leq \fint_{B_{\tau R}} \gamma_m\big(|Du - p_R|_{\epsilon,\delta}\big)dx \\ &\leq \frac{1}{|B_R|}\frac{1}{\tau^n}\int_{B_{\tau R+\epsilon+\delta}} \gamma_m(|Du - p_o|)dx \leq \frac{1}{|B_R|}\frac{1}{\tau^n}D(u, B_R)\end{aligned}$$

hence, integrating over B_R we get

$$\int_{B_R} d(Du - p^*)dx \leq c(1 + \frac{1}{\tau^n})D(u, B_R).$$

Analogously

$$\int_{B_R} \gamma_r(|Du - p^*|)dx \leq c(1 + \frac{1}{\tau^n})E(u, B_R)$$

and this, together with (4.5) proves (4.31). So (4.14) together with (4.27), (4.30) and (4.31) gives (4.2) provided $\mathcal{E}(u, B_R)$ is sufficiently small.

We shall now prove (4.3). We have

$$\psi(v, \frac{R}{2}) = \int_{B_{R/2}} [f(Dv) - f(DH)] =$$

$$(4.32) \qquad = \int_{B_{R/2}} [\bar{F}(M(Dv)) - \bar{F}(M(DH))] + \\ + \int_{B_{R/2}} [f(Dv) - F(M(Dv))]dx \\ + \int_{B_{R/2}} [f(DH) - F(M(DH))].$$

Using Taylor expansion we get

$$f(Dv) - F(M(Dv)) \leq c\,\omega(|Dv - p_R|)|Dv - p_R|^2 + +c|Dv - p_R|^3$$

so

$$\int_{B_{R/2}} [f(Dv) - F(M(Dv))]dx \leq$$

$$\leq c[\omega(\mathcal{E}(u, B_R)) + \mathcal{E}(u, B_R)]E(u, B_R).$$

Similarly, since H is a solution of (3.12)

$$\int_{B_{R/2}} [f(DH) - F(M(DH))]dx \leq$$

$$\leq c[\omega(\mathcal{E}(u, B_R)) + \mathcal{E}(u, B_R))]E(u, B_R))$$

and the first term on the right of (4.32) remains to be estimated. Let s, t be such that $R/2 < t < s < R$. Set

$$\tilde{H} = \begin{cases} H & \text{in} \quad B_{R/2} \\ v & \text{in} \quad B_s \backslash B_{R/2} \end{cases}$$

We have

$$\begin{aligned} \int_{B_{R/2}} [\bar{F}(M(Dv) - \bar{F}(M(DH))]dx = \\ &= \int_{B_s} [\bar{F}(M(Dv)) - \bar{F}(M(D\tilde{H}))] \\ &= \int_{B_s} [\bar{F}(M(Dv)) - \bar{F}(M(Du)_{\epsilon,\delta})]dx \ + \\ &\quad + \int_{B_s} [\bar{F}(M(Du)_{\epsilon,\delta}) - \bar{F}(M(Du))]dx \\ &\quad + \int_{B_s} [F(M(Du)) - \bar{F}(M(D\tilde{H}))]dx. \end{aligned}$$

Now, as in step 1a (4.16)

$$\int_{B_s} [\bar{F}(M(Dv)) - \bar{F}(M(Du)_{\epsilon,\delta})] \leq c\, \mathcal{E}^{\beta} \int_{B_s} \bar{F}(M(Du))dx;$$

using the minimality of u, we obtain

$$\begin{aligned} &\int_{B_s} [\bar{F}(M(Du)) - \bar{F}(M(D\tilde{H}))] \\ &\qquad \leq \int_{B_s} \left[\bar{F}(M(D(\tilde{H} + \eta(\hat{u} - \hat{v}))) - \bar{F}(M(D\tilde{H}))\right] dx \\ &\qquad \leq \int_{B_s \backslash b_t} \bar{F}(M(D[v + \eta(u - v))]dx, \end{aligned}$$

hence, proceeding as in step 1b, we get for osc $u + \epsilon < s - t$

$$\int_{B_s \setminus B_t} [\bar{F}(M(Du)) - \bar{F}(M(D\tilde{H}))]dx = c \int_{B_s \setminus B_t} \bar{F}(M(Du))dx.$$

Summarizing, we conclude with the following estimate

$$\psi(v, R/2) \leq c\, \mathcal{E}^{\beta} \int_{B_R} \bar{F}(M(Du))dx +$$

$$+ c \int_{B_s} [\bar{F}(M(Du)_{\epsilon,\delta}) - \bar{F}(M(Du))]dx$$

$$+ c \int_{(B_s \setminus B_t)_{2\epsilon}} \bar{F}(M(Du))dx \ .$$

Now, arguing exactly as in step 1c and using Lemma 4.2 instead of 4.3, one easily concludes the proof of (4.3).

References

[1] W.K.Allard, *On the first variation of a varifold*, Ann. of Math, **95** (1972), 417-491.

[2] F.J.Almgren Jr.,*Existence and regularity almost everywhere of solutions to elliptic variational problems among surfaces of varying topological type and singularity structure*, Ann. of Math. **87** (1968), 321-391.

[3] G.Anzellotti, M.Giaquinta, *Convex functionals and partial regularity*, Arch. Rat. Mech. Anal., to appear.

[4] E.Bombieri, *Regularity theory for almost minimal currents*, Arch. Rat. Mech. Anal. **78** (1982), 99-130.

[5] E.De Giorgi, *Frontiere orientate di misura minima*, Sem. Mat. Sc. Norm. Sup. Pisa, 1961.

[6] H.Federer, *Geometric measure theory*, Springer Verlag, Berlin 1969.

[7] M.Giaquinta, *Multiple integrals in the calculus of variations and nonlinear elliptic systems*, Annals of Math. Studies n.105, Princeton University Press, Princeton 1983.

[8] M.Giaquinta, *Quasiconvexity, growth conditions and partial regularity*, Preprint 1987.

[9] M.Giaquinta, G.Modica, J.Souček, *Weak diffeomorphisms and nonlinear elasticity*, Arch. Rat. Mech. Anal. **106** (1989), 97-159. *Erratum*, to appear on Arch. Rat. Mech. Anal.

[10] R.Schoen, L.Simon, *A new proof of the regularity theorem for rectifiable currents which minimize parametric elliptic functionals*, Indiana Univ. Math.J. **31**(1982), 415-434.

M.G. & G.M.
Istituto di Matematica Applicata
Università di Firenze
Via S.Marta,3
I-50139 FIRENZE

J.S.
Cekoslovenska Akademie Ved Mathematicky Ustav
Žitna ul.25
PRAHA 1

THEOREME DES MINIMAX LOCAUX ET FONCTIONS TOPOLOGIQUEMENT FERMEES

GABRIELE H. GRECO

Dédié à Ennio de Giorgi pour son soixantième anniversaire

Depuis l'article [10] de E.De Giorgi et T.Franzoni, les Γ-limites sont devenues un outil très important tout en calcul des variations que dans d'autres domaines mathématiques, grâce, notamment, aux propriétés de compacité et de stabilité des points et valeurs de minimum dont dispose la Γ-convergence (i.e. la convergence définie par les Γ-limites).

Il en est de même en théorie des jeux où les points et valeurs de selle disposent de propriétés analogues par rapport à une nouvelle Γ-convergence, introduite par E.Cavazzuti [4], et par rapport à une autre Γ-convergence (plus faible et plus symétrique que celle de Cavazzuti), introduite par H.Attouch et R.Wets [3]. Ces deux Γ- convergences ont en commun l'inconvénient de ne pas être, en général, topologiques; de plus la convergence d'Attouch et Wets n'a pas la propriété d'unicité de la limite.

Pour pallier à ces inconvénients, nous introduisons dans la section 4 une convergence moins fine que les deux précédentes, mais qui est topologique. La topologie associée, baptisée "saddle topology", est compacte et dispose de propriétés de stabilité des points et valeurs de selle, mais elle n'a pas la propriété d'unicité de la limite. Ce défaut d'unicité pour la "saddle topology" pose le problème suivant:

" *La connaissance d'*une *fonction limite d'une suite de fonctions assure-t-elle la connaissance de* toutes *les fonctions limites de la suite?*"

L'introduction de la notion nouvelle de *fonction topologiquement fermée* permettra de donner une réponse au problème d'unicité (voir la deuxième partie de la section 4). Dans le cas, particulièrement important, de fonctions topologiquement fermées et quasi concavo-convexes, "cette response" assurera la métrisablité de la "saddle topology" et, *ipso facto*, celle des Γ-convergences de Cavazzuti et d'Attouch, Wets. On rappelle que la classe des fonctions topologiquement fermées inclue strictement celle des fonctions semi- continues inférieurement dans la variable de minimisation et semi- continues supérieurement dans la variable de maximisation; on mesure l'importance de cette classe au fait qu'elle assure un théorème de minmax [17], dont on trouvera des versions locales dans la section 3. Une partie des résultats contenus dans ce travail a fait l'object de conférences, tenues par l'auteur ces trois dernières années. A la suite de ces conférences, certains collègues, après avoir accordé leur confiance aux théorèmes présentés, souvent sans démonstrations, les ont employés ou cités ou parfois même ameliorés (Cavazzuti [7], Franzoni [12], Guillerme [18]).

1.Définitions et rappels. Dans cet article on emploiera surtout des fonctions à deux variables à valeurs dans la droite achevée $\bar{\mathbf{R}}$. Si $f : X \times Y \to \bar{\mathbf{R}}$ est une fonction donnée, la première variable sera toujours celle correspondant aux opérations de maximisation et la seconde aux opérations de minimisation. Comme habituellement, on associe à la variable de minimisation des conditions de semi-continuité inférieure, d'inf-compacité , de quasi convexité ,...; et à la variable de maximisation les conditions (duales) de semi-continuité supérieure, sup-compacité , quasi-convexité...

Espaces topologiques (Notations). Soient X et Y deux espaces topologiques. Aux points $x \in X$ et $y \in Y$ correspondent leurs systèmes de voisinages notés, respectivement, par $\mathcal{N}(x)$ et $\mathcal{N}(y)$. Si A et$B$ sont des sous-ensembles, respectivement, de Y et X, on désigne par $\mathcal{K}_A$, $\mathcal{K}_B$ (resp. O_A, O_B) les familles des compacts (resp. ouverts) inclus, respectivement, dans A et dans B. Dans la suite on considérera les espaces topologiques dont tous les sous-espaces compacts sont réguliers (en bref: *à compacts réguliers*).

Semi-continuité des fonctions et Γ-transformations (définitions et rappels). On dit qu'une fonction $f : X \times Y \to \bar{\mathbf{R}}$ est *semi-continue inférieurement* (en bref: *s.c.i.*), si pour tout $(x, y) \in X \times Y$

$$f(x,y) = \liminf_{y' \to y} f(x,y').$$

Dualement on dit que f est *semi-continue supérieurement* (en bref: *s.c.s.*), si pour tout $(x, y) \in X \times Y$:

$$f(x,y) = \limsup_{x' \to x} f(x',y).$$

On dit que f est *s.c.s./s.c.i.*, si f est aussi bien s.c.s. que s.c.i. Les quatre Γ-transformations suivantes:

$$\Gamma(X^+),\ \Gamma(Y^-),\ \Gamma(X^+,Y^-),\ \Gamma(Y^-,X^+), \tag{1.1}$$

sont les applications de $\bar{\mathbf{R}}^{X\times Y}$ dans $\bar{\mathbf{R}}^{X\times Y}$ définies par les égalités suivantes:

$$\begin{aligned}
(\Gamma(X^+)f)(x,y) &:= \limsup_{x' \to x} f(x',y) \\
(\Gamma(Y^-)f)(x,y) &:= \liminf_{y' \to y} f(x,y') \\
(\Gamma(Y^-,X^+)f)(x,y) &:= \inf_{U \in \mathcal{N}(x)} \liminf_{y' \to y} \sup_{x' \in U} f(x',y') \\
(\Gamma(X^+,Y^-)f)(x,y) &:= \sup_{V \in \mathcal{N}(y)} \limsup_{x' \to x} \inf_{y' \in V} f(x',y')
\end{aligned}$$

pour tout $(x, y) \in X \times Y$. Les fonctions $\Gamma(Y^-)f$ et $\Gamma(X^+)f$ seront appelées, respectivement, *régularisée semi-continue inférieure et supérieure* de f. On rappelle que les deux autre fonctions $\Gamma(Y^-, X^+)f$ et $\Gamma(X^+, Y^-)f$ sont, respectivement, s.c.s. et s.c.i. Cette semi-continuité et aussi bien la semi-continuité d'autres Γ-transformations introduites dans les paragraphes suivants, peuvent être facilement demontrée, grâce à la propriété suivante (ou à sa duale) [11]: "pour toute fonction d'ensemble $\gamma : \mathcal{P}(Y) \to \bar{\mathbf{R}}$ croissante la fonction g de Y dans $\bar{\mathbf{R}}$, définie pour tout y par $g(y) := \sup\{\gamma(V) : V \in \mathcal{N}(y)\}$ est semi-continue inférieurement".

Points de selle et valeurs de selle (définitions et rappels). Pour une fonction $f : X \times Y \to \bar{\mathbf{R}}$ quelconque on a l'inégalité suivante:

$$\sup_X \inf_Y f \leq \inf_Y \sup_X f;$$

lorsque l'inégalité inverse est également satisfaite, on dit que f a *valeur de selle*; dans ce cas, la valeur commune aux deux parties de l'inégalité, noté par $sv(f)$, est appelée valeur de selle de f. Un point (x_o, y_o) de $X \times Y$ est dit *point de selle* de f, si pour tout $(x, y) \in X \times Y$ on a:

$$f(x, y_o) \leq f(x_o, y_o) \leq f(x_o, y).$$

Si f a un point de selle (x_o, y_o), alors f a aussi valeur de selle et $sv(f) = f(x_o, y_o)$. Il est bien connu que si f a valeur de selle (resp. un point (x_o, y_o), comme point de selle), alors les fonctions $\Gamma(X^+)f, \Gamma(Y^-)f$, $\Gamma(X^+, Y^-)f$ et $\Gamma(Y^-, X^+)f$ ont même valeur de selle que f (resp. le point (x_o, y_o), comme point de selle).

Inf-compacité et inf-compacité locale (de fonctions et de suites; définitions). On dit qu'une fonction $f : X \times Y \to \bar{\mathbf{R}}$ est *inf-compacte* (resp. *sup-compacte*) au point $(x_o, y_o) \in X \times Y$, si pour tout $a \in \mathbf{R}$ on a que l'ensemble $\{f(x_o, \cdot) \leq a\}$ (resp. $\{f(\cdot, y_o) \geq a\}$) est contenu dans un compact fermé de Y (resp. de X). Si une fonction est aussi bien inf-compacte que sup-compacte, on dit que f est *inf/sup-compacte*. Si dans cette définition on remplace "contenu dans un compact fermé " par "contenu dans un localement compact fermé" on obtient les définitions de fonctions *inf-localement compactes, sup-localement compactes et inf/sup localement compactes* en un point. La notion d'inf-compacité locale pour une fonction d'une variable a été introduite et étudiée par J.L.Joly [19] (voir aussi [20]); parmi ces fonctions un rôle privilégié est joué par celles qui sont convexes propres et qui ne contiennent aucune droite dans leur épigraphe (dites:

"saillantes" dans [21]); ces fonctions ont été redécouvertes, dans le cadre des problèmes de Dirchlet, sous une forme équivalente dans [1] où elles sont appelées "demi-coercives".

Enfin on dit qu'une *suite* (généralisée ou non) $\{f_j\}_j$ des fonctions de $X \times Y$ dans $\bar{\mathbf{R}}$ est *inf-compacte* (resp. *sup-compacte*) au point (x_o, y_o), si pour tout $a \in \mathbf{R}$ il existe un j_o tel que les ensembles $\{f_j(x_o, \cdot) \leq a\}$ (resp. $\{f_j(\cdot, y_o) \geq a\}$) sont tous contenus dans un même compact fermé de Y (resp. de X), lorsque j parcourt les indices plus grands que j_o. On a aussi des définitions analogues pour la compacité locale d'une suite.

Quasi concavité et quasi convexité (de fonctions et multi-applications; définitions et rappels). Soient X et Y deux ensembles convexes d'espaces vectoriels topologiques. On dit qu'une fonction $f : X \times Y \to \bar{\mathbf{R}}$ est *quasi concavo-convexe*, si pour tout $a \in \mathbf{R}$ et pour tout $(x_o, y_o) \in X \times Y$ on a que $\{f(x_o, \cdot) \leq a\}$ et $\{f(\cdot, y_o) \geq a\}$) sont convexes. Une multiapplication Ω de X dans Y est dite *concavo-convexe*, si pour tout $x \in X$ l'ensemble Ωx est convexe et pour tout couple x_1, x_2 on a que $\Omega x' \subset \Omega x_1 \cup \Omega x_2$, lorsque x' appartient au segment qui a les point x_1 et x_2 comme extrêmes. Une propriété de ces multiapplications, utile dans la démonstration du théorème des minimax locaux est la *propriété de la sélection constante* [17]: " si Ω est à valeurs fermées dans Y et s'il existe un recouvrement fini et ouvert $B_1, B_2, ..., B_n$ de X, tel que $\cap\{\Omega x : x \in B_i\} \neq \emptyset$ pour $1 \leq i \leq n$, alors les valeurs de Ω ont un point commun". Enfin on rappelle que la *fonction indicatrice* ψ_G d'un sous-ensemble G de $X \times Y$ est, par définition, égale à $-\infty$ dans les points appartenants à G, autrement elle est égale à $+\infty$; l'ensemble G est alors dit concavo-convexe, si sa fonction indicatrice ψ_G est quasi concavo-convexe. Dans le cas où G est le graphe d'une multiapplication Ω, on peut observer que Ω est concavo-convexe, si et seulement si G est concavo-convexe.

Semigroupe engendré par les Γ-transformations. Le semigroupe engendré par les quatre Γ-transformations (1.1) a été étudié dans [12]; il contient ces autres Γ-transformations:

$$\text{(1.2)} \qquad \Gamma(Y^-)\Gamma(X^+),\ \Gamma(X^+)\Gamma(Y^-)\Gamma(X^+)\Gamma(Y^-)$$

$$\Gamma(X^+)\Gamma(Y^-)\Gamma(X^+),\ \Gamma^n(X^+, Y^-)\Gamma^n(Y^-, X^+).$$

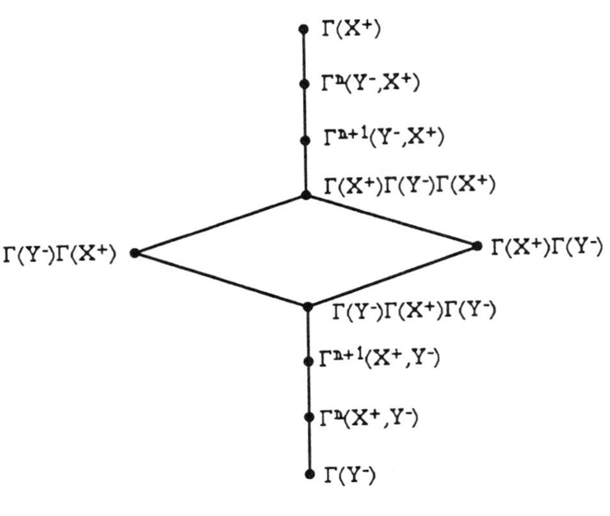

Figure 1

Les premières quatre de celles-ci sont idempotentes. Le diagramme d'ordre entre eux est le suivant:

Toutes les éléments de ce semigroupe transforment les fonctions quasi concavo-convexes, en fonctions quasi concavo-convexes; fonctions inf-compactes en un point en fonctions inf-compactes dans le même point. Et les autres propriétés (sup-compacité, inf-compacité locale et sup-compacité locale) sont aussi invariantes. Il est remarquable que, sauf $\Gamma(X^+)$ et $\Gamma(Y^-)$, toutes les éléments du semigroupe sont des "transformations sellantes" [17]; c'est-à-dire: elles transforment les fonctions quasi concavo-convexes en fonctions qui ont valeur de selle.

Enfin on rappelle que le semigroupe en question est généralement infini [12]; mais, sous des hypothèses de compacité , dans le cadre des fonctions quasi concavo-convexes le semigroupe est fini, parce qu'on a les égalités suivantes [17]:

$$\Gamma^2(X^+, Y^-) = \Gamma(Y^-)\Gamma(X^+)\Gamma(Y^-)$$
$$\Gamma^2(Y^-, X^+) = \Gamma(X^+)\Gamma(Y^-)\Gamma(X^+).$$

Principe d'extension des limitoïdes (rappels). Soient A et B deux ensembles quelconques. Une *limitoïde* [13], [14] est une application T de $\bar{\mathbf{R}}^A$ dans $\bar{\mathbf{R}}^B$ vérifiant les deux propriétés suivantes:

(L.1) $$h \leq g \Rightarrow T(h) \leq T(g)$$

(L.2) $$T(\varphi \circ h) = \varphi \circ T(h),$$

pour tout couple $h, g \in \bar{\mathbf{R}}^A$ et pour toute fonction croissante et continue φ de $\bar{\mathbf{R}}$ dans $\bar{\mathbf{R}}$.

La composition de deux limitoïdes est aussi une limitoïde. Les Γ-transformations précédentes sont des examples de limitoïdes. Même les applications qui à toute fonction f à deux variables associent le nombre "$\inf \sup f$" (ou "$\sup \inf f$") sont des limitoïdes. D'autres limitoïdes peuvent être obtenues par composition de celles-ci.

Soit $\mathbf{H}$ une famille de sous-ensembles de A. On dit qu'une fonction $h : A \to \bar{\mathbf{R}}$ est *dualement* **H**-*mesurable* [15], si pour tout couple $a, b \in \mathbf{R}$ tel que $a < b$ il existe un ensemble $H \in \mathbf{H}$ tel que $\{h \leq a\} \subset H \subset \{h < b\}$. Alors on a le *principe d'extension des limotoïdes* [15]:

"Si T_1 et T_2 sont deux limitoïdes et si pour tout $H \in \mathbf{H}$ on a $T_1(\psi_H) = T_2(\psi_H)$, alors $T_1(h) = T_2(h)$ pour toute fonction h dualement **H**-mesurable".

Ce principe permet aussi d'avoir un analogue pour les inégalités; en effet on en déduit que:

"Si T_1 et T_2 sont deux limitoïdes et si pour tout $H \in \mathbf{H}$ on a $T_1(\psi_H) \leq T_2(\psi_H)$, alors $T_1(h) \leq T_2(h)$ pour toute fonction f dualement **H**-mesurable".

On peut observer que les fonctions quasi concavo-convexes sont les fonctions dualement mesurables par rapport à la famille de tous les ensembles concavo-convexes. Similairement on peut dire que les fonctions inf-compactes, sup-compactes, inf/sup-compactes ou à la fois inf-compact et quasi concavo-convexes... sont dualement mesurables par rapport à la famille des ensembles, dont les fonctions indicatrices ont les mêmes propriétés. Dans tous ces cas on peut appliquer le principe d'extension des limitoïdes pour réduire les démonstrations qu'on rencontrera, à des simples vérifications ensemblistes.

Remarque *(Treillis complets et complétement distributifs).* La droite achevée $\bar{\mathbf{R}}$ est un treillis complet et complétement distributif.

Cette propriété de $\bar{\mathbf{R}}$ est la propriété clef qui fait marcher toutes le démonstrations de cet article. En effect, *mutatis mutandis*, tout ce qui suit est vrai aussi pour les fonctions à valeurs dans un treillis complet et complétement distributif, puisque le principe d'extension des limitoïdes vaut aussi dans le cadre de ces treillis [15].

2. Formules de réduction. Dans les démonstrations qui suivent on a besoin plusieurs fois de quelques formules que nous résumons ici. Ces formules sont appelées *formules de réduction* parce qu'elles permettent de réduir la quantité des extrémisations qu'on trouvera dans les calculs qui suivront. D'après le principe d'extension des limitoïdes, il suffit de les vérifier pour les fonctions indicatrices pour s'assurer de leur validité dans le cas général. Les démonstrations pour les fonctions indicatrices, que nous laisserons au lecteur, ont les avantages d'être très simples, de relever immédiatement les propiétés topologiques nécessaires et de déduire qu'elles sont valables aussi dans le cas de fonctions à valeurs dans les treillis complets complétement distributifs.

1re *Formule de réduction.* Soit Y un espace topologique et soit $\mathcal{K}_Y$ la famille de tous ses sous-ensembles compacts. Donnée une suite généralisés $\{h_j\}_j$ inf-compacte de fonctions de Y dans $\bar{\mathbf{R}}$, alors

$$\inf_{K\in\mathcal{K}_Y} \liminf \inf_K h_j = \liminf \inf_Y h_j. \tag{2.1}$$

2me *Formule de réduction.* Soit Y un espace topologique à compacts réguliers. Donnée une suite généralisée $\{h_j\}_j$ de fonctions de Y dans $\bar{\mathbf{R}}$, si W est un ouvert de Y, alors

$$\inf_{y\in W} \sup_{V\in\mathcal{N}(y)} \inf_{K\in\mathcal{K}_V} \liminf \inf_K h_j = \inf_{K\in\mathcal{K}_W} \liminf \inf_K h_j, \tag{2.2}$$

où $\mathcal{N}(y)$ est le système des voisinages de y et $\mathcal{K}_V$ est l'ensemble de tous les compacts contenus dans V. D'après cette formule on a la propriété suivante [8]: pour tout compact K_o de Y

$$\inf_{y\in K_o} \sup_{V\in\mathcal{N}(y)} \inf_{K\in\mathcal{K}_V} \liminf \inf_K h_j \leq \liminf \inf_{K_o} h_j. \tag{2.2'}$$

La propriété topologique, sousjacent à (2.2′), est la suivante: "tout recouvrement ouvert d'un compact quelconque admet un recouvrement fini plus fin, constitué par des compacts." Un example d'espace topologique ne vérifiant pas cette propriété est donné par la compactification d'Alexandroff de la droite rationnelle [22].

3me *Formule de réduction.* Soient X et Y deux espaces topologiques. Soit $\{f_j\}_j$ une suite généralisée de fonction de $X \times Y$ dans $\bar{\mathbf{R}}$. Alors pour tout point x de X et A contenu en Y on a:

$$\inf_{B\in\mathcal{N}(x)} \inf_{K\in\mathcal{K}_A} \limsup \inf_K \sup_B f_j = \tag{2.3}$$

$$= \sup_{\substack{\{f_{j_k}\}_k \\ sous-suite\ de\ \{f_j\}_j}} \inf_{B\in\mathcal{N}(x)} \inf_{K\in\mathcal{K}_A} \liminf \inf_K \sup_B \{f_{j_k}\}_k.$$

En outre si $\{f_n\}_{n\in\mathbf{N}}$ est une suite usuelle, X un espace métrique, alors pour tout x de X on a:

$$\inf_{B\in\mathcal{N}(x)} \limsup_{n\to\infty} \inf_A \sup_B f_n = \tag{2.3′}$$

$$\sup_{\substack{\{n_k\}_k\subset\mathbf{N} \\ n_k\to+\infty}} \inf_{B\in\mathcal{N}(x)} \liminf_{k\to\infty} \inf_A \sup_B \{f_{n_k}\}_k,$$

et aussi la formule obtenue de (2.3′) en remplacement $\inf_A \sup_B f_n$ par $\sup_B \inf_A f_n$.

4me *Formule de réduction.* Soient X et Y deux espaces topologiques. Soit $\{f_j\}_j$ une suite généralisée de fonctions de $X \times Y$ dans $\bar{\mathbf{R}}$. Si cette suite est inf/sup-localement compacte au point (x_o, y_o) de $X \times Y$, alors on a l'inégalité suivante:

$$\sup_{A\in\mathcal{N}(y_o)} \inf_{B\in\mathcal{N}(x_o)} \inf_{K\in\mathcal{K}_A} \limsup \sup_B \inf_K f_j \leq \tag{2.4}$$

$$\leq \sup_{A\in\mathcal{N}(y_o)} \inf_{B\in\mathcal{N}(x_o)} \sup_{C\in\mathcal{K}_B} \limsup \inf_A \sup_C f_j,$$

et aussi l'inégalité obtenue de (2.4) en remplacement "limsup" par "liminf".

5me *Formule de réduction.* Soit Y un espace topologique. Donnée une suite généralisée $\{h_j\}_j$ de fonctions de Y dans $\bar{\mathbf{R}}$, si K est un compact de Y, alors

$$(2.5) \qquad \inf_{y\in K} \sup_{V\in\mathcal{N}(y)} \liminf \inf_V h_j = \inf_{V\in\mathcal{N}(K)} \liminf \inf_V h_j$$

où $\mathcal{N}(K)$ est la famille des voisinages de K dans Y. Cette formule, apparue dans une première version de [14], a été redécouvert par [9]. D'après (2.5) on a la bien connue et très utile inégalité [10]:

$$(2.5') \qquad \inf_{y\in K} \sup_{V\in\mathcal{N}(y)} \liminf \inf_V h_j \leq \liminf \inf_K h_j .$$

3. Fonctions topologiquement fermées et théorèmes des minimax locaux. En analogie aux fonctions fermées de Rockafellar [24], obtenues par des convexifications, les fonctions qu'on va définir sont appelées topologiquement fermées, parce qu'elles sont obtenues par le seul outil topologique et elles jouent un rôle similaire à celles de Rockafellar. Les fonctions topologiquement fermées sont bien indiquées pour l'étude locale des fonctions; par contre le fonctions fermées de Rockafellar ont un caractère tout global. Une comparaison entre ces deux notions, qui sont indépendantes et qui n'ont rien à voir avec les fonctions fermées, notamment étudiées dans les livres de topologie générale, se trouve dans Cavazzuti [7].

Ce n'est pas difficil de rencontrer des fonctions topologiquement fermées. Par exemple si on considère une fonctions $f : X \to \bar{\mathbf{R}}$ semicontinue supérieurement dans X, alors la fonction marginale "inf f" est égale à "$\inf \Gamma(Y^-)f$", où cetter dernière fonction $\Gamma(Y^-)f$ est, comme on verra, topologiquement fermée.

Même dans les cas où l'on veut avoir des informations supplémentaires sur la quantité "$\sup\inf f$" (lorsqu'on sait que f est semicontinue inférieurement et quasi concavo-convexe) il peut être très utile de savoir que "$\sup\inf f = \inf\sup \Gamma(Y^-)f$", grâce à un théorème de minimax qui vaut pour les fonctions topologiquement fermées et quasi concavo-convexes.

Soient X et Y deux espaces topologiques quelconques. On dira alors qu'une fonction $f : X \times Y \to \bar{\mathbf{R}}$ est *topologiquement fermée* (ou

Γ-*fermée*) si:

$$\liminf_{y\to y_o}\limsup_{x\to x_o} f(x,y) \le f(x_o,y_o) \le \limsup_{x\to x_o}\liminf_{y\to y_o} f(x,y), \tag{3.1}$$

pour tout point $(x_o, y_o) \in X \times Y$; en symboles:

$$\Gamma(Y^-)\Gamma(X^+)f \le f \le \Gamma(X^+)\Gamma(Y^-)f, \tag{3.1'}$$

ou, équivalentement:

$$\Gamma(Y^-)\Gamma(X^+)f = \Gamma(Y^-)f \quad et \quad \Gamma(X^+)f = \Gamma(X^+)\Gamma(Y^-)f.$$

L'ensemble de toutes les fonctions topologiquement fermées est noté par le symbole $C(X, Y)$.

On peut donner des caractérisations de ces fonctions pour tous les goûts. La plus simple est la suivante:

$$\inf_V f = \inf_V \Gamma(X^+)f \quad et \quad \sup_U f = \sup_U \Gamma(Y^-)f, \tag{3.2}$$

pour tout couple U, V d'ouverts, respectivement, de X et de Y. De (3.2) on déduit qu'une fonction f topologiquement fermée a les mêmes points de selle que $\Gamma(Y^-)f$ et $\Gamma(X^+)f$.

Par la méthode des épigraphes et des niveaux [16], on a d'autres caractérisations; en effet la (3.1) est équivalente à chacune des quatre propriétés suivantes:

$$Li^{\iota\times\nu}_{x\to x_o}\overline{\text{épi}f(x,\cdot)} \subset \text{épi}f(x_o,\cdot) \subset \overline{Li^{\iota\times\nu}_{x\to x_o}\text{épi}f(x,\cdot)}, \tag{3.3}$$

$$\text{int}_{X\times\mathbf{R}}cl_{Y\times\mathbf{R}}\ \text{épi}f \subset \text{épi}f \subset cl_{Y\times\mathbf{R}}\text{int}_{X\times\mathbf{R}}\ \text{épi}f, \tag{3.4}$$

$$\begin{gathered}\forall a \in \mathbf{R} \quad \text{int}_X cl_Y\{f \le a\} \subset \{f \le a\} \\ et \\ \{f < a\} \subset cl_Y \text{int}_X \{f \le a\},\end{gathered} \tag{3.5}$$

$$\begin{gathered}\forall a \in \mathbf{R} \quad \text{int}_X cl_Y\{f < a\} \subset \{f \le a\} \\ et \\ \{f < a\} \subset cl_Y \text{int}_X \{f < a\},\end{gathered} \tag{3.6}$$

où ι est la topologie discrète sur Y, ν est la topologie usuelle de $\mathbf{R}$ et "Li" désigne la limite inférieure de Kuratowski. En outre par la méthode des hypographes et des niveaux on peut obtenir les formules duales; c'est-à-dire: on les obtient en remplacement dans la description des quatre propriétés précedentes les symboles suivants "épi, $f(x,\cdot)$, $f(x_o,\cdot)$, $x \to x_o$, X, Y, $\leq$, $<$", respectivement, par "hypo, $f(\cdot,y)$, $f(\cdot,y_o)$, $y \to y_o$, Y, X, $\geq$, $>$".

Afin de reconnaître, de construire, et de chercher des fonctions topologiquement fermées, les propriétés suivantes sont très utiles:

(3.7) *pour tout f, les fonctions* $\Gamma(Y^-)\Gamma(X^+)f$, $\Gamma(X^+)\Gamma(Y^-)f$, $\Gamma(X^+)\Gamma(Y^-)\Gamma(X^+)f$ *et* $\Gamma(Y^-)\Gamma(X^+)\Gamma(Y^-)f$ *sont topologiquement fermées;*

(3.8) *si f est s.c.i. (resp. s.c.s.), alors* $\Gamma(X^+)f$ *(resp.* $\Gamma(Y^-)f$*) sont topologiquement fermées;*

(3.9) *si f et g sont des fonctions, respectivements, s.c.i. et s.c.s. telles que $f \leq g$, alors il existe des fonctions topologiquement fermées comprises entre f et g.*

Toutes les fonctions s.c.s./s.c.i. sont évidentement topologiquement fermées; mais malheureusement la réciproque n'est pas du tout vraie. La liaison entre les fonctions s.c.s./s.c.i. et les fonctions topologiquement fermées est eclaircie dans la proposition suivante, qui permettra d'avoir dans son corollaire une autre caractérisation des fonctions topolgiquement fermées dans le cadre des fonctions quasi concavo-convexes.

Proposition 3.1 (Voir [17]). *Soient X et Y deux sous-ensembles convexes d'espaces vectoriels localement convexes.*
Une fonction $f : X \times Y \to \bar{\mathbf{R}}$ quasi concavo-convexe et inf-localement compacte en tout point, qui vérifie la propriété suivante:

$$f \quad \text{est s.c.i et} \quad \inf_V f \quad \text{est s.c.s.}, \tag{3.10}$$

pour tout ouvert V de Y, est topologiquement fermée. La réciproque est vraie sans aucune hypothèse sur f.

Corollaire 3.2. *Soient X et Y deux sous-ensembles convexes d'espaces vectoriels localement convexes.*

Une fonction $f : X \times Y \to \bar{\mathbf{R}}$ quasi convaco-convexe et inf/sup-localement compacte en tout point, est topologiquement fermée si sont vérifiées les propriétés suivantes:

$$\Gamma(Y^-)\Gamma(X^+)f \leq \Gamma(X^+)\Gamma(Y^-)f, \tag{3.11}$$

$$\sup_U f \quad \textit{est s.c.i. et} \quad \inf_V f \quad \textit{est s.c.s.}, \tag{3.12}$$

pour tout ouvert V de Y et pour tout ouvert U de X. La réciproque est vraie sans aucune hypothèse sur f.

DÉMONSTRATION. La (3.12) implique que les fonctions $\Gamma(Y^-)f$ et $\Gamma(X^+)f$ vérifient (3.10).

Puisque f est par hipothése quasi concavo-convexe et inf/sup-localement compacte, les fonctions $\Gamma(Y^-)f$ et $\Gamma(X^+)f$ sont aussi quasi concavo-convexes et inf/sup-localement compactes. D'après la proposition 3.1 on a donc que les fonctions en question sont toutes les deux topologiquement fermées. Vue la définition (3.1′), ceci implique que

$$\begin{gathered}\Gamma(Y^-)\Gamma(X^+)\Gamma(Y^-)f = \Gamma(Y^-)f \\ \textit{et} \\ \Gamma(X^+)\Gamma(Y^-)\Gamma(X^+)f = \Gamma(X^+)f.\end{gathered} \tag{3.13}$$

D'autre part la (3.11) nous donne les égalités suivantes:

$$\begin{gathered}\Gamma(Y^-)\Gamma(X^+)\Gamma(Y^-)f = \Gamma(Y^-)\Gamma(X^+)f \\ \textit{et} \\ \Gamma(X^+)\Gamma(Y^-)\Gamma(X^+)f = \Gamma(X^+)\Gamma(Y^-)f.\end{gathered} \tag{3.14}$$

De (3.13) et (3.14) on déduit que $\Gamma(Y^-)\Gamma(X^+)f = \Gamma(Y^-)f$ et $\Gamma(X^+)\Gamma(Y^-)f = \Gamma(X^+)f$. La fonction f est donc topologiquement fermée. La réciproque dérive directement de la définition de fonction topologiquement fermée et de la propriété(3.2).∎

Les fonctions topologiquement fermées jouent un rôle remarquable, grâce au théorème de minimax suivant.

Théorème de Minimax (Voir [17]). *Soient X et Y deux sous-ensembles convexes d'espaces vectoriels. Soit $f : X \times Y \to \bar{\mathbf{R}}$ une fonction quasi concavo-convexe et topologiquement fermée. Si f est ou bien inf-compacte ou bien sup-compacte en quelques points, alors f a valeur de selle. Si, en outre, f est inf/sup-compacte en quelques points, alors f a des points de selle.*

Il y a aussi une version locale de ce théorème qui sera employée dans les sections 4 et 5.

Lemme 3.3. *Soit Y un sous-ensemble convexe d'un espace vectoriel topologique et X un sous-ensemble convexe d'un espace vectoriel topologique localement convexe.*
Pour toute fonction $f : X \times Y \to \bar{\mathbf{R}}$ quasi concavo-convexe on a:

$$\inf_V \sup_C \Gamma(X^+)\Gamma(Y^-)f = \sup_C \inf_V \Gamma(X^+)\Gamma(Y^-)f \tag{3.15}$$

lorsque V est un sous-ensemble convexe quelconque de Y et C est un compact convexe de X.

DÉMONSTRATION. Les transformations "$\inf\sup \Gamma(X^+)\Gamma(Y^-)$" et "$\sup\inf \Gamma(X^+)\Gamma(Y^-)$" sont toutes les deux des limitoïdes.

En vertu du principe d'extension des limitoïdes il suffit donc de vérifier (3.15) dans le cas où f est une fonction indicatrice. Il suffit alors de prouver l'implication suivante:

$$\begin{gathered} \sup_C \inf_V \Gamma(X^+)\Gamma(Y^-)\psi_G = -\infty \\ \Downarrow \\ \inf_V \sup_C \Gamma(X^+)\Gamma(Y^-)\psi_G = -\infty, \end{gathered} \tag{3.15'}$$

pout tout $G \subset X \times Y$ concavo-convexe. Supposons que la première partie de l'implication (3.15′) soit vraie. Pour tout $x \in C$ il existe alors un voisinage ouvert B_x de x tel que

$$\bigcap_{x' \in B_x} \overline{Gx'} \cap V \neq \emptyset \tag{3.16}$$

De la compacité de C on déduit qu'il existe un nombre fini de points $x_1, ..., x_n$ tels que C soit contenu dans la réunion des $B_1, ..., B_n$ associés à ces points. On peut donc choisir un sous-ensemble U de X

qui soit ouvert, convexe et tel que $\bigcup\{B_i : 1 \leq i \leq n\} \supset U \supset C$. La (3.16) implique alors que la multiapplication Ω de U dans V, définie pour tout $x \in U$ de la façon suivante:

$$\Omega x := V \cap \overline{Gx}, \tag{3.17}$$

est à valeurs fermées et que le recouvrement fini $\{U \cap B_i\}_{1 \leq i \leq n}$ de U est tel que $\bigcap\{\Omega x : x \in U \cap B_i\} \neq \emptyset$ pour tout i. D'après la propriété de la sélection constante (voir section 1), on déduit donc que

$$\inf_V \sup_U \Gamma(X^+)\Gamma(Y^-)\psi_G = -\infty. \tag{3.18}$$

Mais on a

$$\inf_V \sup_C \Gamma(X^+)\Gamma(Y^-)\psi_G = \inf_{\substack{U \supset C \\ \text{ouvert convex}}} \inf_V \sup_U \Gamma(X^+)\Gamma(Y^-)\psi_G;$$

la démonstration de (3.15′) a donc aboutie.■

Théorème des Minimax locaux. *Soient X et Y deux sous-ensembles convexes d'espaces vectoriels localement convexes. Soit $f : X \times Y \to \bar{\mathbf{R}}$ une fonction quasi concavo-convexe topologiquement fermée. Alors*

$$\inf_A \sup_B f = \sup_B \inf_A f, \tag{3.19}$$

si A et B sont deux sous-ensembles convexes, respectivement, de Y et de X tels que l'un est compact et l'autre ouvert.

DÉMONSTRATION. Supposons que A soit ouvert et B compact. On observe alors que

$$\begin{aligned} \inf_A \sup_B f &\leq \inf_A \sup_B \Gamma(X^+)f = \inf_A \sup_B \Gamma(X^+)\Gamma(Y^-)f \\ &\leq \sup_B \inf_A \Gamma(X^+)\Gamma(Y^-)f = \sup_B \inf_A \Gamma(X^+)f = \sup_B \inf_A f, \end{aligned} \tag{3.19′}$$

où les deux premières égalités sont une conséquence de l' égalité $\Gamma(X^+)\Gamma(Y^-)f = \Gamma(X^+)f$, qui est vraie pour les fonctions topologiquement fermées; la dernière égalité est due à (3.2), puisque A est ouvert; la première inégalité est évidente, parce que $f \leq \Gamma(X^+)f$; enfin l'autre inégalité est déduite du lemme 3.3. De (3.19′) on déduit donc (3.19). On raisonne dualement dans le cas où A est compact et B ouvert. ∎

Remarque 3.A *(Fonctions équivalentes et couples stables).* Dans le but d'étudier les fonctions qui ont les mêmes points de selle, Franzoni, Francaviglia [12] et Cavazzuti [7] ont donné une relation d'équivalence sur les fonctions; la signification "topologique" de cette notion jouera un rôle important dans l'étude de la "saddle topology" (voir §4).

Ils disent que deux fonctions f et g sont *équivalentes*, si elles ont les mêmes régularisations inférieures et supérieures; autrement dit, si $\Gamma(Y^-)f = \Gamma(Y^-)g$ et $\Gamma(X^+)f = \Gamma(X^+)g$.

Cette notion est l'analogue topologique de la relation d'équivalence introduite par Rockafellar [24] au moyen de convexifications.

Afin de comprendre les liaisons entre ce travail et d'autres, on rappelle qu'à côte de cette définition d' équivalence, Franzoni, Francaviglia et Cavazzuti introduisent la notion de *couple stable*; en disant qu'un couple (f,g) de fonctions est stable si $f = \Gamma(Y^-)g$ et $g = \Gamma(X^+)f$; en termes de relation d'équivalence, ceci signifie que l'intervalle de fonctions $[f,g]$ est une classe d'équivalence. L'intervalle $[f,g]$ associé à un couple stable (f,g) est donc composé par des fonctions topologiquement fermées équivalentes; et réciproquement, pour toute fonction f topologiquement fermée le couple $(\Gamma(Y^-)f, \Gamma(X^+)f)$ est stable. Enfin on observe que dans les dernières travaux de Cavazzuti (voir, par exemple [7]) on rencontre souvent deux fonctions f, g telles que:

$$(3.20) \qquad f \textit{ s.c.i.}, \quad g \textit{ s.c.s.}, \quad f \leq g \textit{ et } \Gamma(Y^-)g \leq \Gamma(X^+)f.$$

Cette propriété (3.20) équivaut à dire que $C(X,Y) \cap [f,g]$ est une classe d'équivalence, mais, chose plus importante, la (3.20) entraîne que f et g ont mêmes valeurs de selle, s'ils existent. Evidemment (3.20) est satisfaite si (f,g) est un couple stable.

4. Saddle Topology. Soient X et Y deux espaces topologiques quelconques. Soit $\mathcal{K}_X$ (resp. $\mathcal{K}_Y$) la famille des tous les compacts, non nécessairement séparés, de X (resp. Y) et soit O_X (resp. O_Y) la famille de tout les ouverts de X (resp. Y). La *saddle topology* est la moins fine topologie sur $\bar{\mathbf{R}}_{X\times Y}$ qui a, pour ouverts, les ensembles suivants:

$$\{f \in \bar{\mathbf{R}}^{X\times Y} : \inf_K \sup_U f > \epsilon\} \quad et \quad \{f \in \bar{\mathbf{R}}^{X\times Y} : \sup_C \inf_V f < \epsilon\}$$

où ϵ est un nombre réel fini, $K \in \mathcal{K}_Y, C \in \mathcal{K}_X, V \in O_Y, U \in O_X$. Par rapport à cette topologie on peut dire qu'une suite généralisée $\{f_j\}_j$ *converge vers* f si et seulement si pour chaque couple de compacts $(K,C) \in \mathcal{K}_X \times \mathcal{K}_Y$ et pour chaque couple d'ouverts $(U,V) \in O_X \times O_Y$ on a:

$$\inf_K \sup_U f \leq \liminf \inf_K \sup_U f_j \tag{4.1}$$

$$\sup_C \inf_V f \geq \limsup \sup_C \inf_V f_j \tag{4.2}$$

L'ensemble de toutes les fonctions vers lesquelles $\{f_j\}_j$ converge est appelé *limite* (noté: *Lim* f_j). Si l'un des espaces topologiques X, Y n'est pas discret, la saddle topology n'est ni séparée ni réguliere (réguliere, pour nous, signifie que tout point a une base de voisinages fermés). Mais elle a des propriétés de stabilité des valeurs de selle qui justifient son nom (voir th.4.1. et th. 4.2), et en outre elle est compacte (voir th. 4.3) et dans le cas où les espaces X et Y vérifient le deuxième axiome de dénombrabilité elle est aussi séquenciellement compacte (voir th. 4.4).

Théorème 4.1 (Propriétés de stabilité de valeurs de selle) . *Soit $\{f_j\}_j$ une suite généralisée inf/sup-compacte en quelques points et convergeant vers f par rapport à la saddle topology.*

Si chaque fonction f_j a valeur de selle (notée $sv(f_j)$), alors f a aussi valeur de selle et $sv(f) = \lim sv(f_j)$.

Démonstration. On peut écrire les relations suivantes:

$$\inf_Y \sup_X f = \inf_{K\in\mathcal{K}_Y} \inf_K \sup_X f \leq \inf_{K\in\mathcal{K}_Y} \liminf \inf_K \sup_X f_j =$$

$$\liminf \inf_Y \sup_X f_j \leq \limsup \sup_X \inf_Y f_j = \sup_{C\in\mathcal{K}_X} \limsup \sup_C \inf_Y f_j \leq$$

$$\sup_{C\in\mathcal{K}_X} \sup_C \inf_Y f = \sup_X \inf_Y f.$$

En effet la première et la dernière égalité sont évidentes; d'autre côté la première et la dernière inégalité sont conséquence de (4.1) et (4.2); en outre, la deuxième et l'avant-dernière égalité sont dues à la formule de réduction (2.1); enfin l'inégalité, qui reste, dérive de l'existence des valeurs de selle $sv(f_j)$. D'après les dites relations on a que la valeur de selle de la fonction f existe et qu'elle est la valeur limite de la suite des $sv(f_j)$.■

Théorème 4.2 (Propriétés de stabilité des points de selle) . *Soient X, Y deux espaces topologiques à compacts réguliers. Soit $\{f_j\}_j$ une suite généralisée inf/sup-compacte en quelques points et convergeant vers une fonction f par rapport à la saddle topology. Soit $\{(x_j, y_j)\}_j$ une suite de points telle que (x_j, y_j) soit un point de selle de f_j pour tout j, alors tout son point (x_o, y_o) d'adhérence tel que:*

(4.3) *les fonctions marginales $\sup_X f$ et $\inf_Y f$ soient, respectivement, s.c.i. en y_o et s.c.s. en x_o,*

est un point de selle de f et, en outre, $f(x_o, y_o) = \lim f_j(x_j, y_j)$.

Démonstration. D'après le Théorème 4.1, la fonction f a valeur de selle et, en outre, cette valeur $sv(f)$ est la limite des valeurs $sv(f_j)$ correspondant aux fonctions f_j; observe que $sv(f_j) = f_j(x_j, y_j)$. Maintenant on considere trois cas.
1er cas: $-\infty < sv(f) < +\infty$. Soient alors c, d deux nombres réels tels que $c < sv(f) < d$; il existe, donc, un index j_o tel que

$$c < sv(f_j) < d \quad \textit{pour tout} \quad j \geq j_o \tag{4.4}$$

D'après la inf/sup-compacité de $\{f_j\}_j$, il existe un couple de sous-ensembles compacts C et K, respectivement, de X et de Y et il existe un j_o tel que

$$\{\sup_X f_j < d\} \subset K \ , \ \{\inf_Y f_j > c\} \subset C \quad \textit{pour tout } j \geq j_o. \tag{4.5}$$

De (4.4) et (4.5) on déduit que $x_j \in C$ et $y_j \in K$ pour tous $j \geq j_o$. Tout en restant dans la généralité ,on suppose que la suite $\{(x_j, y_j)\}_j$ converge vers (x_o, y_o).
Donc pour tout $j' \geq j_o$ soit $C_{j'} \times K_{j'}$ le compact donnée par la fermeture de l'ensemble $\{x_j : j \geq j'\} \times \{y_j : j \geq j'\}$ dans $C \times K$.

Le filtre engendré par les ensembles $C_{j'} \times K_{j'}$ converge vers (x_o, y_o). Alors on a:

$$\inf_Y \sup_X f \le \sup_X f(\cdot, y_o) \le \sup_{j'} \inf_{K_{j'}} \sup_X f \le \tag{4.6}$$
$$\sup_{j'} \liminf \inf_{K_{j'}} \sup_X f_j = \lim f_j(x_j, y_j),$$

en effet la première de ces inégalités est évidente; la deuxième dérive de la semi-continuité inférieure de la fonction marginale "$\sup f$" et de la convergence du filtre, engendré par les ensembles $K_{j'}$, vers x_o; la troisième est une conséquence de la propriété (4.1) de la saddle topology; l'égalité qui suit est due au fait que l'ensemble $K_{j'}$ contient y_j pour tout $j \ge j'$. D'une facon analogue on a:

$$\lim f_j(x_j, y_j) = \inf_{j'} \limsup \sup_{C_{j'}} \inf_Y f_j \le \tag{4.7}$$
$$\inf_{j'} \sup_{C_{j'}} \inf_Y f \le \inf_Y f(x_o, \cdot) \le \sup_X \inf_Y f.$$

De (4.6) et (4.7) on déduit que (x_o, y_o) est un point de selle de f et que $f(x_o, y_o) = \lim f_j(x_j, y_j)$.

2me cas: $sv(f) = -\infty$. Ayant choisi un nombre réel $d > sv(f)$, on peut définir l'index j_o, les compacts K et K_j, comme ci-dessus. Ainsi on peut appliquer la (4.6) qui nous donne que (x_o, y_o) est un point de selle de f.

On raisonne dualement dans le troisième cas, où on suppose que $sv(f) = +\infty$.∎

Remarque 4.A. *(Sur les propriétés de stabilité des points et des valeurs de selle).* D'après Cavazzuti (voir [4] et Guillerme [18]) on peut donner des hypothèses légèrement differentes dans le but d'avoir des propriétés de stabilité. *"Soit $\{f_j\}_j$ une suite généralisée convergeant vers f par rapport à la saddle topology. Si toute f_j a valeur de selle et s'il existe deux compacts C et K tels que*

$$sv(f_j) = \inf_K \sup_X f_j = \sup_C \inf_Y f_j \quad \textit{pour tous} \quad j \tag{4.8}$$

alors f a valeur de selle et $sv(f) = \lim sv(f_j)$". A propos de la stabilité de points de selle on a: *"Soient X et Y deux espaces* *topologiques à compacts réguliers. Soit* $\{f_j\}_j$ *une suite généralisée convergeant vers* f *par rapport à la saddle topology. Si toute* f_j *a un point de selle* (x_j, y_j) *et la suite*

(4.9) $\{(x_j, y_j)\}_j$ *est contenue dans un compact de* $X \times Y$

alors tout point (x_o, y_o) *d'adhérence de* $\{(x_j, y_j)\}_j$ *vérifiant (4.3) est un point de selle de* f*, et* $f(x_o, y_o) = \lim f_j(x_j, v_j)$*". ■*

Remarque 4.B *(Sur la condition (4.3)).* On peut observer que la condition (4.3) est vérifiée lorsque f est topologiquement fermée (propriété(3.2)); en particulier, si f est s.c.s./s.c.i..■

Remarque 4.C *(Sur la condition (4.9)).* Dans le cas où $\{f_n\}_n$ est une suite usuelle (c'est-à-dire l'ensemble des indeces n est l'ensemble des nombres naturels, muni de l'ordre usuel) et si la suite $\{(x_n, y_n)\}_n$ est convergente, alors la condition (4.9) est automatiquement satisfaite, et on n'a pas besoin de la regularité ou de la séparation de X et Y. Même dans le cas où X et Y sont localement compacts et la suite $\{(x_j, y_j)\}_j$ est convergente, la (4.9) est satisfaite.■

Remarque 4.D *(Sur le caractère local de la saddle topology).* Si une suite généralisé $\{f_j\}_j$ converge vers f par rapport à la saddle topology, alors pour tout couple d'ouverts U, V respectivement, de X et Y les restrictions $f_j|_{U\times V}$ convergent aussi vers $f|_{U\times V}$. On peut donc obtenir des informations de caractère local à partir des théorèmes précédents, lesquels par contre sont encore valables si on considère la topologie (moins fine que la saddle topology), donnée par la convergence suivante: *"*$\{f_j\}_j$ *converge vers* f *si (4.1) et (4.2) sont vérifiées pour* $U = X, V = Y$ *et pour tout couple* C, K *des compacts". ■*

Théorème 4.3 (compacité). *La saddle topology est compacte.*

DÉMONSTRATION. Il suffit de démontrer que toute suite généralisée ayant un ultrafiltre comme son filtre de section, est convergente. Si $\{f_j\}_j$ est une telle suite, on a que la limite de $\{\inf_A \sup_B f_j\}_j$ existe pour tout couple A, B des sous-ensembles, respectivement, de X et

Y. Donc soit $f : X \times Y \to \bar{\mathbf{R}}$, définie pour tout (x,y) par

$$(4.10) \qquad f(x,y) = \inf_{B \in \mathcal{N}(x)} \sup_{A \in \mathcal{N}(y)} \liminf \sup_{A} \sup_{B} f_j$$

Si $K \in \mathcal{K}_Y$ et $U \in O_X$, d'après (2.5'):

$$\inf_K \sup_U f \leq \inf_{y \in K} \sup_{A \in \mathcal{N}(y)} \liminf \sup_{A} \sup_{U} f_j \leq \liminf \inf_K \sup_U f_j$$

et dualement pour $C \in \mathcal{K}_X$ et $V \in O_Y$ on a:

$$\sup_C \inf_V f \geq \sup_{x \in C} \inf_{B \in \mathcal{N}(x)} \limsup \inf_{B} \inf_{V} f_j \geq \limsup \sup_C \inf_V f_j$$

Donc la suite en question converge vers f par rapport à la saddle topology.■

Théorème 4.4 (Compacité sequentielle). *Si l'un des espaces X, Y vérifie le deuxième axiome de dénombrabilité , tandis que l'autre est un espace de Lindelöff héréditaire, alors la saddle topology est séquentiellement compacte.*

DÉMONSTRATION. Suppose que Y soit l'espace de Lindelöff héréditaire (c'est-à -dire: de toute famille d'ouverts on peut extraire une sous-famille dénombrable ayant la même réunion). Pour des espaces topologiques de ce type on a le théorème de Peirone [23]:
De toute suite $\{h_n\}_{n \in \mathbf{N}}$ de fonctions de Y dans $\bar{\mathbf{R}}$ on peut extraire une sous-suite $\{h_{n_k}\}_{k \in \mathbf{N}}$ telle que pour tout $y \in Y$ on ait:

$$(4.11) \qquad \sup_{A \in \mathcal{N}(y)} \liminf_{k \to \infty} \inf_{A} h_{n_k} = \sup_{A \in \mathcal{N}(y)} \limsup_{k \to \infty} \inf_{A} h_{n_k}.$$

Alors soit $\{f_n\}_{n \in \mathbf{N}}$ une suite de fonctions de $X \times Y$ dans $\bar{\mathbf{R}}$ et soit $\{U_n\}_{n \in \mathbf{N}}$ une base dénombrable de la topologie de X. D'après le théorème de Peirone on choisit des sous-suites $\{f_k^m\}_k$ de $\{f_n\}_n$ telles que (4.11) soit vérifiée, lorsque on remplace h_{n_k} avec $\sup_{U_m} f_k^m$, et telles que $\{f_k^{m+1}\}_k$ soit une sous-suite de $\{f_k^m\}_k$ pour tout m. Alors la (4.11) est aussi vérifiée lorsqu'on remplace h_{n_k} avec la suite diagonale $\sup_{U_m} f_k^k$. La fonction f définie par

$$(4.12) \qquad f(x,y) = \inf_{B\in\mathcal{N}(x)} \sup_{A\in\mathcal{N}(y)} \liminf_{k\to\infty} \inf_A \sup_B f_k^k$$

où $(x,y) \in X \times Y$, vérifie donc

$$(4.12') \qquad f(x,y) = \inf_{B\in\mathcal{N}(x)} \sup_{A\in\mathcal{N}(y)} \limsup_{k\to\infty} \inf_A \sup_B f_k^k$$

Un raisonnement similaire à celui du théorème 4.3 nous permet donc de montrer que la sous-suite diagonale $\{f_k^k\}_k$ converge vers f par rapport à la saddle topology.∎

Ayant vu les propriétés basilaires et de routine de la saddle topology, un problème qui reste à considérer est le *problème d'unicité de la limite*, c'est-à-dire:
"Est-il possible de réconstruire l'ensemble de toutes les fonctions vers lesquelles une suite généralisée donnée converge, une fois qu'on connaît au moins une d'entre elles?".

Une réponse à ce problème nous permettra aussi de spécifier des classes de fonctions sur lesquelles la saddle topology est métrisable.

Evidentement ce problème se pose, parce que la saddle topology n'est pas du tout séparée. En effet si on considere une suite quelconque convergente vers une fonction; appelée celle-ci f, on a que sa **limite** (= l'ensemble des fonctions vers lesquelles la suite converge) contient la fermeture de l'ensemble $\{f\}$; c'est-à-dire:

$$(4.13) \qquad \overline{\{f\}} = [\Gamma(Y^-)f, \Gamma(X^+)f]$$

et cet ensemble contient plus qu'une fonction, lorsque f n'est pas s.c.s./s.c.i. .

Donc pour donner une réponse au problème d'unicité de la limite, il suffirait de trouver, par exemple, une relation d'équivalence par rapport à laquelle la limite d'une suite généralisée quelconque soit une classe d'équivalence. Mais, malhereusement, une telle relation d'équivalence n'existe pas, sauf dans le cas très particulier où les espaces X et Y sont tous les deux munis de la topologie discrète; en effet, seulement dans ce cas, la famille des toutes les limites par rapport à la saddle topology forme une partition de $\bar{\mathbf{R}}^{X\times Y}$.

Alors dans le but de donner une réponse partielle, mais en certain sens satisfaisante, on procédera de la façon suivante. On donnera à priori une relation d'équivalence; et ensuite on cherchera à déterminer les suites généralisées telles que leur limite soit une classe d'équivalence.

La relation d'équivalence qu'on va donner est à mon avis la plus simple et constructive qu'on puisse imaginer. On dit que deux fonctions sont *équivalentes* si elles ont les mêmes voisinages par rapport à la saddle topology. Donc deux fonctions f, g seront équivalentes si et seulement si $\overline{\{f\}} = \overline{\{g\}}$; c'est-à-dire, d'après (4.13), si et seulement si

$$\Gamma(Y^-)f = \Gamma(Y^-)g \quad \text{et} \quad \Gamma(X^+)f = \Gamma(X^+)g. \tag{4.14}$$

Fait remarquable: cette relation d'équivalence, obtenue d'une façon très naturelle dans le cadre de la saddle topology, est exactement celle presentée par Franzoni, Francaviglia [12] et une de celles étudiées par Cavazzuti [7], [5].

On convient de numeroter par (E.1), (E.2)... toutes les propriétés qui concernent cette relation d'équivalence, dont les démonstrations sont laissées au lecteur.

Que signifie l'affirmation que la limite d'une suite est une classe d'équivalence par rapport à la relation d'équivalence (4.14)? Etant donnée une suite $\{f_j\}_j$ convergente, sa limite est évidentement une reunion des classes d'équivalence. Supposons alors que sa limite soit exactement une classe d'équivalence. Ayant pris alors une fonction f dans sa limite, elle converge aussi vers les fonctions $\Gamma(Y^-)f$ et $\Gamma(X^+)f$. Donc ces deux dernières fonctions doivent être équivalentes à f. D'après (4.14), on aura :$\Gamma(Y^-)f = \Gamma(Y^-)\Gamma(X^+)f$ et $\Gamma(X^+)f = \Gamma(X^+)\Gamma(Y^-)f$; c'est-à-dire f est topologiquement fermée. Donc si la limite d'une suite est une classe d'équivalence, alors elle est composée par des fonctions topologiquement fermées.

Alors on dira qu'une suite généralisée a la *propriété d'unicité de la limite* si l'ensemble de toutes les fonctions topologiquement fermées vers lesquelles elle converge est une classe d'équivalence.

Observons que par rapport à la convergence dans la saddle topology il n'est pas trop restrictif de se borner aux fonctions topologiquement fermées. En effet on a la proposition suivante.

Proposition 4.5. *Toute suite généralisée convergente par rap-*

port à la saddle topology converge aussi vers une fonction topologiquement fermée.

DÉMONSTRATION. Si une suite converge vers une fonction f, elle converge aussi vers les fonctions $\Gamma(Y^-)f$ et $\Gamma(X^+)f$, puisque sa limite, étant un ensemble fermé , contient $\overline{\{f\}}$. Donc elle converge aussi vers la fonction $\Gamma(Y^-)\Gamma(X^+)f$ qui est topologiquement fermée (voir (3.7)).■

Pour mieux comprendre la liaison entre le problème et la propriété d'unicité de la limite on a la proposition suivante. On note l'ensemble des fonctions topologiquement fermées par le symbole $C(X,Y)$.

Proposition 4.6. *Soit* $\{f_j\}_j$ *une suite généralisée convergente par rapport à la saddle topology.*
Les propriétés suivantes sont équivalentes:

(4.14) $\{f_j\}_j$ *a la propriete de l'unicite de la limite*;

(4.15) $$\exists\ f \in C(X,Y) \cap Lim f_j \ :\ C(X,Y) \cap Lim f_j = \overline{\{f\}};$$

(4.16) $$C(X,Y) \cap Lim f_j = \overline{\{f\}}\ \ \forall\ f \in C(X,Y) \cap Lim f_j;$$

(4.17) $$C(X,Y) \cap Lim f_j = [\Gamma(Y^-)\Gamma(X^+)\Gamma(Y^-)f, \Gamma(X^+)\Gamma(Y^-)f]$$
pour toute fonction $f \in Lim f_j$;

(4.17') $$C(X,Y) \cap Lim f_j = [\Gamma(Y^-)\Gamma(X^+)f, \Gamma(X^+)\Gamma(Y^-)\Gamma(X^+)f]$$
pour toute fonction $f \in Lim f_j$.

DÉMONSTRATION. Les implications (4.14) $\Rightarrow$ (4.15) $\Rightarrow$ (4.16) et leurs réciproques dérivent des propiétés suivantes:

(E.1) *la classe d'équivalence d'une fonction topologiquement fermée* f *est* $\overline{\{f\}}$;

(E.2) *toute fonction équivalente à une fonction topologiquement fermée est aussi topologiquement fermée.*

(4.16) implique (4.17) et (4.17′), parce que les fonctions topologiquement fermées $\Gamma(X^+)\Gamma(Y^-)f$ et $\Gamma(Y^-)\Gamma(X^+)f$ appartiennent à $Lim f_j$, lorsque $f \in Lim f_j$. Enfin (4.17) ou (4.17′) implique (4.14) parce que les intervalles

$$[\Gamma(Y^-)\Gamma(X^+)\Gamma(Y^-)f, \Gamma(X^+)\Gamma(Y^-)f]$$

et

$$[\Gamma(Y^-)\Gamma(X^+)f, \Gamma(X^+)\Gamma(Y^-)\Gamma(X^+)f],$$

sont, respectivement, les classes d'équivalence des fonctions topolo giquement fermées $\Gamma(X^+)\Gamma(Y^-)f$ et $\Gamma(Y^-)\Gamma(X^+)f$.■

Maintenant on va donner une condition nécessaire et suffisante pour avoir la propriété d'unicité de la limite. Et après on emploiera cette condition sur la classe des fonctions quasi concavo-convexes.

Soit donnée une suite généralisée $\{f_j\}_j$ et soit J l'ensemble des indeces j. Par les symboles $\tilde{\Gamma}_c(J^+, Y^-, X^+)f_j$ et $\tilde{\Gamma}_c(J^-, X^+, Y^-)f_j$ on désigne les fonctions de $X \times Y$ dans $\bar{\mathbf{R}}$ définies par les égalités suivantes:

(4.18)
$$(\tilde{\Gamma}_c(J^+, Y^-, X^+)f_j)(x, y) = \\ = \inf_{B \in \mathcal{N}(x)} \sup_{A \in \mathcal{N}(y)} \sup_{C \in \mathcal{K}_B} \limsup \sup_C \inf_A f_j,$$

(4.19)
$$(\tilde{\Gamma}_c(J^-, X^+, Y^-)f_j(x, y) = \\ = \sup_{A \in \mathcal{N}(y)} \inf_{B \in \mathcal{N}(x)} \inf_{K \in \mathcal{K}_A} \liminf \inf_K \sup_B f_j,$$

où $(x, y) \in X \times Y$. Il est évident que les fonctions $\tilde{\Gamma}_c(J^+, Y^-, X^+)f_j$ et $\tilde{\Gamma}_c(J^-, X^+, Y^-)f_j$ sont respectivement s.c.s. et s.c.i. (voir section 1).

Théorème 4.7 (Sur l'unicité de la limite). *Soient X et Y deux espaces topologiques à compacts réguliers. Une suite généralizée $\{f_j\}_j$ convergente a la propriété d'unicité de la limite si et seulement si $\tilde{\Gamma}_c(J^-, X^+, Y^-)f_j \leq \tilde{\Gamma}_c(J^+, Y^-, X^+)f_j$.*

Corollaire 4.8. *Soient X et Y deux espaces topologiques à compacts régulliers. Si $\{f_j\}_j$ a la propriété d'unicité de la limite, alors*

$$(4.19)\quad C(X,Y) \cap Lim f_j = [\tilde{\Gamma}_c(J^-,X^+,Y^-)f_j, \tilde{\Gamma}_c(J^+,Y^-,X^+)f_j].$$

Réciproquement, si $\{f_j\}_j$ converge et (4.19) vaut, alors $\{f_j\}_j$ a la propriété d'unicité de la limite.

Corollaire 4.9. *Soient X et Y deux espaces topologiques à compacts réguliers. Soit $\{f_j\}_j$ une suite généralisée des fonctions. Si toute sa sous-suite a la propriété d'unicité de la limite, alors*

$$(4.20)\qquad \tilde{\Gamma}_c(J^-,X^+,Y^-)f_j = \tilde{\Gamma}_c(J^+,X^+,Y^-)f_j$$

$$(4.21)\qquad \tilde{\Gamma}_c(J^+,Y^-,X^+)f_j = \tilde{\Gamma}_c(J^-,Y^-,X^+)f_j$$

Réciproquement, si $\tilde{\Gamma}_c(J^+,X^+,Y^-)f_j \leq \tilde{\Gamma}_c(J^-,Y^-,X^+)f_j$, alors toute sous-suite de $\{f_j\}_j$ a la propriété d'unicité de la limite.

Les nouvelles fonctions $\tilde{\Gamma}_c(J^+,X^+,Y^-)f_j$ et $\tilde{\Gamma}_c(J^-,Y^-,X^+)f_j$ sont obtenues, respectivement de (4.18) et (4.19) en changeant dans (4.18) la "liminf" par la "limsup" et dans (4.19) "limsup" par la "liminf".

Pour démontrer le théorème sur l'unicité et ses corollaires on a besoin de quelques lemmes et des fonctions $\Gamma_c(J^-,Y^-,X^+)f_j$ et $\Gamma_c(J^+,X^+,Y^-)f_j$ définies par:

$$(4.22)\qquad (\Gamma_c(J^-,Y^-,X^+)f_j)(x,y) = \\ = \inf_{B\in\mathcal{N}(x)} \sup_{A\in\mathcal{N}(y)} \sup_{K\in\mathcal{K}_A} \liminf \inf_K \sup_B f_j,$$

$$(4.23)\qquad (\Gamma_c(J^+,X^+,Y^-)f_j)(x,y) = \\ = \sup_{A\in\mathcal{N}(y)} \inf_{B\in\mathcal{N}(x)} \sup_{C\in\mathcal{K}_B} \limsup \sup_C \inf_A f_j,$$

où $(x,y) \in X \times Y$. Aussi dans ce cas il est évident que les fonctions $\Gamma_c(J^-,Y^-,X^+)f_j$ et $\Gamma_c(J^+,X^+,Y^-)f_j$ sont respectivement s.c.s. et s.c.i. (voir section 1).

Lemme 4.10. *Soient X et Y deux espaces topologiques à compacts réguliers. Une suite généralisée $\{f_j\}_j$ converge vers toute fonction f telle que $\Gamma_c(J^+, X^+, Y^-)f_j \leq f \leq \Gamma_c(J^-, Y^-, X^+)f_j$. Réciproquement, si $\{f_j\}_j$ converge vers une fonction topologiquement fermée f, alors f vérifie ces inégalités.*

DÉMONSTRATION. Pour tout compact K de Y et tout ouvert U de X on a:

$$\begin{aligned}
&\inf_K \sup_U \Gamma_c(J^-, Y^-, X^+)f_j \leq \\
&\leq \inf_{y \in K} \sup_{A \in \mathcal{N}(y)} \inf_{K' \in \mathcal{K}_A} \liminf \inf_{K'} \sup_U f_j \\
&\leq \liminf \inf_K \sup_U f_j,
\end{aligned}$$

où la première inégalité est due simplement au fait que U est ouvert et l'autre inégalité est une conséquence de la formule de réduction (2.2'). On vérifie dualement que pour tout compact C de X et pour tout ouvert V de Y on a:

$$\sup_C \inf_V \Gamma_c(J^+, X^+, Y^-)f_j \leq \limsup \sup_C \inf_V f_j,$$

Donc la suite $\{f_j\}_j$ converge vers toute fonctions f telle que $\Gamma_c(J^+, X^+, Y^-)f_j \leq f \leq \Gamma_c(J^-, Y^-, X^+)f_j$.

Supposons maintenant que $\{f_j\}_j$ converge vers une fonction topologiquement fermée f. D'après (4.1) on a alors pour tout $(x, y) \in X \times Y$:

$$\inf_{B \in \mathcal{N}(x)} \sup_{A \in \mathcal{N}(y)} \inf_{K \in \mathcal{K}_A} \inf_K \sup_U f \leq (\Gamma_c(J^-, Y^-, X^+)f_j)(x, y);$$

mais le première partie de cette inégalité est la fonction $\Gamma(Y^-, X^+)f$ qui est plus grande que f, puisque f est topologiquement fermée. Donc $f \leq \Gamma_c(J^-, Y^-, X^+)f_j$. On montre dualement l'inégalité $f \geq \Gamma_c(J^+, X^+, Y^-)f_j$. ∎

D'après ce lemme 4.10 et la proposition 4.5 on a le lemme suivant.

Lemme 4.11. *Soient X et Y deux espaces topologiques à compacts réguliers. Une suite généralisée $\{f_j\}_j$ est convergente si et seulement si $\Gamma_c(J^+, X^+, Y^-)f_j \leq \Gamma_c(J^-, Y^-, X^+)f_j$.*

Lemme 4.12. *Soient X et Y deux espaces topologiques à compacts réguliers. Pour toute suite généralisée* $\{f_j\}_j$ *on a*

$$\Gamma(X^+)\Gamma_c(J^+,X^+,Y^-)f_i = \tilde{\Gamma}_c(J^+,Y^-,X^+)f_i$$

et

$$\Gamma(Y^-)\Gamma_c(J^-,Y^{,}X^+)f_i = \tilde{\Gamma}_c(J^-,X^+,Y^-)f_i.$$

DÉMONSTRATION. L'égalité

$$\Gamma(Y^-)\Gamma_c(J^-,Y^-,X^+)f_j = \tilde{\Gamma}_c(J^-,X^+,Y^-)f_j,$$

et sa duale

$$\Gamma(X^+)\Gamma_c(J^+,X^+,Y^-)f_j = \tilde{\Gamma}_c(J^+,Y^-,X^+)f_j,$$

on les déduit directement de la formule de réduction (2.2) et de sa duale.■

Démonstration du théor. 4.7 et du Cor. 4.8. D'après le lemme 4.10 on a que

$$C(X,Y) \cap Lim f_j = C(X,Y) \cap [\Gamma_c(J^+,X^+,Y^-)f_j, \Gamma_c(J^-,Y^-,X^+)f_j]$$

En outre, d'après le lemme 4.11 la fonction s.c.i. $\Gamma_c(J^+,X^+,Y^-)f_j$ est plus petite que la fonction s.c.s. $\Gamma_c(J^-,Y^-,X^+)f_j$, dans le cas où la suite $\{f_j\}_j$ soit convergente. On peut aussi vérifier en toute généralité que si deux fonctions f et g sont, respectivement, s.c.i. et s.c.s. et si $f \leq g$, alors:

(E.3) *les fonctions topologiquement fermées comprises entre* f *et* g *forment une classe d'équivalence si et seulement si*

$$\Gamma(Y^-)g \leq \Gamma(X^+)f.$$

En outre, lorsque cette inégalité est vraie, on a

$$C(X,Y) \cap [f,g] = [\Gamma(Y^-)g, \Gamma(X^+)f].$$

Donc de (E.3) on déduit le théorème 4.7 et son corollaire 4.8, en vertu du lemme 4.12.■

Démonstration du Cor. 4.9. On démontre d'abord la première partie du corollaire. Soit donc $\{f_i\}_i$ une sous-suite de $\{f_j\}_j$.

On a évidentement $C(X,Y) \cap Lim f_i \supset C(X,Y) \cap Lim f_j$; mais par hypothèse les ensembles $C(X,Y) \cap Lim f_j$ et $C(X,Y) \cap Lim f_i$ sont des classes déquivalence; donc ils doivent être egaux. D'après le corollaire 4.8, ça signifie qu'on a:

$$\tilde{\Gamma}_c(J^-, X^+, Y^-) f_j = \tilde{\Gamma}_c(I^-, X^+, Y^-) f_i \tag{4.24}$$

quelle que soit la sous-suite. D'après la formule de réduction (2.3) on a aussi:

$$\tilde{\Gamma}_c(J^+, X^+, Y^-) f_j = \sup_{\substack{\{f_i\}_i \\ sous-suite\, de\, \{f_j\}_j}} \tilde{\Gamma}_c(I^-, X^+, Y^-) f_i. \tag{4.25}$$

De (4.24) et de (4.25) on déduit, l'égalité (4.20). On démontre dualement l'autre égalité . Maintenant on va démontrer la deuxième partie du corollaire. Soit $\{f_i\}_i$ une sous-suite quelconque. De l'inégalité, donnée par hypothèse, on a évidentement:

$$\tilde{\Gamma}_c(I^+, X^+, Y^-) f_i \leq \tilde{\Gamma}_c(I^-, Y^-, X^+) f_i \tag{4.26}$$

Par ailleurs pour tout point $(x, y) \in X \times Y$ on a:

$$\begin{aligned}
&(\Gamma_c(I^+, X^+, Y^-) f_i)(x, y) \leq \\
&\leq \sup_{A \in \mathcal{N}(y)} \inf_{B \in \mathcal{N}(x)} \limsup \sup_B \inf_A f_i \\
&\leq \sup_{A \in \mathcal{N}(y)} \inf_{B \in \mathcal{N}(x)} \limsup \inf_A \sup_B f_i \\
&\leq (\tilde{\Gamma}_c(I^+, X^+, Y^-) f_i)(x, y);
\end{aligned}$$

et analoguement on a aussi: $\tilde{\Gamma}_c(I^-, Y^-, X^+) f_i \leq \Gamma_c(I^-, Y^-, X^+) f_i$. D'après (4.26) et le lemme 4.11, la sous-suite en question est convergente. Étant les inégalités $\tilde{\Gamma}_c(J^-, Y^-, X^+) f_i \leq \tilde{\Gamma}_c(J^+, Y^-, X^+) f_i$ et $\tilde{\Gamma}_c(I^-, X^+, Y^-) f_i \leq \tilde{\Gamma}_c(I^+, X^+, Y^-) f_i$ en tout cas vraies, de (4.26) on déduit que $\{f_i\}_i$a la propriètè d'unicité de la limite, en vertu du théorème sur l'unicité de la limite.■

Remarque 4.E. D'après le lemme 4.12 on a que les fonctions $\tilde{\Gamma}_c(J^-, X^+, Y^-)f_j$ et $\tilde{\Gamma}_c(J^+, Y^-, X^+)f_j$ sont topologiquement fermées, parce qu'elles sont, respectivement, la régularisée inférieure de la fonction s.c.s. $\Gamma_c(J^-, Y^-, X^+)f_j$ et la régularisée supérieure de la fonction s.c.i. $\Gamma(J^+, X^+, Y^-)f_j$. En outre elles appartiennent à la limite de $\{f_j\}_j$, si cette suite est convergente, sans avoir besoin de l'unicité de la limite, en vertu des lemmes 4.10 et 4.11.■

Maintenant on va démontrer l'application envisagée du théorème sur l'unicité de la limite au cas des fonctions quasi concavo-convexes.

Lemme 4.13. *Soit X et Y deux sous-ensembles convexes d'espaces vectoriels localement convexes, jouissant de la propriété que l'enveloppe fermée convexe de tout leur compact est aussi compact. Soit $\{f_j\}_j$ une suite de fonctions quasi concavo-convexes et topologiquement fermées, qui est inf/sup-localement compacte en tout point. Alors on a les égalités suivantes:*

$$\Gamma_c(J^+, X^+, Y^-)f_j = \tilde{\Gamma}_c(J^+, X^+, Y^-)f_j \quad (4.27)$$

et

$$\Gamma_c(J^-, X^+, Y^-)f_j = \tilde{\Gamma}_c(J^-, X^+, Y^-)f_j;$$

$$\Gamma_c(J^+, Y^-, X^+)f_j = \tilde{\Gamma}_c(J^+, Y^-, X^+)f_j \quad (4.28)$$

et

$$\Gamma_c(J^-, Y^-, X^+)f_j = \tilde{\Gamma}_c(J^-, Y^-, X^+)f_j.$$

DÉMONSTRATION. Les relations d'ordre entre les huit fonctions qui apparaissent dans les égalités (4.27) et (4.28) sont schématisées dans le diagramme suivant; ces relations sont toujours vraies, sans aucune hypothèse sur les espaces topologiques X et Y ou sur les fonctions f_j. Donc, la démonstration de (4.27) aura aboutie si on démontre les inégalités suivantes:

$$\Gamma_c(J^+, X^+, Y^-)f_j \geq \tilde{\Gamma}_c(J^+, X^+, Y^-)f_j \quad (4.27')$$

et

$$\Gamma_c(J^-, X^+, Y^-)f_j \geq \tilde{\Gamma}_c(J^-, X^+, Y^-)f_j.$$

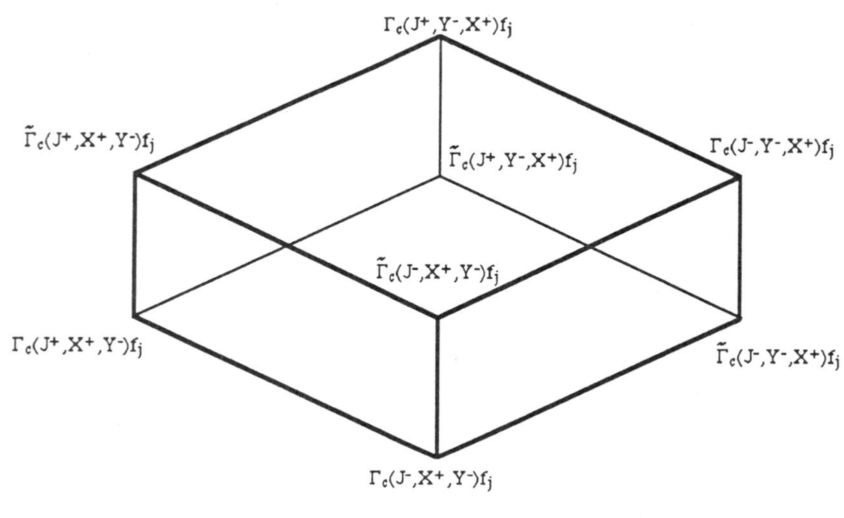

Figure 2

Soient $(x,y) \in X \times Y$, $B \subset X$ et $A \subset Y$. On désigne par les symboles $\mathcal{V}(x), \mathcal{V}(y)$ les systèmes des voisinages convexes et ouverts, respectivement, de x et de y; tandis que $\mathcal{C}_A, \mathcal{C}_B$ désignent les familles des tous les sous-ensembles convexes et compacts, respectivement, de Y et de X. Alors pour tout $(x,y) \in X \times Y$ on a:

$$(\tilde{\Gamma}_c(J^+, X^+, Y^-)f_j)(x,y) = \sup_{V\in\mathcal{V}(y)} \inf_{U\in\mathcal{V}(x)} \inf_{K\in\mathcal{C}_V} \limsup \inf_K \sup_U f_j \leq$$

$$\leq \sup_{V\in\mathcal{V}(y)} \inf_{U\in\mathcal{V}(x)} \inf_{K\in\mathcal{C}_V} \limsup \sup_U \inf_K f_j$$

$$\leq \sup_{V\in\mathcal{V}(y)} \inf_{U\in\mathcal{V}(x)} \sup_{C\in\mathcal{C}_U} \limsup \sup_C \inf_V f_j$$

$$\leq \sup_{V\in\mathcal{V}(y)} \inf_{U\in\mathcal{V}(x)} \sup_{C\in\mathcal{C}_U} \limsup \inf_V \sup_C f_j = (\Gamma_c(J^+,X^+,Y^-)f_j)(x,y),$$

où la première et la dernière égalité sont dues aux propriétés topologiques de X et Y; en outre la première et la dernière inégalité sont une conséquence du théorème des minimax locaux; enfin l'inégalité centrale est la formule de réduction (2.4). Comme cette formule de réduction est aussi valable quand on change la "limsup" avec la "liminf", on peut déduire l'autre inégalité de (4.27′). On peut d'une façon duale démontrer la (4.28).■

Théorème 4.14. *Sous les mêmes hypothèses sur X, Y et sur $\{f_j\}_j$ du lemme (4.13), si $\{f_j\}_j$ converge par rapport à la saddle topology, alors toute sa sous-suite a la propriété d'unicité de la limite et*

$$C(X,Y) \cap Lim f_j = [\Gamma_c(J^+, X^+, Y^-)f_j, \Gamma_c(J^-, Y^-, X^+)f_j].$$

DÉMONSTRATION. D'après le lemme 4.11, la convergence de la suite implique que $\Gamma_c(J^+, X^+, Y^-)f_j \leq \Gamma_c(J^-, Y^-, X^+)f_j$. Mais le lemme 4.13 entraîne $\Gamma_c(J^+, X^+, Y^-)f_j = \tilde{\Gamma}_c(J^+, X^+, Y^-)f_j$ et $\Gamma_c(J^-, Y^-, X^+)f_j = \tilde{\Gamma}_c(J^-, Y^-, X^+)f_j$. En vertu du corollaire 4.9 on a donc que toute sous-suite de $\{f_j\}_j$ a la propriété d'unicité de la limite. Enfin des (4.19), (4.20) et (4.21) on conlue que

$$C(X,Y) \cap Lim f_j = [\Gamma_c(J^+, X^+, Y^-)f_j,\ \Gamma_c(J^-, Y^-, X^+)f_j].\blacksquare$$

Corollaire 4.15. *Sous les mêmes hypothèses sur X, Y et sur $\{f_j\}_j$ du lemme (4.13) les propriétés suivantes sont équivalentes:*

(4.29) $\{f_j\}_j$ *converge par rapport à la saddle topology*

$$\Gamma_c(J^-, Y^-, X^+)f_j = \Gamma_c(J^+, Y^-, X^+)f_j \tag{4.30}$$

$$\Gamma_c(J^+, X^+, Y^-)f_j = \Gamma_c(J^-, X^+, Y^-)f_j \tag{4.31}$$

$$\Gamma_c(J^+, X^+, Y^-)f_j \leq \Gamma_c(J^-, Y^-, X^+)f_j. \tag{4.32}$$

DÉMONSTRATION. D'après le diagramme de la démonstration du lemme 4.13 on a que (4.30) $\Rightarrow$ (4.32) et (4.31) $\Rightarrow$ (4.32). Par le lemme 4.11 on a aussi que (4.32) $\Rightarrow$ (4.29). Et enfin d'après les propriétés (4.20), (4.21), (4.27) et (4.28) on a donc que (4.29) $\Rightarrow$ (4.30) et (4.29) $\Rightarrow$ (4.31), en vertu du Théorème 4.14.$\blacksquare$

Corollaire 4.16. *Sous les mêmes hypothèses sur X, Y et sur $\{f_j\}_j$ du lemme (4.13), si $\{f_j\}_j$ converge par rapport à la saddle* *topology, alors dans l'intervalle* $[\Gamma_c(J^-,X^+,Y^-)f_j, \Gamma_c(J^+,Y^-,X^+)f_j]$ *la fonction* $\Gamma_c(J^-,X^+,Y^-)f_j$ *est la seule fonctions s.c.i. qui lui appartient, tandis que* $\Gamma_c(J^+,Y^-,X^+)f_j$ *est la seule fonction s.c.s.*

DÉMONSTRATION. D'après le théorème 4.14 et son corollaire 4.15 l'intervalle est une classe d'équivalence. Même par la définition de cette équivalence, une classe d'équivalence ne peut pas contenir plus qu'une fonction s.c.i. et plus qu'une fonction s.c.s. Etant donc les fonctions $\Gamma_c(J^-,X^+,Y^-)f_j$ et $\Gamma_c(J^+,Y^-,X^+)f_j$, respectivement, s.c.i. et s.c.s., on en déduit la deuxième partie du corollaire.∎

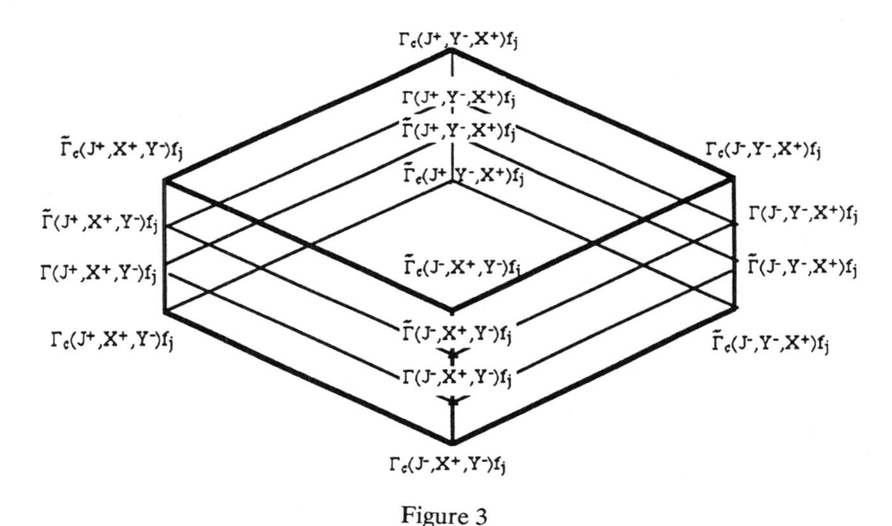

Figure 3

Par exemple on peut appliquer ce corollaire aux huit fonctions:

$$\Gamma(J^+,X^+,Y^-)f_j, \Gamma(J^-,X^+,Y^-)f_j,$$
$$\Gamma(J^+,Y^-,X^+)f_j, \Gamma(J^-,Y^-,X^+)f_j,$$
$$\tilde{\Gamma}(J^+,X^+,Y^-)f_j, \tilde{\Gamma}(J^-,X^+,Y^-)f_j,$$
$$\tilde{\Gamma}(J^-,Y^-,X^+)f_j, \tilde{\Gamma}(J^-,Y^-,X^+)f_j.$$

Les premières quatre de ces fonctions sont très bien connues et très importantes dans l'étude des problèmes liés aux propriétés de stabilité des jeux concavo-convexes (voir les travaux de Cavazzuti et d'Attouch et Wets). Pour ce qui concerne les autre quatre fonctions, jusqu'à ce moment on ne connaissait rien, sauf la définition donnée dans Attouch et Wets [3]. Les définitions de ces fonctions seront rappelées dans le prochain paragraphe. Par rapport aux autres huit fonctions définies ci-dessus, on peut donner un diagramme d'ordre (voir fig.3).

Soit $\psi : Y \to \bar{\mathbf{R}}$ une fonction s.c.i. et inf-localement compacte; et, dualement, soit $\varphi : X \to \bar{\mathbf{R}}$ une fonction s.c.s. et sup-localement compacte. Si $f : X \times Y \to \bar{\mathbf{R}}$ est une fonction quelconque, on désigne par $[f]$ la classe d'équivalence associée à f. Si f est topologiquement fermée (et seulement dans ce cas!), on a $[f] = \overline{\{f\}}$. Soit donc $C_{\psi,\varphi}$ l'espace donné par l'ensemble

$$\{[f] : f \text{ topologiquement fermée, quasi concavo–convexe et } \psi \leq f \leq \varphi\}$$

muni de la topologie quotient; c'est-à-dire $\{[f_j]\}_j$ converge vers $[f]$ si et seulement si $\{f_j\}_j$ converge vers f par rapport à la saddle topology. D'après la compacité de la saddle topology et le théorème 4.14 $C_{\psi,\varphi}$ est compact et séparé . En suivant la même procédure qui a été employée dans Dal Maso [8], on a donc le théorème suivant sur la métrisabilité de $C_{\psi,\varphi}$.

Théorème 4.17. *Si X et Y sont deux espaces métriques séparables par rapport à des topologies vérifiant les hypothèses du lemme 4.13, alors $C_{\psi,\varphi}$ est un espace compact métrisable.*

5. Une comparaison entre la saddle topology et d'autres convergences. Cavazzuti, Attouch et Wets ont défini et étudié des convergences qui ont aussi des propriétés de compacité pour les points et les valeurs de selle. Elles ne sont pas induites par une topologie: c'est là la difference plus remarquable par rapport à la saddle topology.

Soit $f_n : X \times Y \to \bar{\mathbf{R}}$ une fonction pour tout $n \in \mathbf{N}$; soient X et Y deux espaces topologiques quelconques. On définit les huit fonctions de $X \times Y$ dans $\bar{\mathbf{R}}$:

$$\Gamma(\mathbf{N}^+, X^+, Y^-)f_n,\ \Gamma(\mathbf{N}^-, X^+, Y^-)f_n,$$
$$\Gamma(\mathbf{N}^+, Y^-, X^+)f_n,\ \Gamma(\mathbf{N}^-, Y^-, X^+)f_n,$$

$$\tilde{\Gamma}(\mathbf{N}^+, X^+, Y^-) f_n, \ \tilde{\Gamma}(\mathbf{N}^-, X^+, Y^-) f_n,$$
$$\tilde{\Gamma}(\mathbf{N}^+, Y^-, X^+) f_n, \ \tilde{\Gamma}(\mathbf{N}^-, Y^-, X^+) f_n,$$

de la façon suivante. On pose:

$$(\Gamma(\mathbf{N}^{\pm}, X^+, Y^-) f_n)(x,y) := \sup_{A \in \mathcal{N}(y)} \inf_{B \in \mathcal{N}(x)} \left\{ \begin{matrix} \limsup \\ \liminf \end{matrix} \right\} \sup_B \inf_A f_n$$

$$(\tilde{\Gamma}(\mathbf{N}^{\pm}, X^+, Y^-) f_n)(x,y) := \sup_{A \in \mathcal{N}(y)} \inf_{B \in \mathcal{N}(x)} \left\{ \begin{matrix} \limsup \\ \liminf \end{matrix} \right\} \inf_A \sup_B f_n$$

$$(\Gamma(\mathbf{N}^{\pm}, Y^-, X^+) f_n)(x,y) := \inf_{B \in \mathcal{N}(x)} \sup_{A \in \mathcal{N}(y)} \left\{ \begin{matrix} \limsup \\ \liminf \end{matrix} \right\} \inf_A \sup_B f_n$$

$$(\tilde{\Gamma}(\mathbf{N}^{\pm}, Y^-, X^+) f_n)(x,y) := \inf_{B \in \mathcal{N}(x)} \sup_{A \in \mathcal{N}(y)} \left\{ \begin{matrix} \limsup \\ \liminf \end{matrix} \right\} \sup_B \inf_A f_n,$$

où $(x, y) \in X \times Y$. Les premières deux définitions donnent des fonctions s.c.i.; les deux autres des fonctions s.c.s. Comme on a dejà vu, on peut donner le diagramme d'ordre suivant (voir aussi Attouch, Wets [3]):

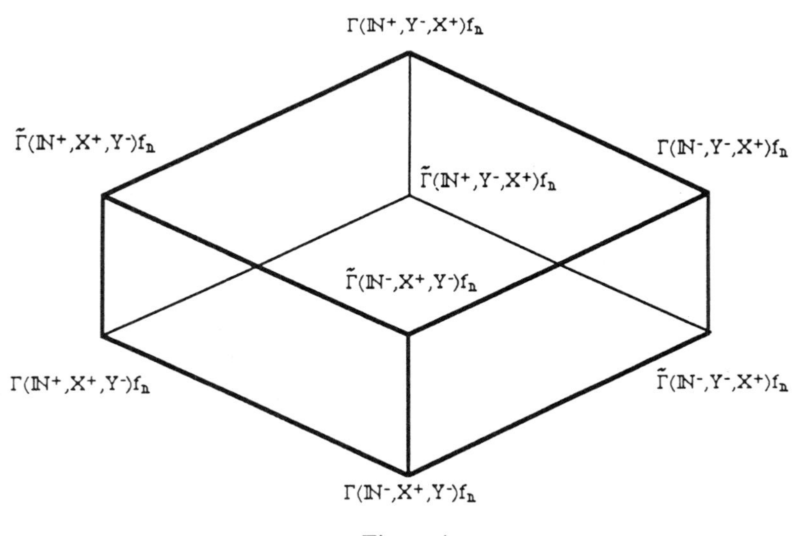

Figure 4

La convergence C^+ (Cavazzuti [4]). On dira qu'une suite $\{f_n\}_n$ de fonctions de $X \times Y$ dans $\bar{\mathbf{R}}$ C^+-*converge* vers la fonction

f, si:

$$(C^+) \qquad f = \Gamma(\mathbf{N}^+, Y^-, X^+) f_n = \Gamma(\mathbf{N}^-, Y^-, X^+) f_n.$$

Cette fonction f sera notée par $Lim^{C^+} f_n$. La convergence duale, appelée C^-, est définie par légalité $f = \Gamma(\mathbf{N}^{\pm}, X^+, Y^-) f_n$. Donc tout ce qu'on dira à propos de la convergence C^+, peut être trasposé à C^-. La limite $Lim^{C^+} f_n$, s'il existe, est une fonction s.c.s.; tandis que $Lim^{C^-} f_n$ est une fonction s.c.i.

La convergence *AW* (Attouch, Wets [3]). On dira qu'une suite $\{f_n\}_n$ de fonctions de $X \times Y$ dans $\bar{\mathbf{R}}$ *AW-converge* vers la fonction f si:

$$(AW) \qquad \Gamma(\mathbf{N}^+, X^+, Y^-) f_n \leq f \leq \Gamma(\mathbf{N}^-, Y^-, X^+) f_n.$$

L'ensemble des fonctions vers lesquelles une suite $\{f_n\}_n$ AW-converge est noté par $Lim^{AW} f_n$.

Les convergences C^+ et AW sont toutes les deux plus fines que la saddle topology; tandis que la convergenve C^+ est plus fine que la convergence AW, grâce au diagramme d'ordre ci-dessus. Le fait que les inégalités de (AW) impliquent (4.1) et (4.2), a été remarqué et employé plusieurs fois par Cavazzuti (voir la remarque 4.A).

La compacité de la saddle topology et de la convergence AW est une conséquence de la bien connue compacité de la convergence C^+ (voir [4]); en effet si on révoir les démonstrations des théorèmes de compacité 4.3 et 4.4, on s'aperçoit qu'en realité nous avons démontré la compacité de C^+.

D'autre part les propriétés de stabilité des points et des valeurs de selle de la saddle topology entraînent les mêmes propriétés pour les convergences plus fines: C^+ et AW.

Le problème d'unicité de la limite n'existe pas pour la convergence C^+; tandis que par rapport à la convergence AW il n'est pas compliqué de donner des conditions nécessaires et suffisantes, d'après le §4 (rappelons-nous aussi les formules de réduction (2.3')).

Proposition 5.1 (Sur l'unicité de la AW-limite). *Soient X et Y deux espaces topologiques. Etant donnée une suite $\{f_n\}_{n \in \mathbf{N}}$ quelconque.*

A) La limite $Lim^{AW} f_n$ *est une classe d' équivalence (nécessairement, des fonctions topologiquement fermées) si et seulement si:*

$$\text{(5.1)}\qquad \Gamma(X^+)\Gamma(\mathbf{N}^+, X^+, Y^-)f_n = \Gamma(\mathbf{N}^-, Y^-, X^+)f_n$$

et

$$\Gamma(Y^-)\Gamma(\mathbf{N}^-, Y^-, X^+)f_n = \Gamma(\mathbf{N}^+, X^+, Y^-)f_n.$$

En outre, si X *et* Y *sont deux espaces métriques et si* $Lim^{AW} f_{n_k}$ *est une classe d'équivalence pour toute sous-suite* $\{f_{n_k}\}_{k\in \mathbf{N}}$ *de* $\{f_n\}_n$*, alors les égalités suivantes sont vraies:*

$$\text{(5.1')}\qquad \begin{aligned}\Gamma(\mathbf{N}^+, X^+, Y^-)f_n &= \Gamma(\mathbf{N}^-, X^+, Y^-)f_n = \\ &= \tilde{\Gamma}(\mathbf{N}^+, X^+, Y^-)f_n = \tilde{\Gamma}(\mathbf{N}^-, X^+, Y^-)f_n,\end{aligned}$$

et

$$\begin{aligned}\Gamma(\mathbf{N}^+, Y^-, X^+)f_n &= \Gamma(\mathbf{N}^-, Y^-, X^+)f_n = \\ &= \tilde{\Gamma}(\mathbf{N}^+, Y^-, X^+)f_n = \tilde{\Gamma}(\mathbf{N}^-, Y^-, X^+)f_n;\end{aligned}$$

B) Si X *et* Y *sont deux espaces métriques, alors*

$$\text{(5.2)}\qquad \Gamma(X^+)\Gamma(\mathbf{N}^+, X^+, Y^-)f_n = \tilde{\Gamma}_c(\mathbf{N}^+, Y^-, X^+)f_n$$

et

$$\Gamma(Y^-)\Gamma(\mathbf{N}^-, Y^-, X^+)f_n = \tilde{\Gamma}_c(\mathbf{N}^-, X^+, Y^-)f_n$$

Donc l'ensemble $C(X,Y)\cap Lim^{AW} f_n$ *des fonctions topologiquement fermées appartenant à la* AW*-limite d'une suite* $\{f_n\}_n$ AW*-convergente est une classe d'équivalence si et seulement si:*

$$\text{(5.3)}\qquad \tilde{\Gamma}_c(\mathbf{N}^-, X^+, Y^-)f_n \leq \tilde{\Gamma}_c(\mathbf{N}^+, Y^-, X^+)f_n$$

et dans ce cas on a que

$$C(X,Y)\cap Lim^{AW} f_n = [\tilde{\Gamma}_c(\mathbf{N}^-, X^+, Y^-)f_n, \tilde{\Gamma}_c(\mathbf{N}^+, Y^-, X^+)f_n].$$

C) Si X *et* Y *sont deux espaces localement compacts, on a:*

$$\text{(5.4)}\qquad \tilde{\Gamma}(\mathbf{N}^-, X^+, Y^-)f_n = \tilde{\Gamma}_c(\mathbf{N}^-, X^+, Y^-)f_n$$

et

$$\tilde{\Gamma}(\mathbf{N}^+, Y^-, X^+)f_n = \tilde{\Gamma}_c(\mathbf{N}^+, Y^-, X^+)f_n.$$

D) Si X et Y sont deux espaces métriques localement compacts, l'inégalité suivante qui est plus forte que la (5.4):

$$\tilde{\Gamma}(\mathbf{N}^+, X^+, Y^-) f_n \leq \tilde{\Gamma}(\mathbf{N}^-, Y^-, X^+) f_n \tag{5.5}$$

est équivalente à dire que

$$C(X,Y) \cap Lim^{AW} f_{n_k} = C(X,Y) \cap Lim^{AW} f_n = [\tilde{\Gamma}(\mathbf{N}^+, X^+, Y^-) f_n, \tilde{\Gamma}(\mathbf{N}^-, Y^-, X^+) f_n] \neq \emptyset$$

pour toute sous-suite $\{f_{n_k}\}_k$; en outre, l'inégalité (5.5) entraîne:

$$\tilde{\Gamma}(\mathbf{N}^+, X^+, Y^-) f_n = \tilde{\Gamma}(\mathbf{N}^-, X^+, Y^-) f_n \quad et \quad \tilde{\Gamma}(\mathbf{N}^-, Y^-, X^+) f_n = \tilde{\Gamma}(\mathbf{N}^+, Y^-, X^+) f_n. \tag{5.6}$$

Dans le but de remarquer encore la différence entre la saddle topology et les convergences C^+ et AW, on peut observer le fonctionnement de celles-ci sur les suites constantes. Il est évident qu'une suite constante associée à une fonction f converge toujours vers f par rapport à la saddle topology; mais elle C^+-converge vers f si et seulement si:

$$f \quad \textit{est s.c.s. et} \quad \sup_U f \quad \textit{est s.c.i.} \tag{5.7}$$

pout tout ouvert U de X. Enfin la suite constante AW-converge vers f si et seulement si:

$$\inf_V f \quad \textit{est s.c.s. et} \quad \sup_U f \quad \textit{est s.c.i.} \tag{5.8}$$

pour tout ouvert V de Y et U de X.

Une propriété qui privilégie la convergence plus fine, c'est-à-dire C^+, mais qui n'est pas jouie par la convergence AW ou par la saddle topology est donnée par la proposition suivante. Sa démonstration, laissée au lecteur, s'appuie sur le théorème de minimax (voir section 3).

Proposition 5.2 (La convergence C^+ et les points de min-sup). *Soient X et Y deux sous-ensembles convexes compacts des espaces topologiques vectoriels. Soit $\{f_n\}_n$ une suite des fonctions s.c.i. quasi concavo-convexes qui C^+-converge vers une fonction f. Alors le* $\min_Y \max_X f$ *existe et*

$$\min_Y \max_X f = \lim_{n\to\infty} \min_Y \sup_X f_n. \tag{5.9}$$

En outre si y_n est un point de min-sup de f_n (c'est-à-dire :$\sup_X f_n(\cdot, y_n) = \min_Y \sup_X f_n$*) et si y_n converge vers y_o, alors y_o est un point de min-sup de f.*

Par ailleurs, il y aussi une propriété qui privilégie les convergences C^+ et AW, mais qui n'est pas géneralement jouie par la saddle topology. Cette propriété a été donnée par l'auteur dans une première version de [14] et, successivement, bien exploitée en [7]. On va la voir dans la proposition suivante, dont la démonstration peut être déduite de la formule de réduction (2.5).

Proposition 5.3 (Une propriété de semicontinuité de la convergence AW). *Si la suite $\{f_n\}_n$ AW-converge vers f, alors pour tout couple de compacts C, K, respectivement, de X et de Y et pour tout couple d'ouverts U, V, respectivement, de X et de Y on a:*

$$\inf_K \sup_U f \leq \sup_{A\in\mathcal{N}(K)} \liminf_{n\to\infty} \inf_A \sup_U f_n \tag{5.10}$$

$$\sup_C \inf_V f \geq \inf_{B\in\mathcal{N}(K)} \limsup_{n\to\infty} \sup_B \inf_V f_n. \tag{5.11}$$

Réciproquement, si pour tous K, C, V, U, comme ci-dessus, (5.10) et (5.11) valent, alors la suite $\{f_n\}_n$ AW-converge vers f.

Après avoir remarqué des différences entre les convergences C^+, AW et la saddle topology, allons déduire du théorème 4.14 et de son corollaire 4.15 le théorème suivant sur leur coincidence dans des cas particuliers.

Théorème 5.4. *Soient X et Y deux sous-ensembles convexes d'espaces vectoriels localement convexes, jouissant de la propriété* *que l'enveloppe fermée convexe de tout leur compact est aussi compact. Si* $\{f_n\}_n$ *est une suite de fonctions quasi concavo-convexes, topologiquement fermées qui est inf/sup-localement compacte en tout point. Alors les propriétés suivantes sont équivalentes:*

$$\{f_n\}_n \text{ converge pa rapport à la saddle topology,} \tag{5.12}$$

$$\{f_n\}_n \ C^+\text{–converge,} \tag{5.13}$$

$$\{f_n\}_n \ C^-\text{–converge,} \tag{5.14}$$

$$\{f_n\}_n \ AW\text{–converge.} \tag{5.15}$$

En outre, dans le cas où une de ces propriétés est vraie, on a:

$$C(X,Y) \cap Lim f_n = Lim^{AW} f_n = [Lim^{C^-} f_n, Lim^{C^+} f_n]. \tag{5.16}$$

$$\Gamma(X^+) Lim^{C^-} f_n = Lim^{C^+} f_n \quad \text{et} \quad \Gamma(Y^-) Lim^{C^+} f_n = Lim^{C^-} f_n. \blacksquare \tag{5.17}$$

Le corollaire suivant nour permet de comprendre mieux la proposition 5.2 et nous donne une relation plus claire entre la convergence C^+ et la saddle topology.

Corollaire 5.5. *Sous les mêmes hypothèses du théorème 5 .4, une suite* $\{f_n\}_n$ *de fonctions s.c.i. qui est inf/sup-localement compacte en tout point,* C^+*-converge vers une fonction* f *si et seulement si* f *est s.c.s. et la suite* $\{\Gamma(X^+)f_n\}_n$ *converge vers* f *par rapport à la saddle topology. Donc dans le cas où la suite* $\{f_n\}_n$ C^+*-converge vers une* f*, alors on a aussi que* f *est topologiquement fermée.*

Des théorèmes 5.4 et 4.4 on peut déduire un théorème de Franzoni et Francaviglia [12] qui, traduit dans notre langage, devient le corollaire suivant.

Corollaire 5.6. *Si X et Y sont deux espaces vectoriels séparés de dimension finie, alors pour toute suite $\{f_n\}_n$ de fonctions quasi concavo-convexes et topologiquement fermées il existe une sous-suite extraite $\{f_{n_k}\}_k$ et une fonction topologiquement fermée f telle que*

$$Lim^{AW} f_{n_k} = [\Gamma(Y^-)f, \Gamma(X^+)f].$$

On doit aussi rappeller qu'il y a des théorèmes de Cavazzuti [6], [7], qui démontrent (sous des hypothèses de compacité plus faibles) que de toute suite $\{f_n\}_n$ de fonctions concavo-convexes propres on peut extraire une sous-suite $\{f_{n_k}\}_k$ telle que $C(X,Y) \cap Lim^{AW}\{f_{n_k}\}_k$ soit une classe d'équivalence.

Enfin, du théorème 4.17 sue la métrisabilité de la saddle topology, on peut déduire le théorème suivant sur la métrisabilité des convergences C^+ et AW.

Théorème 5.7. *Si, sous les mêmes hypothèses du théorème 5.4, X et Y sont deux espaces métriques séparables, alors $C_{\psi,\varphi}$, muni de la convergence AW est fermé , compact et métrisable. En outre l'ensemble des fonctions s.c.s. quasi concavo-convexes et topologiquement fermées qui sont compris entre φ et ψ est fermé , compact et métrisable, lorsque il est muni de la convergence C^+.*

Corollaire 5.8. *Si X et Y sont deux espaces vectoriels séparés de dimension finie, alors l'ensemble de toutes les fonctions quasi concavo-convexes et topologiquement fermées, muni de la convergence AW est fermé , compact et pseudo-métrisable. En outre l'ensemble de toutes les fonctions s.c.s., quasi concavo-convexes et topologiquement fermées, muni de la convergence C^+ est fermé , compact et métrisable.*

Bibliographie

[1] G. Anzellotti, G.Buttazzo, G.Dal Maso, *Dirichlet Problem for Demi-coercive Functionals*, Nonlinear An. Th., Meth.and Appl. **10** (1986), 603-613.

[2] H.Attouch, *Variational Convergence for Functions and Operators*, Pitman, London, 1984.

[3] H.Attouch, R.Wets, *A Convergence Theory for Saddle Functions*, Trans. Am. Math. Soc. **280** (1983), 1-41.

[4] E.Cavazzuti, *Γ-convergenze multiple, convergenze di punti di sella e di max-min*, Boll. U.M.I. (6) **1**-B (1982), 251-274.

[5] E.Cavazzuti, *Alcune caratterizzazioni della Γ-convergenza multipla*, Ann. Mat. Pura ed Appl. (IV) **132** (1982), 69-112.

[6] E.Cavazzuti, N.Pacchiarotti, *Alcune convergenze su spazi di funzioni concavo-convesse: caratterizzazioni duali ed applicazioni*, Boll.U.M.I. (VI) **2** -C (1983), 227-251.

[7] E.Cavazzuti, *Convergence of Equilibria in the Theory of Games*, in Optimization and Related Fields (Erice 1984), Lect. Notes **1190** (1986), 95-130.

[8] D.Dal Maso. *Questioni di topologia legate alla Γ-convergenza*, Ricerche Math. **32** (1983), 135-162.

[9] E.De Giorgi, *On a Definition of Γ-Convergence of Measures*, in Multifunctions and Integrands (Catania 1983), Springer Lect. Notes **1091** (1984), 150-159.

[10] E.De Giorgi, T.Franzoni, *Su un tipo di convergenza variazionale*, Atti Ac. Naz Lincei, (8) **58** (1975), 842-850.

[11] E.De Giorgi, T.Franzoni, *Su un tipo di convergenza variazionale*, Rend. Sem. Mat. Brescia, **3** (1979), 63-101.

[12] T.Franzoni, *Abstract Γ-convergence*, in Optimization and Related Fields (Erice 1984), Lect.Notes **1190** (1986), 229-241.

[13] G.H. Greco, *Limites et fonctions d'ensembles*, Rend.Sem.Mat. Padova, **72** (1984), 89-97.

[14] G.H.Greco, *Limitoidi e reticoli completi*, Ann.Univ.Ferrara **29** (1983), 153-164.

[15] G.H.Greco, *On Measurability of Functions Valued in Completely Distributive Lattices*, à paraître.

[16] G.H.Greco, *Teoria des Semifiltros*, 22° Sem.Bras.d'Analise, Rio de Janeiro 1985, 1-117.

[17] G.H.Greco, *Minimax Theorems and Saddling Transformations*, J. Math. An. and Appl.,à paraître.

[18] J.Guillerme, *Convergence of Approximate Saddle Points*, J. Math. An. and Appl., à paraître.

[19] J.L.Joly, *Une famille de topologies et de convergences sur l'ensemble des fonctionnelles convexes*, Thèse, Grenoble, 1970.

[20] P.J.Laurent, *Approximation et Optimisation*, Hermann, Paris 1972.

[21] C.Lescarret, *Sur la sous-différentiabilité d'une somme de fonctionnelles convexes*, C.R.Acad. Sc.Paris (A) **262** (1966), 443-446.

[22] U.Marconi, communication privée.

[23] R.Peirone, *Sequential Compactness for* Γ*-convergence*, Boll. U.M.I. **5** -A (1986), 413-421.

[24] R.T.Rockafellar, *Convex Analysis*, Princeton University Press, 1970.

Dipartimento di Matematica
Università di Trento
I-38050 POVO (TN)

VARIATIONAL PRINCIPLES WITH LINEAR GROWTH

ROBERT HARDT DAVID KINDERLEHRER

Dedicated to Ennio De Giorgi on his sixtieth birthday

Variational principles which exhibit only linear growth arise in several contexts. As a paradigm for the questions which these principles suggest, we take up here the study of the problem

$$\inf_{v \in \mathcal{A}} \left\{ \int_\Omega \phi(\nabla v) dx - \int_\Omega f v dx \right\}. \tag{0.1}$$

Here $\phi : \mathbf{R}^n \to \mathbf{R}$ is a non-negative convex sufficiently differentiable function satisfying

$$\phi(0) = 0, \ \phi_p(0) = 0, \ \text{and } |p| - \lambda \leq \phi \leq |p| \tag{0.2}$$

for some $\lambda > 0$, subject to

$$\lim_{t \to \infty} \phi(tp)/t|p| = 1. \tag{0.3}$$

The competing v belong to a suitable class $\mathcal{A}$ of functions from Ω to $\mathbf{R}$ where Ω is a bounded domain with sufficiently smooth boundary and f is given.

Our discussion applies in particular to the integrand

$$\phi(p) = \begin{cases} \frac{1}{2}|p|^2 & \text{for } |p| \leq 1 \\ |p| - \frac{1}{2} & \text{for } |p| \geq 1 \end{cases} \tag{0.4}$$

which arises in the study of anti-planar shear in elastic/plastic deformation, [HK_1], [KT], [ST].

A solution to (0.1) may be found direct methods or by resorting to a generalized principle of complementary energy. The solution found by direct methods needs not be unique and, pertinent here, its gradient may be only a measure. The solution of the dual problem of complementary energy is often unique and determines, in case of (0.4), the region associated to elastic and plastic behavior.

Our concerns here will include properties of the generalized stress, the solution of the dual problem, which lead to information about the direct problem and its consequent smoothness.

These issues were brought to our attention by the work of Anzellotti and Giaquinta ([AG_1],[AG_2]) who treated the direct problem, and Strang and Temam [ST] who investigated both the direct problem and the dual problem. As we have indicated, it was partly our desire to achieve a greater unity between these approaches which led us to take up this argument. Our method is transparent and may be summarized in a few words: We have attempted to apply what we have learned from our study of the work of Ennio De Giorgi. It is with special pleasure that we dedicate this paper to him.

Central to our work is the study of the $(n-1)$ density

$$\Theta(a) = \lim_{r \to 0} r^{1-n} \int_{B_r(a)} |Du|$$

for a local solution u of (0.1). In §1 we show that this density exists and is upper semi-continuous in Ω. In §2, we show that the local solution u is continuous at a point $a \in \Omega$ if and only if its $(n-1)$ density vanishes at that point. In §3 the behavior near points of positive density is investigated, and it is shown how the solution is, in a very strong sense, essentially two-sided continuous, with the exception of a set of $n-1$ dimensional measure zero. The curvature of the set of discontinuities, or slip set, is then determined by the inhomogeneous term f.

Some of the present work was announced in $[HK_2]$, the latter containing some discussion of open questions. We wish to point out some more recent work in this general area, $[AG_3]$, and dynamical lems, [RS], [AL].

Implicit in our introduction of the $(n-1)$ density $\Theta(a)$ and our subsequent discussion is the recognition that we shall be obliged to widen our class of admissible funtions to $BV(\Omega)$. To further motivate this, we briefly review the dual problem associated to (0.1).

Assuming that $u_0 \in H^{1,1}(\Omega)$ and $f \in L^\infty(\Omega)$, set

$$\mathcal{A} = \{v \in H^{1,1}(\Omega) : v = u_0 \quad \text{on} \quad \partial\Omega\}.$$

Defining

$$I(v) = \int_\Omega \phi(\nabla v)dx - \int_\Omega f v dx, \quad v \in \mathcal{A}, \tag{0.5}$$

we may extend our functional to $H^{1,1}(\Omega)$ by [ET]

$$I(v) = \int_\Omega \phi(\nabla v)dx + I_{\mathcal{A}}(v) - F(v)$$

where

$$I_{\mathcal{A}}(v) = \begin{cases} 0 & v \in \mathcal{A} \\ \infty & v \notin \mathcal{A} \end{cases}$$

is the indicator function of A and

$$F(v) = \int_\Omega f v dx.$$

Our original variational principle (0.1) now becomes a problem

(P) *To find the solution or characteristic extremals of*

$$i = \inf_{H^{1,1}(\Omega)} \left\{ \int_\Omega \phi(\nabla v)dx + I_{\mathcal{A}}(v) - F(v) \right\}.$$

We may easily calculate the dual functional $I^*(\tau)$ for a vector-field τ. Indeed,

(0.6)

$$I^*(\tau) = \begin{cases} \int_\Omega \phi^*(-\tau)dx + \int_\Omega \tau \cdot \nabla u_0 dx + \int_\Omega f u_0 dx & \text{if } \operatorname{div}\tau = f, \\ +\infty & \text{otherwise.} \end{cases}$$

Several properites of the dual functional ϕ^* are worthy of note. First observe that (0.2) and the convexity of ϕ imply that

$$|\phi_p(p)| \leq 1 \quad \text{whenever} \quad 0 \neq p \in \mathbf{R}^n.$$

Now

$$\phi^*(q) = \begin{cases} p.\phi_p(p) - \phi(p) & \text{whenever } q = \phi_p(p) \text{ and } |q| \leq 1, \\ +\infty & \text{whenever } |q| > 1. \end{cases}$$

Again using (0.2), we obtain the estimate, useful later on,

$$0 \leq \phi^*(q) \leq \lambda \quad \text{if} \quad |q| \leq 1.$$

Since (0.6) implies that

$$\int_\Omega \phi^*(-\tau)dx = +\infty \quad \text{if} \quad \|\tau\|_{L^\infty(\Omega)} > 1,$$

we may further restrict the competitors which will occur in the dual principle. We arrive at the problem

(P^*) *To find or characterize extremals of*

$$(0.7) \quad i^* = \inf_{\operatorname{div}\tau = f, |\tau| \leq 1} \left\{ \int_\Omega \phi^*(-\tau)dx + \int_\Omega \tau \cdot \nabla u_0 dx + \int_\Omega f u_0 dx \right\}.$$

The theory of convex duality ([ET] or [B3],p.8) assures us that whenever

$$\mathcal{A}^* = \{\tau \in L^\infty(\Omega)^n : |\tau| \leq 1 \text{ and } \operatorname{div}\tau = f\}$$

is not empty, there exists a $\sigma \in \mathcal{A}^*$ such that

$$i = i^* = \int_\Omega \phi^*(-\sigma)dx + \int_\Omega \sigma \cdot \nabla u_0 dx + \int_\Omega f u_0 dx.$$

Indeed, (P^*) is a variational inequality [S].

In the special case (0.4) the dual function is

$$\phi^*(\sigma) = \begin{cases} \frac{1}{2}|\sigma|^2 & \text{for } |\sigma| \leq 1 \\ \infty & \text{for } |\sigma| > 1, \end{cases}$$

and the dual functional is

$$I^*(\tau) = \frac{1}{2}\int_\Omega |\tau|^2 dx + \int_\Omega \tau \cdot \nabla u_0 dx + \int_\Omega f u_0 dx \text{ whenever } \tau \in \mathcal{A}^*.$$

However, there need not be $u \in \mathcal{A}$ whose stress achieves this infimum, and thus we are led to search for u in the wider class $BV(\Omega)$ of functions whose gradients are measures.

A second motivation is the absence of a suitable Lagrangian functional which enables us to pair vectorfields τ and v in $H^{1,1}(\Omega)$. Determining the validity of the Lagrangian in an extended sense is one of the subjects treated by Kohn and Temam [KT]. Here we limit ourselves to the relationship between the solution of (0.1) or (0.7) and its own stress vector.

The important question of when $\mathcal{A}^*$ is not empty or when $I(v)$ is bounded below on $\mathcal{A}$ may be resolved in several ways, c.f.[AG_1],[KT], [HK_1]. For example, given Ω there is a constant C_Ω such that

$$\|\zeta\|_{L^1(\Omega)} \leq C_\Omega \|\zeta\|_{BV(\Omega)} \quad \text{for} \quad \zeta \in BV_0(\Omega).$$

Thus $I(v)$ is bounded below whenever

$$\|f\|_{L^\infty(\Omega)} < C_\Omega^{-1}.$$

Methods of limit analysis offer an alternative view, [T_1].

In our motivation we have discussed a variational principle with Dirichlet boundary conditions. Neumann conditions may be considered as well.

Let $\Omega \subset \mathbf{R}^n$ be a bounded domain with Lipschitz boundary. Given $v \in BV(\Omega)$, we decompose its gradient measure Dv into its absolutely continuous and singular parts with respect to Lebesgue measure:

$$Dv = \nabla v dx + D^s v.$$

We define

$$I(v) = \int_\Omega \phi(Dv) = \int_\Omega \phi(\nabla v) dx + \int_\Omega |D^s v|. \tag{0.8}$$

This definition is motivated primarily by the fact that there exist $v_\delta \in C^\infty(\bar{\Omega}), 0 < \delta < 1$, such that (v_δ) is contained in a bounded set in $BV(\Omega)$ and

$$v_\delta \to v \text{ in } L^{\frac{n}{n-1}} \cap L^1(\partial\Omega) \text{ and } I(v_\delta) = \int_\Omega \phi(\nabla v_\delta) dx \to I(v)$$

as $\delta \to 0$, c.f.[AG_1],[HK_1,Th.A.2]. The *stress associated to* v is defined by

$$\sigma = \sigma(v) = \begin{cases} \phi_p(\nabla v) & \text{in } \Omega_a \\ D^s v/|D^s v| & \text{in } \Omega_s \end{cases}$$

where $D^s v/|D^s v|$ is the Radon-Nikodym derivative of $D^s v$ with respect to its total variation measure $|D_s v|$ and $\Omega = \Omega_a \cup \Omega_s$ is the decomposition of Ω with respect the mutually singular measures dx and $|D^s v|$.

We have already noticed that since ϕ is convex and (0.2) holds,

$$|\phi_p(p)| \leq 1 \text{ and } \phi(p) \leq \phi_p(p) \cdot p.$$

It follows that

$$|\sigma(v)| \leq 1 \quad \text{and}$$

$$\phi(Dv) = \phi(\nabla v) + |D^s v| \leq \phi_p(\nabla v) \cdot \nabla v + |D^s v| = \sigma(v) \cdot Dv. \tag{0.9}$$

1. Local solutions and variational formulas. We say that a function $u \in BV(\Omega)$ is a *local solution* in Ω provided that

$$\int_\Omega \phi(Du) - \int_\Omega fu dx \leq \int_\Omega \phi(D(u+\zeta)) - \int_\Omega f(u+\zeta)dx \quad \text{for } \zeta \in BV_0(\Omega).$$

For any $\zeta \in BV_0(\Omega)$ satisfiyng

$$D^s\zeta \ll |D^s u|, \tag{1.1}$$

we have the first variation formula [HK_1,2.4]

$$\int_\Omega \sigma \cdot \nabla\zeta dx + \int_\Omega \sigma \cdot \xi |D^s u| = \int_\Omega f\zeta dx, \tag{1.2}$$

where $\sigma = \sigma(u)$ and ξ is the Radon-Nikodym derivative of $D^s\zeta$ with respect to $|D^s u|$. Indeed, usually we shall write (1.2) as

$$\int_\Omega \sigma \cdot D\zeta dx = \int_\Omega f\zeta dx. \tag{1.3}$$

The absolute continuity (1.1) is necessary for the validity of (1.3) as a simple example illustrates. One may take $n = 1, \Omega = (0,3)$, $u \equiv 0$, and $\zeta = \chi_{(1,2)}$ = characteristic function of$(1,2)$ to see that $\int_\Omega \phi(Du + t\zeta)) = 2|t|$ is not differentiable at $t = 0$ so that the first variation formula makes no sense.

A local variational principle follows from (1.1) for the functional

$$I_{r,a}(v) = \int_{\mathbf{B}_{r(a)}} \phi(Dv) - \int_{\partial \mathbf{B}_r(a)} \sigma \cdot \frac{x}{|x|} v dS - \int_{\mathbf{B}_r(a)} f v dx. \tag{1.4}$$

Proposition 1.1 *The function $u \in BV(\Omega)$ is a local solution if and only if*

$$I_{r,a}(u) \leq I_{r,a}(v) \tag{1.5}$$

for every $v \in BV(\Omega)$ and $r <$ dist$(a, \partial\Omega)$. Moreover,

$$I_{r,a}(u) = - \int_{\mathbf{B}_r(a)} \phi^*(\sigma) dx. \tag{1.6}$$

Here and in the sequel, $\sigma = \sigma(u)$ where u is the local solution under discussion.

The proof is facilitated by an elementary lemma.

Lemma 1.2. *If $v \in BV(\Omega)$ and $D^s v \ll |D^s u|$, then*

$$\int_{\mathbf{B}_r(a)} \sigma \cdot Dv = \int_{\partial \mathbf{B}_r(a)} \sigma \cdot \frac{x}{|x|} v dS + \int_{\mathbf{B}_r(a)} f v dx.$$

Proof. We assume $a = 0$. For fixed $0 < \rho < r$, choose $\zeta = \eta$ where

$$\eta(x) = \begin{cases} 1 & \text{in } \mathbf{B}_\rho \\ (r-\rho)^{-1}(r - |x|) & \text{in } \mathbf{B}_r \sim \mathbf{B}_\rho \\ 0 & \text{in } \Omega \sim \mathbf{B}_r. \end{cases}$$

By (1.1),

$$\int_{\mathbf{B}_r} \eta\sigma \cdot Dv = (r-\rho)^{-1} \int_{\mathbf{B}_r \sim \mathbf{B}_\rho} \sigma \cdot \frac{x}{|x|} v dx =$$

$$= (r-\rho)^{-1} \left\{ \int_{\mathbf{B}_r} \sigma \cdot \frac{x}{|x|} v dx - \int_{\mathbf{B}_\rho} \sigma \cdot \frac{x}{|x|} v dx \right\}.$$

Then, letting $\rho \to r$, the conclusion follows.

Proof of Proposition 1.1. Suppose first that u is a local solution and that $v \in C^1(\Omega)$; hence, $D^s v = 0 \ll |D^s u|$. By (0.2) and Lemma 1.2,

$$\begin{aligned}\int_{\mathbf{B}_r(a)} \phi(Dv) - \int_{\mathbf{B}_r(a)} \phi(Du) &= \\ &= \int_{\mathbf{B}_r(a)} (\phi(\nabla v) - \phi(\nabla u))dx - \int_{\mathbf{B}_r(a)} |D^s u| \\ &\geq \int_{\mathbf{B}_r(a)} \sigma \cdot (\nabla v - \nabla u)dx - \int_{\mathbf{B}_r(a)} |D^s u| \\ &= \int_{\mathbf{B}_r(a)} \sigma \cdot D(v-u) = \int_{\partial\mathbf{B}_r(a)} \sigma \cdot \frac{x}{|x|}(v-u)dS.\end{aligned}$$

Thus $I_{r,a}(u) \leq I_{r,a}(v)$ for $v \in C^1(\Omega)$.

For an arbitrary $v \in BV(\Omega)$, we may find a sequence $v_k \in C^1(\Omega)$ such that, for almost every r,

$$\limsup_{k\to\infty} \int_{\mathbf{B}_r} \phi(Dv_k) \leq \int_{\mathbf{B}_r} \phi(Dv) \quad \text{and} \quad v_k \to v \text{ in } L^1(\partial\mathbf{B}_r).$$

Conversely, suppose that (1.3) holds. If $\zeta \in BV(\Omega)$ has $D^s\zeta \ll |D^s u|$ and supp $\zeta \subset \mathbf{B}_r(a) \subset \Omega$, then

$$\int_{\mathbf{B}_r(a)} \phi(D(u+\zeta)) - \int_{\mathbf{B}_r(a)} \phi(Du) = I_{r,a}(u+\zeta) - I_{r,a}(u) \geq 0;$$

hence, $\int_\Omega \sigma \cdot D\zeta dx = 0$ by our previous reasoning. For an arbitrary $\zeta \in BV_0(\Omega)$ with $D^s\zeta \ll |D^s u|$, we may use a partition of unity to deduce the equation $\int_\Omega \sigma \cdot D\zeta dx = 0$ and then obtain from convexity the desired inequality

$$\int_\Omega \phi(Du) = \int_\Omega \phi(Du) + \int_\Omega \sigma \cdot D\zeta \leq \int_\Omega \phi(Du + D\zeta).$$

The final assertion, the duality relation localized, is a direct calculation.

The duality relation is a useful test for minima.

Theorem 1.3 (Monotonicity). *Let $f \in L^\infty(\Omega)$ and let u be a local solution.*

(i) Then,

$$\rho^{1-n}\int_{\mathbf{B}_\rho(a)}|Du| \le e^{\Lambda(r-\rho)}r^{1-n}\int_{\mathbf{B}_r(a)}|Du| + C_n(r-\rho),$$

for $0<\rho<r<\operatorname{dist}(a,\partial\Omega)$, where $\Lambda = C_n\|f\|_{L^\infty(\Omega)}$ and C_n is a dimensional constant.

(ii) If $u \in L^\infty(\Omega)$, then

$$\rho^{1-n}\int_{\mathbf{B}_\rho(a)}|Du| \le e^{\Lambda(r-\rho)}r^{1-n}\int_{\mathbf{B}_r(a)}|Du| + \\ + C_n(\lambda + \|f\|_{L^\infty(\Omega)}\|u\|_{L^\infty(\Omega)})(r-\rho),$$

(iii) In particular, the $(n-1)$ density

$$\Theta(a) = \lim_{r\to 0} r^{1-n}\int_{\mathbf{B}_r(a)}|Du|$$

exists, is finite, and is uniformly bounded on compact subset of Ω.

We remark immediately that we prove in the next section that any local solution is locally bounded, so that (ii) always holds on compact subset of Ω.

Proof. We again set $a=0$. For $0<r<\operatorname{dist}(0,\partial\Omega)$, define

$$w(x) = \begin{cases} u(r\frac{x}{|x|}) & \text{in } \mathbf{B}_r \\ u(x) & \text{in } \Omega \sim \mathbf{B}_r. \end{cases}$$

For this r, we infer that

$$I_{r,0}(u) \le I_{r,0}(w).$$

Combining this with (0.2) gives that

$$\begin{aligned} \int_{\mathbf{B}_r}|Du| &\le \int_{\mathbf{B}_r}\phi(Du) + \lambda|\mathbf{B}_r| \\ &\le \int_{\mathbf{B}_r}\phi(Dw) - \int_{\mathbf{B}_r} f(w-u)dx + \lambda|\mathbf{B}_r| \qquad (1.7) \\ &\le \int_{\mathbf{B}_r}|Dw| - \int_{\mathbf{B}_r} f(w-u)dx + \lambda|\mathbf{B}_r|. \end{aligned}$$

Now by a well-known calculation, cf.[DCP] for example,

$$\int_{\mathbf{B}_r} |Dw| \le (n-1)^{-1} r \int_{\partial \mathbf{B}_r} |Du|.$$

Now we may estimate the middle term in (1.7) two ways. If u is assumed bounded, then

$$\left| \int_{\mathbf{B}_r} f(w-u)dx \right| \le 2\|f\|_{L^\infty} \|u\|_{L^\infty} |\mathbf{B}_r|.$$

Thus

$$\int_{\mathbf{B}_r} |Du| \le (n-1)^{-1} r \int_{\partial \mathbf{B}_r} |Du| + C_n(\lambda + \|f\|_{L^\infty} \|u\|_{L^\infty}) r^n$$

or

$$(d/dr)\{r^{1-n} \int_{\mathbf{B}_r} |Du| + C_n(\lambda + \|f\|_{L^\infty}\|u\|_{L^\infty})r\} \ge 0,$$

which implies (ii).

If we do not assume that u is bounded, them, since $w - u = 0$ on $\partial \mathbf{B}_r$,

$$\begin{aligned} | \int_{\mathbf{B}_r} f(w-u)dx | &\le \|f\|_{L^\infty} \|w-u\|_{L^1(\mathbf{B}_r)} \\ &\le C_n r \|f\|_{L^\infty} (\int_{\mathbf{B}_r} |Du| + \int_{\mathbf{B}_r} |Dw|), \end{aligned}$$

from which (i) follows.

Corollary 1.4. *The density $\Theta(x)$ is upper semi-continuous on Ω.*

PROOF. We assume that $\mathbf{B}_{2r} \subset \Omega$ and prove upper semi-continuity at 0. First note that $\mathbf{B}_\rho(x) \subset \mathbf{B}_{2r}$ whenever $x \in \mathbf{B}_r$ and $0 < \rho < r$. Thus by 1.3

$$\begin{aligned} \Theta(x) &\le \rho^{1-n} \int_{\mathbf{B}_\rho(x)} |Du| + C_n \rho \\ &\le \rho^{1-n}(\rho + |x|)^{n-1}(\rho + |x|)^{1-n} \int_{\mathbf{B}_{\rho+|x|}} |Du| + C_n(\rho + |x|) \\ &= (1 + \rho^{-1}|x|)^{n-1}(\rho + |x|)^{1-n} \int_{\mathbf{B}_{\rho+|x|}} |Du| + C_n(\rho + |x|). \end{aligned}$$

Holding ρ fixed and letting $x \to 0$ yields that

$$\limsup_{x\to 0} \Theta(x) \le \rho^{1-n} \int_{\mathbf{B}_\rho} |Du| + C_n\rho.$$

Then letting $\rho \to 0$ gives

$$\limsup_{x\to 0} \Theta(x) \le \Theta(0).$$

One may easily derive alternative formulas for the density based on the local functional (1.2) and the minimality of u.

Corollary 1.5. *For u as above and $a \in \Omega$,*

$$\begin{aligned}\Theta(a) &= \lim_{r\to 0} r^{1-n} \int_{\mathbf{B}_r(a)} \phi(Du) \\ &= \lim_{r\to 0} r^{1-n} \int_{\partial\mathbf{B}_r(a)} \sigma \cdot \frac{x}{|x|} u dS = \lim_{r\to 0} r^{1-n} \int_{\mathbf{B}_r(a)} \sigma \cdot Du.\end{aligned}$$

Proof. By (0.1),

$$0 \le r^{1-n} \int_{\mathbf{B}_r(a)} |Du| - r^{1-n} \int_{\mathbf{B}_r(a)} \phi(Du) \le \lambda r^{1-n} |\mathbf{B}_r| \to 0$$

as $r \to 0$, which implies the first equality. By (1.2),(1.4), and the boundedness of σ,

$$\begin{aligned} r^{1-n} \int_{\mathbf{B}_r(a)} \phi(Du) - r^{1-n} \int_{\partial\mathbf{B}_r(a)} \sigma \cdot \frac{x}{|x|} u dS &= \\ = -\frac{1}{2} r^{1-n} \int_{\mathbf{B}_r(a)} \phi^*(\sigma) dx \to 0 \end{aligned}$$

as $r \to 0$, since $0 \le \phi^*(\sigma) \le \lambda$, which implies the second equality. The third equality follows from Lemma 1.2 applied with $v = u$.

Recall that a measurable function u is *approximately continuous* at a if a is a *Lebesgue point* for u, that is, if

$$\lim_{\rho\to 0} \rho^{-n} \int_{\mathbf{B}_\rho(a)} |u - c| dx = 0 \quad \text{for some} \quad c \in \mathbf{R}.$$

Corollary 1.6. *If a local solution is approximately continuous at a, then $\Theta(a) = 0$.*

PROOF. We apply Lemma 1.2 with $v = \eta.(u - c)$ where c is as above and η is as in the proof of Lemma 1.2 with $r = 2\rho$. We conclude that

$$0 = \int_{\mathbf{B}_{2\rho}(a)} \sigma \cdot Dv = \int_{\mathbf{B}_{2\rho}(a)} \sigma \cdot [\eta Du + (u - c)\nabla\eta dx];$$

hence,

$$\int_{\mathbf{B}_{\rho}(a)} |Du| \leq \rho^{-1} \int_{\mathbf{B}_{2\rho}(a)} |u - c| dx.$$

By 1.5,

$$\begin{aligned}\Theta(a) &= \lim_{\rho \to 0} \rho^{1-n} \int_{\mathbf{B}_{\rho}(a)} \sigma \cdot Du \leq \lim_{\rho \to 0} \rho^{1-n} \int_{\mathbf{B}_{\rho}(a)} |Du| \\ &\leq \lim_{\rho \to 0} \rho^{-n} \int_{\mathbf{B}_{2\rho}(a)} |u - c| dx = 0.\end{aligned}$$

2. The local boundedness. We establish here that a local solution is locally bounded and that it is continuous at each point $a \in \Omega$ where the density $\Theta(a) = 0$. Our first step is to estimate the measure of the set where a local solution exceeds a given value. We assume that

u is a local solution in Ω with $f \in L^\infty(\Omega)$, $0 \in \Omega$, and $\theta : \mathbf{R} \to \mathbf{R}$ is a bounded increasing piecewise differentiable function with $\theta(t) \leq 1$ for almost all t.

Suppose that $0 < \rho < r < \min\{1, \operatorname{dist}(0, \partial\Omega)\}$ and let

$$\eta(x) = \begin{cases} 1 & \text{in } \mathbf{B}_\rho \\ (r - \rho)^{-1}(r - |x|) & \text{in } \mathbf{B}_r \sim \mathbf{B}_\rho \\ 0 & \text{in } \Omega \sim \mathbf{B}_r, \end{cases}$$

exactly as in Lemma 1.2. We may apply the variational formula (1.1) with $\zeta = \eta\theta(u - c)$, where c is a constant, to obtain that

$$\begin{aligned}&\int_{\mathbf{B}_r} \eta\sigma.D[\theta(u - c)] = \\ (2.1)\qquad &= (r - \rho)^{-1} \int_{\mathbf{B}_r \sim \mathbf{B}_\rho} \sigma.\frac{x}{|x|}\theta(u - c) dx + \int_{\mathbf{B}_r} \eta\theta(u - c) f dx.\end{aligned}$$

Observe that

$$D[\theta(u-c)] = \theta'(u-c)Du \quad \text{and} \quad |D\theta\,(u-c)| = \theta'(u-c)|Du|.$$

Thus, by (0.2) and (0.8),

$$\begin{aligned}\int_{\mathbf{B}_r} \eta|D[\theta(u-c)]| &\le \int_{\mathbf{B}_r} \eta\theta'(u-c)|Du| \\ &\le \int_{\mathbf{B}_r} \eta\theta'(u-c)\phi(Du) + \lambda\int_{\mathbf{B}_r} \eta\theta'(u-c)dx \\ &\le \int_{\mathbf{B}_r} \eta\theta'(u-c)\sigma\cdot Du + \lambda\int_{\mathbf{B}_r} \eta\theta'(u-c)dx \\ &= \int_{\mathbf{B}_r} \eta\sigma\cdot D[\theta(u-c)] + \lambda\int_{\mathbf{B}_r} \eta\theta'(u-c)dx.\end{aligned}$$

Applying (2.1) and recalling that $|\sigma|\le 1$, we now obtain the basic estimate

$$\begin{aligned}\text{(2.2)}\quad \int_{\mathbf{B}_\rho} \eta|D[\theta(u-c)]| \le (r-\rho)^{-1}\int_{\mathbf{B}_r\sim\mathbf{B}_\rho} |\theta(u-c)|dx + \lambda|A| \\ +\|f\|_{L^\infty(\mathbf{B}_r)}\int_{\mathbf{B}_r}|\theta(u-c)|dx\end{aligned}$$

where $A = \text{supp } \eta\theta(u-c)$.

Now we let $0<h<k<\infty$ and choose

$$\text{(2.3)}\qquad \theta(t) = \begin{cases} 0 & \text{for } t\le k \\ t-k & \text{for } k<t<h \\ h-k & \text{for } t\ge k. \end{cases}$$

Note that

$$\text{supp }\big(\eta\theta(u-c)\big) \subset \mathbf{B}_r\cap\{u-c\ge k\} = A(k,r).$$

Thus (2.2) implies that

$$\text{(2.4)}\quad \int_{\mathbf{B}_\rho} |D[\theta(u-c)]| \le (C+(r-\rho)^{-1})\int_{\mathbf{B}_\rho} |\theta(u-c)|dx+\lambda|A(k,r)|,$$

$$C = \|f\|_{L^\infty(\Omega)}.$$

Let $0 < r_0 < \text{dist}(0, \partial\Omega)$ be given and choose k_0 so large that

$$|A(k_0, \rho)| \le \frac{1}{2}|\mathbf{B}_\rho| \quad \text{for} \quad \frac{1}{2}r_o \le \rho \le r_0.$$

Now $A(k, \rho) \subset A(k_0, \rho)$ for $k > k_0$, so by the isoperimetric inequality, the definition of θ, and (2.4),

$$(h-k)|A(h,\rho)|^{\frac{n-1}{n}} \le \Big(\int_{\mathbf{B}_\rho} |\theta(u-c)|^{\frac{n}{n-1}} dx\Big)^{\frac{n-1}{n}}$$

$$\le c_0 \int_{\mathbf{B}_\rho} |D[\theta(u-c)]|$$

$$\le c_0[C + (r-\rho)^{-1}] \int_{\mathbf{B}_r \sim \mathbf{B}_\rho} |\theta(u-c)| dx + c_0 \lambda |A(k,r)|$$

$$\le c_0[C + (r-\rho)^{-1}](h-k)|A(k,r)| + c_0\lambda|A(k,r)|.$$

Hence,

$$|A(h,\rho)|^{\frac{n-1}{n}} \le c_0(C+\lambda)[1 + (r-\rho)^{-1} + (h-k)^{-1}]|A(k,r)|.$$

Recalling our restriction that $r_0 < 1$, we may rewrite the above as

$$|A(h,\rho)|^{\frac{n-1}{n}} \le \Lambda[(r-\rho)^{-1} + (h-k)^{-1}]|A(k,r)| \tag{2.5}$$

$$\text{for} \quad \frac{1}{2}r_0 \le \rho < r \le r_0 \quad \text{and} \quad h > k \ge k_0,$$

where the $\Lambda = \nu_0 + \nu_1\|f\|_{L^\infty(\Omega)}$ for suitable constants ν_0, ν_1.

The truncation estimate (2.5) differs from the more traditional one of De Giorgi and Stampacchia, cf. for a typical example [KS] p.63, in several notable ways. The first is the sum occuring on the right hand side which is usually a product. The second is the limitation on the range of ρ for which the estimate is valid.

At this point, we state our truncation lemma.

Lemma 2.1. *Suppose that* $\{A(k,r) : \frac{1}{2}r_0 \le r \le r_0, k \ge k_0\}$ *is any collection of subsets of* $\mathbf{B}_{r_0}$ *that satisfies (2.5) and that* δ *is a positive number with*

$$\delta \le \frac{1}{2}|\mathbf{B}| \quad \textit{and} \quad \delta^{\frac{1}{n}} \le c_1^{-1} \quad \textit{where} \quad c_1 = 2^{2n+3}c_0(1+\lambda). \tag{2.6}$$

if

$$|A(k_0, r_0)| \leq \delta r_0^n, \tag{2.7}$$

then

$$|A(k_0 + d, \frac{1}{2} r_0)| = 0 \quad \textit{for} \quad d = c_1 |A(k_o, r_0)|^{\frac{1}{n}}.$$

PROOF. For $i = 1, 2, ...,$ let

$$r_i = \frac{1}{2} r_0 + 2^{-i} r_0 \quad \text{and} \quad k_i = k_0 + d - 2^{-i} d.$$

Then

$$r_i - r_{i+1} = 2^{-i-1} r_0 \quad \text{and} \quad k_{i+1} - k_i = 2^{-i-1} d.$$

With

$$\alpha_i = |A(k_i, r_i)|^{\frac{n-1}{n}},$$

our goal is to prove inductively the estimate

$$\alpha_i \leq 2^{-ni} \alpha_1. \tag{2.8}$$

For this, we first note that

$$\alpha_1 \leq \delta^{\frac{n-1}{n}} r_0^{n-1} \quad \text{and} \quad d = \alpha_1^{\frac{1}{n-1}}.$$

Written in terms of the sequence α_i, the recursion relation (2.5) is

$$\alpha_{i+1} \leq 2^{i+1} c_0 (r_0^{-1} + d^{-1}) \alpha_i^{\frac{n}{n-1}}. \tag{2.9}$$

In particular, by the choise of δ and d,

$$\begin{aligned} \alpha_2 &\leq 4 c_0 (\alpha_1^{\frac{1}{n-1}} r_0^{-1} + \alpha_1^{\frac{1}{n-1}} d^{-1}) \alpha_1 \\ &\leq 4 c_0 (\delta^{\frac{1}{n}} + c^{-1}) \alpha_1 \\ &\leq (2^{-2n-1} + 2^{-2n-1}) \alpha_1 = 2^{-2n} \alpha_1. \end{aligned}$$

Now assume inductively that $\alpha_i \le 2^{-in}\alpha_1$. Then, by (2.9),

$$\begin{aligned}\alpha_{i+1} &\le 2^{i+1}c_o(r_0^{-1}+d^{-1})\alpha_1^{\frac{n}{n-1}}(\alpha_i\alpha_1^{-1})^{\frac{n}{n-1}} \\ &\le 2^i 4c_0(\alpha_1^{\frac{1}{n-1}}r_0^{-1}+\alpha_1^{\frac{1}{n-1}}d^{-1})\alpha_1(2^{-ni})^{\frac{n}{n-1}} \\ &\le 2^i 2^{-2n}(2^{-ni})^{\frac{n}{n-1}}\alpha_1 \le 2^{-(i+1)n},\end{aligned}$$

the latter inequality holding because the inequality

$$i-2n-ni(\frac{n}{n-1}) \le -(i+1)n$$

reduces to simply $n(n-1)+i \ge 0$. This completes the inductive proof of (2.8). The lemma now follows because

$$|A(k_0+d,\frac{1}{2}r_0)| = \lim_{i\to\infty}\alpha_i^{\frac{n}{n-1}} = 0.$$

Note the hypotheses of the lemma implies that

$$d = c_1|A(k_0,r_0)|^{\frac{1}{n}} \le c_1\delta^{\frac{1}{n}}r_0. \tag{2.10}$$

Consequently, whenever (2.5) and (2.7) are satisfied for a local solution u on $\mathbf{B}_{r_0}$,

$$u-c \le k_0 + c_1\delta^{\frac{1}{n}} \text{ in } \mathbf{B}_{r_0/2}.$$

Since we may argue analogously with

$$\theta(t) = \begin{cases} h-k & \text{for } t\le k \\ k-t & \text{for } k<t<h \\ 0 & \text{for } t\ge h,\end{cases}$$

we may conclude that

$$\|u-c\|_{L^\infty(\mathbf{B}_{r_o/2})} \le k_0 + c_1\delta^{1/n}r_0 \tag{2.11}$$

whenever (2.5) and (2.7) are satisfied for the sets

$$A(k,r) = \mathbf{B}_r\cap\{u\ge k\} \text{ and } \tilde{A}(k,r) = \mathbf{B}_r\cap\{u\le k\}.$$

Theorem 2.2. *If u is a local solution in Ω and $\mathbf{B}_{2r}(a) \subset\subset \Omega$, then there is a positive constant M depending only on $\|f\|_{L^\infty}$ so that*

$$\|u - \bar{u}_r\|_{L^\infty(\mathbf{B}_r(a))} \leq M\{r^{1-n} \int_{\mathbf{B}_{2r}(a)} |Du| + r\} \tag{2.12}$$

where

$$\bar{u}_r = |\mathbf{B}_r|^{-1} \int_{\mathbf{B}_r(a)} u dx.$$

Proof. We may assume that $a = 0$. We may also apply (2.11) to $u - \bar{u}_r$, with $r_0 = 2r$. By (2.5) and (2.11) it only remains to estimate k_0.

Suppose that $r \leq \rho \leq 2r < 1$. Then we may average the inequality

$$|\bar{u}_r - \bar{u}_\rho| \leq |u(x) - \bar{u}_r| + |u(x) - \bar{u}_\rho|$$

over the smaller ball $\mathbf{B}_r$ to obtain

$$|\bar{u}_r - \bar{u}_\rho| \leq |\mathbf{B}_r|^{-1} \int_{\mathbf{B}_r} |u - \bar{u}_r| dx + |\mathbf{B}_r|^{-1} |\mathbf{B}_\rho| |\mathbf{B}_\rho|^{-1} \int_{\mathbf{B}_r} |u - \bar{u}_\rho| dx$$

$$\begin{aligned} &\leq c_2\{r^{1-n} \int_{\mathbf{B}_r} |Du| + 2^n \rho^{1-n} \int_{\mathbf{B}_\rho} |Du|\} \\ &\leq c_2\{(1+2^n)^{\Lambda r} \rho^{1-n} \int_{\mathbf{B}_\rho} |Du| + c_n \rho\} \qquad (2.13) \\ &\leq c_3\{\rho^{1-n} \int_{\mathbf{B}_\rho} |Du| + \rho\}, \end{aligned}$$

where we have employed a Poincaré inequality and the monotonicity inequality (Theorem 1.3).

Letting

$$A_r(k, \rho) = \mathbf{B}_\rho \cap \{|u - \bar{u}_r| \geq k\},$$

we use a Sobolev inequality and (2.13) to calculate

$$\begin{aligned} k|A_r(k,\rho)|^{\frac{n-1}{n}} &\leq \|u - \bar{u}_r\|_{L^{\frac{n}{n-1}}(\mathbf{B}_\rho)} \\ &\leq \|u - \bar{u}_\rho\|_{L^{\frac{n}{n-1}}(\mathbf{B}_\rho)} + \|\bar{u}_\rho - \bar{u}_r\|_{L^{\frac{n}{n-1}}(\mathbf{B}_\rho)} \\ &\leq c_4 \int_{\mathbf{B}_\rho} |Du| + c_3 \left\{\rho^{1-n} \int_{\mathbf{B}_\rho} |Du| + \rho\right\} |\mathbf{B}|^{\frac{n-1}{n}} \end{aligned}$$

or

$$k(\rho^{-n}|A_r(k,\rho)|)^{\frac{n-1}{n}} \leq c_5\{\rho^{1-n}\int_{\mathbf{B}_\rho}|Du| + \rho\}.$$

Selecting

$$k_0 = \inf\{\ell : |A_r(k,\rho)| \leq \delta\rho^n \text{ for all } r \leq \rho \leq 2r, k \geq \ell\}$$

now permits us to apply (2.11) with $r_0 = 2r$ as well as to estimate

$$\begin{aligned} k_0 &\leq c_5\delta^{\frac{n}{n-1}} \sup_{r\leq\rho\leq 2r}\{\rho^{1-n}\int_{\mathbf{B}_\rho}|Du| + \rho\} \\ &\leq c_6\{r^{1-n}\int_{\mathbf{B}_r}|Du| + r\}. \end{aligned}$$

This, combined with (2.11), completes the proof.

Corollary 2.3. *The local solution u is continuous at $a \in \Omega$ if and only if*

$$\Theta(a) = \lim_{r\to 0} r^{1-n}\int_{\mathbf{B}_r(a)}|Du| = 0.$$

Moreover,

$$\|u - u(a)\|_{L^\infty(\mathbf{B}_r(a))} \leq N\{r^{1-n}\int_{\mathbf{B}_{2r(a)}}|Du| + r\} \tag{2.14}$$

for some constant N depending only on Ω, f, and a.

Proof. We again take $a = 0$. The necessity of the condition $\theta(a) = 0$ was established in 1.5. To use 2.2 to prove the sufficiency we need to control the averages $\bar{u}_\rho$ as $\rho \to 0$. Using 2.2, we see that, for each positive $\rho \leq r$,

$$\begin{aligned} |\bar{u}_\rho| &\leq |\mathbf{B}_\rho|^{-1}\left|\int_{\mathbf{B}_\rho} u dx\right| \\ &\leq |\bar{u}_r| + |\mathbf{B}_\rho|^{-1}\int_{\mathbf{B}_\rho}|u - \bar{u}_r|dx \\ &\leq |\bar{u}_r| + M\{r^{1-n}\int_{\mathbf{B}_{2r}}|Du| + r\}. \end{aligned}$$

Thus the collection $\{\bar{u}_\rho : 0 < \rho \leq r\}$ is uniformly bounded. Moreover,

$$
\begin{aligned}
|\bar{u}_\rho - \bar{u}_r| &\leq \\
&\leq |\mathbf{B}_\rho|^{-1} \int_{\mathbf{B}_\rho} |u - \bar{u}_\rho| dx + |\mathbf{B}_\rho|^{-1} \int_{\mathbf{B}_\rho} |u - \bar{u}_r| dx \\
&\leq c_2 \rho^{1-n} \int_{\mathbf{B}_\rho} |Du| + +M\{r^{1-n} \int_{\mathbf{B}_{2r}} |Du| + r\}.
\end{aligned}
$$

It follows that $\{\bar{u}_\rho\}$ is Cauchy as $\rho \to 0$ if $\Theta(0) = 0$. In this case, the limit $u(0)$ satisfies

$$|u - u(0)| \leq N\{r^{1-n} \int_{\mathbf{B}_{2r}} |Du| + r\} \text{ in } \mathbf{B}_r.$$

3. Points of positive density. From [F], 4.5.9(3) we recall that a BV function u has, at $\mathcal{H}^{n-1}$ almost all points a of its domain, finite *approximate upper and lower limits* $u^+(a) \geq u^-(a)$ satisfying

$$
\begin{aligned}
u^+(a) &= \inf\{k : \lim_{\rho \to 0+} |\mathbf{B}_\rho|^{-1} |\mathbf{B}_\rho(a) \cap \{u > k\}| = 0\} \text{ and} \\
u^-(a) &= \sup\{k : \lim_{\rho \to 0+} |\mathbf{B}_\rho|^{-1} |\mathbf{B}_\rho(a) \cap \{u < k\}| = 0\}.
\end{aligned}
\tag{3.1}
$$

Moreover, using [F], 4.5.9(27), one finds that these limits differ at $\mathcal{H}^{n-1}$ almost all points a where the upper $n-1$ density

$$\limsup_{r \to 0} r^{1-n} \int_{\mathbf{B}_r(a)} |Du| > 0.$$

In this section we shall limit ourselves to the case where $f = 0$.

Theorem 3.1. *Suppose that $f = 0$ and u is a local solution in Ω, $a \in \Omega$, $\Theta(a) > 0$ and $u^+ > u^-$ where $u^\pm = u^\pm(a)$. For any $\epsilon > 0$, there is a $\sigma > 0$ so that*

$$u^- - \epsilon < u < u^+ + \epsilon \quad \textit{in} \quad \mathbf{B}_\sigma(a).$$

PROOF. We assume $a = 0$. For fixed $0 < \rho < r < \text{dist}(0, \partial\Omega)$, we again let η be as in 1.2, and, for fixed $h > k \geq 0$, let θ be as in (2.3). We now use the test function

$$\zeta(x) = \eta(x) \cdot \theta(u(x) - u^+)$$

in the variational formula (1.1). Since

$$\text{supp}\zeta \subset \mathbf{B}_r \cap \text{supp}(\theta(u - u^+)) = \mathbf{B}_r \cap \{u - u^+ \geq k\},$$

we deduce, exactly as in the proof of (2.4) and using that u is bounded, that

(3.2)

$$\int_{\mathbf{B}_\rho} |D[\theta(u - u^+)]| \leq (r - \rho)^{-1} \int_{\mathbf{B}_r \sim \mathbf{B}_\rho} |\theta(u - u^+)| dx + C|A^+(k, r)|$$

where C is a constant and

$$A^+(k, r) = \mathbf{B}_r \cap \{u - u^+ \geq k\}.$$

By (3.1), we may, for any positive number k_0, choose a positive $r_0 < \min\{k_0, \text{dist}(0, \partial\Omega)\}$ so that

$$|A^+(k_0, r_0)| \leq \min\{\delta, 2^{-n-1}|\mathbf{B}|\} r_0^n \tag{3.3}$$

where δ is the constant from Lemma 2.1. Then, for $\frac{1}{2}r_0 \leq \rho \leq r_0$ and $k \geq k_0$,

$$|A^+(k, \rho)| \leq |A^+(k_0, r_0)| \leq 2^{-n-1}|\mathbf{B}_{r_0}||\mathbf{B}_\rho|^{-1}|\mathbf{B}_\rho| \leq \frac{1}{2}|\mathbf{B}_{r_0}|.$$

Since

$$\mathbf{B}_\rho \cap \text{supp}(\theta(u - u^+)) \subset |A^+(k, \rho)|,$$

we infer from a Sobolev inequality and (3.2) that, for $\frac{1}{2}r_0 \leq \rho < r \leq r_0$ and $h > k \geq k_0$,

$$(h - k)|A^+(h, \rho)|^{\frac{n-1}{n}} \leq \left[\int_{A^+(h,\rho)} |\theta(u - u^+)|^{\frac{n}{n-1}} dx\right]^{\frac{n-1}{n}} \leq$$

$$\leq \left[\int_{\mathbf{B}_\rho} |\theta(u-u^+)|^{\frac{n}{n-1}} dx\right]^{\frac{n-1}{n}} \leq c_0 \int_{\mathbf{B}_r} |D[\theta(u-u^+)]|$$

$$\leq c_0(r-\rho)^{-1} \int_{\mathbf{B}_r \sim \mathbf{B}_\rho} |\theta(u-u^+)| dx + c_0 C |A^+(k,r)|$$

$$\leq c_0(r-\rho)^{-1}(h-k)|A^+(k,r)| + c_0 C|A^+(k,r)|.$$

Thus we find, as in our previous argument in §2 that for a constant C_0,

$$|A^+(h,\rho)|^{\frac{n-1}{n}} \leq C_0\left[(r-\rho)^{-1} + (h-k)^{-1}\right]|A^+(k,r)| \quad \text{for } \frac{1}{2}r_0 \leq \rho < r \leq r_0 \text{ and } h > k \geq k_0. \tag{3.4}$$

By (3.3) and (3.4), we may again apply Lemma 2.1 to conclude that

$$u - u^+ \leq k_0 + d = k_0 + c_1|A(k_0,r_0)|^{\frac{1}{n}} \leq c_7 k_0 \text{ in } \mathbf{B}_{r_o/2}.$$

Similarly, there is a positive s_0 such that

$$u - u^- \geq c_7 k_0 \text{ in } \mathbf{B}_{s_0/2}.$$

Next we seek more precise information on the behavior of u near a point of positive density. From [F],4.5.9(22), we recall that, a BV function u, is, at $\mathcal{H}^{n-1}$ almost all points a of positive upper density, *approximately two-sided continuous* in the sense that there is a unit vector $\nu(a)$ so that

$$\alpha^\pm(\rho) = [\rho^{-n}\int_{\mathbf{B}_\rho^\pm(a)} |u-u^+|^{\frac{n}{n-1}} dx]^{\frac{n-1}{n}} \to 0 \text{ as } \rho \to 0$$

where $\mathbf{B}_\rho^\pm(a) = \mathbf{B}_\rho(a) \cap \{\pm(x-a).\nu(a) > 0\}$.

Theorem 3.2. *Suppose that, in addition to the hypotheses of Theorem 3.1, the local solution u is approximately two-sided continuous at a in the direction ν. For any $0 < \beta < \frac{1}{n-1}$ and any $\epsilon > 0$, there is a $\tau > 0$ so that*

$$|u - u^\pm| \leq \epsilon \quad \textit{in } E^\pm = \mathbf{B}_\tau(a) \cap \{x : \pm(x-a).\nu \geq [\alpha^\pm(2|x-a|)]^\beta |x-a|\} \tag{3.5}$$

where $\alpha^{\pm}$ *is as above. In particular, the two sets*

$$\{x : u(x) > \frac{1}{2}(u^+ + u^-)\} \quad and \quad \{x : u(x) < \frac{1}{2}(u^+ + u^-)\}$$

contain open domains whose boundaries are tangent to the hyperplane $\{(x - a).\nu = 0\}$ *at* a. *Moreover,*

$$\lim_{\rho \to 0} \int_{\mathbf{B}_\rho(a)} Du / \int_{\mathbf{B}_\rho(a)} |Du| = \nu. \tag{3.6}$$

PROOF. We assume $a = 0$. Choose a positive number $s \leq \frac{1}{2}(c_1^{-1}\delta^{-\frac{1}{n}})$ so that $\mathbf{B}_s \subset \Omega$ and

$$[\alpha^+(2t)]^{1-\beta(n-1)} \leq 4^{1-n}(\delta|\mathbf{B}|)^{\frac{n-1}{n}}\epsilon \text{ for all } 0 < t \leq s.$$

Suppose $b \in \mathbf{B}_s$ and $b.\nu \geq [\alpha^+(2|b|)]^\beta |b|$. Setting

$$r_0 = [\alpha^+(2|b|)]^\beta |b| \text{ and } k_0 = \frac{1}{2}\epsilon,$$

we see that

$$k_0^{-1} 4^{n-1} |\mathbf{B}|^{\frac{n}{n-1}} [\alpha^+(2|b|)](r_0^{-1}|b|)^{n-1} \leq \delta^{\frac{n-1}{n}}. \tag{3.7}$$

Next, for $\frac{1}{2}r_0 \leq \rho \leq r_0$ and $k \geq k_0$, we let

$$A_b(k, \rho) = \mathbf{B}_\rho(b) \cap \{u < u^+ - k\},$$

note that $\mathbf{B}_\rho(b) \subset \mathbf{B}^+_{\rho+|b|}$ because $b.\nu \geq \rho$, and use the definition of α^+ and (3.7) to estimate

$$\begin{aligned}
k|A_b(k,\rho)|^{\frac{n-1}{n}} &\leq [\int_{A_b(k,\rho)} |u - u^+|^{\frac{n}{n-1}} dx]^{\frac{n-1}{n}} \\
&\leq [\int_{\mathbf{B}^+_{\rho+|b|}} |u - u^+|^{\frac{n}{n-1}} dx]^{\frac{n-1}{n}} \\
&\leq [\alpha^+(\rho + |b|)](\rho + |b|)^{n-1} \\
&\leq 2^{n-1}[\alpha^+(2|b|)](\rho^{-1}|b|)^{n-1}\rho^{n-1} \\
&\leq 4^{n-1}[\alpha^+(2|b|)](r_0^{-1}|b|)^{n-1}\rho^{n-1} \\
&\leq k_0\delta^{\frac{n-1}{n}}\rho^{n-1} \leq k\delta^{\frac{n-1}{n}}\rho^{n-1}.
\end{aligned}$$

Thus,

$$(3.8) \qquad |A_b(k,\rho)| \le \delta\rho^n \ \text{ for } \ \frac{1}{2}r_0 \le \rho \le r_0 \ \text{ and } \ k \ge k_0.$$

Now for fixed ρ and r with $\frac{1}{2}r_0 \le \rho < r \le r_0$,, we let η be as in 1.2. Also for fixed h and k with $h > k \ge k_0$, we let

$$\theta(t) = \begin{cases} h-k & \text{for } t \le -k \\ -t-k & \text{for } -h < t < -k \\ 0 & \text{for } t \ge -k. \end{cases}$$

Using $\zeta(x) = \eta(x)\cdot\theta(u(x) - u^+)$ in the variational formula (1.1) and noting that

$$\operatorname{supp}(\eta\theta'(u)) \subset \operatorname{supp}\zeta \subset A_b(k,r),$$

we deduce, as in the proof of (2.4), that
(3.9)

$$\int_{\mathbf{B}_\rho} |D[\theta(u-u^+)]| \le (r-\rho)^{-1}\int_{\mathbf{B}_r \sim \mathbf{B}_\rho} |\theta(u-u^+)|dx + C|A_b(k,r)|.$$

Since $\delta \le \frac{1}{2}|\mathbf{B}|$, we may, as in §2, use a Sobolev inequality to find that

$$|A_b(h,\rho)|^{\frac{n-1}{n}} \le \ \text{const.}\ [(r-\rho)^{-1} + (h-k)^{-1}]|A_b(k,r)|$$

for $\frac{1}{2}r_0 \le \rho < r \le r_0$ and $h > k \ge k_0$. Applying 2.1 now gives

$$|A_b(k_0 + d, \frac{1}{2}r_0)| = 0 \ \text{ where } \ d = c_1|A_b(k_0,r_0)|^{\frac{1}{n}}.$$

Thus,

$$u - u^+ \ge -(k_0 + c_1\delta^{\frac{1}{n}}r_0) \ge -(\frac{1}{2}\epsilon + c_1\delta^{\frac{1}{2}}s) \ge \epsilon \ \text{ in } \ \mathbf{B}_{r_o/2}(b).$$

Similarly, for $b \in \mathbf{B}_s$ with $b.\nu \le -[\alpha^-(2|b|)]^\beta|b|$, we use

$$r_0 = [\alpha^-(2|b|)]^\beta|b|$$

and find that

$$u - u^+ \le -\epsilon \ \text{ in } \ \mathbf{B}_{r_o/2}(b).$$

Combining these two inequalities with Theorem 3.1 gives the existence os a suitable τ.

Finally to obtain the formula for the vector ν, we write

$$\int_{\mathbf{B}_\rho} Du = \int_{\partial \mathbf{B}_\rho} u \frac{x}{|x|} dS,$$

where the right hand side is to be understood in the sense of BV trace theory [G], and consider the integrals over the three spherical regions $\partial \mathbf{B}_\rho \cap E^+$, $\partial \mathbf{B}_\rho \cap E^-$, and $\partial \mathbf{B}_\rho \sim (E^+ \cup E^-)$ separately.

To handle the last region, note that by Theorem 3.1, u is bounded on $\mathbf{B}_\rho$ for ρ sufficiently small. Thus,

$$\begin{aligned}
&\left| \int_{\partial \mathbf{B}_\rho \sim (E^+ \cup E^-)} u \frac{x}{|x|} dS \right| \leq \\
&\leq \|u\|_{L^\infty(\mathbf{B}_\rho)} \mathcal{H}^{n-1}(\partial \mathbf{B}_\rho \sim (E^+ \cup E^-)) \\
&\leq c \|u\|_{L^\infty(\mathbf{B}_\rho)} \max\{[\alpha^+(2\rho)]^\beta, [\alpha^-(2|b|)]^\beta\} \rho^{n-1} = o(\rho^{n-1}).
\end{aligned}$$

For the other two regions, we use the first conclusion of 3.2 to see that

$$\begin{aligned}
\Big| \int_{\partial \mathbf{B}_\rho \cap E^\pm} u \frac{x}{|x|} dS &- \int_{\partial \mathbf{B}_\rho \cap E^\pm} u^\pm \frac{x}{|x|} dS \Big| \leq \\
&\leq \int_{\partial \mathbf{B}_\rho \cap E^\pm} |u - u^\pm| dS \\
&\leq \operatorname{ess} \sup_{\partial \mathbf{B}_\rho \cap E^\pm} |u - u^\pm| \mathcal{H}^{n-1}(\partial \mathbf{B}_\rho) = o(\rho^{n-1}).
\end{aligned}$$

Moreover, since each set $E^\pm$ is symmetric in the directions orthogonal to ν,

$$\int_{\partial \mathbf{B}_\rho \cap E^\pm} u^\pm \frac{x}{|x|} dS = (\int_{\partial \mathbf{B}_\rho \cap E^\pm} \frac{x}{|x|} \cdot \nu dS) u^\pm \nu.$$

We conclude that

$$\rho^{1-n} \int_{\mathbf{B}_\rho} Du = \gamma \nu + 0(1) \quad \text{as} \quad \rho \to 0,$$

where $\gamma = [(\int_{\partial \mathbf{B} \cap E^+} \frac{x}{|x|} \cdot \nu dS) u^+ + (\int_{\partial B \cap E^-} \frac{x}{|x|} \cdot \nu dS) u^-]$ is a nonzero scalar. Then

$$\rho^{1-n} \int_{\mathbf{B}_\rho} |Du| = |\gamma| + o(1) \quad \text{as} \quad \rho \to 0,$$

and we may divide and let $\rho \to 0$ to obtain the desired formula.

As a consequence of Theorem 3.2 and the characterization from [F], 4.5.9(22), the set E^+ defined in (3.5) is a set of finite perimeter and (3.6) holds on $\partial E^+ \cap B_\delta(a)$, $\delta < \tau, \mathcal{H}^{n-1}$ a.e. Thus for $f = 0$ and η smooth with supp $\eta \subset B_\delta(a)$,

$$D(\eta\chi_{E^+}) \ll |Du|. \tag{3.10}$$

Working directly with the variational principle for the functional, this leads in a standard way to

Theorem 3.3 *Suppose that $f = 0$ and u is a local solution in Ω, $a \in \Omega$, $\Theta(a) > 0$, and $u^+(a) > u^-(a)$. Suppose in addition that u is two-sided continuous at a in the direction ν. Then there is a neighborhood $\mathbf{B}_\tau(a)$ such that the set*

$$\partial\{x : u(x) > \frac{1}{2}(u^+ + u^-)\} \cap \mathbf{B}_\tau(a)$$

is an area minimizing hypersurface in $\mathbf{B}_\tau(a)$.

We adopt several standard notations: $\mathbf{B}_r(a)$ indicates the open ball $\{x \in \mathbf{R}^n : |x - a| < r\}$ and $\mathbf{B}_r = \mathbf{B}_r(0)$. The Lebesgue outer measure of a set A is written $|A|$, while $\mathcal{H}^{n-1}$ denotes $n - 1$ dimensional Hausdorff measure. When restricted to a sphere $\partial\mathbf{B}_r(a)$, the latter becomes ordinary surface measure and will be, under integrals signs, abbreviated as dS.

This research was supported in part by NSF grants DMS 85-11357 and DMS 87-0672.

References

[A] G.Anzellotti, *On the extremal stress and displacement in Hencky plasticity*, Duke Math. J. **551** (1984), 133-147.

[AG_1] G.Anzellotti and M.Giaquinta, *Existence of the displacement field for an elastoplastic body,* Manus.Math. **32** (1980), 101-136.

[AG_2] G.Anzellotti, M.Giaquinta, *On the existence of the fields of stress and displacement for an elasto-plastic body in static equilibrium*, J.Math. Pures Appl. **61** (1982), 219-244.

[AL] G.Anzellotti, S.Luckhans, *Dynamical evolution of elasto-perfectly plastic bodies,* Appl. Math. Opt. **15** (1987), 121-140.

[B_1] H.Brezis, *Intégrales convexes dans les espaces de Sobolev,* Israel J. Math. **13** (1972), 9-23.

[B_2] H.Brezis, *Multiplicateur de Lagrange en torsion elastoplastique,* Arch. Rat. Mech. Anal. **49** (1972), 32-40.

[B_3] H.Brezis, *Analyse fonctionnelle*, Masson, (1983).

[DCP] E.De Giorgi,F.Colombini, L.Piccinini, *Frontiere orientate di misura minima e questioni collegate*, Pubbl. Scuola Normale, Pisa, (1972).

[ET] I.Ekeland, R.Temam, *Convex duality*, North-Holland, (1976).

[F] H.Federer, *Geometric Measure Theory*, Springer-Verlag, Berlin, Heidelberg and New York, 1969.

[G] E.Gagliardo, *Caratterizzazioni delle tracce sulla frontiera relative ad alcune classi di funzioni in n variabili,* Rend. Sem. Mat. Padova **27** (1957), 284-305.

[HK_1] R.Hardt, D.Kinderlehrer, *Elastic plastic deformation,* Appl. Math. Optim. **10** (1983), 203-246.

[HK_2] R.Hardt, D.Kinderlehrer, *Some regularity results in plasticity,* Appl. Symp. Pure Math. **44** (1986), 239-244.

[KS] D.Kinderlehrer, G.Stampacchia, *Variational Inequalities*, Academic Press, 1980.

[KT] R.Kohn, R.Temam, *Dual spaces of stresses and strains,* Appl. Math. Optim. **10** (1983), 1-35.

[RS] V.Roytburd and M.Slemrod, *Dynamic phase transitions and compensated compactness,* J.Bona, C.Dafermos, J.L.Ericksen, D.Kinderlehrer eds., IMA Volumes in Math. and Appl. 4 (1987), 289-304.

[S] G.Stampacchia, *Formes bilinéaires coércitives sur les ensembles convexes*, CRAS Paris **258** (1964), 4413-4416.

[T] R.Temam, *Plasticity*, Gauthier Villars (1983).

School of Mathematics
University of Minnesota
206 Church St. SE
MINNEAPOLIS, MN 55455

ESTIMATION DE L'ERREUR DANS DES PROBLEMES DE DIRICHLET OU APPARAIT UN TERME ETRANGE

HAYAT KACIMI FRANÇOIS MURAT

Dédié à Ennio De Giorgi pour son soixantième anniversaire

Résumé. On considère le problème de Dirichlet

$$-\Delta u^\varepsilon = f \quad \text{dans} \quad \Omega^\varepsilon, \quad u^\varepsilon = 0 \quad \text{sur} \quad \partial\Omega^\varepsilon,$$

où Ω^ε est obtenu en retirant à Ω de nombreux petits trous, répartis périodiquement avec une période 2ε dans les directions des axes, chaque trou étant obtenu à partir d'un trou modèle par une homothétie de rapport $\varepsilon^{N/(N-2)}$. Dans le problème limite

$$-\Delta u + \mu u = f \quad \text{dans} \quad \Omega, \quad u = 0 \quad \text{sur} \quad \partial\Omega,$$

apparaît un "terme étrange" d'ordre 0.

Désignant par p^ε le potentiel capacitaire de chacun de ces petits trous par rapport à la boule de rayon ε qui l'entoure, nous montrons l'estimation d'erreur

$$\|u^\varepsilon - (1 - p^\varepsilon)u\|_{H_0^1(\Omega)} \leq \text{ cste } \varepsilon.$$

Ce résultat est démontré en deux étapes : la première est une estimation d'erreur abstraite, obtenue dans un cadre qui généralise la situation géométrique considérée ci-dessus ; la deuxième consiste à estimer explicitement certaines quantités, telles que $\|p^\varepsilon\|_{H^1(\Omega)}$, qui apparaissent dans l'estimation d'erreur abstraite.

Abstract. Consider the Dirichlet problem

$$-\Delta u^\varepsilon = f \quad \text{in} \quad \Omega^\varepsilon, \quad u^\varepsilon = 0 \quad \text{on} \quad \partial\Omega^\varepsilon,$$

where Ω^ε is obtained by removing from Ω many small holes T_i^ε, periodically distributed with a period 2ε in the directions of the axes, each of the holes T_i^ε being obtained from a model hole by reducing it at the size $\varepsilon^{N/(N-2)}$. The solutions u^ε converge weakly to the solution of

$$-\Delta u + \mu u = f \quad \text{in} \quad \Omega, \quad u = 0 \quad \text{on} \quad \partial\Omega,$$

where a zero-order term surprisingly appears.

Let p^ε be the capacity potential of each small hole T_i^ε in the ball of radius ε with the same center. We prove in this paper the following error estimate

$$\|u^\varepsilon - (1 - p^\varepsilon)u\|_{H_0^1(\Omega)} \leq \text{ cst } \varepsilon.$$

This result is proved in two steps. In the first, an abstract error estimate is obtained in a framework which generalizes the geometrical situation described above. The second step consists in estimating explicitly quantities like $\|p^\varepsilon\|_{H^1(\Omega)}$,which appear in the abstract error estimate.

1. Position du problème et énoncé du résultat.

Soit Ω un ouvert borné de $\mathbb{R}^N$, $N \geq 2$. On définit un ouvert perforé

$$\Omega^\varepsilon = \Omega \setminus \bigcup_{i=1}^{n(\varepsilon)} T_i^\varepsilon \tag{1.1}$$

en creusant dans Ω des trous T_i^ε que l'on fabrique à partir d'un trou modèle T par une homothétie de rapport a^ε et que l'on répartit périodiquement dans Ω avec une période 2ε dans les directions des axes ; pour être précis, on se donne une base $e_1, e_2, \ldots, e_N$ de $\mathbb{R}^N$ et un fermé borné T de $\mathbb{R}^N$; T n'est pas nécessairement régulier, mais si $N = 2$ on supposera que T contient une boule centrée en 0 (voir à ce sujet la Remarque 2.2) ; pour $a^\varepsilon > 0$ donné, on définit de façon précise Ω^ε par :

$$\text{(1.2a)} \qquad \begin{aligned} &\Omega^\varepsilon = \Omega \cap Q^\varepsilon \\ &Q^\varepsilon = \mathbb{R}^N \setminus \bigcup_{k \in \mathbf{Z}^N} T_k^\varepsilon \end{aligned}$$

$$\text{(1.2b)} \qquad T_k^\varepsilon = a^\varepsilon T + \varepsilon \sum_{\ell=1}^{N} 2k_\ell \, e_\ell$$

où k est le multi-indice $k = (k_1, k_2, \ldots, k_N) \in \mathbf{Z}^N$. Soit B la boule unité de $\mathbb{R}^N$:

$$B = \{x \in \mathbb{R}^N |\ |x| < 1\} \ ;$$

on pose

$$\text{(1.2c)} \qquad B_k^\varepsilon = \varepsilon B + \varepsilon \sum_{\ell=1}^{N} 2 \, k_\ell \, e_\ell .$$

Si a^ε est choisi tel que

$$\text{(1.3)} \qquad a^\varepsilon T \subset \varepsilon B$$

(ce que l'on supposera toujours dans la suite), les trous T_k^ε sont déconnectés, chacun étant contenu dans "sa" boule B_k^ε de rayon ε.

On se donne $f \in H^{-1}(\Omega)$, i.e.

$$\text{(1.4)} \qquad f = -\operatorname{div} g \, , \quad g \in (L^2(\Omega))^N,$$

et on considère le problème de Dirichlet dans Ω^ε : trouver u^ε tel que

$$\text{(1.5)} \qquad \begin{cases} -\Delta u^\varepsilon = f & \text{dans} \quad \Omega^\varepsilon, \\ u^\varepsilon = 0 & \text{sur} \quad \partial\Omega^\varepsilon, \end{cases}$$

qui est équivalent à la formulation variationnelle : trouver u^ε tel que

$$\text{(1.6)} \qquad \begin{cases} \displaystyle\int_{\Omega^\varepsilon} \operatorname{grad} u^\varepsilon \operatorname{grad} v = \int_{\Omega^\varepsilon} g \operatorname{grad} v \quad \forall v \in H_0^1(\Omega^\varepsilon) \\ u^\varepsilon \in H_0^1(\Omega^\varepsilon). \end{cases}$$

Le lemme de Lax-Milgram assure que (1.6) a une solution unique et que

$$\text{(1.7)} \qquad \|\operatorname{grad} u^\varepsilon\|_{(L^2(\Omega^\varepsilon))^N} \leq \|g\|_{(L^2(\Omega^\varepsilon))^N} \leq \|g\|_{(L^2(\Omega))^N}.$$

Si v^ε est une fonction de $L^2(\Omega^\varepsilon)$, on note $\tilde{v}^\varepsilon$ son prolongement à $L^2(\Omega)$ défini par

$$\text{(1.8)} \qquad \tilde{v}^\varepsilon = v^\varepsilon \quad \text{dans} \quad \Omega^\varepsilon, \quad \tilde{v}^\varepsilon = 0 \quad \text{dans} \quad \bigcup_{i=1}^{n(\varepsilon)} T_i^\varepsilon = \Omega \backslash \Omega^\varepsilon.$$

Ce prolongement par 0 vérifie

$$\|\tilde{v}^\varepsilon\|_{L^2(\Omega)} = \|v^\varepsilon\|_{L^2(\Omega^\varepsilon)}.$$

De plus, si $v^\varepsilon \in H_0^1(\Omega^\varepsilon)$ on a $\tilde{v}^\varepsilon \in H_0^1(\Omega)$ avec

$$\operatorname{grad} \tilde{v}^\varepsilon = (\operatorname{grad} v^\varepsilon)^\sim, \quad \|\operatorname{grad} \tilde{v}^\varepsilon\|_{(L^2(\Omega))^N} = \|\operatorname{grad} v^\varepsilon\|_{(L^2(\Omega^\varepsilon))^N}.$$

De (1.7), on déduit que $\|\operatorname{grad} \tilde{u}_\varepsilon\|_{(L^2(\Omega))^N}$ est borné, et donc que $\tilde{u}^\varepsilon$ est borné dans $H_0^1(\Omega)$. Quitte à extraire une sous-suite, que l'on note encore avec l'indice ε, on a donc

$$\text{(1.9)} \qquad \tilde{u}^\varepsilon \rightharpoonup u \quad \text{dans} \quad H_0^1(\Omega) \text{ faible}.$$

En fait l'extraction de cette sous-suite s'avérera inutile (voir Remarque 1.5).

Définissons maintenant dans $\mathbb{R}^N$ la fonction $w^\varepsilon = 1 - p^\varepsilon$ où p^ε est le potentiel capacitaire du trou T_k^ε dans la boule B_k^ε, de rayon ε et de même centre ; de façon précise on définit w^ε par

$$\text{(1.10)} \qquad \begin{cases} w^\varepsilon \in H^1_{\text{loc}}(\mathbb{R}^N) & \\ w^\varepsilon = 0 & \text{dans } \displaystyle\bigcup_{k \in \mathbf{Z}^N} T_k^\varepsilon \\ w^\varepsilon = 1 & \text{dans } \mathbb{R}^N \backslash \displaystyle\bigcup_{k \in \mathbf{Z}^N} B_k^\varepsilon \\ \Delta w^\varepsilon = 0 & \text{dans } \displaystyle\bigcup_{k \in \mathbf{Z}^N} (B_k^\varepsilon \backslash T_k^\varepsilon) \end{cases}$$

Chaque trou T_k^ε étant contenu dans la boule B_k^ε d'après l'hypothèse (1.3), la fonction w^ε est parfaitement définie par (1.10).

Le but de cet article est de montrer le résultat suivant :

Théorème 1.1. *On suppose que* $N \geq 2$ *et que* a^ε *est donné par :*

$$(1.11) \qquad \begin{cases} a^\varepsilon = C_0\ \varepsilon^{N/(N-2)} & si \quad N \geq 3 \\ a^\varepsilon = \exp\left(-C_0/\varepsilon^2\right) & si \quad N = 2 \end{cases}$$

où $C_0 > 0$ *est un réel donné.*

On suppose d'autre part que la limite u *définie par* (1.9) *est telle que*

$$(1.12) \qquad u \in W^{2,\infty}(\Omega).$$

On a alors l'estimation :

$$(1.13) \quad \limsup_{\varepsilon \to 0} \frac{1}{\varepsilon} \|u^\varepsilon - w^\varepsilon u\|_{H_0^1(\Omega^\varepsilon)} = \limsup_{\varepsilon \to 0} \frac{1}{\varepsilon} \|\tilde{u}^\varepsilon - w^\varepsilon u\|_{H_0^1(\Omega)} \leq C,$$

où C *est une constante qui ne dépend que de la dimension* N, *de l'ouvert* Ω, *du trou* T, *de la constante* C_0 *qui apparaît dans* (1.11) *et de la norme de* u *dans* $W^{2,\infty}(\Omega)$.

Remarque 1.2. Le problème du passage à la limite dans des problèmes de Dirichlet du type de celui que nous avons décrit ci-dessus a été étudié par de nombreux auteurs : il est difficile de donner une bibliographie complète, mais on peut citer par exemple [1] [4] [5] [6] [8] [10] [11] [12] [13] [19] [20] [22] [23] [24] [25] [28] [29] [30] [31] [32] [33] [34].

Un problème très lié au problème précédent est celui du passage à la limite dans des inéquations variationnelles avec des obstacles ψ^ε qui oscillent rapidement, l'exemple modèle étant $\psi^\varepsilon = 0$ dans Ω^ε et $\psi^\varepsilon = \psi$ sur les trous T_k^ε définis par (1.2) ; un problème équivalent est celui de la minimisation d'une fonctionnelle sur les convexes unilatéraux variables $\{v \in H_0^1(\Omega) | v \geq \psi^\varepsilon \text{ dans } \Omega\}$. Ce problème, proposé par E. De Giorgi [15] a été la source de nombreux travaux [1]

[3] [5] [6] [8] [9] [10] [12] [13] [17]. Son étude a fourni de belles illustrations de la puissante théorie de la Γ-convergence introduite par E. de Giorgi [14] [15] [16] [18].

Remarque 1.3. Sous les hypothèses du Théorème 1.1, on sait que

$$\tilde{u}^\varepsilon - w^\varepsilon u \to 0 \quad \text{dans} \quad H_0^1(\Omega) \text{ fort ;}$$

ce résultat, démontré dans [6], Théorème 3.4 et Remarque 3.3, est rappelé au Théorème 2.5 ci-dessous. Le résultat du présent article est plus précis puisqu'il donne une *estimation d'erreur.*

D'autres estimations d'erreurs, obtenues par des techniques différentes, sont données dans [19] [20] [29] [30] [31].

Remarque 1.4. Une autre façon de présenter l'estimation (1.13) est d'utiliser le potentiel capacitaire p^ε défini par $p^\varepsilon = 1 - w^\varepsilon$, et d'écrire

$$\tilde{u}^\varepsilon = u + p^\varepsilon u + r^\varepsilon \quad \text{avec} \quad \|r^\varepsilon\|_{H_0^1(\Omega)} \leq C\varepsilon \quad \text{pour } \varepsilon \text{ assez petit.}$$

Cette présentation fait apparaître $\tilde{u}^\varepsilon$ (prolongement de u^ε par 0, voir (1.8)) comme une somme de trois termes : la limite u, un terme $p^\varepsilon u$ qui converge vers 0 dans $H_0^1(\Omega)$ faible mais pas fort (cf. (2.3) et la Remarque 2.8 ci-dessous) et qui est le *correcteur du premier ordre* de la limite, enfin le reste r^ε qui est d'ordre supérieur. Cette présentation se rapproche de la théorie des couches limites (cf. par exemple [27]) ; on notera qu'ici, la couche limite est répartie dans tout l'ouvert Ω (et non pas concentrée), ce qui est assez naturel puisque sa cause, la frontière $\bigcup_{i=1}^{n(\varepsilon)} \partial T_i^\varepsilon$ des trous, est répartie dans tout Ω.

Remarque 1.5. Dans la situation géométrique que nous avons décrite dans ce paragraphe, où les trous sont obtenus par homothétie à partir d'un trou modèle, et répartis périodiquement, il n'est pas nécessaire d'extraire de sous-suite dans (1.9) car la suite $\tilde{u}^\varepsilon$ toute entière converge vers u.

On sait en effet (voir les références citées dans la Remarque 1.2 et le Théorème 2.5 du présent article), que si a^ε est donné par (1.11), la fonction u définie par (1.9) est la solution *unique* du problème

$$(1.14) \qquad \begin{cases} -\Delta u + \mu u = f & \text{dans} \quad \Omega \\ u \in H_0^1(\Omega) \end{cases}$$

où μ est une constante positive définie à partir de la capacité de T dans $\mathbb{R}^N$ quand $N \geq 3$ (voir (2.5) ci-dessous).

Si les trous sont plus petits que la taille a^ε définie par (1.11), alors on a encore (1.14) mais avec $\mu = 0$; si par contre les trous sont plus gros, on a $u \equiv 0$ dans Ω (voir Remarque 2.8 ci-dessous).

Remarque 1.6. Comme d'habitude dans les problèmes d'estimation d'erreur (voir par exemple [27]), l'estimation (1.13) n'est démontrée que si la limite u est assez régulière ; ici u doit appartenir à $W^{2,\infty}(\Omega)$ (hypothèse (1.12)), alors que la limite u des $\tilde{u}^\varepsilon$ appartient, en principe, seulement à $H_0^1(\Omega)$. Notons cependant que si f est assez régulière et si le bord $\partial\Omega$ de l'ouvert est assez régulier, la fonction u, qui est solution du problème elliptique (1.14) a la régularité voulue.

La démonstration du Théorème 1.1 que nous présentons dans cet article s'articule en 2 étapes.

La *première étape* consiste à plonger la situation géométrique décrite par (1.2) et (1.11) (trous de taille a^ε donnée par (1.11) répartis périodiquement dans Ω à des distances de l'ordre de 2ε) dans le "cadre abstrait" introduit dans [6] : nous rappelons ci-dessous (paragraphe 2) ce cadre et les principaux résultats qu'il permet d'obtenir.

Dans ce "cadre abstrait", il est facile d'obtenir (paragraphe 3) une estimation de l'erreur qui est une *généralisation naturelle* de (1.13) voir Théorème 3.1. Cette estimation de l'erreur "abstraite" donne une majoration de $\|u^\varepsilon - w^\varepsilon u\|_{H_0^1(\Omega^\varepsilon)}$ en fonction des deux quantités $\|w^\varepsilon - 1\|_{L^2(\Omega)}$ et $\|\mu^\varepsilon - \mu\|_{H^{-1}(\Omega)}$.

Il suffit alors, dans une *deuxième étape* (paragraphe 4) de majorer explicitement ces deux quantités par $C\varepsilon$ dans le cas où la géométrie est décrite par (1.2), (1.11), pour obtenir le Théorème 1.1.

2. Un cadre abstrait pour certains problèmes de Dirichlet dans un ouvert variable. Rappel de résultats. Dans ce paragraphe, nous rappelons le cadre introduit dans [6] et les résultats qu'il permet d'obtenir. Cette façon d'étudier certains problèmes de Dirichlet dans des ouverts variables est directement inspirée de la méthode mise au point par L. Tartar [35] pour étudier les problèmes d'homogénéisation ; elle est basée sur la fabrication et l'usage de bonnes fonctions test.

Commencons par énoncer une propriété "concrète" :

Théorème 2.1. *Les fonctions w^ε définies par* (1.10) *sont telles que*

$$-\Delta w^\varepsilon = \mu^\varepsilon - \gamma^\varepsilon \quad dans \quad \mathcal{D}'(\mathbb{R}^N) \tag{2.1a}$$

où μ^ε et γ^ε sont des éléments de $H^{-1}_{\text{loc}}(\mathbb{R}^N)$ définis par :

$$\left\{ \begin{aligned} &< \mu^\varepsilon, \varphi > = \sum_{k \in \mathbf{Z}^N} \int_{\partial B_k^\varepsilon} \frac{\partial w^\varepsilon}{\partial n} \varphi \, ds, \\ &< \gamma^\varepsilon, \varphi > = \sum_{k \in \mathbf{Z}^N} \int_{\partial T_k^\varepsilon} \frac{\partial w^\varepsilon}{\partial n} \varphi \, ds \quad \forall \varphi \in \mathcal{D}(\mathbb{R}^N). \end{aligned} \right. \tag{2.1b}$$

Pour tout ouvert borné Ω de $\mathbb{R}^N$, on a, en désignant par $\tilde{v}^\varepsilon$ le prolongement (défini en (1.8)*) de v^ε par* 0 *:*

$$< \gamma^\varepsilon, \tilde{v}^\varepsilon > = 0 \quad si \quad v^\varepsilon \in H^1_0(\Omega^\varepsilon). \tag{2.2}$$

Si a^ε est donné par (1.11)*, on a en outre*

$$w^\varepsilon \rightharpoonup 1 \quad dans \quad H^1_{\text{loc}}(\mathbb{R}^N) \text{ faible} \tag{2.3}$$

$$\mu^\varepsilon \to \mu \quad dans \quad H^{-1}_{\text{loc}}(\mathbb{R}^N) \text{ fort} \tag{2.4}$$

où μ est défini par

$$\left\{ \begin{aligned} &\mu = \frac{C_0^{N-2}}{2^N} Cap\, T \quad si\ N \geq 3 \\ &\mu = \frac{\pi}{2C_0} \quad\quad\quad\quad si\ N = 2. \end{aligned} \right. \tag{2.5}$$

Dans cet énoncé, *Cap* T désigne la capacité du fermé T dans $\mathbb{R}^N$, i.e.

$$\text{(2.6)} \qquad Cap\, T = \inf_{\substack{\varphi \in \mathcal{D}(\mathbb{R}^N) \\ \varphi = 1 \text{ sur } T}} \int_{\mathbb{R}^N} |\text{grad}\, \varphi|^2 dx \ .$$

On a désigné par $<\ ,\ >$ la dualité entre $H_0^1(\Omega)$ et $H^{-1}(\Omega)$. Dans (2.1b) n désigne la normale extérieure à la couronne $B_k^\varepsilon \backslash T_k^\varepsilon$ et on a noté comme des intégrales les dualités entre $H^{-1/2}$ et $H^{1/2}$ sur ∂B_k^ε et ∂T_k^ε ; cette notation est justifiée pour $\int_{\partial B_k^\varepsilon} \frac{\partial w^\varepsilon}{\partial n} \varphi \, ds$ qui est bien une intégrale, car w^ε est harmonique dans la couronne $B_k^\varepsilon \backslash T_k^\varepsilon$ et vaut 1 sur la sphère ∂B_k^ε : $\frac{\partial w^\varepsilon}{\partial n}$ est donc une fonction très régulière sur ∂B_k^ε ; il n'en va pas de même pour $\int_{\partial T_k^\varepsilon} \frac{\partial w^\varepsilon}{\partial n} \varphi \, ds$ si ∂T_k^ε n'est pas régulier ; de façon correcte, on doit poser (en tenant compte du fait que w^ε est harmonique dans $B_k^\varepsilon \backslash T_k^\varepsilon$)

$$\text{(2.7)} \qquad \int_{\partial T_k^\varepsilon} \frac{\partial w^\varepsilon}{\partial n} \varphi \, ds = \int_{\mathbb{R}^N \backslash T_k^\varepsilon} \text{grad}\, w^\varepsilon \text{grad}\, \phi \, dx$$

où ϕ est un élément de $\mathcal{D}(\mathbb{R}^N)$ tel que $\phi = \varphi$ sur T_k^ε et $\phi \equiv 0$ en dehors de B_k^ε ; le second membre de (2.7) définit un élément de $H^{-1/2}(\partial T_k^\varepsilon)$ quelle que soit la régularité de T_k^ε.

Remarque 2.2. L'expression de μ diffère selon que $N = 2$ ou $N \geq 3$. Si $N \geq 3$, μ est égal (à une constante multiplicative près) à la capacité du trou T. En dimension $N = 2$, μ est indépendant de T : cela provient des propriétés spéciales du potentiel Newtonien en dimension 2, qui font que pour tout fermé borné T, on a *Cap* $T = 0$ quand $N = 2$.

Soulignons le fait qu'en dimension 2, nous avons supposé que T contenait une boule B_ρ centrée à l'origine. Cette hypothèse n'est probablement pas nécessaire ; nous ne l'utiliserons que pour démontrer que μ est égal à $\pi/2C_0$ en dimension 2, et dans la démonstration du lemme 4.6. Il est possible, sans faire cette hypothèse, de montrer que l'on a (2.3) et qu'il existe un μ et une sous-suite en ε telle que l'on ait (2.4) : il suffit de procéder par comparaison ([6], Théorème 2.7 et Lemme 2.8) avec une boule centrée en 0 et contenant T. Ce raisonnement montre aussi que l'on a nécessairement $0 \leq \mu \leq \pi/2C_0$. Il reste alors à identifier μ. Nous ne savons pas résoudre ce problème si T est quelconque ; mais quand T contient une boule B_ρ, on obtient facilement (encore par comparaison) que $\mu \geq \pi/2C_0$, ce qui démontre alors (2.5) (sans extraire de sous-suite).

On pourrait d'ailleurs se contenter de supposer que T contient un segment de droite centré à l'origine ; il suffirait alors d'employer la Proposition de l'Appendice 3 de [1]. Mais cette dernière hypothèse n'est pas non plus satisfaisante puisqu'elle ne permet pas de donner la valeur de μ quand T ne contient pas de segment, par exemple quand T est un ensemble de Cantor ; il est probable que même dans ce cas, on a $\mu = \pi/2C_0$.

Démonstration du Théorème 2.1. La propriété (2.1) résulte immédiatement du fait que w^ε est nulle sur les trous T_k^ε, est harmonique dans les couronnes $B_k^\varepsilon \backslash T_k^\varepsilon$, et est égale à 1 en dehors de l'union des boules B_k^ε. La propriété (2.2) est également évidente, puisque γ^ε ne charge que les trous.

Les propriétés (2.3) et (2.4) sont les affirmations les plus intéressantes du Théorème 2.1 ; ce sont d'ailleurs les seules choses à démontrer et les seuls points où intervient l'hypothèse (1.11) sur la valeur de a^ε. Si T est une boule, (2.3) et (2.4) résultent de calculs explicites en coordonnées polaires : voir [6], Théorème 2.2 et Lemme 2.3 ; si T est un fermé borné quelconque, on démontre (2.3) et (2.4) par comparaison ([6], Théorème 2.7 et Lemme 2.8).

Il reste alors à démontrer (2.5), c'est-à-dire à identifier μ ; on utilise pour cela la Proposition 1.1 de [6]. Si $N = 2$ on utilise le fait que la capacité est une fonction croissante et on procède par comparaison : comme nous avons supposé que T contient une boule et est contenu dans une autre boule, et comme toutes les boules conduisent, indépendemment de leur rayon, à $\mu = \pi/2C_0$, on en déduit (2.5). Si $N \geq 3$, on effectue un changement de variable $x \to x/a^\varepsilon$ pour ramener le trou $a^\varepsilon T$ à sa taille originelle T. Il reste alors à faire un raisonnement analogue (mais en fait plus simple) à celui que nous ferons pour démontrer le Lemme 4.6 (voir aussi [1], Lemme 2.4).

L'examen du Théorème 2.1 et de la définition (1.10) de w^ε montrent que, lorsque a^ε est donné par (1.11), la situation géométrique décrite au paragraphe 1 (trous de taille a^ε répartis périodiquement dans Ω, à des distances de 2ε) rentre dans le "cadre abstrait" suivant :

Hypothèse (H). *Soit Ω un ouvert borné de $\mathbb{R}^N$ et soit $\Omega^\varepsilon = \Omega \setminus \bigcup_{i=1}^{n(\varepsilon)} T_i^\varepsilon$ l'ouvert obtenu en lui retirant des fermés T_i^ε de $\mathbb{R}^N$. On suppose que la géométrie est telle qu'il existe $w^\varepsilon, \mu^\varepsilon, \gamma^\varepsilon$ et μ qui vérifient :*

$$(2.8) \qquad \begin{cases} w^\varepsilon \in H^1(\Omega) & \\ w^\varepsilon = 0 & \textit{dans} \quad T_i^\varepsilon, \quad 1 \le i \le n(\varepsilon) \\ w^\varepsilon \rightharpoonup 1 & \textit{dans} \quad H^1(\Omega) \textit{ faible} \\ \mu^\varepsilon, \gamma^\varepsilon, \mu \in H^{-1}(\Omega) & \\ -\Delta w^\varepsilon = \mu^\varepsilon - \gamma^\varepsilon & \textit{dans} \quad \mathcal{D}'(\Omega) \\ \mu^\varepsilon \to \mu & \textit{dans} \quad H^{-1}(\Omega) \textit{ fort} \\ < \gamma^\varepsilon, \tilde{v}^\varepsilon >= 0 & \forall v^\varepsilon \in H_0^1(\Omega^\varepsilon). \end{cases}$$

Remarque 2.3. L'hypothèse (H) n'est pas satisfaite seulement dans la situation géométrique du paragraphe 1 : on trouvera dans [6], paragraphe 2, plusieurs exemples où cette hypothèse est vérifiée et conduit à l'apparition d'un "terme étrange" μu dans l'équation limite : trous répartis périodiquement sur un hyperplan, cylindres disposés "en forêt" ou "en rideau", "treillis tridimensionnels" par exemple. On trouvera dans [25], chapitre 1, l'exemple (en dimension $N \ge 3$) de trous ellipsoïdaux, de révolution autour de leur grand axe, répartis dans Ω avec une période 2ε dans les directions des axes de $\mathbb{R}^N$, et dont les $(N-1)$ petits axes ℓ^ε et le grand axe L^ε ont des ordres de grandeur différents :

$$(2.9) \qquad \begin{cases} \ell^\varepsilon = C_0 \varepsilon^\alpha, & L^\varepsilon = C_1 \varepsilon^\beta \\ 1 \le \beta \le \alpha, & \beta + (N-3)\alpha = N \end{cases}$$

quand $N \geq 4$; là encore, l'hypothèse (H) est vérifiée.

Le "cadre abstrait" de l'hypothèse (H) permet également de traiter des problèmes de Dirichlet avec des opérateurs d'ordre supérieur [5], ainsi que des problèmes d'obstacles qui oscillent rapidement ([6], paragraphe 4, et [5]) (pour ces problèmes voir aussi [13]).

Remarque 2.4. L'hypothèse (H) n'est autre que le "cadre abstrait" (H1) (H2) (H3) (H4) (H5) (ou plus exactement (H5)$'$) de [6], à un détail près cependant : on a ici supposé seulement que μ appartient à $H^{-1}(\Omega)$, alors que dans [6] on avait supposé $\mu \in W^{-1,\infty}(\Omega)$. Nous allons voir que $\mu \in H^{-1}(\Omega)$ suffit.

En effet, il est facile de voir, en utilisant comme fonction test φw^ε avec $\varphi \in \mathcal{D}(\Omega)$ dans $-\Delta w^\varepsilon = \mu^\varepsilon - \gamma^\varepsilon$ que

$$
(2.10) \qquad < \mu, \varphi > = \lim_{\varepsilon \to 0} \int_\Omega \varphi |\mathrm{grad}\, w^\varepsilon|^2 \quad \forall \varphi \in \mathcal{D}(\Omega)
$$

(cf. [6], Proposition 1.1). Ainsi μ est nécessairement une *mesure positive* de $H^{-1}(\Omega)$.

Un résultat de J. Deny [7] (voir aussi [2]) assure alors que toute fonction $v \in H_0^1(\Omega)$ est mesurable pour la mesure μ et appartient à $L^1(\Omega; d\mu)$:

$$
(2.11) \qquad \begin{cases} \mu \in H^{-1}(\Omega), \quad \mu \geq 0 \quad \text{et} \quad v \in H_0^1(\Omega) \\ \Longrightarrow v \in L^1(\Omega; d\mu) \quad \text{et} \quad < \mu, v > = \int_\Omega v \, d\mu. \end{cases}
$$

Dans le cadre (H), on obtient les résultats suivants :

Théorème 2.5. *On se place sous l'hypothèse* (H), *et on considère, pour* $f \in H^{-1}(\Omega)$ *donné, les solutions* u^ε *des problèmes de Dirichlet :*

$$
(2.12) \qquad \begin{cases} -\Delta u^\varepsilon = f \quad \textit{dans} \quad \mathcal{D}'(\Omega^\varepsilon) \\ u^\varepsilon \in H_0^1(\Omega^\varepsilon). \end{cases}
$$

Alors la suite $\tilde{u}^\varepsilon$ *(toute entière) des prolongements par* 0 *de* u^ε *(voir* (1.8)*) vérifie*

$$
(2.13) \qquad \tilde{u}^\varepsilon \rightharpoonup u \quad \textit{dans} \quad H_0^1(\Omega) \textit{ faible,}
$$

où u est la solution unique de

$$
(2.14) \qquad \begin{cases} -\Delta u + \mu u = f & dans \quad \mathcal{D}'(\Omega) \\ u \in H_0^1(\Omega) \cap L^2(\Omega; d\mu) \end{cases}
$$

et l'on a

$$
(2.15) \qquad \begin{cases} \tilde{u}^\varepsilon = w^\varepsilon u + r^\varepsilon \\ avec\ r^\varepsilon \to 0 \quad dans \quad W_0^{1,1}(\Omega)\ fort. \end{cases}
$$

Si en outre u appartient à $W_0^{1,p}(\Omega)$ où $p > N$, alors $r^\varepsilon \to 0$ dans $H_0^1(\Omega)$ fort.

Théorème 2.6. *On se place sous l'hypothèse* (H), *et on considère une suite z^ε telle que*

$$
(2.16) \qquad \begin{cases} z^\varepsilon \in H_0^1(\Omega) \\ z^\varepsilon = 0 \quad sur \quad T_i^\varepsilon, \quad 1 \le i \le n(\varepsilon) \\ z^\varepsilon \rightharpoonup z \quad dans \quad H_0^1(\Omega)\ faible. \end{cases}
$$

Alors (noter que d'après (2.11) *on sait déjà que $z \in L^1(\Omega; d\mu)$) on a*

$$
(2.17) \qquad \begin{cases} z \in L^2(\Omega; d\mu) \\ \liminf_{\varepsilon \to 0} \int_\Omega |\mathrm{grad}\, z^\varepsilon|^2 dx \ge \int_\Omega |\mathrm{grad}\, z|^2 dx + \int_\Omega z^2 d\mu. \end{cases}
$$

Si on suppose de plus que

$$
(2.18) \qquad \int_\Omega |\mathrm{grad}\, z^\varepsilon|^2 dx \to \int_\Omega |\mathrm{grad}\, z|^2 dx + \int_\Omega z^2 d\mu
$$

on obtient

$$
(2.19) \qquad \begin{cases} z^\varepsilon = w^\varepsilon z + r^\varepsilon \\ où\ r^\varepsilon \to 0 \quad dans \quad W_0^{1,1}(\Omega)\ fort\ ; \end{cases}
$$

en outre r^ε tend vers 0 dans $H_0^1(\Omega)$ fort si z appartient à $W_0^{1,p}(\Omega)$ où $p > N$. Enfin, pour tout $\varphi \in \mathcal{D}(\Omega)$ on a

$$
(2.20) \qquad \begin{cases} \varphi w^\varepsilon \rightharpoonup \varphi \quad dans\ H_0^1(\Omega)\ faible \\ \varphi w^\varepsilon = 0 \quad sur\ T_i^\varepsilon,\ 1 \le i \le n(\varepsilon) \\ \int_\Omega |\mathrm{grad}\,(\varphi w^\varepsilon)|^2 \to \int_\Omega |\mathrm{grad}\, \varphi|^2 dx + \int_\Omega \varphi^2 d\mu. \end{cases}
$$

Remarque 2.7. Si on note

$$1_{T^\varepsilon}(u) = \begin{cases} 0 & \text{si} \quad u = 0 \quad \text{sur} \quad T_i^\varepsilon, \quad 1 \leq i \leq n(\varepsilon) \\ +\infty & \text{sinon} \end{cases}$$

écrire (2.17) et (2.20) est équivalent à affirmer la Γ-convergence, notion très féconde introduite par E. De Giorgi ([14] [18]), des fonctionnelles

$$F^\varepsilon(u) = \int_\Omega |\text{grad}\, u|^2 dx + 1_{T^\varepsilon}(u)$$

vers $F(u) = \int_\Omega |\text{grad}\, u|^2 dx + \int_\Omega u^2 d\mu.$

Démonstrations des Théorèmes 2.5 et 2.6. Dans le cas où l'on suppose $\mu \in W^{-1,\infty}(\Omega)$, les Théorèmes 2.5 et 2.6 sont démontrés dans [6], Théorèmes 1.2 et 3.4, Propositions 3.1 et 3.2 et Remarque 3.3. Pour démontrer ces théorèmes quand μ appartient seulement à $H^{-1}(\Omega)$, il suffit de suivre les démonstrations de [6] en utilisant le résultat de J. Deny (2.11), et en approchant z par des fonctions z_n telles que :

$$\begin{cases} z_n \in H_0^1(\Omega) \\ |z_n(x)| \leq \inf\{|z(x)|, n\} \\ z_n \to z \quad \text{dans} \quad H_0^1(\Omega) \quad \text{et} \quad L^1(\Omega; d\mu) \text{ fort.} \end{cases}$$

Les fonctions obtenues par troncature de z à la hauteur n conviennent parfaitement pour cela ; on peut également utiliser les approximations de L. Hedberg [21] (voir aussi [2], paragraphe 2), qui, elles, opèrent dans les espaces $W^{m,p}$ et permettent d'obtenir les analogues des Théorèmes 2.5 et 2.6 pour des opérateurs d'ordre supérieur [5] (voir aussi [13]).

Remarque 2.8. Revenons pour finir ce paragraphe à la situation géométrique décrite au paragraphe 1 (trous de taille a^ε répartis dans Ω avec une période 2ε), et examinons la situation en fonction des valeurs de a^ε.

Nous venons de voir que si a^ε vérifie (1.11), les solutions des problèmes de Dirichlet (1.5) convergent dans $H_0^1(\Omega)$ faible vers u

solution de

$$\begin{cases} -\Delta u + \mu u = f & \text{dans} \quad \Omega \\ u \in H_0^1(\Omega) \end{cases}$$

où μ est la constante définie par (2.5).

Si les trous T_i^ε sont *"plus petits"* que la taille définie par (1.11), i.e. si

$$(2.21) \qquad \begin{cases} a^\varepsilon / \varepsilon^{N/(N-2)} \to 0 & \text{si} \quad N \geq 3 \\ \varepsilon^2 \operatorname{Log} a^\varepsilon \to -\infty & \text{si} \quad N = 2 \end{cases}$$

on peut majorer $\|\operatorname{grad} w^\varepsilon\|_{(L^2_{\text{loc}}(\mathbb{R}^N))^N}$ par un calcul explicite où l'on remplace le trou T_i^ε par une boule $a^\varepsilon B_r$; on montre facilement que les w^ε définis par (1.10) vérifient $\operatorname{grad} w^\varepsilon \to 0$ dans $(L^2_{\text{loc}}(\mathbb{R}^N))^N$ fort ; c'est dire que

$$w^\varepsilon \to 1 \quad \text{dans} \quad H^1_{\text{loc}}(\mathbb{R}^N) \text{ fort.}$$

L'hypothèse (H) est alors vérifiée avec $\mu = 0$, et l'on a

$$(2.22) \qquad \tilde{u}^\varepsilon \to u \quad \text{dans} \quad H_0^1(\Omega) \text{ fort}$$

où u est la solution du problème de Dirichlet (classique puisque $\mu = 0$)

$$(2.23) \qquad \begin{cases} -\Delta u = f & \text{dans } \Omega \\ u \in H_0^1(\Omega). \end{cases}$$

Si au contraire, les trous T_i^ε sont *"plus gros"* que la taille définie par (1.11), i.e. si

$$(2.24) \qquad \begin{cases} a^\varepsilon / \varepsilon^{N/(N-2)} \to +\infty & \text{si} \quad N \geq 3 \\ \varepsilon^2 \operatorname{Log} a^\varepsilon \to 0 & \text{si} \quad N = 2 \end{cases}$$

on utilise (2.16) et (2.17) pour des trous, plus petits, de taille $C_0\ \varepsilon^{N/(N-2)}$ ou $\exp(-C_0/\varepsilon^2)$ qui sont contenus dans les T_i^ε ; on obtient $\liminf_{\varepsilon \to 0} \int_\Omega |\operatorname{grad} \tilde{u}^\varepsilon|^2 dx \geq \int_\Omega u^2 d\mu$, où μ est donné par (2.5) ; il suffit alors de faire tendre C_0 vers l'infini pour voir que les solutions u^ε des problèmes de Dirichlet (1.5) vérifient dans ce cas

$$(2.25) \qquad \tilde{u}^\varepsilon \rightharpoonup 0 \quad \text{dans} \quad H_0^1(\Omega) \text{ faible.}$$

L'étude détaillée de la situation (2.24) est effectuée dans [25], chapitre 3 : on y montre par exemple que si $a^\varepsilon = C_0\varepsilon^\alpha$ avec $1 < \alpha < N/(N-2)$ en dimension $N \geq 3$, on a

$$\frac{\tilde{u}^\varepsilon}{\varepsilon^{N-\alpha(N-2)}} \to \frac{f}{\mu} \quad \text{dans} \quad L^2(\Omega) \text{ fort} \tag{2.26}$$

où μ est donnée par (2.5).

Dans le *cas de la dimension* $N = 1$, on est toujours dans la situation (2.25) : l'injection compacte de $H^1(\Omega)$ dans $C^0(\bar{\Omega})$ en dimension 1 et le fait que $\tilde{u}^\varepsilon$ est nul sur des trous T_i^ε répartis dans Ω, entrainent qu'à la limite $u = 0$, aussi petits que soient les trous T_i^ε.

3. Estimation de l'erreur dans le cadre abstrait. Dans ce paragraphe, on démontre le théorème suivant, qui généralise le Théorème 1.1 dans le cadre de l'hypothèse (H) du paragraphe 2 (voir 2.8).

Théorème 3.1. *On se place sous l'hypothèse* (H), *et on considère les problèmes de Dirichlet* (2.12). *On suppose que la limite u des solutions, qui est définie par* (2.14), *vérifie*

$$u \in W^{2,\infty}(\Omega). \tag{3.1}$$

On a alors l'estimation :

$$\left\{\begin{array}{l} \|\text{grad}\,(u^\varepsilon - w^\varepsilon u)\|_{(L^2(\Omega^\varepsilon))^N} = \|\text{grad}\,(\tilde{u}^\varepsilon - w^\varepsilon u)\|_{L^2(\Omega))^N} \\ \leq [2\,\|\text{grad}\,u\|_{(L^\infty(\Omega))^N} + C_\Omega\,\|\Delta u\|_{L^\infty(\Omega)}]\ \|w^\varepsilon - 1\|_{L^2(\Omega)} \\ +[\ \|u\|_{L^\infty(\Omega)} + C_\Omega\,\|\text{grad}\,u\|_{(L^\infty(\Omega))^N}]\ \|\mu^\varepsilon - \mu\|_{H^{-1}(\Omega)}. \end{array}\right. \tag{3.2}$$

Dans cet énoncé $\tilde{u}^\varepsilon$ est la prolongement par 0 défini en (1.8), C_Ω désigne la constante de Poincaré de l'ouvert Ω, i.e. la plus petite constante telle que

$$\|v\|_{L^2(\Omega)} \leq C_\Omega\,\|\text{grad}\,v\|_{(L^2(\Omega))^N} \quad \forall v \in H_0^1(\Omega) \tag{3.3}$$

et on a choisi pour norme dans $H_0^1(\Omega)$ la norme du gradient, de sorte que

$$\|f\|_{H^{-1}(\Omega)} = \sup_{v \in H_0^1(\Omega)} \frac{< f, v >}{\|\text{grad}\, v\|_{(L^2(\Omega))^N}}. \tag{3.4}$$

DÉMONSTRATION. Comme u appartient à $W^{2,\infty}(\Omega)$ (il suffirait pour cela d'avoir $u \in W^{1,\infty}(\Omega)$) on a $\tilde{u}^\varepsilon - w^\varepsilon u \in H_0^1(\Omega)$, où $\tilde{u}^\varepsilon$ est le prolongement par 0 défini en (1.8). Définissons $\lambda^\varepsilon \in H^{-1}(\Omega)$ par

$$\lambda^\varepsilon = -\Delta(\tilde{u}^\varepsilon - w^\varepsilon u). \tag{3.5}$$

L'idée est d'écrire λ^ε d'une autre manière ; les calculs suivants sont licites (au sens de $\mathcal{D}'(\Omega)$) car u est assez régulière :

$$\begin{aligned}\lambda^\varepsilon &= -\Delta\tilde{u}^\varepsilon + w^\varepsilon \Delta u + 2\,\text{grad}\, w^\varepsilon\ \text{grad}\, u + u\ \Delta w^\varepsilon \\ &= -\Delta\tilde{u}^\varepsilon + 2\ \text{div}((w^\varepsilon - 1)\text{grad}\, u) - (w^\varepsilon - 1)\Delta u + \Delta u + u\ \Delta w^\varepsilon.\end{aligned}$$

Mais, d'après l'hypothèse (H), $-\Delta w^\varepsilon = \mu^\varepsilon - \gamma^\varepsilon$, et d'après le Théorème 2.5, $-\Delta u + \mu u = f$ dans $\mathcal{D}'(\Omega)$; si on définit $\eta^\varepsilon \in H^{-1}(\Omega)$ par

$$\eta^\varepsilon = -\Delta\tilde{u}^\varepsilon - f \tag{3.6}$$

on obtient donc

$$\lambda^\varepsilon = 2\ \text{div}((w^\varepsilon - 1)\text{grad}\, u) - (w^\varepsilon - 1)\Delta u - (\mu^\varepsilon - \mu)u + \gamma^\varepsilon u + \eta^\varepsilon. \tag{3.7}$$

Montrons que $\eta^\varepsilon \in H^{-1}(\Omega)$ ne charge que les trous T_i^ε, i.e. que

$$< \eta^\varepsilon, \tilde{v}^\varepsilon >= 0 \quad \text{si} \quad v^\varepsilon \in H_0^1(\Omega^\varepsilon) \ ; \tag{3.8}$$

ce résultat est naturel puisque, formellement, η^ε est égal à la somme de la restriction de f aux trous T_i^ε et de la mesure $\dfrac{\partial u^\varepsilon}{\partial n}$ concentrée sur les bords ∂T_i^ε ; pour démontrer (3.8) de façon précise, on utilise (3.6), (1.4) et les propriétés du prolongement par 0 : on écrit que pour tout $v \in H_0^1(\Omega)$

$$\begin{aligned}< \eta^\varepsilon, v > &= \int_\Omega \text{grad}\, \tilde{u}^\varepsilon \text{grad}\, v - \int_\Omega g\ \text{grad}\, v \\ &= \int_{\Omega^\varepsilon} \text{grad}\, u^\varepsilon\ \text{grad}\, v - \int_{\Omega^\varepsilon} g\ \text{grad}\, v - \sum_{i=1}^{n(\varepsilon)} \int_{T_i^\varepsilon} g\ \text{grad}\, v\ dx\end{aligned}$$

qui est nul d'après (1.6) si $v = \tilde{v}^\varepsilon$ où $v^\varepsilon \in H_0^1(\Omega^\varepsilon)$.

Multipliant (3.5) et (3.7) par $\tilde{u}^\varepsilon - w^\varepsilon u \in H_0^1(\Omega)$ et intégrant par parties, on obtient :

(3.9)
$$\int_\Omega |\mathrm{grad}\,(\tilde{u}^\varepsilon - w^\varepsilon u)|^2 = -2\int_\Omega (w^\varepsilon - 1)\mathrm{grad}\,u\ \mathrm{grad}\,(\tilde{u}^\varepsilon - w^\varepsilon u)$$
$$- \int_\Omega (w^\varepsilon - 1)\Delta u(\tilde{u}^\varepsilon - w^\varepsilon u) - < \mu^\varepsilon - \mu, u(\tilde{u}^\varepsilon - w^\varepsilon u) >$$
$$+ < \gamma^\varepsilon, u(\tilde{u}^\varepsilon - w^\varepsilon u) > + < \eta^\varepsilon, \tilde{u}^\varepsilon - w^\varepsilon u > .$$

Mais les deux derniers termes sont nuls, à cause de l'hypothèse (H) et de (3.8), car $\tilde{u} - w^\varepsilon u = 0$ sur les trous T_i^ε. Pour obtenir (3.2) il suffit alors de majorer le second membre de (3.9) par l'inégalité de Cauchy-Schwarz ; on notera que, à cause de (3.3), on a si $v \in H_0^1(\Omega)$ et $u \in W^{1,\infty}(\Omega)$:

$$\|uv\|_{H_0^1(\Omega)} \leq [\ \|u\|_{L^\infty(\Omega)} + C_\Omega \|\mathrm{grad}\,u\|_{(L^\infty(\Omega))^N}]\ \|\mathrm{grad}\,v\|_{(L^2(\Omega))^N}.$$

4. Démonstration du Théorème 1.1. Le Théorème 1.1 est une conséquence immédiate de l'estimation (3.2) donnée par le Théorème 3.1, (noter que les fonctions w^ε définies par (1.10) vérifient l'hypothèse (H) quand a^ε est donné par (1.11), voir Théorème 2.1), et des deux Propositions suivantes :

Proposition 4.1. *On se place dans la situation* (1.2) *et on suppose que* a^ε *est donné par* (1.11). *Alors, pour tout cube* Q *de* $\mathbb{R}^N$, *les fonctions* w^ε *définies par* (1.10) *vérifient*

$$\limsup_{\varepsilon\to 0} \frac{1}{\varepsilon}\|w^\varepsilon - 1\|_{L^2(Q)} = 0. \tag{4.1}$$

On verra dans la démonstration de cette Proposition qu'on peut en fait obtenir une estimation meilleure que (4.1), à savoir

$$\|w^\varepsilon - 1\|_{L^2(Q)} \leq C\ |Q|^{1/2}\ \varepsilon^p \qquad \text{pour } \varepsilon \text{ assez petit} \tag{4.2}$$

avec $p > 1$, où p ne dépend que de la dimension N (voir (4.9)), où $|Q| = \int_Q dx$ désigne la mesure N-dimensionnelle du cube Q, et où C est une constante qui ne dépend que de la dimension N, du trou T et de la constante C_0 qui apparaît dans (1.11).

Proposition 4.2. *On se place dans la situation* (1.2) *et on suppose que* a^ε *est donné par* (1.11). *Alors, si* μ *est la constante définie par* (2.5), *les distributions* μ^ε *définies par* (2.1) *vérifient pour tout cube* Q *de* $\mathbb{R}^N$

$$\limsup_{\varepsilon \to 0} \frac{1}{\varepsilon} \|\mu^\varepsilon - \mu\|_{H^{-1}(Q)} \leq C|Q|^{1/2} \tag{4.3}$$

où C *est une constante qui ne dépend que de la dimension* N, *du trou* T *et de la constante* C_0 *qui apparaît dans* (1.11).

Comme on le verra à la Remarque 4.7, l'estimation (4.3) est "optimale", car $\|\mu^\varepsilon - \mu\|_{H^{-1}(Q)}$ est de l'ordre de ε quand T est une boule.

Démonstration de la Proposition 4.1. La fonction $w^\varepsilon - 1$ est périodique de période 2ε dans chaque direction d'axe, c'est-à-dire est périodique de période P^ε

$$P^\varepsilon = \{x \in \mathbb{R}^N \mid |x_\ell| < \varepsilon, \quad \ell = 1, 2, \ldots N\}; \tag{4.4}$$

d'autre part, le cube Q est recouvert par (et contient) un nombre de périodes P^ε qui est équivalent à $|Q|/(2\varepsilon)^N$; on a donc pour ε assez petit

$$\|w^\varepsilon - 1\|^2_{L^2(Q)} = \int_Q |w^\varepsilon - 1|^2 dx \leq c\frac{|Q|}{(2\varepsilon)^N} \int_{P^\varepsilon} |w^\varepsilon - 1|^2 dx \tag{4.5}$$

où le facteur c peut être pris aussi proche de 1 que l'on veut.

Soit B_r une boule de $\mathbb{R}^N$ de centre 0, de rayon r, telle que $T \subset B_r$ et soit $\bar{w}^\varepsilon$ défini sur P^ε par

$$\begin{cases} \bar{w}^\varepsilon \in H^1(P^\varepsilon) & \\ \bar{w}^\varepsilon = 0 & \text{dans} \quad a^\varepsilon B_r \\ \bar{w}^\varepsilon = 1 & \text{dans} \quad P^\varepsilon \backslash \varepsilon B \\ \Delta \bar{w}^\varepsilon = 0 & \text{dans} \quad \varepsilon B \backslash a^\varepsilon B_r. \end{cases} \tag{4.6}$$

Si on compare cette définition avec (1.10), on constate que $\bar{w}^\varepsilon$ (qui n'est défini que sur P^ε, mais que l'on peut définir par périodicité sur $\mathbb{R}^N$ tout entier) est l'analogue (relatif au trou sphérique $a^\varepsilon B_r$) de w^ε (qui est relatif au trou $a^\varepsilon T$). Comme $a^\varepsilon T \subset a^\varepsilon B_r$, on a, par le principe du maximum

$$0 \leq 1 - w^\varepsilon \leq 1 - \bar{w}^\varepsilon \quad \text{dans} \quad P^\varepsilon,$$

ce qui avec (4.5) montre que pour ε assez petit

$$\|w^\varepsilon - 1\|^2_{L^2(Q)} \leq c \frac{|Q|}{(2\varepsilon)^N} \int_{P^\varepsilon} |\bar{w}^\varepsilon - 1|^2 \, dx. \tag{4.7}$$

Mais $\bar{w}^\varepsilon$ se calcule explicitement en coordonnées polaires : on a

$$\begin{cases} \bar{w}^\varepsilon(x) = \dfrac{\text{Log } |x| - \text{ Log } (ra^\varepsilon)}{\text{Log } \varepsilon - \text{ Log } (ra^\varepsilon)} & \text{si} \quad N = 2 \\[2ex] \bar{w}^\varepsilon(x) = \dfrac{\left(1/|x|\right)^{N-2} - \left(1/ra^\varepsilon\right)^{N-2}}{\left(1/\varepsilon\right)^{N-2} - \left(1/ra^\varepsilon\right)^{N-2}} & \text{si} \quad N \geq 3. \end{cases} \tag{4.8}$$

Un calcul explicite, facile mais un peu long, montre alors que si a^ε vérifie (1.11), on a :

$$\frac{1}{(2\varepsilon)^N} \int_{P^\varepsilon} |\bar{w}^\varepsilon - 1|^2 \sim \begin{cases} \dfrac{2\pi}{16} \dfrac{1}{C_0^2} \varepsilon^4 & \text{si} \quad N = 2 \\[2ex] \dfrac{S_3}{24} C_0^2 r^2 \varepsilon^4 & \text{si} \quad N = 3 \\[2ex] \dfrac{S_4}{16} C_0^4 r^4 \varepsilon^4 |\text{Log } \varepsilon| & \text{si} \quad N = 4 \\[2ex] \dfrac{S_N}{2^N} \dfrac{2(N-2)}{N(N-4)} C_0^N r^N \varepsilon^{2N/(N-2)} & \text{si} \quad N \geq 5 \end{cases} \tag{4.9}$$

où l'on a désigné par $S_N = \int_{|x|=1} ds$ la mesure $(N-1)$-dimensionnelle de la sphère unité de $\mathbb{R}^N$ ($S_2 = 2\pi$).

De (4.7) et (4.9) on déduit l'estimation (4.2) qui entraine (4.1).

Démonstration de la Proposition 4.2. Elle est un peu plus compliquée que celle de la Proposition 4.1, et nous allons la diviser en trois étapes.

1ère étape. Le but de cette étape est d'établir l'estimation (qui sera enoncée de façon plus précise au Lemme 4.3)

$$\|h(\frac{x}{\varepsilon})\|_{H^{-1}(Q)} \leq C\ \varepsilon$$

quand h est une distribution de $H^{-1}_{\text{loc}}(\mathbb{R}^N)$, périodique et de moyenne nulle. Cette estimation est due à R.V. Kohn & M. Vogelius [26].

Pour être précis, soit h une distribution de $H^{-1}_{\text{loc}}(\mathbb{R}^N)$ qui est périodique de période 2 dans chaque direction d'axe, c'est-à-dire qui est périodique de période P

$$P = \{x \in \mathbb{R}^N |\ |x_\ell| < 1, \quad \ell = 1, 2, \ldots, N\} \tag{4.10}$$

où P doit en fait être considéré comme un tore.

Définissons la distribution h^ε par :

$$< h^\varepsilon, \varphi >= \varepsilon^N < h(x), \varphi(\varepsilon x) > \quad \forall \varphi \in \mathcal{D}(\mathbb{R}^N)\ ; \tag{4.11}$$

formellement, h^ε n'est autre que la distribution $h(x/\varepsilon)$; en effet, si h appartient à $L^1_{\text{loc}}(\mathbb{R}^N)$, on peut écrire les dualités comme des intégrales, et on a d'une part

$$< h^\varepsilon, \varphi >= \int_{\mathbb{R}^N} h^\varepsilon(x)\varphi(x)dx$$

et d'autre part, par le changement de variable $x \to x/\varepsilon$:

$$\begin{aligned} < h^\varepsilon, \varphi > &= \varepsilon^N < h(x), \varphi(\varepsilon x) >= \varepsilon^N \int_{\mathbb{R}^N} h(x)\varphi(\varepsilon x)dx \\ &= \int_{\mathbb{R}^N} h\left(\frac{x}{\varepsilon}\right)\varphi(x)dx. \end{aligned}$$

Considérons maintenant le problème : trouver v tel que

$$(4.12) \qquad \begin{cases} v \in H^1_{\text{loc}}(\mathbb{R}^N), & v \text{ périodique de période } P, \\ -\Delta v = h & \text{dans } \mathcal{D}'(\mathbb{R}^N). \end{cases}$$

Si l'on définit $H^1_p(P)$ comme l'espace des fonctions de $H^1(P)$ qui sont les restrictions à P de fonctions de $H^1_{\text{loc}}(\mathbb{R}^N)$, périodiques de période P, on démontre que résoudre (4.12) est équivalent à résoudre le problème variationnel suivant : trouver v tel que

$$(4.13) \qquad \begin{cases} v \in H^1_p(P) \\ \int_P \operatorname{grad} v \operatorname{grad} \varphi \, dx = < h, \varphi >_P \quad \forall \varphi \in H^1_p(P) \end{cases}$$

où $< \, , \, >_P$ désigne la dualité entre $H^1_p(P)$ et son dual. On démontre aussi que (4.13) (et donc (4.12)) a une solution si et seulement si h est de moyenne nulle sur P, i.e. si et seulement si

$$(4.14) \qquad < h, 1 >_P = 0 \; ;$$

formellement (4.14) n'est autre que $\int_P h(x)dx = 0$; la fonction v est alors déterminée à une constante additive près, qui est fixée si on impose

$$(4.15) \qquad \int_P v(x)dx = 0 \; ;$$

dans tout ce qui précède, P doit être considéré comme un tore.

On a alors le lemme suivant, dû à R.V. Kohn & M. Vogelius [26] :

Lemme 4.3. *Soit $h \in H^{-1}_{\text{loc}}(\mathbb{R}^N)$ périodique de période P et vérifiant (4.14). On définit v par (4.12), (4.15). Alors, si h^ε est défini par (4.11), on a, pour tout cube Q de $\mathbb{R}^N$*

$$(4.16) \qquad \limsup_{\varepsilon \to 0} \frac{1}{\varepsilon} \|h^\varepsilon\|_{H^{-1}(Q)} \leq \left(\frac{|Q|}{2^N} \right)^{1/2} \|\operatorname{grad} v\|_{(L^2(P))^N}.$$

Remarque 4.4. En fait la démonstration montre que si $h \neq 0$ est périodique, régulière et de moyenne nulle (cf. (4.14)), $\|h(x/\varepsilon)\|_{H^{-1}(Q)}$ est exactement de l'ordre de ε ; ce résultat complète les résultats bien connus qui affirment que $\|h(x/\varepsilon)\|_{L^2(Q)}$ est de l'ordre de 1, et que $\|h(x/\varepsilon)\|_{H^1(Q)}$ est de l'ordre de $1/\varepsilon$.

Démonstration du Lemme 4.3. Posons

$$v^\varepsilon(x) = v\left(\frac{x}{\varepsilon}\right).$$

Cette fonction est périodique, de période $P^\varepsilon = \varepsilon P$. Comme un cube Q de $\mathbb{R}^N$ est recouvert par (et contient) un nombre de périodes P^ε qui est équivalent à $|Q|/(2\varepsilon)^N$, on a, par le changement de variable $x \longrightarrow x/\varepsilon$:

$$\begin{aligned}
\|\operatorname{grad} v^\varepsilon\|^2_{(L^2(Q))^N} &= \int_Q |\operatorname{grad} v^\varepsilon|^2(x)dx \\
&\leq c\frac{|Q|}{(2\varepsilon)^N}\int_{P^\varepsilon} |\operatorname{grad} v^\varepsilon|^2(x)dx \\
&= c\frac{|Q|}{(2\varepsilon)^N}\int_{P^\varepsilon} |\frac{1}{\varepsilon}\operatorname{grad} v|^2\left(\frac{x}{\varepsilon}\right)dx \\
&= c\frac{|Q|}{2^N}\frac{1}{\varepsilon^2}\int_P |\operatorname{grad} v|^2(x)dx,
\end{aligned}$$

où le facteur c peut être pris aussi proche de 1 que l'on veut quand ε est assez petit. Ceci entraîne que

$$\text{(4.17)} \quad \limsup_{\varepsilon\to 0} \|\varepsilon \operatorname{grad} v^\varepsilon\|_{(L^2(Q))^N} \leq \left(\frac{|Q|}{2^N}\right)^{1/2} \|\operatorname{grad} v\|_{(L^2(P))^N}.$$

D'autre part, on vérifie facilement à partir de (4.12) et (4.11) que l'on a

$$-\Delta(v^\varepsilon) = \frac{1}{\varepsilon^2}h^\varepsilon \quad \text{dans} \quad \mathcal{D}'(\mathbb{R}^N),$$

ce que l'on peut encore écrire sous la forme

$$\frac{1}{\varepsilon}h^\varepsilon = -\operatorname{div}(\varepsilon \operatorname{grad} v^\varepsilon) \quad \text{dans} \quad \mathcal{D}'(\mathbb{R}^N).$$

De cette égalité et de la majoration (4.17), on déduit immédiatement (4.16), ce qui achève la démonstration du Lemme 4.3 et la 1ère étape.

2ème étape. Considérons les fonctions w^ε et les distributions μ^ε définies par (1.10) et (2.1), et définissons, pour chaque ε, un réel m^ε par

$$m^\varepsilon = \frac{1}{(2\varepsilon)^N} \int_{P^\varepsilon} |\mathrm{grad}\, w^\varepsilon|^2(x)\, dx. \tag{4.18}$$

A partir du Lemme 4.3, nous allons démontrer le lemme suivant :

Lemme 4.5. *On suppose que a^ε est donné par* (1.11). *Alors pour tout cube Q de $\mathbb{R}^N$ on a*

$$\limsup_{\varepsilon \to 0} \frac{1}{\varepsilon} \|\mu^\varepsilon - m^\varepsilon\|_{H^{-1}(Q)} \leq C|Q|^{1/2} \tag{4.19}$$

où C est une constante qui ne dépend que de T, de la dimension N et de la constante C_0 qui apparaît dans (1.11).

Démonstration du Lemme 4.5.
a) On introduit la fonction z^ε définie sur P par

$$\begin{cases} z^\varepsilon \in H^1(P) & \\ z^\varepsilon = 0 & \text{dans } \dfrac{a^\varepsilon}{\varepsilon} T \\ z^\varepsilon = 1 & \text{dans } P \backslash B \\ \Delta z^\varepsilon = 0 & \text{dans } B \backslash \dfrac{a^\varepsilon}{\varepsilon} T \end{cases} \tag{4.20}$$

que l'on prolonge par périodicité en une fonction de $H^1_{\mathrm{loc}}(\mathbb{R}^N)$, périodique de période P. On définit ensuite, pour ε fixé, la distribution g^ε par :
(4.21)

$$< g^\varepsilon, \varphi > = \sum_{k \in \mathbf{Z}^N} \frac{1}{\varepsilon^2} \int_{\partial B_k} \frac{\partial z^\varepsilon}{\partial n} \varphi \, ds - \frac{1}{2^N \varepsilon^2} \left(\int_{\partial B} \frac{\partial z^\varepsilon}{\partial n} ds \right) \int_{\mathbb{R}^N} \varphi(x) dx \quad \forall \varphi \in \mathcal{D}(\mathbb{R}^N)$$

où B_k est la boule de $\mathbb{R}^N$ de rayon 1 et de centre $\sum_1^N 2k_\ell\, e_\ell$ (noter que les intégrales ont un sens puisque z^ε est harmonique dans la couronne $B\backslash\frac{a^\varepsilon}{\varepsilon}T$ et vaut 1 sur ∂B). La distribution g^ε appartient à $H^{-1}_{\text{loc}}(\mathbb{R}^N)$ et est périodique de période P : elle est définie sur sa période P par :

$$g^\varepsilon = \frac{1}{\varepsilon^2}\frac{\partial z^\varepsilon}{\partial n}\bigg|_{\partial B}\delta_{\partial B} - \frac{1}{2^N\varepsilon^2}\left(\int_{\partial B}\frac{\partial z^\varepsilon}{\partial n}ds\right)$$

où $\delta_{\partial B}$ est la masse de Dirac concentrée sur la sphère $|x| = 1$. On a évidemment (noter que $|P| = 2^N$) :

$$< g^\varepsilon, 1 >_P = \int_P g^\varepsilon dx = 0.$$

C'est dire que g^ε vérifie (4.14). Pour chaque ε il existe donc (cf.(4.12), (4.15)) une fonction q^ε telle que :
(4.22)

$$\begin{cases} q^\varepsilon \in H^1_{\text{loc}}(\mathbb{R}^N), & q^\varepsilon \text{ périodique de période P}, \quad \int_P q^\varepsilon(x)dx = 0 \\ -\Delta q^\varepsilon = g^\varepsilon & \text{dans} \quad \mathcal{D}'(\mathbb{R}^N). \end{cases}$$

Si on définit la distribution h^ε par

$$(4.23) \qquad < h^\varepsilon, \varphi > = \varepsilon^N < g^\varepsilon(x), \varphi(\varepsilon x) > \qquad \forall\varphi \in \mathcal{D}(\mathbb{R}^N)$$

c'est-à-dire (voir (4.11)) si on pose formellement

$$h^\varepsilon(x) = g^\varepsilon\left(\frac{x}{\varepsilon}\right),$$

on a, d'après une variante du Lemme 4.3

$$(4.24) \quad \limsup_{\varepsilon\to 0}\frac{1}{\varepsilon}\|h^\varepsilon\|_{H^{-1}(Q)} \le \left(\frac{|Q|}{2^N}\right)^{1/2}\limsup_{\varepsilon\to 0}\|\text{grad}\, q^\varepsilon\|_{(L^2(P))^N}.$$

Utilisant z^ε comme fonction test dans (4.20), on voit que

$$\int_{\partial B}\frac{\partial z^\varepsilon}{\partial n}ds = \int_{B\backslash\frac{a^\varepsilon}{\varepsilon}T}|\text{grad}\, z^\varepsilon|^2 dx,$$

et comme z^ε est défini par (4.20) et w^ε par (1.10), on a

$$w^\varepsilon(x) = z^\varepsilon\left(\frac{x}{\varepsilon}\right).$$

Le changement de variable $x \to x/\varepsilon$ montre alors que h^ε défini par (4.23) et (4.21) vérifie

$$h^\varepsilon = \mu^\varepsilon - m^\varepsilon \tag{4.25}$$

où la distribution μ^ε est définie par (2.1) et le réel m^ε par (4.18).

b) Majorons maintenant $\|\operatorname{grad} q^\varepsilon\|_{(L^2(P))^N}$, où q^ε est la solution de (4.22).

En utilisant q^ε comme fonction test dans la formulation variationnelle équivalente à (4.22) (voir (4.12) et (4.13)), et le fait que q^ε est de moyenne nulle, on a d'après la définition (4.21) de g^ε :

(4.26)
$$\begin{aligned}\int_P |\operatorname{grad} q^\varepsilon|^2 dx &= < g^\varepsilon, q^\varepsilon >_P \\ &= \frac{1}{\varepsilon^2}\int_{\partial B}\frac{\partial z^\varepsilon}{\partial n} q^\varepsilon ds - \frac{1}{2^N\varepsilon^2}(\int_{\partial B}\frac{\partial z^\varepsilon}{\partial n}\, ds)\int_P q^\varepsilon dx \\ &= \frac{1}{\varepsilon^2}\int_{\partial B}\frac{\partial z^\varepsilon}{\partial n} q^\varepsilon ds.\end{aligned}$$

On majore le second membre de (4.26) par

$$\frac{1}{\varepsilon^2}\|\frac{\partial z^\varepsilon}{\partial n}\|_{L^\infty(\partial B)}\|q^\varepsilon\|_{L^1(\partial B)} \leq \frac{1}{\varepsilon^2}\|\frac{\partial z^\varepsilon}{\partial n}\|_{L^\infty(\partial B)} C_N \|\operatorname{grad} q^\varepsilon\|_{(L^2(P))^N}$$

où C_N est la norme (qui ne dépend que de N) de l'application qui à une fonction de $H^1(P)$, de moyenne nulle, associe sa trace dans $L^1(\partial B)$. On déduit de (4.26) que

$$\|\operatorname{grad} q^\varepsilon\|_{(L^2(P))^N} \leq C_N \frac{1}{\varepsilon^2}\|\frac{\partial z^\varepsilon}{\partial n}\|_{L^\infty(\partial B)}. \tag{4.27}$$

Majorons le second membre de (4.27). Pour cela on considère une boule B_r de $\mathbb{R}^N$, de centre 0 et de rayon r, telle que $T \subset B_r$; on définit $\bar{z}^\varepsilon$ sur P par

$$\begin{cases} \bar{z}^\varepsilon \in H^1(P) & \\ \bar{z}^\varepsilon = 0 & \text{dans} \quad \dfrac{a^\varepsilon}{\varepsilon} B_r \\ \bar{z}^\varepsilon = 1 & \text{dans} \quad P \backslash B \\ \Delta \bar{z}^\varepsilon = 0 & \text{dans} \quad B \backslash \dfrac{a^\varepsilon}{\varepsilon} B_r \end{cases}$$

que l'on prolonge à $H^1_{\text{loc}}(\mathbb{R}^N)$ par périodicité de période P. La fonction $\bar{z}^\varepsilon$ est l'analogue de z^ε défini par (4.20), où le trou T a été remplacé par la boule B_r.

D'après le principe du maximum on a

$$0 \leq \bar{z}^\varepsilon \leq z^\varepsilon \leq 1 \quad \text{dans} \quad B \backslash \frac{a^\varepsilon}{\varepsilon} B_r$$

(puisque $z^\varepsilon \geq 0$ sur $\dfrac{a^\varepsilon}{\varepsilon} \partial B_r$) ; comme $\bar{z}^\varepsilon = z^\varepsilon = 1$ sur ∂B, on a donc

$$\frac{\partial \bar{z}^\varepsilon}{\partial n} \geq \frac{\partial z^\varepsilon}{\partial n} \geq 0 \quad \text{sur} \quad \partial B \tag{4.28}$$

(noter que $\bar{z}^\varepsilon$ et z^ε sont harmoniques, donc régulières au voisinage de ∂B).

Mais $\left.\dfrac{\partial \bar{z}^\varepsilon}{\partial n}\right|_{\partial B}$ se calcule explicitement par un calcul en coordonnées polaires ; c'est une constante donnée par :

$$\begin{cases} \left.\dfrac{\partial \bar{z}^\varepsilon}{\partial n}\right|_{\partial B} = 1/\operatorname{Log}\left(\dfrac{\varepsilon}{r a^\varepsilon}\right) & \text{si} \quad N = 2 \\[2ex] \left.\dfrac{\partial \bar{z}^\varepsilon}{\partial n}\right|_{\partial B} = \dfrac{N-2}{(\varepsilon / r a^\varepsilon)^{N-2} - 1} & \text{si} \quad N \geq 3. \end{cases}$$

Si a^ε est donné par (1.11) on a alors

$$\begin{cases} \dfrac{1}{\varepsilon^2} \left.\dfrac{\partial \bar{z}^\varepsilon}{\partial n}\right|_{\partial B} \to \dfrac{1}{C_0} & \text{si} \quad N = 2 \\[2ex] \dfrac{1}{\varepsilon^2} \left.\dfrac{\partial \bar{z}^\varepsilon}{\partial n}\right|_{\partial B} \to (N-2) C_0^{N-2} r^{N-2} & \text{si} \quad N \geq 3. \end{cases} \tag{4.29}$$

En rassemblant (4.27), (4.28) et (4.29) on voit que si $N \geq 3$ on a, quand a^ε est donné par (1.11)

$$\limsup_{\varepsilon \to 0} \|\operatorname{grad} q^\varepsilon\|_{(L^2(P))^N} \leq C_N(N-2)C_0^{N-2} r^{N-2} \tag{4.30}$$

et un résultat analogue pour $N = 2$.

On déduit alors immédiatement le Lemme 4.5 de (4.25), (4.24) et (4.30).

3ème étape. Pour démontrer la Proposition 4.2, il suffit maintenant, vu le Lemme 4.5, d'écrire

$$\mu^\varepsilon - \mu = \mu^\varepsilon - m^\varepsilon + m^\varepsilon - \mu$$

et de majorer $|m^\varepsilon - \mu|_{\mathbb{R}}$ par un terme en ε ; on a en effet pour tout cube Q de $\mathbb{R}^N$ et tout réel t

$$\begin{aligned} \|t\|_{H^{-1}(Q)} &= \sup_{\varphi \in H_0^1(Q)} \frac{|t| \int_Q |\varphi(x)| dx}{\|\operatorname{grad} \varphi\|_{(L^2(Q))^N}} \\ &= C_N |Q|^{(N+2)/2N} |t|_{\mathbb{R}} \end{aligned} \tag{4.31}$$

où C_N est une constante qui ne dépend que de N et qui est liée à la norme de l'injection de $H_0^1(P)$ dans $L^1(P)$.

La démonstration du lemme suivant terminera donc la démonstration de la Proposition 4.2.

Lemme 4.6. *Si* a^ε *est donné par* (1.11), *les réels* m^ε *et* μ *définis par* (4.18) *et* (2.5) *vérifient*

$$\limsup_{\varepsilon \to 0} \frac{1}{\varepsilon} |m^\varepsilon - \mu|_{\mathbb{R}} = 0. \tag{4.32}$$

On verra dans la démonstration de ce lemme qu'on peut en fait obtenir une estimation meilleure que (4.32), à savoir, pour ε assez petit

$$\begin{cases} |m^\varepsilon - \mu| \leq C\, \varepsilon^2 & \text{si} \quad N \geq 3 \\ |m^\varepsilon - \mu| \leq C\, \varepsilon^2 |\operatorname{Log} \varepsilon| & \text{si} \quad N = 2\ , \end{cases} \tag{4.33}$$

où C est une constante qui ne dépend que de la dimension N, du trou T et de la constante C_0 qui apparaît dans (1.11).

Démonstration du Lemme 4.6. Pour démontrer ce lemme, il est commode de faire le changement de variable $x \to x/a^\varepsilon$ qui ramène le trou $T^\varepsilon = a^\varepsilon T$ au trou modèle T. On introduit donc la fonction y^ε définie dans $\mathbb{R}^N$ par

$$\begin{cases} y^\varepsilon \in H^1_{\text{loc}}(\mathbb{R}^N) & \\ y^\varepsilon = 0 & \text{dans } T \\ y^\varepsilon = 1 & \text{dans } \mathbb{R}^N \backslash \dfrac{\varepsilon}{a^\varepsilon} B \\ \Delta y^\varepsilon = 0 & \text{dans } \dfrac{\varepsilon}{a^\varepsilon} B \backslash T\,. \end{cases} \tag{4.34}$$

Il est alors facile de voir que m^ε défini par (4.18) s'écrit

$$m^\varepsilon = \frac{(a^\varepsilon)^{N-2}}{(2\varepsilon)^N} \int_{\mathbb{R}^N} |\operatorname{grad} y^\varepsilon|^2 dx. \tag{4.35}$$

La suite de la démonstration diffère si $N = 2$ ou si $N \leq 3$.

Etudions d'abord le cas $N = 2$. Si on suppose que T est une boule de rayon r, il est facile de calculer explicitement la solution y^ε de (4.34) et m^ε donné par (4.35). On trouve

$$m^\varepsilon = \frac{2\pi}{(2\varepsilon)^2} \frac{1}{\text{Log } (\varepsilon / r a^\varepsilon)}$$

qui, lorsque $a^\varepsilon = \exp(-C_0/\varepsilon^2)$ conduit à

$$m^\varepsilon - \mu = \frac{\pi}{2C_0} \frac{\varepsilon^2 \text{ Log } (r/\varepsilon)}{C_0 - \varepsilon^2 \text{ Log } (r/\varepsilon)},$$

ce qui implique (4.32) quand T est une boule de centre 0 et de rayon r. Comme nous avons supposé que T est contenu dans une boule de rayon r et contient une boule de rayon ρ, on obtient facilement le Lemme 4.6 par comparaison puisque, pour chaque ε, m^ε est un nombre qui est croissant avec T.

Passons maintenant au cas $N \geq 3$. Quand a^ε est donné par (1.11), on a $\dfrac{(a^\varepsilon)^{N-2}}{(2\varepsilon)^N} = \dfrac{C_0^{N-2}}{2^N}$, de sorte que

$$m^\varepsilon - \mu = \frac{C_0^{N-2}}{2^N}\left[\int_{\mathbb{R}^N} |\text{grad}\, y^\varepsilon|^2 dx - Cap\ T\right]. \tag{4.36}$$

Mais quand $N \geq 3$, il existe une fonction $p \in D^{1,2}$ (l'espace obtenu par complétion de $\mathcal{D}(\mathbb{R}^N)$ pour la norme $(\int_{\mathbb{R}^N} |\text{grad}\, \varphi|^2 dx)^{1/2}$) dont l'énergie est égale à la capacité de T, i.e. telle que

$$\begin{cases} p \in D^{1,2}(\mathbb{R}^N), \quad p = 1 \quad \text{sur} \quad T \\ \displaystyle\int_{\mathbb{R}^N} |\text{grad}\, p|^2 dx = Cap\ T\ ; \end{cases} \tag{4.37}$$

la fonction p est le potentiel capacitaire de T ; elle est harmonique dans $\mathbb{R}^N \backslash T$. Posons

$$y = 1 - p. \tag{4.38}$$

La fonction y^ε étant nulle sur T, et harmonique dans $\dfrac{\varepsilon}{a^\varepsilon} B\backslash T$, on a par intégration par parties :

$$\int_{\mathbb{R}^N} |\text{grad}\, y^\varepsilon|^2 dx = \int_{\frac{\varepsilon}{a^\varepsilon} B\backslash T} |\text{grad}\, y^\varepsilon|^2 dx = \int_{\frac{\varepsilon}{a^\varepsilon}\partial B} \frac{\partial y^\varepsilon}{\partial n} 1\ ds.$$

D'autre part, la fonction y étant harmonique dans $\mathbb{R}^N \backslash T$ et $y - y^\varepsilon$ étant nul sur ∂T et à l'infini, on a par intégration par parties

$$\int_{\mathbb{R}^N \backslash T} \text{grad}\, y\ \text{grad}\,(y - y^\varepsilon) dx = 0\ ;$$

on en déduit que

$$\int_{\mathbb{R}^N} |\text{grad}\, y|^2 dx = \int_{\mathbb{R}^N \backslash T} |\text{grad}\, y|^2 dx = \int_{\mathbb{R}^N \backslash T} \text{grad}\, y\ \text{grad}\, y^\varepsilon\ dx$$
$$= \int_{\frac{\varepsilon}{a^\varepsilon} B\backslash T} \text{grad}\, y\ \text{grad}\, y^\varepsilon\ dx.$$

Une nouvelle intégration par parties, tenant compte du fait que y^ε est harmonique dans $\frac{\varepsilon}{a^\varepsilon}B\backslash T$ et que y est nulle sur le trou T donne

$$\int_{\frac{\varepsilon}{a^\varepsilon}B\backslash T} \operatorname{grad} y \operatorname{grad} y^\varepsilon \, dx = \int_{\frac{\varepsilon}{a^\varepsilon}\partial B} \frac{\partial y^\varepsilon}{\partial n} \, y \, ds.$$

De ces calculs on déduit que

(4.39)

$$\begin{aligned} &\int_{\mathbb{R}^N} |\operatorname{grad} y^\varepsilon|^2 dx - Cap\ T \\ &= \int_{\mathbb{R}^N} |\operatorname{grad} y^\varepsilon|^2 dx - \int_{\mathbb{R}^N} |\operatorname{grad} y|^2 dx = \int_{\frac{\varepsilon}{a^\varepsilon}\partial B} \frac{\partial y^\varepsilon}{\partial n}(1-y) ds. \end{aligned}$$

Soit maintenant B_r une boule de centre 0 et de rayon r telle que $T \subset B_r$, et soient $\bar{y}^\varepsilon$ et $\bar{y}$ les analogues relatifs à B_r et non plus à T, de y^ε et y (définis par (4.34) et (4.37), (4.38)). On a $y^\varepsilon = \bar{y}^\varepsilon = 1$ sur $\frac{\varepsilon}{a^\varepsilon}\partial B$ et $y^\varepsilon \geq 0 = \bar{y}^\varepsilon$ sur ∂B_r ; les deux fonctions étant harmoniques dans $\frac{\varepsilon}{a^\varepsilon}B\backslash B_r$, on a donc $1 \geq y_\varepsilon \geq \bar{y}^\varepsilon$ dans cette couronne, d'où

$$0 \leq \frac{\partial y^\varepsilon}{\partial n} \leq \frac{\partial \bar{y}^\varepsilon}{\partial n} \quad \text{sur} \quad \frac{\varepsilon}{a^\varepsilon}\partial B.$$

Un raisonnement analogue montre que

$$0 \leq 1 - y \leq 1 - \bar{y} \quad \text{dans} \quad \mathbb{R}^N.$$

Mais les fonctions $\bar{y}^\varepsilon$ et $\bar{y}$ se calculent explicitement : on a

$$\frac{\partial \bar{y}^\varepsilon}{\partial n} = \frac{(N-2)}{(1/r)^{N-2} - (a^\varepsilon/\varepsilon)^{N-2}} \frac{1}{|x|^{N-1}} \qquad 1 - \bar{y}(x) = (r/|x|)^{N-2}$$

et (4.39) implique donc que

$$\begin{aligned} 0 &\leq \int_{\mathbb{R}^N} |\operatorname{grad} y^\varepsilon|^2 dx - Cap\ T \leq \int_{\frac{\varepsilon}{a^\varepsilon}\partial B} \frac{\partial \bar{y}^\varepsilon}{\partial n}(1-\bar{y}) ds \\ &\leq \frac{(N-2)S_N}{(1/r)^{N-2} - (a^\varepsilon/\varepsilon)^{N-2}} \left(\frac{r\, a^\varepsilon}{\varepsilon}\right)^{N-2}. \end{aligned} \tag{4.40}$$

Lorsque a^ε est donné par (1.11), ce dernier nombre est équivalent à $(N-2)\ S_N\ r^{2N-4}\ C_0^{N-2}\ \varepsilon^2$, ce qui, joint à (4.36), démontre le Lemme 4.6 quand $N \geq 3$.

Remarque 4.7. En considérant le cas où T est une boule, il est facile de montrer que l'estimation (4.3) de la Proposition 4.2 est optimale, i.e. que $\|\mu^\varepsilon - \mu\|_{H^{-1}(Q)}$ est exactement de l'ordre de ε.

En effet, quand T est la boule de rayon 1 et quand a^ε est donné par (1.11), des calculs explicites en coordonnées polaires (voir si nécessaire [6], paragraphe 2) montrent que, pour $N \geq 3$

$$\mu = \frac{S_N(N-2)C_0^{N-2}}{2^N}$$

$$m^\varepsilon = \frac{1}{1 - C_0^{N-2}\varepsilon^2}\mu$$

$$\mu^\varepsilon = \frac{1}{1 - C_0^{N-2}\varepsilon^2}(N-2)C_0^{N-2}\sum_i \varepsilon\ \delta_i^\varepsilon$$

où δ_i^ε désigne la masse de Dirac portée par la sphère ∂B_i^ε, de rayon ε.

Ecrivons

$$\mu^\varepsilon - \mu = \mu^\varepsilon - m^\varepsilon + m^\varepsilon - \mu$$

et notons d'une part que $m^\varepsilon - \mu$ est un réel de l'ordre de ε^2, et d'autre part que

$$\mu^\varepsilon - m^\varepsilon = \frac{(N-2)C_0^{N-2}}{1 - C_0^{N-2}\varepsilon^2}\left(\sum_i \varepsilon\ \delta_i^\varepsilon - \frac{S_N}{2^N}\right)\ ;$$

on voit alors que montrer que $\|\mu^\varepsilon - \mu\|_{H^{-1}(Q)}$ est exactement de l'ordre de ε est équivalent à montrer que $\|\sum_i \varepsilon\delta_i^\varepsilon - \frac{S_N}{2^N}\|_{H^{-1}(Q)}$ est exactement de l'ordre de ε.

Mais si $v \in H^1_{\text{loc}}(\mathbb{R}^N)$ est la fonction périodique, de période P, solution de l'équation (cf. (4.12))

$$-\Delta v = \sum_i \delta_i - \frac{S_N}{2^N} \quad \text{dans} \quad \mathcal{D}'(\mathbb{R}^N)$$

où δ_i désigne la masse de Dirac portée par la sphère ∂B_i de rayon 1, et si $v^\varepsilon(x) = v(\frac{x}{\varepsilon})$, on a (voir la démonstration du Lemme 4.3)

$$\sum_i \varepsilon\, \delta_i^\varepsilon - \frac{S_N}{2^N} = -\varepsilon \operatorname{div}(\varepsilon \operatorname{grad} v^\varepsilon) \quad \text{dans} \quad \mathcal{D}'(\mathbb{R}^N) ;$$

le fait que $\|\varepsilon \operatorname{grad} v^\varepsilon\|_{(L^2(Q))^N}$ est exactement de l'ordre de 1 montre le résultat désiré.

Bibliographie

[1] H.Attouch, C.Picard, *Asymptotic analysis of variational problems with constraints of obstacle type*, J. Funct. Anal. **15** (1983), 329-386.

[2] H.Brézis, F.E.Browder, *Some properties of higher order Sobolev spaces*, J. Math. pures et appl. **61** (1982), 245-259.

[3] L.Carbone, F.Colombini, *On convergence of functionals with unilateral constraints*, J. Math. pures et appl. **59** (1980), 465-500.

[4] G.Choquet, *Potentiels sur un ensemble de capacité nulle. Suites de potentiels*, C.R. Acad. Sc. Paris **244** (1957), 1707-1710.

[5] D.Cioranescu, L.Hedberg, F.Murat, *Le bilaplacien et le tapis du fakir,* (1983), non publié.

[6] D.Cioranescu, F.Murat, *Un terme étrange venu d'ailleurs I & II,* in *Nonlinear partial differential equations and their applications, Collège de France Seminar, Vol. II & III,* ed. by H. Brézis & J.L. Lions, Res. Notes in Math. **60** & **70**, Pitman, London, (1982), 98-138 & 154-178.

[7] J.Deny, *Les potentiels d'énergie finie*, Acta Math. **82** (1950), 107-183.

[8] G.Dal Maso, *Asymptotic behaviour of minimum problems with bilateral obstacles,* Ann. Mat. pura ed appl. **129** (1981), 327-366.

[9] G.Dal Maso, *Limits of minimum problems for general integral functionals with unilateral obstacles,* Atti Accad. Naz. Lincei, Rend. Cl. Sci. Fis. Mat. Natur. **74** (1983), 56-61.

[10] G.Dal Maso, *Γ-convergence and μ-capacities,* Ann. Sc. Norm. Sup. Pisa **14** (1987), 423-464.

[11] G.Dal Maso, A.Defranceschi, *Limits of nonlinear Dirichlet problems in varying domains,* Preprint SISSA, Trieste, (1987).

[12] G.Dal Maso, P.Longo, *Γ-limits of obstacles,* Ann. Mat. pura ed appl. **128** (1981), 1-50.

[13] G.Dal Maso, G.Paderni, *Variational inequalities for the biharmonic operator with variable obstacles,* Preprint SISSA, Trieste, (1987).

[14] E.De Giorgi, *Sulla convergenza di alcune successioni di integrali del tipo dell' area,* Rendiconti di Mat. **8** (1975), 277-294.

[15] E.De Giorgi, *Convergence problems for functionals and operators,* in *Proceedings of the international meeting on recent methods in non linear analysis, (Rome, may 1978),* ed. by E. De Giorgi, E. Magenes, U. Mosco, Pitagora Editrice, Bologna, (1979), 131-188.

[16] E.De Giorgi, *G-operators and Γ-convergence,* in *Proceedings of the International Congress of Mathematicians, (August 1983, Warszawa),* P.W.N. Polish Scientific Publishers, Warszawa, & North-Holland, Amsterdam, (1984), 1175-1191.

[17] E.De Giorgi, G.Dal Maso, P.Longo, *Γ-limiti di ostacoli,* Atti Accad. Naz. Lincei, Rend. Cl. Sci. Fis. Mat. Natur. **68** (1980), 481-487.

[18] E.De Giorgi, T.Franzoni, *Su un tipo di convergenza variazionale,* Atti Accad. Naz. Lincei, Rend. Cl. Sci. Mat. Fis. Natur. **58** (1975), 842-850, & Rend. Sem. Mat. Brescia **3** (1979), 63-101.

[19] R.Figari, E.Orlandi, S.Teta, *The Laplacian in regions with many small obstacles : fluctuations around the limit operator,* J. Stat. Phys. **41** (1985), 465-488.

[20] R.Figari, G.Papanicolaou, J.Rubinstein, *The point interaction approximation for diffusion in regions with many small holes,* in *Stochastic Methods in Biology,* ed. by M. Kimura et al., Lecture Notes in Biomathematics **70**, Springer-Verlag, Berlin, (1987), 202-220.

[21] L.Hedberg, *Spectral synthesis in Sobolev spaces and uniqueness of solutions of the Dirichlet problem,* Acta Math. **147** (1981), 237-264.

[22] E.Ja.Hrouslov, *The method of orthogonal projections and the Dirichlet problem in domains with a fine-grained boundary,* Math. USSR Sb. **17** (1972), 37-59.

[23] E.Ja.Hrouslov, *The first boundary value problem in domains with a complicated boundary for higher order equations*, Math. USSR Sb. **32** (1977), 535-549.

[24] M.Kac, *Probabilistic methods in some problems of scattering theory*, Rocky Mountain J. Math. **4** (1974), 511-538.

[25] H.Kacimi, *Homogénéisation de problèmes de Dirichlet avec de petits trous*, Thèse 3ème cycle, Université Paris VI, (1987).

[26] R.V.Kohn, M. Vogelius, *A new model for thin plates with rapidly varying thickness. II : a convergence proof*, Quat. Appl. Math., **43** (1985), 1-22.

[27] J.L.Lions, *Perturbations singulières dans les problèmes aux limites et en contrôle optimal*, Lecture Notes in Math. **323**, Springer-Verlag, Berlin, (1973).

[28] V.A.Marcenko, E. Ja. Hrouslov, *Problèmes aux limites dans des domaines avec frontières finement granulées*, (en russe), Naukova Dumka, Kiev, (1974).

[29] S.Ozawa, *On an elaboration of M. Kac's theorem concerning eigenvalues of the Laplacian in a region with randomly distributed small obstacles*, Comm. Math. Phys. **91** (1983), 473-487.

[30] S.Ozawa, *Random media and eigenvalues of the Laplacian*, Comm. Math. Phys. **94** (1984), 421-437.

[31] S.Ozawa, *Fluctuation of spectra in random media*, in *Proceedings of the Taniguchi Symposium "Probabilistic Methods in Mathematical Physics"*, ed. by N. Ikeda & K. Ito, Kinokuniya, Tokyo, (1987), 335-361.

[32] G.C.Papanicolaou, S.R.S.Varadhan, *Diffusion in regions with many small holes,* in *Stochastic differential systems, filtering and control, Proceedings of the IFIP WG 7/1 working conference, (Vilnius, USSR, 1978)*, ed. by B. Grigelionis, Lecture Notes in Control and Information Sciences **25**, Springer-Verlag, (1980), 190-206.

[33] J.Rauch, M.Taylor, *Potential and scattering theory on wildly perturbed domains*, J. Funct. Anal. **18** (1975), 27-59.

[34] E.Sanchez-Palencia, *Boundary value problems containing perforated walls*, in *Nonlinear partial differential equations and their applications, Collège de France Seminar, Vol. III*, ed. by H. Brézis & J.L. Lions, Research Notes in Maths **70**, Pitman, London, (1982), 309-325.

[35] L.Tartar, *Cours Peccot, Collège de France (mars 1977),* par-

tiellement rédigé dans : F. Murat, *H-convergence*, Séminaire d'analyse fonctionnelle et numérique de l'Université d'Alger, (1977-78), ronéoté, 34 p.

Laboratoire d'Analyse Numérique, Université Paris VI
Tour 55-65, 5ème étage
F-75252 PARIS cedex 05

ON ATYPICAL VARIATIONAL PROBLEMS

HANS LEWY †

Dedicated to Ennio De Giorgi on his sixtieth birthday

In this short note I describe some minimum problems of the Calculus of Variations which were encountered in work on index problems of plane vector fields. Although the integrands involved are essentially positive, yet the Eulerian fails to have elliptic character in spite of the second variation being positive. A more extensive treatment and its application to index questions I hope to present at a later date.

1. Let Ω be a strictly convex domain of R^2 with C^2 boundary of arc length S_2. On $\Omega \cup \partial\Omega$ consider a C^2 real function $u(x,y)$ of given boundary values $u(s)$ with $du/ds > 0$ for $0 < s < S_1$, and $du/ds < 0$ for $S_1 < s < S_2$, $u(0) = m = \min u(x,y)$, $u(S_1) = M = \max u(x,y)$, $(x,y) \in \Omega \cup \partial\Omega$, $m < u(x,y) < M$ for $(x,y) \in \Omega$. Moreover assume $u_x^2 + u_y^2 \neq 0$ in Ω and for any constant c in $m < c < M$ the curve $u(x,y) = c$ is a Jordan arc joining the points with

† Died suddenly on August 23th, 1988.

$u(s) = c$ of $\partial\Omega$. The class of these functions u call U.

Denote by $\sigma(c)$ the straight segment joining the points $u(s) = c$. Put

$$F[u] = \int\int_\Omega \sqrt{u_x^2 + u_y^2} dx\, dy, \quad u \in U.$$

Theorem. *There is a unique $u \in U$, yielding a relative minimum for $F[u]$. It is that u which is constant on each $\sigma(c)$. For all $u \in U$ the second variation of F is positive.*

PROOF. Denote by ζ a $C^2(\Omega)$ real function of compact support. With

$$z = x + iy, \quad \bar{z} = x - iy,$$

$$\partial/\partial x + i\partial/\partial y = 2\partial/\partial \bar{z}, \quad \partial/\partial x - i\partial/\partial y = 2\partial/\partial z,$$

$$F[u] = i \int\int_\Omega \sqrt{u_z u_{\bar{z}}}\, dz \wedge d\bar{z}$$

we find

$$u + \epsilon\zeta \in U \quad \text{if} \quad \epsilon > 0 \quad \text{is small, and}$$

$$F'([u]) = \frac{d}{d\epsilon} F[u + \epsilon\zeta]|_{\epsilon=0} =$$

$$= \frac{i}{2} \int\int_\Omega \left(\sqrt{\frac{u_{\bar{z}}}{u_z}} \zeta_z + \sqrt{\frac{u_z}{u_{\bar{z}}}} \zeta_{\bar{z}} \right) dz \wedge d\bar{z}.$$

Putting

$$\sqrt{\frac{u_{\bar{z}}}{u_z}} = e^{i\theta}; \quad u_z e^{i\theta} - u_{\bar{z}} e^{-i\theta} = 0 \quad (*)$$

$$F'([u]) = \frac{i}{2} \int\int_\Omega (e^{i\theta}\zeta_z + e^{-i\theta}\zeta_{\bar{z}}) dz \wedge d\bar{z} =$$

$$= -\frac{i}{2} \int\int_\Omega \zeta (e^{i\theta} i\theta_z - e^{-i\theta} i\theta_{\bar{z}}) dz \wedge d\bar{z}\,.$$

Hence any extremum of $F[u]$, $u \in U$, requires

$$e^{i\theta}\theta_z - e^{-i\theta}\theta_{\bar{z}} = 0\,. \tag{1}$$

In view of (*) we obtain

$$u_z \theta_{\bar{z}} - u_{\bar{z}} \theta_z = 0. \tag{2}$$

At any $z \in \Omega$ we have a direction dz satisfying

$$0 = du = u_z \, dz + u_{\bar{z}} d\bar{z}.$$

Compare this with (*). We get

$$dz : d\bar{z} = e^{i\theta} : -e^{-i\theta}.$$

Thus $e^{i\theta}$ is a vector perpendicular to the curve $u =$ const. and (2) yields, since $u_z u_{\bar{z}} \neq 0$, that throughout Ω

$$\theta = f(u),$$

proving that the vector $ie^{i\theta}$ must be constant on $u = c =$ const., $m < c < M$.

The second variation of $F[u]$ is easily verified as

$$\begin{aligned} F''[u] &= \frac{\partial^2}{\partial \epsilon^2} F[u + \epsilon\zeta]|_{\epsilon=0} = \\ &= \frac{i}{4} \int\int_\Omega -(\zeta_z u_{\bar{z}} - \zeta_{\bar{z}} u_z)^2 (u_z u_{\bar{z}})^{-3/2} dz \wedge d\bar{z} \end{aligned}$$

Since $\zeta_z u_{\bar{z}} - \zeta_{\bar{z}} u_z$ is a pure imaginary, its square is negative and $F''[u] > 0$ unless, identically in Ω, $\zeta_z u_{\bar{z}} - \zeta_{\bar{z}} u_z = 0$ and we have ζ equal a function of u; because of the boundary condition ζ vanishes identically.

Notice that the positivity of $F''[u]$ holds for all u in U. In particular the function u which has value c on $\sigma(c)$ yields a relative minimum of $F[u]$ as u belongs in U.

2. The next example is a functional over some domain Ω of $\mathbf{R}^2$

$$G[u] = i \int\int_\Omega (u_{zz} u_{\bar{z}\bar{z}})^{1/2} dz \wedge d\bar{z}$$

on C^4 real functions u of non-vanishing u_{zz}. It is even stranger than the previous example as the search for a relative minimum leads to Eulerian equations forming a system of two hyperbolic equations with identical characteristics.

The first variation of $G[u]$ is

$$G'[u] = \frac{i}{2} \int\int \left(\left(\frac{u_{\bar z \bar z}}{u_{zz}} \right)^{1/2} \zeta_{zz} + \left(\frac{u_{zz}}{u_{\bar z \bar z}} \right)^{1/2} \zeta_{\bar z \bar z} \right) dz \wedge d\bar z$$

or with

$$\sqrt{\frac{u_{\bar z \bar z}}{u_{zz}}} = e^{2i\theta} \quad \text{or} \quad u_{zz} e^{2i\theta} - u_{\bar z \bar z} e^{-2i\theta} = 0 \quad (**)$$

$$\begin{aligned} G'[u] =& \frac{i}{2} \int_{\partial\Omega} \left(-\left(e^{2i\theta}\right)_z \zeta + \theta e^{2i\theta} \zeta_z \right) d\bar z + \\ & - \left(-\left(e^{-2i\theta}\right)_{\bar z} \zeta + e^{-2i\theta} \zeta_{\bar z} \right) dz + \\ & + \frac{i}{2} \int\int_{\Omega} \zeta \left[\left(e^{2i\theta}\right)_{zz} + \left(e^{-2i\theta}\right)_{\bar z \bar z} \right] dz \wedge d\bar z. \end{aligned}$$

$G'[u] = 0$ requires

$$\left(e^{2i\theta}\right)_{zz} + \left(e^{-2i\theta}\right)_{\bar z \bar z} = 0, \quad z \in \Omega$$

which is a real differential equation for θ

$$0 = e^{2i\theta} \theta_{zz} - e^{-2i\theta} \theta_{\bar z \bar z} + 2i \left(e^{2i\theta} \theta_z^2 + e^{-2i\theta} \theta_{\bar z}^2 \right)$$

This together with (**) is the Eulerian system necessary for the vanishing of the first variation of $G[u]$.

The second variation is

$$G'[u] = \frac{i}{4} \int\int_{\Omega} -(\zeta_{zz} u_{\bar z \bar z} - \zeta_{\bar z \bar z} u_{zz})^2 (u_{zz} u_{\bar z \bar z})^{-3/2} dz \wedge d\bar z;$$

it is positive unless $\zeta_{zz} u_{\bar z \bar z} - \zeta_{\bar z \bar z} u_{zz}$ vanishes identically in Ω. This holds for all functions u under consideration. In particular if u is chosen such as to annull the contribution of the boundary in $G'[u]$, any solution of the above Eulerian system yields a relative minimum.

The hyperbolic nature of the above system is seen by observing that the directions of the characteristics satisfy

$$e^{2i\theta} (d\bar z)^2 - e^{-2i\theta} (dz)^2 = 0$$

or

$$dz = |dz|^{\theta} e^{i\theta} \quad \text{and} \quad dz = |dz| i e^{i\theta}$$

The hyperbolicity of the Eulerian of $G[u]$ imposes severe restriction on possible forms of $\partial\Omega$ and boundary conditions which may be prescribed on u and derivatives on $\partial\Omega$ in order that there exist a relative minimum of $G[u]$.

Department of Mathematics
University of California
BERKELEY, CA 94720

SUR LA CONTROLABILITE EXACTE ELARGIE

JACQUES-LOUIS LIONS

Dédié à Ennio De Giorgi pour son soixantième anniversaire

1. Introduction. Soit Ω un ouvert borné de $\mathbf{R}^n$, de frontière régulière Γ. On considère dans le domaine $\Omega \times (0, +\infty)$, l'équation des ondes

$$\frac{\partial^2 y}{\partial t^2} - \Delta y = 0. \tag{1.1}$$

L'état du système est défini par la solution de (1.1), avec les conditions initiales

$$y(x,0) = y^o(x), \ \frac{\partial y}{\partial t}(x,0) = y^1(x) \ \text{ dans } \ \Omega \tag{1.2}$$

et avec la condition aux limites

$$y = v \ \text{ sur } \ \Gamma \times (0, +\infty) \tag{1.3}$$

où v est la fonction contrôle.

Nous ne précisons pas, pour l'instant, le cadre fonctionnel. Soit donc, de façon formelle (pour l'instant), $y(x,t;v) = y(v)$ la solution de (1.1), (1.2), (1.3).

Le problème de la *contrôlabilité exacte* (CE) est le suivant: soit $T > 0$ donné. Peut-on, pour chaque couple y^0, y^1 (pris dans un espace fonctionnel "convenable" qu' il faudra préciser), trouver un contrôle v tel que

$$y(x,T;v) = \frac{\partial y}{\partial t}(x,T;v) = 0 \quad \text{dans} \quad \Omega \ ? \tag{1.4}$$

Si la réponse est positive, on dit qu' il y a CE. Il y a alors *une infinité* de contrôles v qui réalisent (1.4). Il sera naturel de chercher "le meilleur" – par ex. au sens: celui des contrôles qui *minimise la norme de l'espace où l'on prend* v (cela est précisé ci-après).

Le problème de la *contrôlabilité exacte élargie* (CEE) est le suivant: *on se donne encore* $T > 0$. On se donne en *outre un espace fonctionnel* G *convenable*, formé de *couples de fonctions* $\{g^0, g^1\}$.

On dit qu' il y a CEE (et, plus précisément, CEE *par rapport à* G), si, pour chaque couple $\{y^0, y^1\}$ (dans un espace fonctionnel convenable), on peut trouver un contrôle v telque

$$\{y(\cdot,T;v), \frac{\partial y}{\partial t}(\cdot,T;v)\} \in G. \tag{1.5}$$

La notion dépend évidemment de G.

Si $G = \{0,0\}$, on retrouve la notion de CE.

Si G est *tout l'espace* parcouru par $\{y(\cdot,T;v), \frac{\partial y}{\partial t}(\cdot,T;v)\}$, la notion est vide.

S'il y a CEE (par rapport à G), alors il existe une infinité de contrôles v qui réalisent (1.5) et, là encore, on cherchera "le meilleur" des v.

Remarque 1.1. S'il y a CEE alors il y a CEE par rapport à n'importe quel espace G. Mais la CEE par rapport à un espace G n'entraîne pas le CE.

Remarque 1.2. A cause de la vitesse finie de propagation des ondes, on ne peut avoir CE *que pour* T *assez grand.*

Donnons maintenant (toujours de manière formelle) quelques exemples.

Exemple 1.1. Soit ω un ouvert contenu dans Ω.
On cherche v (le "meilleur" v) tel que

$$y(x,T;v) = 0 \quad dans \quad \omega. \tag{1.6}$$

C'est le problème CEE avec pour G l'espace

$$\begin{aligned} &G = G_0 \times \{espace\ \ entier\} \\ &G_0 = espace\ de\ fonctions\ (ou\ distributions)\ nulles\ sur\ \omega. \end{aligned} \tag{1.7}$$

Une variante évidente est

Exemple 1.2. On cherche v avec

$$\frac{\partial y}{\partial t}(x,T;v) = 0 \quad dans \quad \omega. \tag{1.8}$$

C'est le problème CEE avec
(1.9)

$$G = \{espace\ \ entier\} \times G_1,$$

$$G_1 = espace\ de\ fonctions\ (ou\ de\ distributions)\ nulles\ sur\ \omega.$$

Exemple 1.3. Soient $w_1, w_2, \ldots$ les fonctions propres de l'opérateur $-\Delta$ pour la condition de Dirichlet

$$\begin{aligned} &-\Delta w_j = \lambda_j w_j, \quad w_j = 0 \quad sur\ \Gamma, \int_\Omega w_j^2 dx = 1, \\ &0 < \lambda_1 \le \lambda_2 \le \ldots \end{aligned} \tag{1.10}$$

On cherche v tel que

$$\begin{aligned} &\int_\Omega y(x,T;v) w_j(x) dx = 0, \\ &\int_\Omega \frac{\partial y}{\partial t}(x,T;v) w_j(x) dx = 0, \quad 1 \le j \le m. \end{aligned} \tag{1.11}$$

Alors G est l'espace des couples orthogonaux aux w_j, $1 \leq j \leq m$.

Remarque 1.3. On peut poser les mêmes problèmes avec

$$y = \begin{cases} v & \text{sur } \Sigma_0 \subset \Sigma = \Gamma \times (0,T) \\ 0 & \text{sur } \Sigma \setminus \Sigma_0 \end{cases} \tag{1.12}$$

où Σ_0 est une partie "convenable" de Σ.

Remarque 1.4. Nous avons pris la situation (1.1), (1.2), (1.3) comme *exemple modèle*. Les méthodes que nous allons donner sont *générales* et s'appliquent à toutes sortes d'autres situations: autres opérateurs (en particulier opérateurs de type Petrowsky, non nécessairement hyperboliques),

autres conditions aux limites,

autres actions sur le système que par le bord — et notamment contrôle ponctuel (dont un exemple sera donné à la fin).

Remarque 1.5. Le problème de la CE est un problème classique. Un exposé complet des diverses méthodes qui avaient pour objet la CE *sans* étudier la question supplémentaire de la recherche "du meilleur v" est donné dans D.L.Russel [1].

En rajoutant la question de la recherche du "meilleur v" on est conduit à une méthode nouvelle, beaucoup plus générale, introduite dans J.L.Lions [1], Méthode dite HUM (pour Hilbert Uniqueness Method). Cette méthode a fait l'objet d'un exposé (assez) systématique dans J.L.Lions [2].

Le problème de la CEE a été introduit dans J.L.Lions [2], Chap.1, n.9. On y a adapté la méthode HUM.

On va proposer ici une solution un peu différente, toujours basée sur les mêmes principes (que nous allons rappeler), mais utilisant aussi l'idée de RHUM (Reverse HUM) introduite (pour d'autres raisons) dans J.L.Lions [3].

Le plan est le suivant:

2. Position fonctionnelle précise du problème.
3. Méthode générale pour la CEE.
4. CEE et pénalisation.
5. Examples.
6. Variantes.

Bibliographie.

2. Position fonctionnelle précise du problème. Nous allons préciser a priori *un* cadre fonctionnel — celui qui parait le plus simple. Mais il y en a *beaucoup* d'autres (on considérera pour cela J.L.Lions [2] [3] et les travaux de C.Bardos, G.Lebeau, J.Rauch [1], A.Haraux [1]).

Dans toutes ces questions, il y a des choix de deux types:

i) le choix de l'espace fonctionnel dans lequel on va considérer le contrôle v;
ii) le choix de l'espace fonctionnel où l'on prend $\{y^0, y^1\}$.

Ces choix doivent être cohérents, bien sûr. Mais l'on voit bien qu'il peut-être plus difficile de conduire à l'équilibre (ou dans un état donné dans l'espace G) un système qui a, à l'instant $t = 0$, des données initiales *singulières* qu'un système ayant des données initiales *très régulières*.

Une fois les choix ci-dessus faits, alors G doit être un sous espace (fermé) d'un espace fonctionnel qui est fixé.

Nous allons nous placer (cela n'ajoute pas de complications sérieuses dans la formulation des problèmes) dans la cadre de la Remarque 1.3. Plus précisément on introduit

$$\Sigma_0 = \Gamma_0 \times (0,T), \qquad \Gamma_0 \subset \Gamma \tag{2.1}$$

et on *choisit*

$$v \in L^2(\Sigma_0), \tag{2.2}$$

avec la convention que $v = 0$ dans $\Sigma \backslash \Sigma_0$.

On introduit alors les espaces de Sobolev $H_0^1(\Omega)$ (espace des fonctions $\varphi \in L^2(\Omega)$ avec $\partial\varphi/\partial x_i \in L^2(\Omega)$ et $\varphi = 0$ sur Γ et $H^{-1}(\Omega)$, dual de $H_0^1(\Omega)$.

On considère

$$y^0 \in L^2(\Omega), \qquad y^1 \in H^{-1}(\Omega). \tag{2.3}$$

Le problème (1.1) (1.2) (1.3) admet alors une solution unique, $y(v)$, telle que

(2.4)
$$\{y(v), y'(v) = \frac{\partial y}{\partial t}(v)\} \quad \textit{est continue de} \quad [0,T] \to L^2(\Omega) \times H^{-1}(\Omega).$$

Ici et dans la suite, on a remplacé $\frac{\partial y}{\partial t}$ par y', donc $\frac{\partial^2 y}{\partial t^2}$ sera remplacé par y''. On introduit alors

$$G = \textit{sous espace vectoriel fermé de} \quad L^2(\Omega) \times H^{-1}(\Omega). \tag{2.5}$$

Le problème de la CEE (par rapport à G) est ainsi défini sans ambiguïté.

Remarque 2.1. La démonstration de (2.4) sous les hypothèses (2.2) (2.3) est donnée dans J.L. Lions [2], Chap. 1, n° 4.

3. Méthode générale pour la CEE. On introduit a priori la méthode–adaptée de RHUM que nous prososons. On va la justifier. On indiquera au n^o suivant comment la méthode de pénalisation *conduit* à la méthode proposée et, en même temps, démontrer que le contrôle obtenu est *le meilleur.*

On introduit

$$G^0 \subset L^2(\Omega) \times H_0^1(\Omega), \qquad \textit{espace polaire de } G. \tag{3.1}$$

Donc

$$\begin{aligned} &\varrho = \{\rho^1, \varrho^0\} \in G^0 \qquad \textit{si et seulement si} \\ &(\varrho^1, g^0) + (\varrho^0, g^1) = 0 \;\; \forall\, g = \{g^0, g^1\} \in G. \end{aligned} \tag{3.2}$$

Dans (3.2) (ϱ^1, g^0) désigne le produit scalaire dans $L^2(\Omega)$, et (ρ^0, g^1) désigne le produit scalaire entre $H_0^1(\Omega)$ et $H^{-1}(\Omega)$. Soit $\{\varrho^0, \varrho^1\}$ t.q.

$$\{\varrho^1, -\varrho^0\} \in G^0. \tag{3.3}$$

On résout le problème *rétrograde*

$$\begin{array}{ll} \varrho'' - \Delta\varrho = 0 & \textit{dans } \ \Omega \times (0,T) \\ \varrho(T) = \varrho^0 \ , \ \varrho'(T) = \varrho^1 & \textit{dans } \ \Omega \\ \varrho = 0 & \textit{sur } \ \Sigma = \Gamma \times (0,T). \end{array} \tag{3.4}$$

Ce problème admet une solution unique. On a en outre la propriété (cf. J.L. Lions [2] pour les détails des démonstrations)

$$\frac{\partial \rho}{\partial \nu} \in L^2(\Sigma). \tag{3.5}$$

On peut donc considérer le problème

$$\begin{array}{ll} \sigma'' - \Delta\sigma = 0 & \textit{dans } \ \Omega \times (0,T) \\ \sigma(0) = y^0 \ , \ \sigma'(0) = y^1 & \textit{dans } \ \Omega \\ \sigma = +\partial\varrho/\partial\nu & \textit{sur } \ \Sigma_0 \\ \quad 0 & \textit{sur } \ \Sigma\backslash\Sigma_0. \end{array} \tag{3.6}$$

On obtient ainsi σ telle que

$$\{\sigma, \sigma'\} \ \textit{est continue de} \ [0,T] \to L^2(\Omega) \times H^{-1}(\Omega). \tag{3.7}$$

Si l'on peut trouver ρ^0, ρ^1 (*les inconnues*) avec (3.3) tels que

$$\{\sigma(T), \sigma'(T)\} \in G \tag{3.8}$$

alors

$$v = \frac{\partial \varrho}{\partial \nu} \ \textit{sur} \ \Sigma_0 \tag{3.9}$$

donne *un* contrôle donnant CEE. On a alors

$$y(v) = \sigma. \tag{3.10}$$

Pour exprimer (3.8) par une équation, il est naturel d'introduire

$$\pi = \textit{projection orthogonale } \big(\textit{dans } L^2(\Omega) \times H^{-1}(\Omega)\big) \textit{ sur l'espace } G^\perp \textit{ orthogonal de } G. \tag{3.11}$$

On définit alors

$$M\{\rho^1, -\rho^0\} = \pi\{\sigma(T), \sigma'(T)\}. \tag{3.12}$$

On a ainsi défini un opérateur M affine de G^0 sur $G^\perp$. Tout revient à résoudre l'équation

$$M\{\varrho^1, -\varrho^0\} = \{0, 0\}. \tag{3.13}$$

Décomposons M en sa partie linéaire + partie constante.
Introduisons y_0 solution de

$$\begin{aligned} & y_0'' - \Delta y_0 = 0 && \textit{dans } \Omega \times (0,T) \\ & y_0(0) = y^0 \ , \ y_0'(0) = y^1 && \textit{dans } \Omega \\ & y_0 = 0 && \textit{sur } \Sigma \end{aligned} \tag{3.14}$$

Puis introduisons τ solution de

$$\begin{aligned} & \tau'' - \Delta\tau = 0 && \textit{dans } \Omega \times (0,T) \\ & \tau(0) = \tau'(0) = 0 && \textit{dans } \Omega \\ & \tau(0) = +\partial\varrho/\partial\nu && \textit{sur } \Sigma_0 \\ & \quad\quad 0 && \textit{sur } \Sigma \backslash \Sigma_0. \end{aligned} \tag{3.15}$$

On a alors

$$M\{\varrho^1, -\varrho^0\} = M_0\{\varrho^1, -\varrho^0\} + \pi\{y_0(T), y_0'(T)\} \tag{3.16}$$

où

$$M_0\{\varrho^1, -\varrho^0\} = \pi\{\tau(T), \ \tau'(T)\}. \tag{3.17}$$

Tout revient donc à résoudre l'équation

$$M_0\{\varrho^1, -\varrho^0\} = -\pi\{y_0(T), y_0'(T)\}. \tag{3.18}$$

L'opérateur M_0 vérifie:

$$M_0 \in \mathcal{L}(G^0; G^\perp) \ \ (G^\perp \text{ s'identifiant à } (G^0)'). \tag{3.19}$$

On vérifie en outre que

$$M_0^* = M_0. \tag{3.20}$$

Calculons — c'est le point essentiel — le produit scalaire

$$\begin{gathered} < M_0\{\varrho^1, -\varrho^0\} \ , \ \{\varrho^1, -\varrho^0\} >= \mu \\ \{\varrho^1, -\varrho^0\} \in G^0, \end{gathered} \tag{3.21}$$

le crochet désignant la dualité entre $G^\perp$ et G^0. Par définition, on a:

$$< \pi\{\tau^0, \tau^1\} - \{\tau^0, \tau^1\}, \tilde{g} >= 0 \quad \forall \tilde{g} \in G^0. \tag{3.22}$$

On applique (3.22) avec $\tilde{g} = \{\rho^1, -\rho^0\}$, $\tau^0 = \tau(T)$ et $\tau^1 = \tau'(T)$. Alors

$$\mu =< \{\tau^0, \tau^1\}, \tilde{g} >= (\tau(T), \varrho^1) - (\tau'(T), \varrho^0). \tag{3.23}$$

Pour calculer cette dernière expression, multiplions la 1ère équation (3.15) par ρ. Après intégrations par parties (qui sont justifiées par *définition* même de τ, solution faible définie par transposition, comme dans J.L.Lions et E.Magenes [1]), il vient:

$$\begin{gathered} (\tau'(T), \rho(T)) - (\tau'(0), \rho(0)) - (\tau(T), \rho'(T)) + \\ + (\tau(0), \rho'(0)) + \int_\Sigma \tau \frac{\partial \rho}{\partial \nu} d\Sigma = 0. \end{gathered} \tag{3.24}$$

Tenant compte des conditions initiales et aux limites dans (3.15), on déduit de (3.24) et (3.23) que

$$\mu = \int_{\Sigma_0} (\frac{\partial \rho}{\partial \nu})^2 d\Sigma. \tag{3.25}$$

Le point essentiel est maintenant le suivant: la quantité

$$(\int_{\Sigma_0} (\frac{\partial \rho}{\partial \nu})^2 d\Sigma)^{1/2} \tag{3.26}$$

est une seminorme sur G^0.

Supposons que cette quantité soit une norme équivalente à celle de G^0.

Alors

$$\text{(3.27)} \qquad \mu \ \textit{est un isomorphisme de} \ G^0 \ \textit{sur} \ G^\perp$$

et l'équation (3.18) admet une solution unique.

Donc

Théorème 3.1. *On se donne $G \subset L^2(\Omega) \times H^{-1}(\Omega)$. On suppose que (3.26) est une norme sur G^0 équivalente à la norme induite par $L^2(\Omega) \times H_0^1(\Omega)$. Alors il y a CEE* (relative à G).
Le contrôle $v = \dfrac{\partial \rho}{\partial \nu}$ est donné par la résolution de (3.18).

Remarque 3.1. Naturellement la difficulté essentielle est maintenant la vérification de l'hypothèse dans le Théorème 3.1. Notons que dire que (3.26) est une *norme* est un *théorème d'unicité*: si ρ est solution de

$$\text{(3.28)} \qquad \begin{array}{ll} \varrho'' - \Delta\varrho = 0 & \textit{dans } \Omega \times (0,T) \\ \varrho = 0 & \textit{sur } \Sigma \\ \dfrac{\partial \varrho}{\partial \nu} = 0 & \textit{sur } \Sigma_0 \end{array}$$

et si ϱ vérifie

$$\text{(3.29)} \qquad \{\varrho'(T), \ -\varrho(T)\} \in G^0$$

alors

$$\text{(3.30)} \qquad \varrho \equiv 0 \ \text{ dans } \ \Omega \times (0,T).$$

La CE ordinaire correspond au cas où $G = \{0,0\}$, donc G^0 est sans objet. On a alors affaire à un théorème d'unicité du type de Holmgren. La condition nécessaire et suffisante sur Σ_0 pour que l'on ait unicité est alors donnée dans C.Bardos, G.Lebeau et J.Rauch [1].

Mais on peut avoir (3.30) avec un Σ_0 *plus petit* (i.e. n'assurant pas la CE), si l'on ajoute la condition (3.29). C'est le cas *si par exemple G^0 est un espace de dimension finie.*

Remarque 3.2. La terminologie RHUM commence à être justifée: on construit une norme *hilbertienne* à partir d'un théorème d'*unicité* pour un problème *rétrograde.*

Remarque 3.3. On n'a pas encore démontré que le contrôle $v = \frac{\partial \varrho}{\partial \nu}$ ainsi construit correspond à celui de *norme minimum.* Cela sera conséquence du n° 4 suivant.

Remarque 3.4. La méthode proposée est constructive. Dans le cas de la CE (usuelle), les méthodes numériques (qui demamdent quelques précautions) sont présentées dans R.Glowinski, C.H.Li et J.L.Lions [1]. D'autres résultats, pour des contrôles ponctuels notamment, sont donnés dans A.El Jai et A.Gonzales [1].

Remarque 3.5. La résolution de (3.18) équivaut à la minimisation de

$$J(\varrho) = \frac{1}{2} < M_0\{\varrho^1, -\varrho^0\}, \{\varrho^1, -\varrho^0\} > + \\ + < \pi\{y_0(T), y_0'(T)\}, \{\varrho^1, -\varrho^0\} > .$$

Mais

$$< \pi\{y_0(T), y_0'(T)\}, \{\varrho^1, -\varrho^0\} > = (y_0(T), \varrho') - (y_0'(T), \varrho^0).$$

Multipliant la première equation (3.14) par ϱ, il vient

$$(y_0(T)), \varrho') - (y_0'(T), \varrho^0) = (y^0, \varrho'(0)) - (y^1, \varrho(0))$$

de sorte que

$$J(\varrho) = \frac{1}{2} \int_{\Sigma_0} \left(\frac{\partial \varrho}{\partial \nu}\right)^2 d\Sigma + (y^0, \varrho'(0)) - (y^1, \varrho(0)). \tag{3.31}$$

Le problème *est alors équivalent à*

$$\inf J(\varrho) \;, \qquad \{\varrho^1, -\varrho^0\} \in G^0. \tag{3.32}$$

(C'est la forme duale du problème initial, en un sens qui peut être précisé, comme dans J.L.Lions [2], Chap. 8).

4. CEE et pénalisation. Nous allons maintenant reprendre le problème de façon différente , à partir de la *pénalisation.* Nous prenons v, y^0, y^1 comme dans (2.2) (2.3) et G comme dans (2.5).

On désigne par $\mathcal{U}$ l'ensemble des $v \in L^2(\Sigma_0)$ tels que si $y(v)$ est la solution de (1.1) (1.2) (1.3) on ait

$$\{y(T;v), y'(T;v)\} \in G. \tag{4.1}$$

On suppose cet ensemble non vide, i.e. on suppose que l'on a CEE (relative à G). On va en déduire — via la pénalisation — des conséquences qui a) conduisent de manière naturelle à la méthode présentée a priori au n° 3; b) montrent que

$$\begin{aligned} &\text{le contrôle } v = \frac{\partial \varrho}{\partial \nu} \text{ sur } \Sigma_0 \textit{ fourni par la méthode du n° 3} \\ &\textit{est celui qui minimise } \int_{\Sigma_0} v^2 d\Sigma \textit{ dans l'ensemble } \mathcal{U}. \end{aligned} \tag{4.2}$$

Soit $\epsilon > 0$ fixé. On pose

$$J_\epsilon(v,y) = \frac{1}{2}\int_{\Sigma_0} v^2 d\Sigma + \frac{1}{2\epsilon}\int_{\Omega\times(0,T)} (y'' - \Delta y)^2 dx\, dt \tag{4.3}$$

où

$$\begin{aligned} &y, y'' - \Delta y \in L^2(Q)\ , \quad Q = \Omega \times (0,T), \\ &y(0) = y^0\ ,\ y'(0) = y^1, \\ &y = v \ \ sur \ \ \Sigma_0\ ,\ y = 0 \ \ sur \ \ \Sigma\backslash\Sigma_0 \\ &\{y(T), y'(T)\} \in G. \end{aligned} \tag{4.4}$$

L'ensemble des couples y, v avec (4.4) n'est pas vide. Il suffit de prendre une fonction y (assez régulière) telle que

$$y(0) = y^0,\ y'(0) = y^1,\ \{y(T), y'(T)\} \in G,\ y = 0 \ \ sur \ \ \Sigma\backslash\Sigma_0$$

puis de *définir* v par $v = y$ sur Σ_0. Il n'y a pas de difficulté, au moins si y^0, y^1 sont assez régulières.

On considère alors le problème

$$\begin{aligned} &\inf J_\epsilon(v, y), \\ &v, y \text{ vérifiant } (4.4). \end{aligned} \tag{4.5}$$

Soit $\{u_\epsilon, y_\epsilon\}$ la solution de (4.5). Posons

$$p_\epsilon = -\frac{1}{\epsilon}(y_\epsilon'' - \Delta y_\epsilon). \tag{4.6}$$

Puisque $\mathcal{U}$ n'est pas vide (par hypothèse), on a

$$J_\epsilon(v, y) \leq \frac{1}{2}\int_{\Sigma_0} v^2 d\Sigma \quad si \quad v \in \mathcal{U} \tag{4.7}$$

(en prenant $y = y(v)$). On en déduit facilement que

$$u_\epsilon \to u \qquad \text{dans } L^2(\Sigma_0) \tag{4.8}$$

où u est la solution de

$$\frac{1}{2}\int_{\Sigma_0} u^2 d\Sigma = \inf_{v \in \mathcal{U}} \frac{1}{2}\int_{\Sigma_0} v^2 d\Sigma. \tag{4.9}$$

L'équation d'Euler relative au problème (4.5) est

$$\int_{\Sigma_0} u_\epsilon v d\Sigma - \int_{\Omega\times(0,T)} p_\epsilon(\eta'' - \Delta\eta) dx\, dt = 0 \tag{4.10}$$

pour tout couple $\{v, \eta\}$ tel que

$$\begin{aligned} &\eta(0) = \eta'(0) = 0, \\ &\eta = \nu \ \ sur \ \ \Sigma_0, \quad \eta = 0 \ \text{ sur } \ \Sigma \backslash \Sigma_0, \\ &\{\eta(T)\ ,\ \eta'(T)\} \in G. \end{aligned} \tag{4.11}$$

On en déduit que p_ϵ vérifie

$$\begin{aligned} &p_\epsilon'' - \Delta p_\epsilon = 0 \quad && dans \ \ \Omega \times (0, T), \\ &p_\epsilon = 0 && sur \ \ \Sigma \\ &\frac{\partial}{\partial\nu} p_\epsilon = u_\epsilon && sur \ \ \Sigma_0 \end{aligned} \tag{4.12}$$

et que

$$(4.13)\qquad \begin{array}{l}(p'_\epsilon(T),\eta(T)) - (p_\epsilon(T),\eta'(T)) = 0\\ \forall\eta \;\; avec \;\; \{\eta(T),\eta'(T)\}\in G.\end{array}$$

Donc

$$(4.14)\qquad \{p'_\epsilon(T),-p_\epsilon(T)\}\in G^0.$$

Si l'on suppose que l'on a l'inégalité suivante:

$$(4.15)\qquad \int_{\Sigma_0}(\frac{\partial p_\epsilon}{\partial\nu})^2 d\Sigma \geq c[\|p_\epsilon(T)\|^2_{H^1_0(\Omega)} + \|p'_\epsilon(T)\|^2_{L^2(\Omega)}]$$

alors on peut passer à la limite en $\epsilon\to 0$.

Noter que (4.15) *équivaut* aux hypothèses faites au n° 3.

On obtient alors à la limite — s'il y a CEE relative à G — le *système d'optimalité* suivant:

$$(4.16)\qquad \begin{array}{ll}y''-\Delta y = 0 & dans \;\; \Omega\times(0,T)\\ y(0)=y^0, y'(0)=y^1 & dans \;\; \Omega,\\ y=u \;\; sur \;\; \Sigma_0, \;\; y=0 & sur \;\; \Sigma\backslash\Sigma_0,\\ p''-\Delta p=0 & dans \;\; \Omega\times(0,T),\\ p=0 & sur \;\; \Sigma,\\ \{p'(T),-p(T)\}\in G^0, & \{y(T),y'(T)\}\in G\\ \dfrac{\partial p}{\partial\nu}=u & sur \;\; \Sigma_0\end{array}$$

Ce système d'optimalité peut être interprété de manière à conduire à la méthode donné au n° 3.

En effet, on prend $\{p'(T),-p(T)\}$ comme *"inconnue de base"*.

On introduit ainsi ϱ solution de (3.4) avec (3.3).

On *définit* alors σ par (3.6) — ce qui correspond à $y=\frac{\partial p}{\partial\nu}$ dans (4.16) — et on aura alors $\sigma = y$ si

$$(4.17)\qquad \{\sigma(T),\sigma'(T)\}\in G.$$

C'est la méthode donnée au n° 3, qui montre que l'équation (3.17) en $\{\varrho^1,-\varrho^0\}$ *admet une solution unique.* On a donc montré le caractère "naturel" de la méthode introduite au n° 3 *et on a en outre montré* (4.2).

Remarque 4.1. On peut interpréter différemment le système (4.16) en prenant $\{p'(0), -p(0)\}$ comme *"inconnue de base"*. C'est ce que l'on a fait dans J.L.Lions [2], Chap.1, n.8. Le choix donné ici est plus simple.

Remarque 4.2. On voit bien que le procédé *est général.*

5. Exemples. On reprend, dans le même ordre, les Exemples considérés au n.1.

Exemple 5.1 (suite de l'Exemple 1.1). On a donc

$$\begin{aligned} &G = G_0 \times H^{-1}(\Omega), \\ &G_0 = \textit{espace des fonctions de } L^2(\Omega) \textit{ nulles sur } \omega. \end{aligned} \tag{5.1}$$

Alors

$$G^0 = L^2(\omega) \times \{0\} \tag{5.2}$$

où $L^2(\omega)$ désigne l'espace des fonctions de $L^2(\omega)$ prolongées par 0 en dehors de ω.

La présentation de la méthode se simplifie. On a comme "*inconnue de base*" la fonction

$$\varrho^1 \in L^2(\omega).$$

Identifions ϱ^1 à sa restriction à ω. On a donc comme inconnue de base $\varrho^1 \in L^2(\omega)$. On résout

$$\begin{aligned} &\varrho'' - \Delta\varrho = 0 && \textit{dans } \ \Omega \times (0,T), \\ &\varrho(T) = 0 \ , \ \varrho'(T) = \varrho^1 && \textit{dans } \ \omega, \ \ \varrho'(T) = 0 \ \ \textit{dans } \Omega\backslash\omega \\ &\varrho = 0 && \textit{sur } \ \Sigma. \end{aligned} \tag{5.3}$$

Puis on résout (3.6).

On définit ensuite

$$M_1\varrho^1 = \sigma(T)|_\omega \quad \text{(restriction de } \sigma(T) \text{ à } \omega). \tag{5.4}$$

On définit ainsi un opérateur affine continu de $L^2(\Omega) \to L^2(\omega)$.

Tout revient à résoudre

$$M_1\varrho^1 = 0. \tag{5.5}$$

(On vérifie sans peine que ce qui précède n'est que la traduction dans le cas particulier présent de la méthode générale du n° 3).

L'équation (5.5) admet une solution unique si l'on a l'inégalité

$$\int_{\Sigma_0} \left(\frac{\partial\varrho}{\partial\nu}\right)^2 d\Sigma \geq c \int_\omega (\varrho^1)^2 dx \quad , \quad c > 0. \tag{5.6}$$

La condition nécessaire et suffisante portant sur Σ_0 pour que (5.6) ait lieu peut probablement être obtenue par les méthodes de C.Bardos, G.Lebeau et J.Rauch, [1], mais cela reste à faire.

Un résultat précis est le suivant: soit x^0 un point quelconque de $\mathbf{R}^n$ et $\Gamma(x^0) = \{x | x \in \Gamma, \nu(x)(x - x^0) \geq 0,\ \nu(x) = \text{vecteur unitaire normal à } \Gamma \text{ dirigé vers l'extérieur de } \Omega\}$. On prend

$$\Gamma_0 = \Gamma(x^0). \tag{5.7}$$

On a alors une inégalité du type (5.6) si

$$\begin{aligned} &T > R(x^0), \\ &R(x^0) = \sup \nu(x)(x - x^0),\ x \in \Gamma. \end{aligned} \tag{5.8}$$

Remarque 5.1. On a donc CEE relative à $L^2(\Omega\backslash\omega) \times H^{-1}(\Omega)$ si $T > R(x^0)$ alors que l'on a CE si $T > 2R(x^0)$. (cf. J.L.Lions [2], Chap.1, n.8).

Remarque 5.2. On doit pouvoir améliorer (5.8) avec une constante $T > T_0$ *où* T_0 *dépend de* ω. Il est évident en effet que si ω est vide, il y a CEE pour un temps arbitrairement petit, puisque l'on n'impose *aucune* condition sur $\{y(T), y'(T)\}$! Le meilleur choix de v est alors $v = 0$.

Exemple 5.2 (suite de l'Exemple 1.2.).
On a cette fois

$$G = L^2(\Omega) \times G_1, \tag{5.9}$$
$$G_1 \textit{ sous espace des éléments de } H^{-1} \textit{ nuls sur } \omega.$$

Donc

$$G^0 = \{0\} \times H_0^1(\omega). \tag{5.10}$$

On prend donc comme 'inconnues de base" $\varrho^0 \in H_0^1(\omega)$ et l'on résout

$$\begin{aligned} &\varrho'' - \Delta\varrho = 0 && \textit{dans } \Omega \times (0,T), \\ &\varrho(T) = \varrho^0 && \text{(prolongement de } \varrho^0 \text{ par 0 hors de } \omega), \\ &\varrho'(T) = 0, && \\ &\varrho = 0 && \textit{sur } \Sigma. \end{aligned} \tag{5.11}$$

Puis on définit σ par (5.6) et on définit ensuite

$$M_2\varrho^0 = \sigma'(T)|_\omega = \text{restriction de } \sigma'(T) \text{ à } \omega. \tag{5.12}$$

On a ainsi défini un opérateur affine continu de $H_0^1(\omega)$ dans $H^{-1}(\omega)$. Tout revient à résoudre

$$M_2\varrho^0 = 0. \tag{5.13}$$

Cette équation admet une solution unique si

$$\int_{\Sigma_0} \left(\frac{\partial \varrho}{\partial \nu}\right)^2 d\Sigma \geq c\|\varrho^0\|^2_{H_0^1(\omega)}. \tag{5.14}$$

Cette inégalité est vérifiée si $\Gamma_0 = \Gamma(x^0)$ et si (5.8) a lieu.

Exemple 5.3 (suite de l'Exemple 1.3.).
Dans ce cas, l'espace G consiste en les couples g^0, g^1 avec

$$\begin{aligned} &g^0 \in L^2(\Omega),\ (g^0, w_j) = 0\ ,\ 1 \leq j \leq m, \\ &g^1 \in H^{-1}(\Omega),\ (g^1, w_j) = 0\ ,\ 1 \leq j \leq m. \end{aligned}$$

Donc G^0 est l'espace engendré par h^0, h^1 combinaisons linéaires des $w_j, 1 \leq j \leq m$.

On résout donc

$$
(5.15) \qquad \begin{aligned}
&\varrho'' - \Delta\varrho = 0,\\
&\varrho(T) = \sum_{j=1}^{m} \varrho_j^0 w_j \ , \ \varrho'(T) = \sum_{j=1}^{m} \varrho_j^1 w_j,\\
&\varrho = 0 \ \ sur \ \Sigma
\end{aligned}
$$

les "inconnues de base" étant $\{\varrho_j^0, \varrho_j^1\} \in \mathbf{R}^{2m}$.

On résout ensuite (3.6) et on doit résoudre les équations linéaires algébriques

$$
(5.16) \qquad (\sigma(T), w_j) = 0 \ , \ (\sigma'(T), w_j) = 0 \ , \ 1 \leq j \leq m.
$$

C'est un système linéaire dans $\mathbf{R}^{2m}$. On a existence si l'on a unicité, i.e. si

$$
(5.17) \qquad \frac{\partial\varrho}{\partial\nu} = 0 \ sur \ \Sigma_0 \ entra\hat{i}ne \ \varrho \equiv 0.
$$

Mais

$$
(5.18) \qquad \begin{aligned}
&\frac{\partial\varrho}{\partial\nu} = \sum_{j=1}^{m} a_j(t) \frac{\partial w_j}{\partial\nu},\\
&a_j(t) = \varrho_j^0 \cos(T-t)\sqrt{\lambda_j} + \frac{\varrho_j^1}{\sqrt{\lambda_j}} \ \sin(T-t)\sqrt{\lambda_j}.
\end{aligned}
$$

Faisons donc *l'hypothèse*

(5.19) les fonctions $\dfrac{\partial w_j}{\partial\nu}$, $1 \leq j \leq m$, sont *linéairement indépendantes* sur Γ_0.

Alors (5.17) implique que

$$
(5.20) \qquad a_j(t) = 0 \ \ sur \ \ (0,T)
$$

donc

$$
\varrho_j^0 = \varrho_j^1 = 0
$$

d'où la CEE.

6. Variantes. Nous indiquons maintenant quelques variantes sur des exemples. Tous ces exemples sont eux mêmes susceptibles de rentrer dans des cadres plus généraux.

Exemple 6.1. Nous reprenons les situations des n^{os} précédents, mais avec un contrôle qui s'exerce *par la condition de Neumann.*

Donc l'état est donné (formellement, tant qu'on ne précise pas les espaces fonctionnels où sont prises les données) par

$$\text{(6.1)} \qquad \begin{array}{ll} y'' - \Delta y = 0 & \textit{dans } \Omega \times (0,T), \\ \dfrac{\partial y}{\partial \nu} = v & \textit{sur } \Sigma = \Gamma \times (0,T), \\ y(0) = y^0 \ , \ y'(0) = y^1 & \textit{dans } \Omega. \end{array}$$

Remarque 6.1. On pourrait également considérer le cas où v est à support dans

$$\Sigma_0 = \Gamma_0 \times (0,T) \ , \ \Gamma_0 \subset \Gamma.$$

Si ω est un ouvert $\subset \Omega$, on cherche la CEE au sens suivant: $T > 0$ est donné; peut-on, pour chaque y^0, y^1 (dans un espace fonctionnel convenable) trouver v (dans un espace fonctionnel convenable) tel que

$$\text{(6.2)} \qquad y(T;v) = y'(T;v) = 0 \textit{ dans } \omega.$$

Remarque 6.2. Bien sûr si $\omega = \Omega$ on retrouve la CEE habituelle, pour laquelle nous référons à J.L.Lions [1] [2] et à la Bibliographie de ces travaux.

On opère par analogie avec la méthode (du type RHUM) donnée au n.3. On considère $\{\varrho^0, \varrho^1\}$ comme "inconnues de base" et l'on résout

$$(6.3)\qquad \begin{array}{ll} \varrho'' - \Delta\varrho = 0 & \textit{dans } \Omega \times (0,T), \\ \varrho(T) = \varrho^0, \varrho'(T) = \varrho^1 & \textit{dans } \Omega, \\ \dfrac{\partial\varrho}{\partial\nu} = 0 & \textit{sur } \Sigma \end{array}$$

avec

$$(6.4)\qquad \varrho^0, \varrho^1 \textit{ sont nuls en dehors de } \bar{\omega}.$$

Admettons le problème (6.3) résolu. Nous considérons alors l'équation

$$(6.5)\qquad \begin{array}{ll} \sigma'' - \Delta\sigma = 0 & \textit{dans } \Omega \times (0,T), \\ \sigma(0) = y^0, \sigma'(0) = y^1 & \textit{dans } \Omega, \\ \dfrac{\partial\sigma}{\partial\nu} = +\varrho & \textit{sur } \Sigma \end{array}$$

et nous définissons l'opérateur *affine* M par

(6.6)
$$M\{\varrho^0, \varrho^1\} = \{\sigma'(T), -\sigma(T)\}_\omega = \text{restriction de } \{\sigma'(T), -\sigma(T)\} \text{ à } \omega.$$

Remarque 6.3. Dans le n^{os} précédents, on avait défini $M\{\varrho^1, \varrho^0\} = \{\sigma(T), -\sigma'(T)\}$. Il s'agit dans (6.6) d'une variante évidente! On était parti de $\{\varrho^1, \varrho^0\} \in G^0$ d'où le choix fait au $n^o 3$. Si l'on introduit ${}^vG^0$ image de G^0 par l'application $\{h^0, h^1\} \to \{h^1, -h^0\}$, on peut partir de $\{\varrho^0, \varrho^1\} \in {}^vG^0$ dans le cas général. C'est un détail de présentation.

Le problème de CEE est maintenant: trouver (si possible) $\{\varrho^0, \varrho^1\}$ tel que

$$(6.7)\qquad M\{\varrho^0, \varrho^1\} = 0.$$

On calcule

$$(6.8)\qquad < M\{\varrho^0, \varrho^1\}, \{\varrho^0, \varrho^1\} >= (\sigma'(T), \varrho^0) - (\sigma(T), \varrho^1).$$

Si l'on multiplie (6.5) par ϱ, il vient (par définition de la solution (faible) de (6.5), définie par transposition):

$$(6.9)\quad (\sigma'(T), \varrho^0) - (\sigma(T), \varrho^1) = \int_\Sigma \frac{\partial\sigma}{\partial\nu} \varrho d\Sigma - (y^0, \varrho'(0)) + (y^1, \varrho(0)).$$

On déduit de (6.9) (6.8) et (6.5) que

$$(6.10) \quad < M\{\varrho^0, \varrho^1\}, \{\varrho^0, \varrho^1\} >= \int_\Sigma \varrho^2 d\Sigma - (y^0, \varrho'(0)) + (y^1, \varrho(0)).$$

On fait alors l'hypothèse

$$(6.11) \quad (\int_\Sigma \varrho^2 d\Sigma)^{1/2} \text{ définit une norme sur l'espace des fonctions } \varrho^0, \varrho^1 \text{ assez régulières qui vérifient (6.4).}$$

Remarque 6.4. On a donné dans J.L.Lions [2] notamment — et dans l'Appendice 2 dû à C.Bardos, C.Lebeau et J.Rauch — des exemples où (6.11) a lieu en prenant seulement l'intégrale sur $\Sigma_0 \subset \Sigma$ et *sans condition sur* ϱ^0, ϱ^1. Evidemment ces cas *impliquent* (6.11).

Mais (6.11) peut être vérifié, lorsque ω est assez petit, sans que l'on ait CE.

On désigne alors par F_ω l'espace complété des fonctions ϱ^0, ϱ^1 nulles hors de $\bar{\omega}$ pour la norme (6.11).

La *caractérisation* de cet espace n'est pas facile. On a déjà introduit ces espaces lorsque $\omega = \Omega$ dans J.L.Lions loc. cit. Des précisions sur cet espace (avec $\omega = \Omega$) sont données dans I.Lasiecka et R.Triggiani [1].

On fait l'hypothèse que

$$(6.12) \quad \{\varrho^0, \varrho^1\} \to -(y^0, \varrho'(0)) + (y^1, \varrho(0)) \text{ est une forme linéaire continue sur } F_\omega.$$

Alors (6.10) *montre que (6.7) admet une solution unique.*

Remarque 6.5. Le problème revient à la recherche de

$$(6.13) \quad \inf_{\{\varrho^0, \varrho^1\} \in F_\omega} \left[\frac{1}{2} \int_\Sigma \varrho^2 d\Sigma - (y^0, \varrho'(0)) + (y^1, \varrho(0))\right].$$

Exemple 6.2. Nous considérons maintenant un exemple avec *contrôle ponctuel.*

On considère l'état donné par

$$
(6.14) \qquad \begin{array}{ll} y'' - \Delta y = v(t)\delta(x-b) & \textit{dans } \Omega \times (0,T) \\ y = 0 & \textit{sur } \Sigma, \\ y(0) = y^0, \ y'(0) = y^1 & \textit{dans } \Omega. \end{array}
$$

Dans (6.14), b est donné dans Ω et $\delta(x-b)$ désigne la masse de Dirac au point b. Soit par ailleurs ω donné ouvert $\subset \Omega$.

Soit $T > 0$ donné. On cherche, pour chaque couple y^0, y^1 (dans un espace fonctionnel convenable), une fonction contrôle v (si elle existe, et dans un espace fonctionnel convenable) telle que

$$
(6.15) \qquad y(T;v) = y'(T;v) = 0 \qquad \textit{dans } \omega.
$$

C'est un problème de CEE relatif à ω. On suit la même démarche que dans les n^{os} précédents. On part donc de ϱ solution de

$$
(6.16) \qquad \begin{array}{ll} \varrho'' - \Delta\varrho = 0 & \textit{dans } \Omega \times (0,T) \\ \varrho(T) = \varrho^0 \ , \ \varrho'(T) = \varrho^1 & \textit{dans } \Omega, \\ \varrho = 0 & \textit{sur } \Sigma \end{array}
$$

avec encore (6.4).

On considère ensuite (formellement pour l'instant) l'équation

$$
(6.17) \qquad \begin{array}{ll} \sigma'' - \Delta\sigma = \varrho(b,t)\delta(x-b) & \textit{dans } \Omega \times (0,T) \\ \sigma(0) = y^0, \ \sigma'(T) = y^1 & \textit{dans } \Omega, \\ \sigma = 0 & \textit{sur} \Sigma \end{array}
$$

et l'on définit

$$
(6.18) \qquad M\{\varrho^0, \varrho^1\} = \{\sigma'(T), -\sigma(T)\}_\omega
$$

Prenant le produit scalaire de (6.17) avec ϱ, on trouve que

(6.19)

$$
(\sigma'(T), \varrho^0) - (\sigma(T), \varrho^1) = \int_0^T \varrho(b,t)^2 dt - (y^0, \varrho'(0)) + (y^1, \varrho(0)).
$$

Donc

(6.20)

$$
< M\{\varrho^0, \varrho^1\}, \{\varrho^0, \varrho^1\} > \geq \int_0^T \varrho(b,t)^2 dt - (y^0, \varrho'(0)) + (y^1, \varrho(0)).
$$

Pour ϱ^0, ϱ^1 données assez régulières avec (6.4), on pose

$$\|\{\varrho^0, \varrho^1\}\|_{F_B} = \left(\int_0^T \varrho(b,t)^2 dt \right)^{1/2}. \tag{6.21}$$

On définit ainsi une *semi*-norme.

On fait *l'hypothèse qu'il s'agit d'une norme.*

Remarque 6.6. Décider si (6.21) est, ou non, une norme n'est pas une question simple. Supposons, pour un peu simplifier, que le spectre de $-\Delta$ dans Ω pour les conditions de Dirichlet est simple. Soient w_j les fonctions propres.

Alors, pour avoir une *norme*, il est *nécessaire* que le point b soit *stratégique*, i.e. que

$$w_j(b) \neq 0 \quad \forall j. \tag{6.22}$$

Cette condition — assez curieusement... — n'est pas suffisante, un contre exemple (absolument non trivial !) étant donné par Y.Meyer [1], dans le cas $\omega = \Omega$. Evidemment les "chances" d'avoir une norme augmentent lorsque ω est plus petit!

Désignons par F_b l'espace complété pour la norme (6.21). La structure de cet espace est compliquée. C'est un espace très grand, qui contient des éléments qui ne sont plus des distributions mais des ultra-distributions, cf. (pour le cas $\omega = \Omega$) A.Haraux [1].

On obtient alors pour M un opérateur affine continu de F_b dans F_b' et dont la partie linéaire est un isomorphisme. Si donc

$$\{\varrho^0, \varrho^1\} \to -(y^0, \varrho'(0)) + (y^1, \varrho(0))$$

est une forme linéaire continue sur F_b, on a CEE.

Remarque 6.7. Dans le cas où $\omega = \Omega$, une étude numérique intéressante est donnée dans A.El Jai et A.Gonzalez [1].

Terminons par deux remarques supplémentaires:

Remarque 6.8. Tout ce qui vient d'être fait n'est pas restreint aux problèmes *hyperboliques.* On peut adapter tout ce qui a été fait à des opérateurs du type Petrowsky, des opérateurs à mémoire etc., comme ceux considérés dans J.L.Lions [3].

Remarque 6.9. Une extension *non linéaire* très intéressante de la méthode HUM a été donnée par E.Zuazua [1]. La méthode de Zuazua est susceptibile de s'adapter aux problèmes de CEE *non linéaires.*

Bibliographie

C.Bardos, G.Lebeau, J.Rauch [1], Appendice II de J.L.Lions [2].

A.El Jai, A.Gonzales [1], *Actionneurs frontières pour l'exacte contrôlabilité de systèmes hyperboliques,* Rapport Université de Perpignan, Décembre 1987.

R.Glowinski, C.H.Li, J.L.Lions [1], *A numerical approach to the exact boundary controllability of the wave equation* (I) Dirichlet controls: description of the numerical methods, Japan Journal of Applied Math. 1989.

A.Haraux [1], *Contrôlabilité exacte d'une membrane rectangulaire au moyen d'une fonctionnelle analytique localisée.* C.R.A.S. 1988.

I.Lasiecka, R.Triggiani [1], *Exact controllability for the wave equation with Neumann boundary control,* à paraître.

J.L.Lions [1], *Exact controllability, stabilization and perturbations for distributed systems.* J.Von Neumann Lecture, Boston 1986, SIAM Rev. March 1988. [Developpements de la note C.R.A.S. Paris **302** (1986), 471-475].

J.L.Lions [2], *Contrôlabilité exacte des systèmes distribués. Vol.1. Contrôlabilité exacte,* R.M.A. Masson, 1988 (Notes du cours au Collège de France, rédigées par E.Zuazua).

J.L.Lions [3], *Contrôlabilité exacte des systèmes distribués, Vol. 2 Perturbations,* R.M.A. Masson, 1988

J.L.Lions, E.Magenes[1], *Problèmes aux limites non homogènes et applications,* Vol. 1 et 2, Dunod, Paris,1968.

Y.Meyer [1], Travail à paraître.

D.L.Russel [1], *Controllability and stabilization theory for linear partial differential equations. Recent progress and open questions,* SIAM Rev. **20** (1978), p.639-739.

E.Zuazua [1], *Contrôlabilité exacte de systèmes d'évolution non linéaires.* C.R.A.S. Paris,1988.

Collège de France
11, Pl. Marcelin Berthélot
F-75005 PARIS

LOW AND HIGH FREQUENCY VIBRATION IN STIFF PROBLEMS

MIGUEL LOBO-HIDALGO   ENRIQUE SANCHEZ-PALENCIA

Dedicated to Ennio De Giorgi on his sixtieth birthday

1. Introduction. Stiff problems are partial differential equation problems with very different values of the coefficients in two regions of the domain Ω of the space variables. Very different means that the ratio of coefficients is described by a parameter $\epsilon \to 0$. Stiff problems originated in Lions [4]. The behavior of the corresponding eigenvalue problems depends highly on the form of the H-like space (in the standard framework $V \subset H$ with dense and compact imbedding) and in particular on the fact that the corresponding coefficients do depend or not on ϵ. We refer to Lions [5], Geymonat et al. [1,2], Panasenko [7], Gibert [3] and Sanchez-Palencia [8] for the study of several kinds of eigenvalue stiff problems. It turns out that the case when the coefficients in front of the H-like space do not depend on ϵ was not very widely explored, as the "low frequency" region does not provide a good insight on the vibration problems all over Ω. We consider here such problems, in particular the high frequency region. In order to obtain properties of the spectral family we use a Fourier transform technique starting from the solutions of the corresponding hyperbolic problems involving second order derivatives with respect to time. (We refer to Lobo and Sanchez Palencia [6] and Sanchez-Palencia [8], chap. 12, sect. 3 for other problems using this method).

Let us consider a model problem for the Laplacian. Let Ω be a bounded domain of $\mathbf{R}^N$, containing the origin, divided in two parts Ω_o and Ω_1 by the interface Γ (= the hyperplane $x_1 = 0$):

$$\Omega = \Omega_o \cup \Omega_1 \cup \Gamma.$$

Let ϵ be a positive small parameter. We consider a standard vibration problem in the spaces $V = H_o^1(\Omega)$ and $H = L^2(\Omega)$:

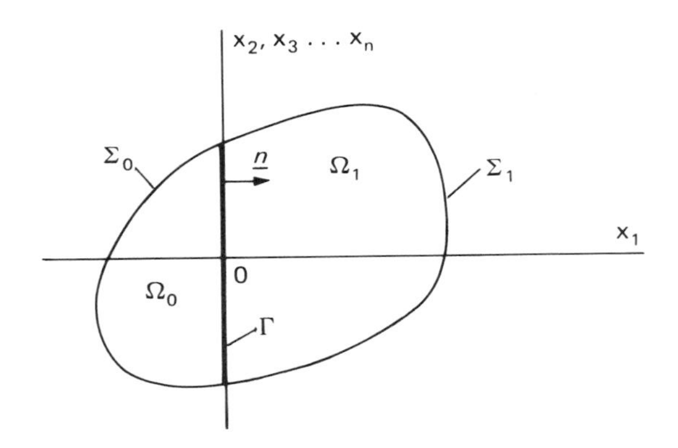

Figure 1

Find $u^\epsilon(t)$ with values in V satisfying

$$(1.1) \qquad (\ddot{u}^\epsilon, v) + a_o(u^\epsilon, v) + \epsilon a_1(u^\epsilon, v) = 0 \quad \forall v \in V$$

$$(1.2) \qquad u^\epsilon(0) = \alpha; \qquad \dot{u}^\epsilon(0) = \beta$$

where

$$(1.3) \qquad (u, v) = \int_\Omega u \; v \; dx$$

$$(1.4) \qquad a_i(u, v) = \int_{\Omega_i} \text{grad } u \cdot \text{ grad } v \; dx, \qquad i = 0, 1.$$

Clearly (1.1) amounts to

$$\ddot{u}_o^\epsilon - \Delta u_o^\epsilon = 0 \quad \text{in } \Omega_o \tag{1.5}$$

$$\ddot{u}_1^\epsilon - \epsilon\Delta u_1^\epsilon = 0 \quad \text{in } \Omega_1 \tag{1.6}$$

$$u_o^\epsilon = 0 \ \text{ on } \ \Sigma_o \ \text{ and } \ u_1^\epsilon = 0 \ \text{ on } \ \Sigma_1 \tag{1.7}$$

$$u_o^\epsilon = u_1^\epsilon \text{ on } \Gamma \tag{1.8}$$

$$\frac{\partial u_o^\epsilon}{\partial n} = \epsilon \frac{\partial u_1^\epsilon}{\partial n} \text{ on } \Gamma \tag{1.9}$$

where the indices 0, 1 denote the restriction of the functions to the subdomains Ω_o, Ω_1.

Remark 1.1 It will prove useful to define the limit vibration problem (1.2), (1.3) with $\epsilon = 0$. In this case, H_o^1 must be replaced by

$$V_* = \{u \in L^2(\Omega);\ u_o \in H^1(\Omega_o), u_o = 0 \text{ on } \Sigma_o\}. \tag{1.10}$$

This is not a standard vibration problems as the embedding of V_* into $H = L^2$ is no longer compact. The classical formulation is (1.5)-(1.7) and (1.9) with $\epsilon = 0$. The condition (1.8) of the case $\epsilon > 0$ desappears, as the trace on Γ of u_1 does not make sense. Of course, (1.9) becomes a Neumann boundary conditionfor u_o, and there are not boundary conditions for u_1. Clearly, $V = H_o^1$ is dense in V_*.

The spectral problem associated (1.2) is obviously:

$$a^o(u,v) + \epsilon a^1(u,v) = \lambda(u,v). \tag{1.11}$$

For reasons which will be clear later, we shall write sometimes $\lambda+1 = \mu$, and (1.11) reads:

$$a^o(u,v) + \epsilon a^1(u,v) + (u,v) = \mu(u,v) \tag{1.12}$$

which corresponds to the evolution problem

$$(\ddot{u}^\epsilon, v) + a^o(u,v) + \epsilon a^1(u,v) + (u,v) = 0 \tag{1.13}$$

instead of (1.2).

2. Low frequency vibrations. We seek for eigenvalues of order ϵ, i.e.:

$$\lambda = \mathcal{E}\zeta; \quad \zeta = O(1) \tag{2.1}$$

which were considered by Panasenko [7], Gibert [3], Lions [5]; we only give some indications hereafter. The formulation of the problem (1.11) in terms of equations and boundary conditions after the rescaling (2.1) is

$$-\Delta u_o^\epsilon = \epsilon\zeta u_o^\epsilon \quad \text{on } \Omega_o \tag{2.2}$$

$$-\Delta u_1^\epsilon = \zeta u_1^\epsilon \quad \text{on } \Omega_1 \tag{2.3}$$

$$u_o^\epsilon = 0 \text{ on } \Sigma_o \text{ and } u_1^\epsilon = 0 \text{ on } \Sigma_1 \tag{2.4}$$

$$u_o^\epsilon = u_1^\epsilon \quad \text{on } \Gamma \tag{2.5}$$

$$\frac{\partial u_o^\epsilon}{\partial n} = \epsilon \frac{\partial u_1^\epsilon}{\partial n} \quad \text{on } \Gamma. \tag{2.6}$$

Let us seek for the expansions:

$$u^\epsilon = u^o + \epsilon u^1 + \dots \tag{2.7}$$

$$\zeta^\epsilon = \zeta^o + \epsilon\zeta^1 + \dots \tag{2.8}$$

The substitution of this into (2.2)-(2.6) gives at order O(1):

$$-\Delta u_o^o = 0 \quad \text{on } \Omega_o \tag{2.9}$$

$$u_1^o = 0 \quad \text{on} \quad \Sigma_1 \tag{2.10}$$

$$\frac{\partial u_1^o}{\partial n} = 0 \quad \text{on} \quad \Gamma \tag{2.11}$$

for u_1^o which implies $u_1^o = 0$, and then

$$-\Delta u_1^o = \zeta^o u_1^o \quad \text{on} \quad \Omega_1 \tag{2.12}$$

$$u_1^o = 0 \quad \text{on} \quad \Sigma_o \quad \text{and} \quad \Gamma \tag{2.13}$$

which implies that u_1^o and ζ^o are eigenvector and eigenvalue of the Dirichlet problem in Ω_1. Prescribing the normalization condition

$$(u_1^\epsilon, u_1^o)_{L^2(\Omega_1)} = 1 \quad \Leftrightarrow \quad (u_1^i, u_1^o)_{L^2(\Omega_1)} = \delta_{i1} \tag{2.14}$$

and assuming that the eigenvalue is simple, we may consider u_1^o, ζ^o as known. Then, at order ϵ, we have

$$-\Delta u_o^1 = 0 \quad \text{on} \quad \Omega_o \tag{2.15}$$

$$u_o^1 = 0 \quad \text{on} \quad \Sigma_o \tag{2.16}$$

$$\frac{\partial u_o^1}{\partial n} = \frac{\partial u_1^o}{\partial n} \quad \text{on} \quad \Gamma \tag{2.17}$$

which defines u_o^1 uniquely. For u_1^1 we have

$$(-\Delta - \zeta^o) u_1^1 = \zeta^1 u_1^o \quad \text{on} \quad \Omega_1 \tag{2.18}$$

$$u_1^1 = 0, \quad \text{on} \quad \Sigma_1; \quad u_1^1 = u_o^1 \quad \text{on} \quad \Gamma \tag{2.19}$$

where ζ^1 is not known. It is determined from the compatibility condition; then u_1^1 is uniquely determined by the orthogonality condition (2.14). The other terms of the expansion are determined in the same way.

The justification of the expansions (2.7),(2.9) may be done by methods analogous to those of Geymonat and al., [1,2].

3. High frequency vibration in Ω_0. Coming back to (1.11) or (1.12), we are studying now the eigenvalues with λ or μ of order 1; they are called of "high frequency" by opposition to the preceding ones; in fact the term "medium frequencies" would be more appropriate. For reasons which will be clear later (Remark 4.6) we are not able to give an asymptotic expansion for the eigenvalues and eigenfunctions. Instead we shall give the limit behavior of the spectral family.

Let us denote by A_ϵ the operator associated with the form $a^o + \epsilon a^1$ via the Lax-Milgram theorem, and let $B_\epsilon = A_\epsilon + I$ so that (1.13) reads

$$\ddot{u}^\epsilon + B_\epsilon u^\epsilon = 0 \tag{3.1}$$

and similarly let A_* be the operator associated with the form a^o, and $B_* = A_* + I$. This operator is coercive on V_*, defined in (1.10), which is the appropriate space for the limit problem (1.13) with $\epsilon = 0$, which is

$$\ddot{u} + B_* u^* = 0. \tag{3.2}$$

Then, we have

Lemma 3.1. *Let $v \in L^2(\Omega)$, and let u^ϵ, u^* be the solutions of (3.1), (3.2) with the initial conditions*

$$u^\epsilon(0) = 0,\ \dot{u}^\epsilon(0) = v; \qquad u^*(0) = 0,\ \dot{u}^*(0) = v \tag{3.3}$$

respectively. Then,

$$\begin{cases} u^\epsilon \to u^* & \textit{weakly* in } L^\infty(-\infty,+\infty;V^*) \\ \dot{u}^\epsilon \to \dot{u}^* & \textit{weakly* in } L^\infty(-\infty,+\infty;H). \end{cases} \tag{3.4}$$

PROOF. The energy equality for (3.1),(3.3) reads:

$$\|\dot{u}^\epsilon(t)\|_H^2 + a_o(u^\epsilon(t)) + \epsilon a_1(u^\epsilon(t)) + \|u^\epsilon(t)\|_H^2 = \|v\|_H^2 \tag{3.5}$$

and we may extract subsequences (in fact the sequence itself as the limit will be proved to be unique) converging in the sense of (3.4). Moreover

$$\epsilon a_1(u^\epsilon(t)) \leq C, \qquad t \in]-\infty, +\infty[\tag{3.6}$$

and it follows from (3.3),(3.4) and the trace theorem for $t = 0$ that $u^*(0) = 0$. We must prove that the limit u^* is the solution of (3.2), (3.3). To this end, we use the characterization of u^ϵ of Lobo and Sanchez [6] or Sanchez [8], chap 12, sect.3:

$$\int_0^T \{[a_o(u^\epsilon, w) + \epsilon a_1(u^\epsilon, w) + (u^\epsilon, w)_H] \ \varphi(t) - (u^\epsilon, w)_H \dot{\varphi}(t) = \tag{3.7}$$

$$= (v, w)_H \ \varphi(0)$$

for $w \in V$ (or a dense subset of) and $\varphi \in C^1([0,T])$, with $\varphi(0) = 0$. Fixing $w \in V$ and passing to the limit according to (3.4) and (3.6) we obtain for u^* an expression analogous to (3.7) without the term ϵa^1. As V is dense in V_* (Remark 1.1), this is the characterization of the solution of (3.2), (3.3).

Now, denoting by $\mathcal{E}(B_\epsilon, \lambda)$, $\mathcal{E}(B_*, \lambda)$ the spectral families of B_ϵ and B_*, let us take the inner product of ($)_2$ with $w \in H$ and then the Fourier transform. The solution u^ϵ reads

$$u^\epsilon(t) = A^{-1/2} \sin(tA^{1/2})v$$

$$\dot{u}^\epsilon(t) = \cos(tA^{1/2})v.$$

Taking the Fourier transform of the second of (3.4) and on account of the fact that the spectral family of $A^{1/2}$ vanishes for $\lambda \leq 0$, we easily obtain

$$\frac{d}{d\lambda}(\mathcal{E}(B_\epsilon^{1/2}, \lambda)v, w)_H \quad \rightarrow \quad \frac{d}{d\lambda}(\mathcal{E}(B_*^{1/2}, \lambda)v, w)_H \tag{3.8}$$

in $\mathcal{S}'(-\infty, +\infty)$. Then, using the obvious estimate

$$|(\mathcal{E}(B, \lambda)v, w)_H| \leq \|v\|_H \|w\|_H$$

we may pass in (3.8) from the derivatives to the functions, and even to the spectral families of B instead of $B^{1/2}$ (see the proof of theorem

IX, 4.2 for details). Finally, by a unit shift of the variable λ we pass to the spectral families of $A = B - I$ and obtain

Theorem 3.2. *The spectral family of A_ϵ converges to that of A_* as $\epsilon \to 0$ in the sense*

$$(\mathcal{E}(A_\epsilon, \lambda)v, w)_H \quad \to \quad (\mathcal{E}(A_*, \lambda)v, w)_H$$

for any $v, w \in H$, weakly in $L^\infty(-\infty, +\infty)$.*

As we pointed out in Remark 1.1, in the limit problem there is no coupling between Ω_o and Ω_1. In fact, the operator in Ω_1 is the null operator (and then, $\lambda = 0$ is eigenvalue with the entire space $L^2(\Omega_1)$ as eigenspace), and the operator in Ω_o is $-\Delta$ with Dirichlet and Neumann boundary conditions on Σ_o and Γ, respectively. Such is the limit behavior of the spectral family in Ω_o, but not in Ω_1, where all the spectral family appears flattened to 0. Another asymptotic process is needed to furnish a deeper insight on the vibration in Ω_1. Before going on, let us consider Fig. 2 and 3 which describe the spectral family in the region Ω_1 in a somewhat allegorical way . In Fig. 2 we have the eigenvalues $\lambda = \epsilon\zeta$, $\zeta = O(1)$ which provide a good decription of the low frequency phenomena. Of course, as $\zeta \to +\infty$ the spectral family tends to the asymptote $\mathcal{E} = I$. Thus, when studying such a function in the variable $\lambda = \epsilon\zeta$, the function is flattened to zero, and we only see a step at the origin (Fig.3). This is the reason why the study of part b furnished a good description of the region Ω_o but not of Ω_1. Now, let us think about the fact the family of eigenvalues tends to $+\infty$ (for each fixed ϵ); then, in Ω_1 there are vibrations (the high frequency vibrations) with difference between the curve and the asymptote. In order to see these vibrations, which have a short wave lenght, we shall dilate the space variable x, and of course we may do a new normalization in order to see the eigenfunctions (Fig. 3). This we proceed to do:

4. High frequency vibrations in Ω_1. Geometrical acoustics.

According to the preceding considerations we perform the di-

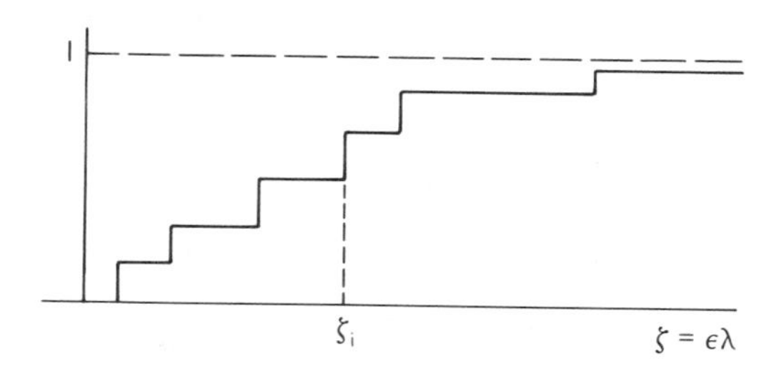

Figure 2

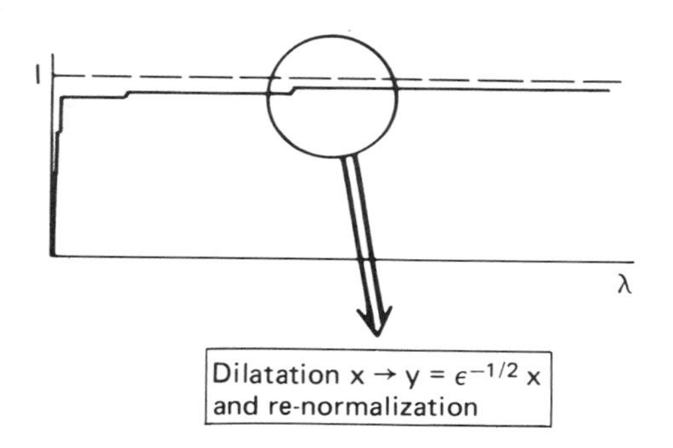

Figure 3

latation

$$x \rightarrow y = x\epsilon^{-1/2} \tag{4.1}$$

and we denote the transformed domains with the index ϵ :

$$\Omega_\epsilon = \epsilon^{-1/2}\Omega; \qquad \Omega_{j\epsilon} = \epsilon^{-1/2}\Omega_j; \qquad j = 0,1 \tag{4.2}$$

which are obviously "large" domains, tending to the space $\mathbf{R}^N$ or the half space as $\epsilon \rightarrow 0$. The equations (1.5) and (1.6) become

$$\ddot{u}_o^\epsilon - \epsilon^{-1}\Delta_y u_o^\epsilon = 0 \quad \text{on} \quad \Omega_{o\epsilon} \tag{4.3}$$

$$\ddot{u}_1^\epsilon - \Delta_y u_1^\epsilon = 0 \quad \text{on} \quad \Omega_{1\epsilon} \tag{4.4}$$

and (1.7)-(1.9) remain unchanged on the transformed boundaries.

We also define bilinear forms on the transformed domains:

$$a_{j\epsilon}(u,v) = \int_{\Omega_{j\epsilon}} \underline{\mathrm{grad}}_y u \cdot \underline{\mathrm{grad}}_y v dy, \qquad j = 0,1. \tag{4.5}$$

Then, the evolution problem analogous to (1.13) reads

$$(\ddot{u}^\epsilon, v) + \epsilon^{-1} a_{o\epsilon}(u^\epsilon, v) + a_{1\epsilon}(u^\epsilon, v) + (u^\epsilon, v) = 0 \quad \forall v \in H_o^1(\Omega_\epsilon) \tag{4.6}$$

or, with an obvious notation,

$$\ddot{u}^\epsilon + \tilde{B}_\epsilon = 0, \ \tilde{B}_\epsilon = \tilde{A}_\epsilon + I. \tag{4.7}$$

Let us define a problem which will appear later as the limit of the preceding one. Let

$$\mathbf{R}^n_\pm = \{y \in \mathbf{R}^N; \ \pm y_1 > 0\} \tag{4.8}$$

and let consider the evolution problem (the initial conditions will be fixed later):
(4.9)

$$\begin{cases} \text{Find } u^*(t) \text{ with values in } H_o^1(\mathbf{R}^N_+) \text{ such that } \forall v \in H_o^1(\mathbf{R}^N_+) \\ \int_{\mathbf{R}^N_+} \ddot{u}^* v dy + \int_{\mathbf{R}^N_+} (\underline{\mathrm{grad}}_y u^* \cdot \underline{\mathrm{grad}}_y v + u^* v) dy = 0 \end{cases}$$

which is obviously equivalent to

$$\ddot{u}^* + \tilde{B}^* u^* = 0; \quad \tilde{B}^* = \tilde{A}^* + I; \quad \tilde{A}^* = -\Delta_y \tag{4.10}$$

on $H_o^1(\mathbf{R}^N_T)$; we then emphasize that the Dirichlet boundary condition $u = 0$ is satisfied on $x_N = 0$. We then have:

Lemma 4.1. *Let $v \in L^2(\mathbf{R}^N)$ and denote by v_ϵ and v_+ its restrictions to Ω_ϵ and $\mathbf{R}^N_+$ respectively. Moreover, let u^ϵ and u^* be the solutions of (4.7) and (4.10) with the initial conditions*

$$u^\epsilon(0) = 0, \ \dot{u}^\epsilon(0) = v_\epsilon \tag{4.11}$$

$$u^*(0) = 0, \ \dot{u}^*(0) = v_+ \tag{4.12}$$

respectively. We consider u^ϵ, u^* extended by 0 to $\mathbf{R}^N \backslash \Omega_\epsilon$ and $\mathbf{R}^N_-$ respectively. Then,

$$u^\epsilon \to u^* \quad \textit{weakly in} \quad L^\infty(-\infty, +\infty; H^1(\mathbf{R}^N)) \tag{4.13}$$

$$\dot{u}^\epsilon \to \dot{u}^* \quad \textit{weakly in} \quad L^\infty(-\infty, +\infty; L^2(\mathbf{R}^N)) \tag{4.14}$$

PROOF. The energy equality for (4.7), (4.11), with an obvious notation, reads:

$$\|\dot{u}^\epsilon\|^2_{\Omega_\epsilon} + \epsilon^{-1} a_{o\epsilon}(u^\epsilon) + a_{1\epsilon}(u^\epsilon) + \|u^\epsilon\|^2_{\Omega_\epsilon} = \|v_\epsilon\|^2_{\Omega_\epsilon} \leq \|v\|^2_{\mathbf{R}^N} \tag{4.15}$$

Then, by extraction of a subsequence (the whole sequence) we have (4.13) and (4.14) for some u^* belonging to the space quoted in these relations. In addition, the trace thoerem at $t = 0$ shows that $u^*(0) = 0$. Moreover, on account of (4.13) and of the coefficient ϵ^{-1} in (4.15),

$$\underline{\operatorname{grad}}_y\ u^\epsilon \to \underline{\operatorname{grad}}_y\ u^* = 0 \text{ in } L^\infty(-\infty, +\infty; \mathbf{R}^N_-)$$

i.e. u^* is constant on $\mathbf{R}^N_-$ for each t; as it belongs to $H^1(\mathbf{R}^N)$, it vanishes on $\mathbf{R}^N_-$, as well as its derivative with respect to time. Consequently, u^* takes values in $H^1_o(\mathbf{R}^N_+)$, continued with value 0 to $\mathbf{R}^N_-$. Let us check that u^* is the solution of (4.10),(4.12). To this end, we use, as previously in this section, the characterization of u^ϵ :

$$\begin{aligned} \int_0^T \{[\epsilon^{-1} a_{o\epsilon}(u^\epsilon, w) + a_{1\epsilon}(u^\epsilon, w) + (u^\epsilon, w)]\varphi(t) - \\ - (\dot{u}^\epsilon, w)\dot{\varphi}(t)\}dt = (v, w)\varphi(0) \end{aligned} \tag{4.16}$$

valid for any w belonging to a dense set of $H^1_o(\Omega_\epsilon)$ (extended to $\mathbf{R}^N$ with value 0) and φ of class $C^1([0, T])$ with $\varphi(0) = 0$. Then, choosing $w \in \mathcal{D}(\mathbf{R}^N_+)$ continued with value 0 to $\mathbf{R}^N_-$, the term in ϵ^{-1} of (4.16) disappears, and the integral on its right side becomes obviously an inner product in $L^2(\mathbf{R}^N_+)$, then, passing to the limit $\epsilon \to 0$ with (4.13), (4.14) and noting that $\mathcal{D}(\mathbf{R}^N_+)$ is dense in $H^1_o(\mathbf{R}^N_+)$ we see that u^* satisfies the characterization of the (4.10), (4.12).

The limit behavior of the spectral family is obtained by the same routine as Theorem 3.2. We note in this respect that the fact of extending u^ϵ and u^* from Ω_ϵ and $\mathbf{R}^N_+$ to $\mathbf{R}^N_+$ with value zero imply that the corresponding spectral families are also extended with value 0. In fact we have:

Theorem 4.2. *Let $\mathcal{E}(\tilde{A}_\epsilon, \lambda)$ and $\mathcal{E}(\tilde{A}^*, \lambda)$ be the spectral families of the operators $\tilde{A}_\epsilon, \tilde{A}^*$ defined in (4.7) (4.10). They converge in the sense*

$$(4.17) \qquad (\mathcal{E}(\tilde{B}_\epsilon, \lambda)v_\epsilon, w_\epsilon)_{L^2(\Omega_\epsilon)} \to (\mathcal{E}(\tilde{B}^*, \lambda)v_+, w_+)_{L^2(\mathbf{R}^N_+)}$$

for any $v, w \in L^2(\mathbf{R}^N)$, weakly in $L^\infty(-\infty, +\infty)$.*

Here the indices $\epsilon, +$ denote the restrictions to Ω_ϵ and $\mathbf{R}^N_+$, as in Lemma 4.1.

Remark 4.3. We note that $\tilde{B}_\epsilon$ is the transmission operator on the dilated domain in Ω_ϵ; it has a compact resolvent and consequently its spectral family is a step function. Oppositely, $\tilde{B}^*$ is merely $-\Delta_y$ in $\mathbf{R}^N$ with a Dirichlet boundary condition on $y_N = 0$. The corresponding spectral family is classical; it is continuous, and is described by plane waves with arbitary direction and wavelenght but satisfying $u = 0$ on $y_N = 0$, i.e. satisfying the reflection laws on the boundary. It may be said that the limit behavior of the waves is given by the geometrical acoustics in $\mathbf{R}^N_+$ with boundary condition $u = 0$ on $y_N = 0$.

Remark 4.4. It is clear that Theorem 4.2 gives information on the limit behavior after the homothety (4.1); consequently it has some local character (i.e. for x close to $x = 0$). In order to have information on the limit spectral family at another point $0' \neq 0$, we may perform the change $y = \epsilon^{-1/2}(x - 0')$. Obviously the results are anlogous for $0' \in \Gamma$. On the other hand, if $0'$ is an interior point to Ω_1, we have $\mathbf{R}^N$ instead of $\mathbf{R}^N_+$ in theorem, i.e. the limit spectral family is merely that of $-\Delta_y$ in $\mathbf{R}^N$, without boundary condition. In the same way, in the case of a boundary with a corner (Fig. 4) the limit spectral family is that of $-\Delta_y$ in an infinite angular sector with $u = 0$ on its boundary.

Remark 4.5. We recall the part of the spectrum considered in sect. 3 and 4 is $\lambda = O(1)$. It appears from sect. 3 (resp. sect.

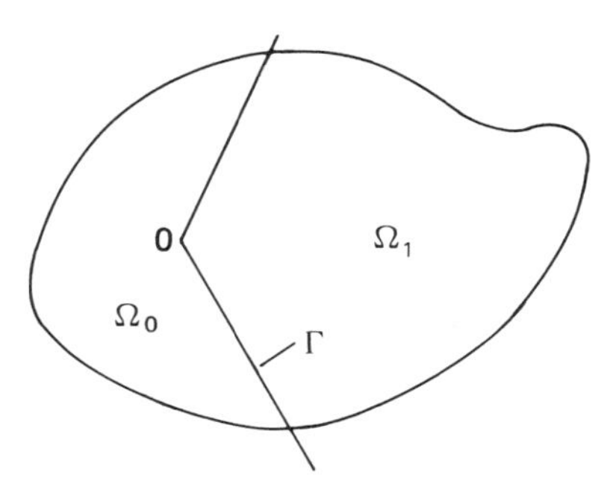

Figure 4

4) that the limit behavior in the region Ω_o (resp. Ω_1) is a discrete (resp. continuous) spectrum. This shows that we cannot "solve" the problem in Ω_1 to get a modified problem in Ω. This situation is very different from that of sect. 2 and explains why we are only able to describe the limit behavior, but not to exhibit an asymptotic expansion.

Remark 4.6. We emphasize that, according to theorem 4.2, the limit behavior of the vibrations in Ω_1 with $\lambda = O(1)$ satisfies a Dirichlet condition on $y_1 = 0$, i.e. on the interface Γ. This is natural for the values of the limit problem in Ω_o (see the preceding Remark if necessary) as the corresponding vibration vanishes in Ω_o. Nevertheless, for the λ of the preceding set, there is certainly some coupling between the vibration in Ω_o and Ω_1, and the latter does not satisfy $u|_\Gamma = 0$. This intuitive assertion is not in contradiction with Theorem 4.2., as the convergence in it holds in the weak* topology of L^∞ of the variable λ. This is a very poor convergence, which is consistent with the existence of narrow (as $\epsilon \to 0$) regions of the variable λ where other phenomena appear. This question seems to be open.

References

[1] G.Geymonat, M.Lobo-Hidalgo, E.Sanchez-Palencia, *Spectral properties of certain stiff problems in elasticity and acoustic,* Math. Appl. Sci. **4** (1982), 291-306.

[2] G.Geymonat, E.Sanchez-Palencia, *Spectral properties of certain stiff problems in elasticity and acoustics,* Part II, Proceedings of the Centre for the Mathematical Analysis, Australian National Univ. **5** (1984), 15-38.

[3] P.Gibert, *Les basses et moyennes frequences dans les structures fortement hétérogènes,* Compt. Rend. Acad. Sci. Paris **295** (1982), 951-954.

[4] J.L.Lions *Perturbations singulières dans les problèmes aux limites et en contrôle optimal,* Lect. Notes in Math. **323**, Springer (1973).

[5] J.L.Lions, *Remarques sur les problèmes d'homogénéisation dans les milieux à structure périodique et sur quelques problèmes raides,* in Les Méthodes de l'Homogénéisation, Eyrolles, Paris (1985), 129-228.

[6] M.Lobo-Hidalgo, E.Sanchez-Palencia, *Sur certaines propriétés spectrales des perturbations de domaine dans les problèmes aux limites,* Comm. Part. Diff. Eq. **4** (1979), 1085-1098.

[7] G.P.Panasenko, *Asymptotics of the solutions and eigenvalues of elliptic equations with strongly varying coefficients,* Dokl. Akad. Nauk. **252** (1980), 1320-1324 (Soviet Math. Doklady **21**, 1980, 942-947).

[8] E.Sanchez-Palencia, *Non-homogeneous media and vibration theory,* Springer Lect. Notes in Physics **127** (1980).

Departamiento de Ecuaciones Funcionales
Facultad de Ciencias
Avenida de los Castros
SANTANDER - Spain

Laboratoire Modélisation Mécanique
Université Pierre et Marie Curie
4, Place Jussieu
F-75252 PARIS

A TIME-DISCRETIZATION SCHEME APPROXIMATING THE NON-LINEAR EVOLUTION EQUATION $u_t + ABu = 0$

ENRICO MAGENES

Dedicated to Ennio De Giorgi on his sixtieth birthday

The aim of this paper is to extend a time discretization scheme to the abstract differential equation in the Hilbert spaces

$$(I) \qquad u_t \ + \ ABu = 0 \quad , \quad u(0) = u_0$$

where A and B are suitable operators with A linear and B monotone. The scheme has been introduced in [4] and more recently studied in [21], [22], [23] for the case of the partial differential parabolic equation of the Stefan-type.

The algorithm is formally the following:

$$(II) \qquad \begin{cases} U^0 = u_0 \\ \Theta^n + \frac{\tau}{\mu} A\Theta^n = BU^{n-1} \\ U^n = U^{n-1} + \mu\{\Theta^n - BU^{n-1}\} \quad , \quad n = 1, 2, \dots \end{cases}$$

where τ is the time-step and μ s a relaxation parameter. It allows to calculate the unknowns U^n and Θ^n, approximating respectively the functions $u(t)$ and $\theta(t) = Bu(t)$, through two simpler equations containing separately the operators A and B. In the applications both these functions u and θ represent significant physical quantities.

The outline of the paper is the following: Section 1 is devoted to state the assumptions and the results. Sections 2 and 3 are devoted to prove the results. Section 4 contains the application to many interesting concrete equations, as the parabolic equations of Stefan-type, certain linear and non-linear pseudoparabolic (or Sobolev) equations.

The author express his aknowledgements to C.Verdi and A.Visintin for the helpful discussions on the subject to this paper.

1. Notations, assumptions and results. Let V and H two real Hilbert spaces with

$$V \subset H, \text{ with continuous injection; } V \text{ dense in } H; \tag{1.1}$$

we denote by $<,>$ either the inner product on H or the pairing between V' and V and by $|\cdot|$, $\|\cdot\|$, $\|\cdot\|_*$ the norms respectively in H, V and V'.

Let

$$\begin{cases} a(u,v) \text{ a bilinear, symmetric, continuous form on} \\ V \times V, \text{ coercive on } V, \text{ i.e. there exist } \alpha > 0 \\ \text{such that } a(v,v) \geq \alpha\|v\|^2 \quad \forall v \in V. \end{cases} \tag{1.3}$$

Let us denote by A the isomorphism of V onto V' defined by

$$< Au, v >= a(u,v) \quad , \quad \forall u,v \in V. \tag{1.4}$$

Let B an operator of H into H satisfying the two following conditions:

$$\begin{cases} B0 = 0 \quad \text{and there exist } \bar{\mu} > 0 \quad \text{such that} \\ \bar{\mu}|Bu - Bv|^2 \leq < Bu - Bv, u - v > \quad \forall u,v \in H \end{cases} \tag{1.5}$$

$$\begin{cases} \text{there exists a function } \Phi \text{ from } H \text{ to } \mathbf{R} \\ \text{Gateaux} - \text{differentiable in } H \text{ with } \partial\Phi = B\,. \end{cases} \tag{1.6}$$

Then we can deduce the following remarks:

a) By (1.5) we have that B is a *monotone* and *Lipschitz-continuous* operator in H. Consequently B is also *maximal monotone*. Moreover, as a consequence of (1.6) (cf. e.g. [12]) we have that Φ is *convex* and *bounded from below* because $\Phi(u) \geq \Phi(0)+ < B0, u >= \Phi(0)$ $\forall u \in H$. Therefore, without lost of generality, we can suppose $\Phi(0) = 0$. Still by (1.6) we have

$$< Bu - Bv, u - v > \leq \frac{1}{\bar{\mu}} |u - v|^2 \quad \forall u, v \in H \tag{1.7}$$

Then the operator

$$B_\mu = I - \mu B \quad , \quad 0 < \mu \leq \bar{\mu} \tag{1.8}$$

is also *monotone* and *Lipschitz-continuous* in H. Denoting by Φ_μ the function

$$\Phi_\mu(u) = \frac{1}{2}|u|^2 - \mu\Phi(u) \quad , \quad 0 < \mu \leq \bar{\mu} \tag{1.9}$$

Φ_μ is Gateaux-differentiable in H and

$$\partial\Phi_\mu = B_\mu \tag{1.10}$$

Then Φ_μ is also *convex*.

b) Under the hypothesis (1.5), the assumption (1.6) can be replaced by the following one

$$B \text{ is cyclically monotone} \tag{1.11}$$

i.e. $\sum_{i=1}^{m} < u_i - u_{i-1}, Bu_i > \geq 0$ for every $u_0, u_1, ..., u_m$ in H with $u_o = u_m$.

Let us suppose indeed B cyclically monotone. Then (cf. [7]) B is the sub-differential $\partial\Phi$ of a convex function Φ. We shall prove now that Φ is also Gateaux-differentiable. The function Φ can be defined indeed (cf. e.g. [7]) as follows: given $u_0 \in H$ we set $\forall u \in H$

$$\begin{aligned} \Phi(u) = \sup\{ &< u - u_n, Bu_n > + < u_n - u_{n-1}, Bu_{n-1} > + \\ &+ < u_{n-1} - u_{n-2}, Bu_{n-2} > + ... + < u_1 - u_0, Bu_0 > \} \end{aligned}$$

where the sup is taken with respect to $u_1, u_2, ..., u_n$ in H. Since B is cyclically monotone we have

$$\{...\} = [\{...\} + < u_0 - u, Bu >] - < u_o - u, Bu > \leq - < u_o - u, Bu > .$$

Then by (1.5) we obtain

$$\Phi(u) \leq \frac{1}{\bar{\mu}}(|u_0|\ |u|\ +\ |u|^2)\ .$$

Therefore $\Phi(u)$ is bounded from above in the neighbourhood of every point of H. Then Φ is continuous in H (cf. e.g. [12]). Since B is a single-valued operator, we obtain (cf. e.g. [12]) that Φ is Gateaux-differentiable. Viceversa if (1.6) is satisfied B is cyclically monotone (cf. again [7]).

Now, given $T > 0$, we can state the problem (P)

$$(P) \quad \begin{cases} \text{given } u_o \in V', \text{ find } t \to u(t) \text{ absolutely continuous in} \\ [0,T] \text{ with values in } V' \text{ such that} \\ \dfrac{du(t)}{dt} + ABu(t) = 0\ ,\ Bu(t) \in V\ ,\ \text{a.e. in } [0,T] \\ u(0) = u_o \end{cases}$$

Setting $\theta(t) = Bu(t)$ and denoty by C the inverse B^{-1} of B (C can be multi-valued), the problem (P) is frequently written as follows

$$(P') \qquad \frac{du(t)}{dt} + A\theta(t) = 0\ ,\ u(t) \in C\theta(t)\ \text{ a.e. in } [0,T], \qquad u(0) = u_o\ .$$

It is well known that, (P) can be studied from different points of view and by different techniques; many existence and uniqueness results have been obtained (cf. e.g. the books [2], [7], [8], [9], [16], [18] and the papers [1], [3], [5], [6], [10]). We shall assume now the following hypothesis

$$(1.12) \qquad \begin{cases} \forall u_0 \in H \text{ the problem } (P) \text{ has one} \\ \text{and only one solution } u \text{ belonging to} \\ \text{the space } H^1(0,T;V') \cap L^2(0,T;H) \text{ with} \\ Bu \in L^2(0,T;V) \end{cases}$$

(for the notations cf. e.g. [18]). We shall see in Section 4 many interesting cases for which the assumption (1.2) is satisfied.

Let us state the time-discrete algorithm for (P).

Let $0 < \mu \leq \bar{\mu}$ be a fixed number; let N be a fixed positive integer. We denote by $\tau = T/N$ the time step. and we set $t^n = n\tau$, $I^n =]t^{n-1}, t^n]$, $n = 1, ..., N$. The time-discret algorithm is the following

$$(1.13) \qquad \begin{cases} U^0 = u_0 \\ \Theta^n + \frac{\tau}{\mu} A\Theta^n = BU^{n-1}, \Theta^n \in V \\ U^n = U^{n-1} + \mu\{\Theta^n - BU^{n-1}\}, \; n = 1, ..., N. \end{cases}$$

It is well defined because the second equation in (1.13) has a unique solution Θ^n and because U^n defined by the third line of (1.13) belongs to H.

We shall prove in Section 2 and 3 the following theorems, under the assumptions (1.1), (1.3), (1.5), (1.6), (1.12) and for $T > 0$, $o < \mu \leq \bar{\mu}$ fixed.

Theorem 1 (stability). *There exists a positive number K, depending on $|u_0|$, such that*

$$(1.14) \qquad \max_{n=1,...,N} \Phi(U^n) + \sum_{n=1}^{N} |U^n - U^{n-1}|^2 + \tau \sum_{n=1}^{N} \|\Theta^n\|^2 \leq K.$$

Moreover setting

$$(1.15) \qquad e_u = u(t) - U^n; \; \theta(t) = Bu(t); \; e_\theta = \theta(t) - \Theta^n, \quad t \in I^n$$

Theorem 2 (error estimates). *There exists a positive number K, depending on $|u_0|$ such that*

(1.16)

$$\|e_u\|_{L^\infty(0,T;V')} + \|e_\theta\|_{L^2(0,T;H)} + \| \int_0^t e_\theta ds \|_{L^\infty(0,T;V)} \leq K\tau^{1/4}.$$

As we said, the algorithm (1.13) is suggested by [4]. It is also related from one side with the "Laplace modified forward Galerkin method" of [11], the "alternating phase truncation method" of [24] and the "discret-time phase relaxation method" of [27], introduced for the partial differential parabolic equations, and for the other side with the "local correction" of the finite differernce schemes for abstract differential equations due to the Soviet School (cf. e.g. [25], [14], [15]).

Following the same technique as in [21], it is easy to extend the results of the present paper to the non-homogeneous equation $u_t + ABu = f$, under suitable assumptions on f.

2. Proof of the stability. By (1.13) we have

$$< U^n - U^{n-1}\phi > + \tau < A\Theta^n, \phi >= 0 \quad \forall \phi \in V. \tag{2.1}$$

Let us take $\phi = \Theta^n$ and add from 1 to m, $1 \leq m \leq N$:

$$\sum_{n=1}^{m} < U^n - U^{n-1}, \Theta^n > + \tau \sum_{n=1}^{m} < A\Theta^n, \Theta^n >= 0. \tag{2.2}$$

By (1.13), (1.8) we obtain

$$\begin{aligned} &\frac{1}{2}BU^n + \frac{1}{2\mu}\{B_\mu U^n - B_\mu U^{n-1}\} + \frac{1}{2\mu}U^n - \frac{1}{2\mu}BU^{n-1} = \\ &\qquad = \frac{U^n}{\mu} - \frac{U^{n-1}}{\mu} + BU^{n-1} = \Theta^n. \end{aligned} \tag{2.3}$$

Then, using (2.2), we have

(2.4)

$$\begin{aligned} &\sum_{n=1}^{m} < U^n - U^{n-1}, \Theta^n >= \sum_{n-1}^{m} < U^n - U^{n-1}, \frac{1}{2}BU^n > + \\ &+ \sum_{n=1}^{m} < U^n - U^{n-1}, \frac{1}{2\mu}\{B_\mu U^n - B_\mu U^{n-1}\} > \\ &+ \sum_{n=1}^{m} < U^n - U^{n-1}, \frac{1}{2\mu}U^n > - \sum_{n=1}^{m} < U^n - U^{n-1}, \frac{1}{2\mu}B_\mu U^{n-1} > \\ &= I + II + III + IV. \end{aligned}$$

Since Φ is convex and $B = \partial\Phi$, we have

$$I \geq \frac{1}{2}\sum_{n=1}^{m}\{\Phi(U^n) - \Phi(U^{n-1})\} = \frac{1}{2}\Phi(U^m) - \frac{1}{2}\Phi(U^0).$$

Since B_μ is monotone we have

$$II \geq 0.$$

Moreover, we have $\forall\phi$ and $\psi \in H$

$$2 < \phi - \psi, \phi >= |\phi|^2 - |\psi|^2 + |\phi - \psi|^2. \tag{2.5}$$

Consequently

$$III = \frac{1}{4\mu}\{|U^m|^2 - |U^0|^2 + \sum_{n=1}^{m}|U^n - U^{n-1}|^2\}.$$

Finally since Φ_μ is convex and $\partial\Phi_\mu = B_\mu$ we have

$$\begin{aligned} IV \geq & \frac{1}{2\mu}\sum_{n=1}^{m}\{\Phi_\mu(U^n) - \Phi_\mu(U^{n-1})\} = \\ & -\frac{1}{2\mu}\{\Phi_\mu(U^m) - \Phi_\mu(U^0)\}. \end{aligned}$$

Therefore, using (1.9) and $U^0 = u_0$, we obtain

$$\begin{aligned} \sum_{n-1}^{m} < U^n - U^{n-1}, \Theta^n > \ & \geq \frac{1}{2}\Phi(U^m) - \frac{1}{2}\Phi(u_0) + \\ & + \frac{1}{4\mu}|U^m|^2 - \frac{1}{4\mu}|u_0|^2 + \frac{1}{4\mu}\sum_{n=1}^{m}|U^n - U^{n-1}|^2 + \\ & - \frac{1}{4\mu}|U^m|^2 + \frac{1}{2}\Phi(U^m) + \frac{1}{4\mu}|u_0|^2 - \frac{1}{2}\Phi(u_o) = \\ & = \Phi(U^m) - \Phi(u_0) + \frac{1}{4\mu}\sum_{n=1}^{m}|U^n - U^{n-1}|^2. \end{aligned} \tag{2.6}$$

Then from (2.2) it follows

$$(2.7)\quad \Phi(U^m)+\frac{1}{2\mu}\sum_{n=1}^{m}|U^n-U^{n-2}|^2+\tau\sum_{n-1}^{m}< A\Theta^n,\Theta^n > \leq \Phi(u_0).$$

The assertion (1.14) follows now from (2.7), (1.3) and (1.4).

Remark. Since $\Phi(u) \geq 0$ (cf.§1 a)), (1.14) is a real "stability" condition for (1.13).

By (1.14) we have

$$(2.8)\qquad 0 \leq \Phi(U^n) \leq K, \qquad n = 1, ..., N$$

$$(2.9)\qquad \sum_{n=1}^{N}|U^n - U^{n-1}|^2 \leq K$$

$$(2.10)\qquad \tau\sum_{n-1}^{N}\|\Theta^n\|^2 \leq K.$$

(2.9) and (2.10) may be regarded as estimates in the discrete $H^{\frac{1}{2}}(0,T;H)$ and $L^2(0,T;V)$ norms.

3. Proof of the error estimates. First let us introduce the following notations: if $t \to v(t)$ is a function integrable in $[0,t]$ with values in H (or V, or V') we shall set

$$(3.1)\qquad \bar{v}^n = \frac{1}{\tau}\int_{I^n} v(t)\, dt\ , \qquad n = 1,\ldots,N\ ;$$

if v is continuous we shall set

$$(3.2)\qquad v^n = v(t^n) \quad n = 1, ..., N$$

The following relation is easily verified, since $a(u, v)$ is bilinear and symmetrie:

$$(3.3)\qquad \begin{aligned} a(\sum_{i=1}^{m} v_i, \sum_{i=1}^{m} v_i) + \sum_{i=1}^{m} a(v_i, v_i) = \\ =2\sum_{i=1}^{m} a(\sum_{n=1}^{i} v_n, v_i) \quad \forall v_1, ..., v_m \in V. \end{aligned}$$

Let u be the solution of (P). Then for all most every $t \in [0,T]$, we have

$$(3.4) \qquad < \frac{du}{dt}, \phi > + a(\theta, \phi) = 0 \quad \forall \phi \in V.$$

Integrating on I^n, and using (1.12) we obtain

$$(3.5) \qquad < u^n - u^{n-1}, \phi > + \tau a(\bar{\theta}^n, \phi) \quad \forall \phi \in V.$$

Let us remark that (2.2) can be written in the following form

$$(3.6) \qquad < U^n - U^{n-1}, \phi > + \tau a(\Theta^n, \phi) = 0 \quad \forall \in V.$$

Take the difference between (3.5) and (3.6) and sum over n from 1 to $i \leq N$:

$$(3.7) \qquad < u^i - U', \phi > + \tau a(\sum_{n=1}^{i} (\bar{\theta}^n - \Theta^n), \phi) = 0 \quad \forall \phi \in V.$$

Take in (3.7) $\phi = \tau(\bar{\theta}^i - \Theta^i)$ and sum over i from 1 to $m \leq N$; we obtain

$$(3.8) \qquad \sum_{i=1}^{m} \tau < u^i - U^i, \bar{\theta}^i - \Theta^i > + \tau^2 \sum_{i=1}^{m} a(\sum_{n=1}^{i} (\bar{\theta}^n - \Theta^n), \bar{\theta}^i - \Theta^i) = = I + II = 0.$$

Using (3.3) and (1.3) and the notations (1.15), it follows:

$$(3.9) \qquad \begin{aligned} II = & \frac{1}{2}\tau^2 a(\sum_{i=1}^{m} (\bar{\theta}^i - \Theta^i), \sum_{i=1}^{m} (\bar{\theta}^i - \Theta^i)) + \\ & + \frac{1}{2}\tau^2 \sum_{i=1}^{m} a(\bar{\theta}^i - \Theta^i, \bar{\theta}^i - \Theta^i) \geq \\ \geq & \frac{1}{2} a(\sum_{i=1}^{m} \int_{I^i} (\theta(t) - \Theta^i) dt, \sum_{i=1}^{m} \int_{I^i} (\theta(t) - \Theta^i) dt) \\ \geq & \frac{1}{2} a(\int_0^{t^m} e_\theta(t) dt, \int_0^{t^m} e_\theta(t) dt) \geq \frac{\alpha}{2} \| \int_0^{t^m} e_\theta(t) dt \|^2. \end{aligned}$$

Moreover we have

(3.10)

$$
\begin{aligned}
I &= \sum_{i=1}^{m} \int_{I^i} < u^i - U^i, e_\theta(t) > dt \\
&= \sum_{i=1}^{m} \int_{I^i} < u^i - u(t),, e_\theta(t) > dt + \sum_{i=1}^{m} \int_{I^i} < u(t) - U^i, e_\theta(t) dt \\
&= \sum_{i=1}^{m} \int_{I^i} < u^i - u(t), e_\theta(t) > dt + \sum_{i=1}^{m} \int_{I^i} < e_u(t), e_\theta(t) > dt \\
&= III + IV.
\end{aligned}
$$

Now

$$
\begin{aligned}
|III| &= |\sum_{i=1}^{m} \int_{I^i} < \int_t^{t^i} \frac{du}{ds} ds, e_\theta(t) > dt| \\
&\leq \tau \| \frac{du}{dt} \|_{L^2(0,t^m;V')} \| e_\theta \|_{L^2(0,t^m;V)}.
\end{aligned}
\tag{3.11}
$$

By (2.10) and by $\frac{du}{dt} \in L^2(0,T;V')$ and $\theta \in L^2(0,T;V)$ we obtain (K denoting again a positive number depending only on $|u_0|$)

$$
|III| \leq K\tau. \tag{3.12}
$$

Now from (1.13) we have

$$
\begin{cases} e_\theta = Bu - BU^{n-1} - \frac{1}{\mu}(U^n - U^{n-1}) \\ e_u - \mu e_\theta = u - U^{n-1} - \mu(Bu - BU^{n-1}) \quad t \in I^n. \end{cases} \tag{3.13}
$$

Then, again by (1.12), we obtain

(3.14)

$$
\begin{aligned}
IV &= \sum_{i=1}^{m} \mu \int_{I^i} |e_\theta(t)|^2 dt + \sum_{i=1}^{m} \int_{I^i} < e_u(t) - \mu e_\theta(t), e_\theta(t) > dt = \\
&= \mu \| e_\theta \|^2_{L^2(0,t^m;H)} + \\
&+ \sum_{i=1}^{m} \int_{I^i} < u(t) - U^{i-1} - \mu(Bu(t) - BU^{i-1}), Bu(t) - BU^{i-1} > dt +
\end{aligned}
$$

$$-\frac{1}{\mu}\sum_{i=1}^{m}\int_{I^i} < u(t), U^i - U^{i-1} > dt + \frac{1}{\mu}\sum_{i=1}^{m}\int_{I^i} < U^i, U^i - U^{i-1} > dt +$$

$$+\sum_{i=1}^{m}\int_{I^i} < e_\theta(t), U^i - U^{i-1} > dt = V + VI + VII + VIII + IX.$$

Since $\mu \leq \bar{\mu}$ by (1.5) we have

$$VI \geq 0. \tag{3.15}$$

Moreover, using also (2.9), we have

$$|VII| \leq \|u\|_{L^2(0,T;H)}[\tau \sum_{i=1}^{m} |U^i - U^{i-1}|^2]^{\frac{1}{2}} \leq K\tau^{\frac{1}{2}}. \tag{3.16}$$

From (2.5) it follows

$$VIII \geq -\frac{1}{\mu}\frac{\tau}{2}|u_0|^2 \geq -K\tau. \tag{3.17}$$

Again by (2.9) we have

$$\begin{aligned} |IX| &\leq \frac{\mu}{2}\|e_\theta\|^2_{L^2(0,t^m;H)} + \frac{2}{\mu}\tau\sum_{i=1}^{m}|U^i - U^{i-1}|^2 \\ &\leq \frac{\mu}{2}\|e_\theta\|^2_{L^2(0,T^m;H)} + K\tau. \end{aligned} \tag{3.18}$$

Recalling (3.8), (3.9), (3.10), (3.11), (3.12), (3.14), (3.15), (3.16), (3.17), (3.18) and (1.12) we obtain finally

$$\frac{1}{2}\|e_\theta\|^2_{L^2(0,T;H)} + \|\int_0^t e_\theta ds\|^2_{L^\infty(0,T;V)} \leq K\tau^{1/2}. \tag{3.19}$$

Now we have only to deduce the estimate for $\|e_u\|_{L^\infty(0,T;V')}$. Let us take $\phi = A^{-1}(u^i - U^i)$ in (3.7); then for $i \leq N$:
(3.20)

$$< U^i - U^i, A^{-1}(u^i - U^i) > +\tau a(\sum_{n=1}^{i}(\bar{\theta}^n - \Theta^n), A^{-1}(u^i - U^i)) = 0.$$

By (1.3) we have first

$$(3.21)\qquad \begin{aligned} < u' - U^i, A^{-1}(u^i - U^i) > &\geq \alpha \|A^{-1}(u^i - U^i)\|^2 \\ &\geq \alpha' \|u^i - U^i\|_*^2 \end{aligned}$$

where α' is a positive constant independent on $u^i - U^i$. Moreover,
(3.22)

$$\tau \left| a\left(\sum_{n-1}^{i} (\bar{\theta}^n - \Theta^n), A^{-1}(u^i - U^i)\right) \right| =$$

$$= \left| a\left(\int_o^{t^i} e_\theta(t)dt, A^{-1}(u^i - U^i)\right) \right| \leq \alpha'' \| \int_o^{t^i} e_\theta(t)dt\| . \|u^i - U^i\|_*$$

where α'' is a positive constant independent on $u^i - U^i$. Therefore we have

$$(3.23)\qquad \|u^i - U^i\|_* \leq \frac{\alpha'}{\alpha''} \| \int_0^{t^i} e_\theta(t)dt\|, \quad i = 1, ..., N$$

from which, using (3.19), we obtain

$$(3.24)\qquad \max_{i=1...N} \|u^i - U^i\|_* \leq K\tau^{1/4}$$

Finally (1.16) follows from (3.24) and (3.19), using also (1.12).

4. Applications. We show now some families of equations to which the previous results can be applied.

a) *Non linear parabolic equations of Stefan-type.*

This family of equations is certainly the most important. As we said, the algorithm (1.13) has been deeply studied in this case (of [4], [20], [21] [22], [23], [24]). We obtain it from (1.13) taking $H = L^2(\Omega)$, where Ω is a bounded open set, sufficiently smooth, in $\mathbf{R}^d, d \geq 1$;

$$V = H_0^1(\Omega), \quad a(u, v) = \int_\Omega \nabla u \nabla v dx, \quad (Bu)(x) = \beta(u(x)),$$

where $s \to \beta(s)$ is a non decreasing and Lipschitz-continuous function from $\mathbf{R}$ to $\mathbf{R}$, with $\beta(0) = 0$. Formally the problem (P) becomes

$$(4.1) \qquad \begin{cases} \frac{\partial u}{\partial t} - \Delta\beta(u) = 0 & \text{in } \Omega\times]0,T[\\ \beta(u) = 0 & \text{on } \partial\Omega\times]0,T[\\ u(0) - u_0. \end{cases}$$

The classical model of the two-phase Stefan problem in the enthalpy formulation is a particular case of (4.1), The hypothesis (1.1) (1.3), (1.5) (1.6) (1.12) are satisfied (cf. e.g. [13]. [14] and the referenceres there quotel). The extension to the other boundary conditions and to more general elliptic operator instead of the Laplacian Δ, has been emphasized in the quoted papers.

b) *Some problems with B linear.*

A second family of applications, which contains some pseudoparabolic linear equations is the following. Let V and H be again two real Hilbert spaces with (1.1) and let $a(u,v)$ verify (1.3). Let $c(\theta,\eta)$ be still a bilinear, symmetric, continuous and coercive from (i.e. verifying (1.13)). Let C be the isomorphism of V onto V' defined by $c(\theta,\eta)$, i.e.

$$< C\theta,\eta >= c(\theta,\eta) \quad \forall\theta,\eta \in V.$$

Let us set

$$(4.2) \qquad B = \textit{restriction to } H \textit{ of } C^{-1}.$$

It is easy to verify that B satisfies (1.5). We have inded

$$(4.3) \qquad \begin{cases} < Bu,u >=< C^{-1}u, CC^{-1}u >= c(Bu,Bu) \geq \\ \geq c_1\|Bu\|^2 \geq c_2|Bu|^2 \qquad \forall u \in H \end{cases}$$

where c_1 and c_2 are suitable positive constant. The condition (1.5) follows now from (4.3) since B is linear.

Let us set $\Phi(u) = \frac{1}{2} < Bu,u >$. Since $< Bu,v >=< u,Bv >$ $\forall u,v \in H$, it is easy to verify that Φ is Gateaux-differentiable in H and $\partial\Phi = B$. So (1.6) is satisfied. Moreover the operator AC^{-1} is linear and continuous from V' into V'. Then the problem

$$(4.4) \qquad \frac{du}{dt} + AC^{-1}u = 0$$

$$(4.5) \qquad u(0) = u_0 \quad , \quad u_o \in V'$$

has one and only one solution u Lipschitz-continuous in $[0, T]$ with values in V'. It is intersting to write the problem (4.4)-(4.5) in the following form

$$(4.6) \qquad \begin{cases} \frac{d(C\theta)}{dt} + A\theta = 0, \quad \theta(t) \in V, \quad C\theta(0) = u_0 \\ u(t) = C\theta(t). \end{cases}$$

Obviously in this abstract formulation we don't have in general the property (1.12), but, in the applications to the partial differential equations, (1.12) is a consequence of the regularity properties of the coefficients of the operators A and C, frequently verified. A simple example is given by the following linear pseudo-parabolic equation: take

$$\Omega = \text{bounded, open and sufficiently smooth set in } \mathbf{R}^d, \ d \geq 1$$
$$H = L^2(\Omega), \quad V = H_0^1(\Omega)$$

and

$$(4.7) \qquad \begin{cases} a(u,v) = \int_\Omega (\sum_{i,j=1}^d a_{ij} \frac{\partial u}{\partial x_i} \frac{\partial v}{\partial x_j} + a_0 uv) dx \\ \text{where } a_{ij} \in C^1(\bar{\Omega}), a_{ij} = a_{ij} \\ \sum_{i,j=1}^d a_{ij}\xi_i\xi_j \geq \alpha \sum_{i=1}^d |\xi_i|^2 \quad \forall \xi \in \mathbf{R}^d, \ \alpha \ \text{ constant} > 0 \\ a_0 \in L^\infty(\Omega), \ a_0 \geq 0 \text{ in } \Omega; \end{cases}$$

$$(4.8) \qquad \begin{cases} c(\theta, \eta) = \int_\Omega (\sum_{i,j=1}^d c_{ij} \frac{\partial \theta}{\partial x_j} \frac{\partial \eta}{\partial x_j} + c_0 \theta\eta) dx \\ \text{where } c_{ij} \in C^1(\bar{\Omega}), c_{ij} = c_{ij} \\ \sum_{i,j=1}^d c_{ij}\xi_i\xi_j \geq \gamma \sum_{i=1}^d |\xi|^2 \quad \forall \xi \in \mathbf{R}^d, \gamma \ \text{ constant} \\ c_0 \in L^\infty(\Omega), c_0 \geq 0 \text{ in } \Omega. \end{cases}$$

Then it is well known that $B(= C^{-1})$ is a linear continuous operator from H into $V \cap H^2(\Omega)$ and A is a linear continuous operator from $H^2(\Omega)$ into H.

Then AB is linear continuous from H into H, and the problem (4.4)-(4.5) with $u_0 \in H$ has one and only one solution Lipschitz-continuous with values in H. Therefore (1.12) is satisfied.

Other similar problems with more general coefficients of A and C can be considered, proving (1.12) by some "regularization" techniques as in the following example d).

c) *Some non linear problems, I.*

Take again:

$$\begin{cases} H = L^2(\Omega), \quad V = H_0^1(\Omega) \quad \text{bounded open and sufficiently} \\ \text{smooth set in } \mathbf{R}^d, \ a(u,v) \text{ satisfing (4.7).} \end{cases}$$

It is possible to consider operator C of the type of the "calculus of variations":

$$C\theta = -\sum_{i=1}^{d} \frac{\partial}{\partial x_i} c_i(x, \theta, \nabla\theta) + c_0(x, \theta, \nabla\theta)$$

where the functions c_i, $i = 0, 1, ..., d$ verifie suitable conditions such that the problem $C\theta = u$ has one and only one solution in V (cf. e.g. [17], chap. 2, n.26).

For sake of simplicity we shall consider here only a one-dimensional case. We take

$$\begin{cases} d = 1, \Omega =]0,1[, H = L^2(\Omega), V = H_0^1(\Omega); \\ a(u,v) = \int_\Omega a \frac{du}{dx}\frac{dv}{dx} dx \text{ where } a \in L^\infty(\Omega), \ a \geq \alpha, \ \alpha \text{ pos. const.} \\ C\theta = -\frac{d}{dx}(c(\frac{d\theta}{dx})), \text{ where c is a Lipschitz continuous function} \\ \qquad \text{from } \mathbf{R} \text{ into } \mathbf{R} \text{ and } c(0) = 0, \ c' \geq \gamma \text{ a.e. in } \mathbf{R}, \text{ with } \gamma \\ \qquad\qquad \text{positive constant.} \end{cases}$$

Then the problem $C\theta = u$ has one and only one solution θ in V, $\forall u \in V'$ and the inverse operator C^{-1} is Lipschitz continuous from

V' into V. Take indeed u and v in V' and set $\theta = C^{-1}u, \eta = C^{-1}v$. We have

$$C\theta - C\eta = u - v$$

Therefore we obtain, in the pairing between V' and V,

$$<u - v, \theta - \eta> = < C\theta - C\eta, \theta - \eta> =$$

$$= \int_\Omega [c(\frac{d\theta}{dx}) - c(\frac{d\eta}{dx})](\frac{d\theta}{dx} - \frac{d\eta}{dx})dx \geq \gamma \int_\Omega (\frac{d\theta}{dx} - \frac{d\eta}{dx})^2 dx \tag{4.9}$$

Then

$$\|\theta - \eta\| \leq K\|u - v\|_* \text{ , } K \text{ constant,} \tag{4.10}$$

i.e. C^{-1} is Lipschitz-continuous.

Let us define now

$$B = \text{restriction to } H \text{ of } C^{-1}.$$

Since $|\theta| \leq \|\theta\| \quad \forall \theta \in V$, the assumption (1.5) follows from (4.9).

In order to verify (1.6) we can (cfr. §1), (1.11)) prove that B is cyclically monotone (cf. also the following Remark 1). Let $u_0, u_1, ..., u_m$ be in H, with $u_0 = u_m$. Setting $\theta_i = Bu_i$ we obtain

$$\sum_{i=1}^m < u_i - u_{i-1}, Bu_i > = \sum_{i=1}^m < C\theta_i, -C\theta_{i-1}, \theta_i > =$$

$$= \sum_{i=1}^m \int_\Omega [c(\frac{d\theta_i}{dx}) - c(\frac{d\theta_{i-1}}{dx})] \frac{d\theta_i}{dx} dx \geq 0$$

because we have $\sum_{i=1}^m [c(s_i) - c(s_{i-1})]s_i \geq 0$ whenever $s_o, s_1, ..., s_m = s_0$ are real nombers as we shall prove.

Since $c'(s) \geq \gamma > 0, C^{-1}$ is a single-valued and strictly increasing; then the function $\varphi(\xi) = \int_0^\xi c^{-1}(s)ds$ is convex and continuous and we have

$$\varphi(\xi) - \varphi(\eta) \leq c^{-1}(\xi)(\xi - \eta) \quad \forall \xi, \eta \in \mathbf{R};$$

therefore, setting $\xi_i = c(s_i)$, we have

$$\sum_{i=1}^m |c(s_i) - c(s_{i-1})|s_i = \sum_{i=1}^m (\xi_i - \xi_{i-1})c^{-1}(\xi_i) \geq$$

$$\geq \sum_{i=1}^m (\varphi(\xi_i) - \varphi(\xi_{i-1})) = \varphi(\xi_m) - \varphi(\xi_0) = 0.$$

We have now to verify (1.12). First of all the operator AC^{-1} is Lipschitz-continuous from V' into V' and therefore the problem

$$\frac{du}{dt} + AC^{-1}u = 0, \quad u(0) = u_0 \in V' \tag{4.11}$$

has one and only one solution Lipschitz-continuous in $[0,T]$ with values in V'. Let us now introduce the new assumption on the derivative of c :

$$c'(s) \text{ is Lipschitz continuous in } \mathbf{R}. \tag{4.12}$$

Under this hypothesis AC^{-1} is Lipschitz-continuous from H into H. Let indeed u and v be in H and set again $\theta = C^{-1}u, \eta = C^{-1}v$. We have

$$\frac{d^2\theta}{dx^2} - \frac{d^2\eta}{dx^2} = -\frac{1}{c'(\frac{d\theta}{dx})}u + \frac{1}{c'(\frac{d\eta}{dx})}v. \tag{4.13}$$

Then, since $\frac{1}{c'}$ is also Lipschitz-continuous and $\geq \delta, \delta$ positive constant:

$$|\frac{d^2\theta}{dx^2} - \frac{d^2\eta}{dx}| = |(\frac{1}{c'(\frac{d\theta}{dx})} - \frac{1}{c'(\frac{d\eta}{dx})})u + \frac{1}{c'(\frac{d\eta}{dx})}(u - v)| \leq$$

$$\leq K\{|u|\,|\frac{d\theta}{dx} - \frac{d\eta}{dx}| + |u - v|\}, \quad K \text{ suitable constant.}$$

Moreover from (4.9) we have

$$|\frac{d\theta}{dx} - \frac{d\eta}{dx}| \leq K|u - v|. \tag{4.14}$$

Then

$$\|\theta - \eta\|_{H^2(\Omega)} \leq K|u - v|. \tag{4.15}$$

Finally C^{-1} is Lipschitz-continuous from H into $V \cap H^2(\Omega)$ and AC^{-1} from H into H. Under the new assumption (4.12) we have so proved (1.12).

d) *Some non linear problems: II.*

We shall consider now another non linear problem, related again to the diffusion theory (cf. e.g. [8], [16] and the references there quoted).

Take $H = L^2(\Omega)$, $V = H^1(\Omega), \Omega$ bounded open set in $\mathbf{R}^d$, sufficiently smooth,

$$a(u,v) = \int_\Omega (\nabla u \nabla v + \lambda uv)dx \tag{4.16}$$

where λ is a fixed positive number. Then we have $A = -\Delta + \lambda$. Let us consider the operator C, formally defined by

$$C\theta = \theta + \mathcal{H}(\theta) - \Delta\theta \tag{4.17}$$

where $\mathcal{H}$ is the Heaviside maximal monotone graph. It is known from the theory of the monotone operators and of the elliptic equations that $\forall u \in L^2(\Omega)$ the problem

$$C\theta \ni u, \quad \theta \in H^2(\Omega), \quad \frac{\partial\theta}{\partial\nu} = 0 \quad \text{on } \Gamma \tag{4.18}$$

where ν is the normal on Γ, has one and only one solution. Therefore we can set $B = C^{-1}$. The condition (1.5) is easy verified since $\mathcal{H}$ is monotone: setting $\theta = Bu$ and $\eta = Bv, u, v \in H$, we have indeed, since $\frac{\partial\theta}{\partial\nu} = \frac{\partial\eta}{\partial\nu} = 0$ on Γ,

$$\begin{aligned} &< Bu - Bv, u - v > = < \theta - \eta, , C\theta - c\eta > = \\ &= |\theta - \eta|^2 + < \theta - \eta, \mathcal{H}(\theta) - \mathcal{H}(\eta) > + |\nabla(\theta - \eta)|^2 \\ &\geq |\theta - \eta|^2 = |Bu - Bv|^2. \end{aligned} \tag{4.19}$$

We shall prove that B is cyclically monotone in H (cf. also the following Remark 1). Let $u_0, u_1, \dots, u_m$ be in H with $u_0 = u_m$ and define $\theta_i = Bu_i$, $i = 0, \dots, m$.

Denote by β the function (non decreasing and Lipschitz-continuous from $\mathbf{R}$ into $\mathbf{R}$) which is the inverse of the graph $I + \mathcal{H}$. Let $v_i \in H$, $i = 0, \dots, m$ be a fixed function such that $v_i(x) \in \theta_i(x) + \mathcal{H}(\theta_i(x))$ a.e. in Ω.

We can choose $\nu_0 = v_m$ a.e. in Ω because $\theta_0 = \theta_m$ a.e. in Ω.

Then we have

$$\tag{4.20} \begin{aligned} &\sum_{i=1}^{m} < u_i - u_{i-1}, Bu_i >= \sum_{i=1}^{m} < C\theta_i - C\theta_{i-1}, \theta_i >= \\ &= \sum_{i=1}^{m} < v_i - v_{i-1}, \beta(v_i) > + \sum_{i=1}^{m} \int_\Omega \nabla(\theta_i - \theta_{i-1}) \nabla \, \theta_i \, dx. \end{aligned}$$

Setting $\psi(\xi) = \int_0^\xi \beta(s) ds$, ψ is conmvex and continuous and then we have

$$\begin{aligned} &\sum_{i=1}^{m} < v_i - v_{i-1}, \beta(v_i) >\geq \sum_{i=1}^{m} \int_\Omega (\psi(v_i) - \psi(v_{i-1}) dx = \\ &= \int_\Omega \psi(v_m(x)) dx - \int_\Omega \psi(v_0(x)) dx = 0. \end{aligned}$$

In a similar way, since $\theta_0 = \theta_m$ we have

$$\begin{aligned} \sum_{i=1}^{m} \int_\Omega \nabla(\theta_i - \theta_{i-1}) \nabla\theta_i, dx = \frac{1}{2} \int_\Omega |\nabla\theta_m|^2 dx - \frac{1}{2} \int_\Omega (\nabla\theta_0|^2 dx + \\ + \frac{1}{2} \sum_{i=1}^{m} \int_\Omega |\nabla\theta_i - \nabla\theta_{i-1}|^2 dx \geq 0 \end{aligned}$$

and finally $\sum_{i=1}^m < u_i - u_{i-1}, Bu_i >\geq 0$; then B is cyclically monotone and (1.6) is verified.

We shall prove now (1.12). First let us remark that the problem

$$\tag{4.21} \frac{du}{dt} + ABu = 0, \quad u(0) = u_0$$

written in the equivalent form ($\theta = Bu$)

$$\begin{cases} \frac{d}{dt}(I + \mathcal{H} - \Delta)\theta - \Delta\theta + \lambda\theta \ni 0, & \frac{\partial\theta(t)}{\partial\nu} = 0 \text{ on } \Gamma, \ 0 < t < T \\ (I + \mathcal{H} - \Delta)\theta(0) \ni u_0 \end{cases}$$

belongs to a family of problems already studied (cf. e.g. [10]); then $\forall u_0 \in H$ there exist one and only one solution $u \in H^1(0,T;V')$ with $Bu \in L^2(0,T;V)$. It is possible to obtain the property (1.12), using a regolarization "technique". More precisely $\forall\epsilon > 0$ let $\mathcal{H}_\epsilon$ be a function such that

$$\mathcal{H}_\epsilon \in C^2(\mathbf{R}),\ \ \mathcal{H}_\epsilon(\xi) = 0 \ \text{ if } \ \xi \leq 0,\ \ \mathcal{H}_\epsilon(\xi) = 1 \ \text{ if } \ \xi \geq 1$$
$$\mathcal{H}'_\epsilon(\xi) \geq 0 \quad \forall \xi \in \mathbf{R}.$$

Let

$$C_\epsilon\theta = \theta + \mathcal{H}_\epsilon(\theta) - \nabla\theta$$

Then $\forall u \in H^1(\Omega)$ the problem $C_\epsilon\theta_\epsilon = u$, $\theta_\epsilon \in H^3(\Omega)$, $\frac{\partial\theta_\epsilon}{\partial\nu} = 0$ on Γ has one and only one solution and the inverse operator $B_\epsilon = C_\epsilon^{-1}$ is Lipschitz-continuous from $H^1(\Omega)$ into $H^3(\Omega)$ and it verifies not only (1.5) and (1.6) but also the property that AB_ϵ is Lipschitz-continuous from $H^1(\Omega)$ into $H^1(\Omega)$. Then using well known techniques (cf. e.g. [10]), we consider the problem

$$\frac{du_\epsilon}{dt} + AB_\epsilon u_\epsilon = 0, \quad u_\epsilon(0) = u_{0,\epsilon} \tag{4.22}$$

where $u_{0,\epsilon} \in H^1(\Omega)$ and $\lim\limits_{\epsilon\to 0} = u_0$ in H. This problem has only and only one solution u_ϵ Lipschitz-continuous in $[0,T]$ with values in $H^1(\Omega)$. We can multiply (4.22) by $B_\epsilon u_\epsilon(t)$ and integrate from 0 to $t, 0 < t \leq T$. Then we obtain the estimate

$$\|B_\epsilon u_\epsilon\|_{L^2(0,T;V)} \subseteq K \tag{4.23}$$

where K is independent on ϵ. Moreover we can multiply (4.22) also by $u_\epsilon(t)$ since $u_\epsilon(t)$ belongs to $H^1(\Omega)$. Since $\frac{\partial B_\epsilon u_\epsilon(t)}{\partial\nu} = 0$ on Γ and $\mathcal{H}_\epsilon$ is monotone, if we integrate from 0 to $t, 0 < t \leq T$, we obtain the other estimate

$$|u_\epsilon(t)|^2 \leq K, \quad 0 < t \leq T \tag{4.24}$$

where K is again independent on ϵ. From (4.23) and (4.24), using compactness and monotonicity properties we obtain that $u \in L^\infty(0,T;H)$ and (1.12) is verified.

Remark 1. The proof of the cyclically monotonicity of B in the examples c) and d) are essentially equivalent to the proofs of the cyclically monotonicity of C, because B is the inverse of C.

Remark 2. Many other interesting non linear problem considered in [10], can be approximate by (1.13) and studied in similar way as for the examples a), b), c), d).

Remark 3. The error estimate of the Theorem 2 is not "optimal". In particular in the examples given, it could be improved, under more restrictive hypothesis on u_0 (cf. e.g. the results obtained in [21] for the Stefan problem).

References

[1] C.Baiocchi, *Sulle equazioni differenziali astratte lineari del primo e secondo ordine negli spazi di Hilbert,* Ann. Mat. Pura e Appl. **76** (1967), 233-304.

[2] V.Barbu, *Non linear semigroups and differential equations in Banach spaces,* Noordhoff Int. Pub., Leyden, 1976.

[3] Ph.Benilan, *Opérateurs accretifs et semi-groupes dans les espaces $L^p(1 \leq p \leq \infty)$*, Funct. Anal. and Numerical Anal. (H. Fujita Ed.), Japan Soc. for the Promotion of Sciences (1978), 15-53.

[4] A.E.Berger, H.Brezis, J.C.W.Rogers, *A numerical method for solving the problem $u_t - \Delta f(u) = 0$*, R.A.I.R.O. Anal. Numer. **13** (1979), 297-312.

[5] H.Brezis, *On some degenerate non-linear parabolic equations,* in Non-linear Functionals Analysis (F.E.Browder Ed.), A.M.S. **18** (1970), 28-38.

[6] H.Brezis, *Monotonicity methods in Hilbert space and some applications to non linear differential equations,* Contribution to the non linear functional analysis, Acad. Press, New York (1971), 101-155.

[7] H.Brezis, *Operateurs maximaux monotones et semigroupes de contractions dans les espaces de Hilbert,* North Holland, Amsterdam, 1973.

[8] R.W.Carrol, R.E.Showalter, *Singular and degenerate Cauchy problems* , Academic Press, New York, 1976.

[9] G.Da Prato, *Applications croissantes et équation d'évolution dans les espaces de Banach,* Academic Press, New York, 1976.

[10] E.Di Benedetto, R.E.Showalter, *Implicit degenerate evolution equations and applications*, SIAM J. Math. Anal. **12** (1981), 731-751.

[11] J.Douglas, T.Dupont, *Alternating-direction Galerkin methods on rectangles*, in Numerical solutions of partial differential equations (B.Hubbard Ed.), vol. II, Academic Press, New York (1971), 133-214.

[12] I.Ekeland, R.Temam, *Analyse convexe et problèmes variationnels*, Dunod-Gauthier-Villars, Paris, 1974.

[13] J.W.Jerome, *Approximation of nonlinear evolution system*, Academic Press, New York (1983).

[14] A.Lapin, *On the local correction of non linear finite difference schemes* , (in russian) Izvestijia Vuzov Math. **9** (1972), 48-53.

[15] A.Lapin, *Sur la "correction locale" du schéma aux différences. Applications aux problèmes non linéaires*, Rapport **254** de Math. Appl., Univ. de Grenoble, 1976.

[16] J.L.Lions, *Equations différentielles opérationnelles et problèmes aux limites*, Springer Verlag, Berlin, 1961.

[17] J.L.Lions, *Quelques méthodes de resolution des problèmes aux limites non lineaires*, Dunod-Gauthier-Villars, Paris, 1969.

[18] J.L.Lions, E.Magenes, *Non-homogeneous boundary value problems and applications*, vol.I, Springer-Verlag, Berlin, 1972.

[19] E.Magenes, *Problemi di Stefan bifase in più variabili spaziali*, V S.A.F.A., Le Matematiche **36** (1981), 65-108.

[20] E.Magenes, C.Verdi, *On the semigroup approach to the two-phase Stefan problem with non linear flux condition*, in "Free Boundary Problems. Applications and Theory" (A.Bossavit, A.Damlamian, M.Fremond Eds.), Pitman, 1985, 28-39.

[21] E.Magenes, R.H.Nochetto, C.Verdi, *Energy error estimates for a linear scheme to approximate non linear parabolic problems*, M^2AN, Model. Math. et Anal. Numer. **21** (1987), 655-678.

[22] R.H.Nochetto, C.Verdi, *An efficient linear scheme to approximate parabolic free boundary problems: Error estimates and implementation*, Math. of Comp. **51** (1988), 27-53.

[23] M.Paolini, G.Sacchi, C.Verdi, *Finite element approximations of singular parabolic problems*, Int. J. Math. Num. Meth. in Eng. **28** (1988), 1989-2007.

[24] J.C.Rogers, A.E. Berger. M.Ciment, *The alternating phase truncation method for numerical solution of Stefan problem*,

SIAM J. Numer. Anal. **16** (1979), 563-587.

[25] A.Samarsky, *Introduction to the difference schemes theory* (in russian), Nauka, Mosca, 1971.

[26] C.Verdi, *On the numerical approach to a two-phase Stefan problem with nonlinear flux*, Calcolo **22** (1985), 351-381.

[27] C.Verdi, A. Visintin, *Error estimates for a semi-explicit numerical scheme for Stefan-type problems*, Numer. Math. (1988), **52** 165-185.

Dipartimento di Matematica e
Istituto di Analisi Numerica del C.N.R.
Università di Pavia
Strada Nuova 65
I-27100 PAVIA

THE STORED-ENERGY FOR SOME DISCONTINUOUS DEFORMATIONS IN NONLINEAR ELASTICITY

PAOLO MARCELLINI

Dedicated to Ennio De Giorgi on his sixtieth birthday

Sommario. In questo lavoro viene applicata all'elasticità non lineare un'idea che Ennio De Giorgi ha dimostrato essere feconda in contesti diversi, ad esempio nello studio delle superfici minime o della Γ-convergenza. Si tratta della cosiddetta "estensione dello spazio ambiente": è opportuno ricercare a priori superfici cartesiane di area minima nella classe BV delle funzioni a variazione limitata, piuttosto che tra le funzioni di classe C^1, come pure in problemi di Γ-convergenza o di esistenza del minimo di integrali del calcolo delle variazioni è opportuno estendere lo spazio ambiente C^1 fino ad uno spazio di Sobolev $H^{1,p}$, od anche perfino ad L^p.

L'idea di base, già sperimentata da De Giorgi e da altri, ha dato buoni frutti. Il resto non so. L'una e l'altro gli sono dedicati con affetto in occasione del suo 60° compleanno.

1. Introduction. We consider an elastic body that occupies a bounded open set $\Omega \subset \mathbf{R}^n$ in a reference configuration. We denote by $u(x)$ a deformation of Ω; that is, a particle $x \in \Omega$ is displaced to $u(x) \in \mathbf{R}^n$. By Du we denote the $n \times n$ matrix of the deformation gradient. We are concerned with *hyperelastic* materials having stored-energy function $W(\xi), \xi \in \mathbf{R}^{n\times n}$. That is, the total stored-energy E is given by

$$E(u) = \int_\Omega W(Du(x))dx. \tag{1.1}$$

If the material is frame-indifferent and isotropic, the energy function W can be represented by

$$W(\xi) = \phi(v_1, \ldots, v_n), \tag{1.2}$$

where $v_1, \ldots, v_n$ are the eigenvalues of the symmetric matrix $(\xi^T \xi)^{1/2}$ assuming that $\det \xi$ (the determinant of ξ) is positive.

One of the most interesting problems in this field is to find appropriate (both from the mathematical and the physical point of view) assumptions on the behaviour of W; that is, to describe the largest number of properties of the stored-energy that are common to a given class of materials and that are useful in a mathematical approach.

Some models have been proposed and have been studied in *incompressible elasticity*, in which the deformation $u(x)$ is subjected to the pointwise constraint $\det Du(x) = 1$. There are also papers that relate the proposed theoretical expression of the stored-energy with measures in experiments (see the references in Ball [1], [2]).

In the incompressible case, the stored energy $\hat{W}(\xi)$, defined for ξ with $\det \xi = 1$, can be extended to every ξ with $\det \xi > 0$, by setting

$$W_1(\xi) = \hat{W}\left(\frac{\xi}{(\det \xi)^{1/n}}\right); \tag{1.3}$$

this kind of extension has been studied in a recent paper by Charrier, Dacorogna, Hanouzet, Laborde [6].

Up to now the most considered form of stored-energy for *compressible materials* is of the type:

$$W(\xi) = W_1(\xi) + g(\det \xi), \tag{1.4}$$

where g is a given real function.

Odgen in [8] proposed for rubberlike solids a stored-energy of the form (1.4); more precisely, he proposed an expression in terms of v_i of the form

$$W(\xi) = \phi(v_1, \ldots, v_n) = \sum_{i,j} a_j v_i^{\alpha_j} + g(v_1 \cdot v_2 \ldots v_n). \tag{1.5}$$

Some of the exponents α_j can be negative; of course the product $v_1 \cdot v_2 \ldots v_n$ is the determinant of the matrix ξ.

The contribution to the energy, corresponding to the function g, expressed by the integral

$$\int_{\Omega} g(\det Du(x))dx, \tag{1.6}$$

takes into account the part of the energy that depends on changes in volume. We assume that the energy goes to $+\infty$ if we expand the solid ($\det Du \rightarrow +\infty$) or if we compress the solid to a point ($\det Du \rightarrow 0$). More precisely, we assume that $g = g(t)$, defined for $t > 0$, goes to $+\infty$ both as $t \rightarrow +\infty$ and as $t \rightarrow 0^+$. We assume also that g is smooth, so that g has a minimum, say at $t = 1$.

In the following we quote some considerations by Odgen [18]. First, there are physical reasons to think that $g(t)$ is decreasing for $t < 1$ and increasing for $t > 1$. Secondly, an inequality considered by Odgen is:

$$(tg')' = g' + tg'' > 0 \quad , \quad \forall t > 0. \tag{1.7}$$

For $t < 1$, since $g' \leq 0$, we have $g'' > 0$, thus g is convex in $(0,1)$. The conclusion is not the same for $t > 1$, where the above inequality can also be satisfied if g is concave.

In relation to the concavity of g we have two possible situations, schematized in figures 1 and 2.

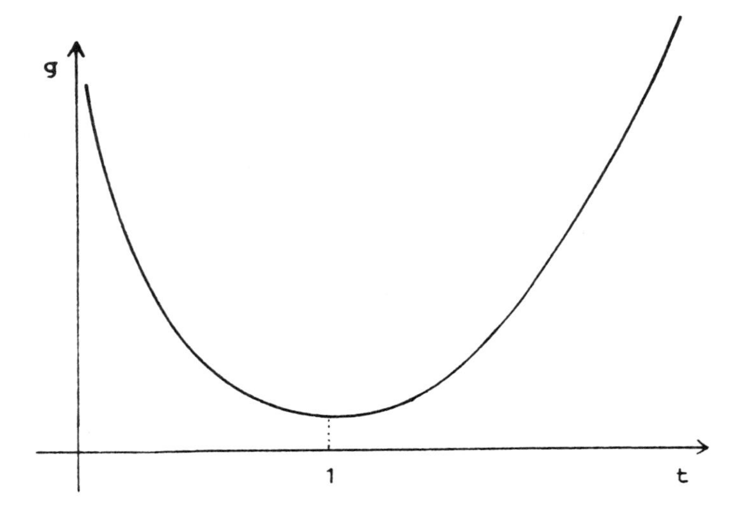

Fig.1. Graph of a convex g.

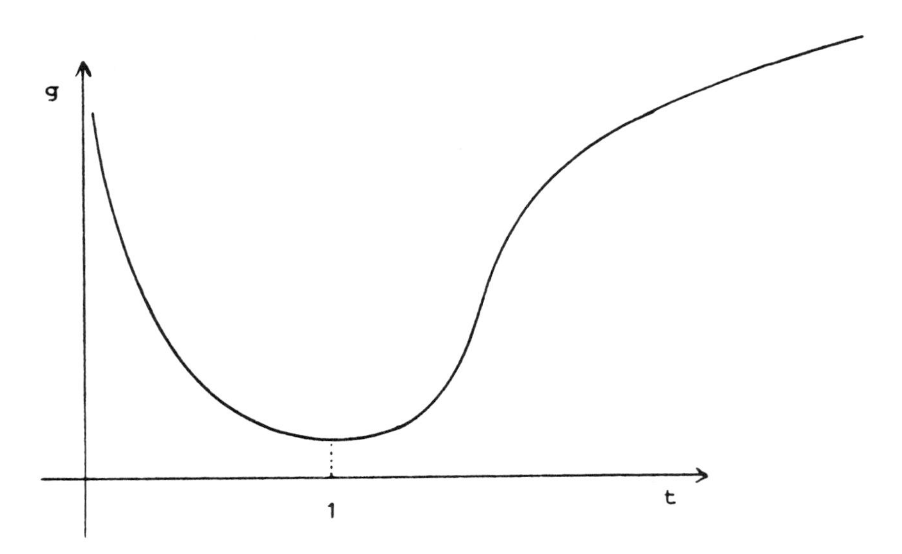

Fig.2. Graph of a non convex g.

Finally let us quote from Odgen [18], pages 572, 573: *"Little has been determined experimentally about situations in which* $t > 1$"; *"It seems likely (though it is not certain) that, so long as the material remains elastic, an increase in all-round tension would always be required to effect a volume increase; we would then have* $g'' > 0$ *for all* t (> 0). *Once the* elastic limit *is reached the possibility of* g'' *being less than zero cannot be ruled out"*.

From a mathematical point of view, in particular from the point of view of the existence of an energy minimum, the situation schematized in figure 1 is easier for at least two reasons.

First, the convexity of g plays a role in the existence of equilibrium solutions. In fact Ball [1] pointed out the importance in this theory of the assumption of *quasi-convexity* (see (5.2)) of $W(\xi)$, and recently Ball and Murat ([4], lemma 4.3) proved that quasiconvexity of a function of the type

$$W(\xi) = |\xi|^p + g(\det \xi) \tag{1.8}$$

implies convexity of g, if $1 < p < 2n$.

A second reason to say that the situation schematized in figure 1 is easier for existence of minima, depends on the coercivity condition:

$$g(t) \geq \text{ constant } \cdot t^r \quad , \quad t > 0 \tag{1.9}$$

for an exponent $r > 1$. Such an r may exist in the case of figure 1, but it cannot exist in the case of figure 2, if g is concave for large t.

The validity of (1.9) for some $r > 1$ is one of the assumptions in the well known existence theorem by Ball [1]. Recently, new existence results for some integrals of elasticity have been obtained by Ball and Murat ([4], theorem 6.1) and by Marcellini ([14], theorem 1).

Here we present a *mathematical argument* to show that the case $r = 1$ in (1.9) is one of the most relevant in this theory. In some sense, $r = 1$ is an "intermediate" case between the scheme considered in figure 1 (with g convex) and the scheme in figure 2 (with g concave for large t), since, if $r = 1$, then g is "linear at infinity".

The mathematical argument is based on studying an hyperelastic material for which the phenomenon of *cavitation* may occur. We describe the phenomenon of cavitation in the next section. We present an approach substantially different from Ball's approach on cavitation [2]. We will show that, under our approach, cavitation may occur only if g is not convex and if it is linear at infinity. Moreover, if g is linear at $+\infty$ and cavitation occurs, then the slope $g'(+\infty)$ should be equal to the radial component of the Cauchy stress at the surface of the cavity. Thus we propose an indirect method to find experimentally the behaviour of the function $g(t)$ for large values of t.

On reading this paper a referee pointed out to me the references [27], [28], [29], where it is studied the "Blatz-Ko material". The energy function W, proposed by Blatz and Ko for $n = 3$, with respect to $t = \det Du$ is of the form (see formula (2.21) of [29])

$$W(..,..,t) = 2\beta t^{-1} + (1-\beta)(t^{-2} + 2t). \tag{1.10}$$

The parameter β, determinated experimentally under three distinct homogeneous deformations, takes the values:

$$\text{(1.11)}\quad (a) \quad \beta = 0.13 \qquad (b) \quad \beta = 0.07 \qquad (c) \quad \beta = -0.19.$$

In our context it is very interesting to notice that $g(t)$ is linear at infinity and, in the cases (a), (b), g is convex like in figure 1, while in the case (c) g is not convex and behaves like the function in figure 2.

2. The phenomenon of cavitation.

The phenomenon of cavitation has been first studied by Ball [2]. The reader interested in cavitation is also referred to the papers by Stuart [25], [26], Podio Guidugli, Vergara Caffarelli, Virga [19], [20], Sivaloganathan [23], [24].

The idea of cavitation is as follows: we consider a body that occupies the unit ball $\Omega = \{|x| < 1\}$ of $\mathbf{R}^n$, with $n \geq 2$. We expand the body with deformation $u(x) = \lambda x$ at $|x| = 1$, for some $\lambda > 1$ (that is, we impose the boundary condition that the deformed surface of the body is a sphere of radius λ). We expect that, if λ is too large, then for some materials a hole forms inside the body.

To describe mathematically the phenomenon of cavitation, following Ball [2], we consider radial deformations

$$u(x) = v(r)\frac{x}{r} \quad , \qquad \text{with} \quad r = |x|, \tag{2.1}$$

where $v(r)$ is a function defined for $r \in [0,1]$, such that $v \geq 0$, $v' \geq 0$.

A computation shows that the eigenvalues v_i of $(Du^T Du)^{1/2}$ are

$$v_1 = v'(r) \quad , \quad v_i = \frac{v(r)}{r} \quad \text{for} \quad i = 2, \ldots, n. \tag{2.2}$$

Thus the determinant of Du, i.e. the product of the v_i for $i = 1, \ldots, n$, is equal to

$$\det Du = v'(r) \left(\frac{v(r)}{r}\right)^{n-1} . \tag{2.3}$$

By using polar coordinates, the stored-energy (1.1), (1.2) takes the form:

$$E = \omega_n \int_o^1 r^{n-1} \phi \left(v', \frac{v}{r}, \ldots, \frac{v}{r}\right) dr, \tag{2.4}$$

where, as usual, ω_n is the $(n-1)$-measure of the surface of the unit sphere in $\mathbf{R}^n$.

The transformation $u(x)$ defined in (2.1) is a map of the unit sphere to the sphere of radius $v(1)$. If $v(0) > 0$, then in this deformation a *cavity* forms at the center, with radius $v(0)$. In this case it is easy to see that u belongs to the Sobolev space $H^{1,p}(\Omega; \mathbf{R}^n)$ for every $p < n$, but u does not belong to $H^{1,n}(\Omega; \mathbf{R}^n)$.

If $v(0) > 0$, then $u(x) = v(r)x/r$ is a singular transformation at $x = 0$, and the corresponding energy needs to be defined carefully. We propose a definition in the next section.

3. Definition of the stored-energy for discontinuous deformations. From this point we follow a different approach from Ball's approach in [2]. For fixed $p > 1$, we consider the set

$$\mathcal{A} = \{v \in H^{1,p}_{loc}(0,1) : v(0) = 0, \quad v \geq 0, \quad v' \geq 0 \quad \text{a.e.}\}; \tag{3.1}$$

where by the notation "$v \geq 0$" we mean $v(r) \geq 0$ for every $r \in (0,1)$; and by "$v' \geq 0$ a.e." we mean $v'(r) \geq 0$ for almost every $r \in (0,1)$. We recall that every function $v \in H^{1,p}_{loc}(0,1)$ is continuous in $(0,1)$.

Since we consider functions $v = v(r)$ that are increasing with respect to r, we can define v at the endpoints of $(0,1)$, by defining, for example at $x = 0$:

$$v(0) = \inf\{v(r) : r \in (0,1)\} = \lim_{r \to 0^+} v(r). \tag{3.2}$$

Thus the value $v(0)$ in (3.1) is defined by (3.2).

It is easy to see that the set $\mathcal{A}$ is dense, with respect to the strong topology of $H^{1,p}_{loc}(0,1)$, in the set $\bar{\mathcal{A}}$ defined by

$$\bar{\mathcal{A}} = \{v \in H^{1,p}_{loc}(0,1) : v \geq 0, \quad v' \geq 0 \quad \text{a.e.}\}. \tag{3.3}$$

By its convexity, the set $\bar{\mathcal{A}}$ is closed both in the strong and in the weak topology of $H^{1,p}_{loc}(0,1)$.

With abuse of notation we denote the energy either by $E(u)$ or by $E(v)$, where $u(x)$ and $v(r)$ are associated by (2.1). The integral E in (2.4) is well defined in $\mathcal{A}$, since the integrand is assumed to be positive; in fact E is the supremum (with respect to a, b) of the corresponding integrals on subintervals $[a,b] \subset (0,1)$.

We extend E from $\mathcal{A}$ to $\bar{\mathcal{A}}$. We denote the extension by F. The idea is to define the energy $F(v)$ for $v \in \bar{\mathcal{A}}$ by continuity:

$$F(v) = \lim_k E(v_k), \tag{3.4}$$

where v_k is a sequence in $\mathcal{A}$ that converges (we will consider either the strong or the weak topology) to v. To be sure that the definition is independent of the particular sequence v_k, we proceed as follows.

As usual we use the letter s to denote the strong topology, and the letter w to denote the weak topology of $H^{1,p}_{loc}(0,1)$. For every $v \in \bar{\mathcal{A}}$ we define:

$$F_s(v) = \inf\{\liminf_k E(v_k) : v_k \in A, v_k \xrightarrow{s} v\}, \tag{3.5}$$

$$F_w(v) = \inf\{\liminf_k E(v_k) : v_k \in A, v_k \xrightarrow{w} v\}.$$

The scheme of the above definitions is classical. It was introduced by Lebesgue in his thesis [11], and then considered again by De Giorgi, Giusti, Miranda (see e.g. [10], [16]), Serrin [22], and recently by many others (see for example [5], [7], [9]). In this context this scheme was introduced by the author in [13].

It is easy to see that F_s is lower semicontinuous in the strong topology of $H^{1,p}_{loc}(0,1)$ and that (under coercivity conditions) F_w is lower semicontinuous in the weak topology. It is less easy to derive representation formulas for F_s and F_w; we consider this problem in the next section.

4. Representation formulas for the stored-energy. With the aim of giving a characterization of F_s, F_w, we state our assumptions on the integrand ϕ in (2.4).

(4.1) $\phi(\xi,\eta,\ldots,\eta)$ *is a continuous function for* $\xi > 0$ *and* $\eta > 0$.

(4.2) *There exist an exponent* $q < n$, *some positive constants* c, ξ_o *and a convex function* $h : [0,+\infty) \to [0,+\infty)$ *such that :*

$$\phi(\xi,\eta,\ldots,\eta) \geq h(\xi\eta^{n-1}), \quad \forall \xi \geq 0, \quad \forall \eta \geq 0; \tag{4.2a}$$

$$\phi(\xi,\xi,\ldots,\xi) \leq c(1+\xi^p) + h(\xi^n), \quad \forall \xi \geq \xi_o. \tag{4.2b}$$

Note that we do not require that $\phi = \phi(\xi,\eta,...,\eta)$ is convex with respect to ξ, neither do we require that ϕ is bounded from above as $\xi \to 0^+$.

Assumptions (4.1), (4.2) are very general and natural in the theory of nonlinear elasticity by Ball [1], [2]. Of course they can be satisfied by integrands of the type (1.8) with $p < n$ and g as in figures 1 or 2.

Theorem 1. *Let E, F_s, F_w be defined respectively by (2.4), (3.5), (3.6). Under assumptions (4.1), (4.2) the following representation formula holds:*

$$F_s(v) = \omega_n \int_o^1 r^{n-1}\phi\left(v', \frac{v}{r}, ..., \frac{v}{r}\right) dr + \tilde{h}\frac{\omega_n}{n}[v(0)]^n, \tag{4.3}$$

for every $v \in \bar{A}$, where the constant $\tilde{h} \in [0, +\infty]$ is given by

$$\tilde{h} = \lim_{t \to +\infty} h(t)/t = \lim_{t \to +\infty} h'(t). \tag{4.4}$$

Moreover, if $\phi = \phi(\xi, \eta, ..., \eta)$ is convex with respect to ξ, the above representation formula holds also for F_w, i.e. $F_w(v) = F_s(v)$ for every $v \in \bar{A}$.

Note that in principle F_s and F_w in (3.5), (3.6) depend on p; but, under the assumptions of the above theorem, F_s and F_w are actually independent of p.

Before giving the proof of theorem 1 we state in the following lemma 2 a known result about convex functions.

For a convex function $h : \mathbf{R} \to \mathbf{R} \cup \{+\infty\}$ the inequality of convexity can be written by:

$$\frac{1}{\sum \lambda_i} \sum \lambda_i h(\xi_i) \geq h\left(\frac{\sum \lambda_i \xi_i}{\sum \lambda_i}\right), \tag{4.5}$$

where $\xi_i \in \mathbf{R}$ and $\lambda_i \geq 0$, with $\sum \lambda_i \neq 0$. By approximating L^1-functions $\lambda(r), \xi(r)$ by step functions, each of them assuming a finite number of values λ_i, ξ_i, we easily obtain the following form of Jensen's inequality (the usual Jensen's inequality is obtained for $\lambda(r)$ = constant):

Lemma 2. *Let $h : \mathbf{R} \to \mathbf{R} \cup \{+\infty\}$ be a convex function and let λ, ξ be L^1 functions, with $\lambda \geq 0, \lambda \not\equiv 0$.*

$$\frac{\int_a^b \lambda(r) h(\xi(r)) dr}{\int_a^b \lambda(r) dr} \geq h\left(\frac{\int_a^b \lambda(r)\xi(r) dr}{\int_a^b \lambda(r) dr}\right). \tag{4.6}$$

Proof of Theorem 1. If $v(0) = 0$ formula (4.3), i.e. $F_s = E$, follows from the lower semicontinuity of E in the strong topology of $H^{1,p}_{loc}(0,1)$ (by Fatou's lemma) and, if $\phi(\xi, \eta, ..., \eta)$ is convex with respect ξ, then $F_w = F_s = E$ follows from the lower semicontinuity of E in the weak topology of $H^{1,p}_{loc}(0,1)$.

Thus let us consider $v(0) > 0$ and let v_k be a sequence in $\mathcal{A}$ that converges to v in the strong topology of $H^{1,p}_{loc}(0,1)$. We extract a subsequence of v_k, that we still denote by v_k, with the properties: (i) the subsequence of real numbers $E(v_k)$ admits a limit and this limit is equal to the limes inferior of the original sequence; (ii) $v_k(r)$ converges to $v(r)$ for every $r \in (0,1)$.

Let M be an upper bound for $v_k(1/2)$, so that

$$v_k(1/2) \leq M \quad , \quad \forall k. \tag{4.7}$$

For every $r \in (0,1)$ we have

$$\lim_k v_k(r) = v(r) \geq v(0); \tag{4.8}$$

thus, for every natural ν, by choosing $r = 1/\nu$, there exists k_ν such that

$$v_{k_\nu}(1/\nu) > v(0) - 1/\nu. \tag{4.9}$$

If we define $w_\nu(r) = (v(0)\nu - 1)r$ and if ν is sufficiently large ($\nu > (2M+1)/v(0)$), by (4.7) and (4.9) we obtain

$$w_{k_\nu}(1/2) > M \geq v_{k_\nu}(1/2) \quad , \quad w_{k_\nu}(1/\nu) < v_{k_\nu}(1/\nu). \tag{4.10}$$

Therefore, if ν is sufficiently large, there exists $r_\nu \in (1/\nu, 1/2)$ such that $v_{k_\nu}(r_\nu) = w_{k_\nu}(r_\nu) = (v(0)\nu - 1)r_\nu$; moreover the following relations hold:

$$\lim_\nu r_\nu = 0 \quad ; \quad \liminf_\nu v_{k_\nu}(r_\nu) \geq v(0). \tag{4.11}$$

In fact r_ν converges to zero since, by (4.7), we have

$$\frac{1}{\nu} < r_\nu = \frac{v_{k_\nu}(r_\nu)}{v(0)\nu - 1} \leq \frac{v_{k_\nu}(1/2)}{v(0)\nu - 1} \leq \frac{M}{v(0)\nu - 1}; \tag{4.12}$$

while the second relation in (4.11) holds by (4.9), since $v_{k_\nu}(r_\nu) \geq v_{k_\nu}(1/\nu) > v(0) - 1/\nu$.

With the aim of finding a lower bound for F_s in (3.5), we use assumption (4.2a) to obtain

$$\begin{aligned} \liminf_k \ E(v_k) &= \lim_\nu \ E(v_{k_\nu}) \geq \\ &\geq \liminf_\nu \ \omega_n \int_{r_\nu}^1 r^{n-1} \phi\left(v'_{k_\nu}, \frac{v_{k_\nu}}{r}, \ldots, \frac{v_{k_\nu}}{r}\right) dr + \\ &+ \liminf_\nu \ \omega_n \int_o^{r_\nu} r^{n-1} h\left(v'_{k_\nu} \left(\frac{v_{k_\nu}}{r}\right)^{n-1}\right) dr. \end{aligned} \tag{4.13}$$

We estimate the second term on the right hand side by using Jensen's inequality (4.6) with $\lambda(r) = r^{n-1}$ and $\xi(r) = v'_{k_\nu}(v_{k_\nu}/r)^{n-1}$. Since $v^n_{k_\nu}/n$ is a primitive of $v'_{k_\nu} v^{n-1}_{k_\nu}$, we have

$$
\begin{aligned}
&\int_o^{r_\nu} r^{n-1} h\left(v'_{k_\nu}\left(\frac{v_{k_\nu}}{r}\right)^{n-1}\right) dr \geq \\
(4.14)\qquad &\geq \frac{r^n_\nu}{n} h\left(\frac{n}{r^n_\nu} \cdot \frac{v^n_{k_\nu}(r_\nu)}{n}\right) \\
&= \frac{v^n_{k_\nu}(r_\nu)}{n} \frac{h(t_\nu)}{t_\nu},
\end{aligned}
$$

where we have posed $t_\nu = (v_{k_\nu}(r_\nu)/r_\nu)^n$. By (4.11) $t_\nu \to +\infty$. By again using (4.11) and definition (4.4), we obtain

$$
\begin{aligned}
(4.15)\qquad &\liminf_\nu \omega_n \int_o^{r_\nu} r^{n-1} h\left(v'_{k_\nu}\left(\frac{v_{k_\nu}}{r}\right)^{n-1}\right) dr \geq \\
&\geq \tilde{h}\frac{\omega_n}{n}[v(0)]^n.
\end{aligned}
$$

Let us go back to (4.13). For every fixed ν_o, if $r_\nu < r_{\nu_o}$, by Fatou's lemma we have:

$$
\begin{aligned}
(4.16)\qquad &\liminf_k E(v_k) \geq \\
&\geq \omega_n \int_{r_{\nu_o}}^1 r^{n-1}\left(v', \frac{v}{r}, \ldots, \frac{v}{r}\right) dr + \tilde{h}\frac{\omega_n}{n}[v(0)]^n.
\end{aligned}
$$

As $\nu_o \to +\infty$ we obtain

$$
(4.17)\qquad F_s(v) \geq \omega_n \int_o^1 r^{n-1}\phi\left(v', \frac{v}{r}, .., \frac{v}{r}\right) dr + \tilde{h}\frac{\omega_n}{n}[v(0)]^n.
$$

To get the opposite inequality we can compare a fixed function $v \in \bar{A}$, having $v(0) > 0$, with the sequence $w_k(r) = kr$. For every $k > v(1/2)$ we can choose $r_k \in (0, 1/2)$ so that $v(r_k) = kr_k$. Since $v(r_k) \leq v(1/2)$, then r_k converges to zero as $k \to +\infty$. let us define v_k by:

$$
(4.18)\qquad v_k(r) = \begin{cases} kr & \text{if } 0 \leq r \leq r_k \\ v(r) & \text{if } r_k < r < 1. \end{cases}
$$

Since $v_k' = v_k/r = k$ for $r \in (0, r_k)$, we have

$$
\begin{aligned}
F_s(v) &\leq \limsup_k \omega_n \int_o^1 r^{n-1}\phi\left(v_k', \frac{v_k}{r}, \ldots, \frac{v_k}{r}\right) dr \cdot \\
&\leq \omega_n \int_o^1 r^{n-1}\phi\left(v', \frac{v}{r}, \ldots, \frac{v}{r}\right) dr \cdot \qquad (4.19)\\
&\cdot \limsup_k \frac{\omega_n}{n} r_k^n \phi(k, k, \ldots, k).
\end{aligned}
$$

We estimate the last term on the right hand side using assumption (4.2b). Since $kr_k = v(r_k)$, and since $v(r_k)$ converges to $v(0)$, we have

$$
\begin{aligned}
&\limsup_k r_k^n \phi(k, k, ..., k) \leq \\
&\leq \lim_k r_k^n [c(1 + k^q) + h(k^n)] \qquad (4.20)\\
&= \lim_k k^{-n}[c(1 + k^q) + h(k^n)][v(r_k)]^n = \tilde{h}[v(0)]^n.
\end{aligned}
$$

Here we have used the assumption $q < n$.

Since $v_k(r) = v(r)$ for $r > r_k$ and since $r_k \to 0$, the sequence v_k converges to v in $H_{loc}^{1,p}(0,1)$. Thus the opposite inequality to (4.17) follows from (4.19), (4.20).

The statement relating to F_w follows analogously. The only difference is that (4.16) then follows from the lower semicontinuity of the integral with respect to the weak topology of $H_{loc}^{1,p}(0,1)$ whenever $\phi(\xi, \eta, ..., \eta)$ is convex with respect to ξ.

By combining theorem 1 with theorem 3.8 of Marcellini and Sbordone [15] (see also [8]) it is possible to prove the following further characterization result when $\phi(\xi, \eta, \ldots, \eta)$ is not convex with respect to ξ.

Theorem 3. *For every $\nu > 0$ let $\phi^{**}(\xi, \eta)$ be the greatest function convex with respect to $\xi > 0$ and less or equal to $\phi(\xi, \eta, \ldots, \eta)$, with ϕ satisfying (4.1), (4.2). Let $\tilde{h}$ be the constant given by (4.4) and let F_w be the functional defined in (3.6). Then, for every $v \in \bar{A}$ we have*

$$
F_w(v) = \omega_n \int_o^1 r^{n-1}\phi^{**}\left(v', \frac{v}{r}, \ldots, \frac{v}{r}\right) dr + \tilde{h}\frac{\omega_n}{n}[v(0)]^n. \qquad (4.21)
$$

5. Interpretation and consequences of the representation formulas. A first consequence of the representation formula (4.3) is stated in the following:

Corollary 4. *Let us assume that (4.1), (4.2) hold and that the energy F_s is represented by (4.3). Then the phenomenon of cavitation may occur only if $h(t)$ is "(sub-)linear at infinity". More precisely, if $h(t) \geq ct^r$ for some $r > 1$ and $c > 0$, then $F_s(v) = +\infty$ for every $v \in \bar{A}$ with $v(0) > 0$.*

A second consequence can be easier described if we consider again the general situation without radial symmetry, with the energy integrand $W(\xi)$ of the form (1.4), that is:

$$W(\xi) = W_1(\xi) + g(\det \xi), \tag{5.1}$$

$W_1(\xi)$ being a quasiconvex function in the sense of the following definition (5.2).

We already pointed out that the linearity of $g(t)$ at ∞ is of interest in our approach. The function $g(t)$ must be convex or concave for large values of t?

If $g(t)$ is convex then it is well known that $W(\xi)$ satisfies the *quasiconvexity condition* by Morrey [17]:

$$\int_\Omega W(\xi + D\phi(x))dx \geq \int_\Omega W(\xi)dx = W(\xi)|\Omega|, \quad \forall \phi \in C_o^1(\Omega; \mathbf{R}^n). \tag{5.2}$$

J.Ball pointed out that this mathematical condition by Morrey has an interpretation in nonlinear elasticity; we quote from Ball [1], pages 338, 339: "*... for ... a body that admits as a possible displacement a homogeneous strain, we require that this homogeneous strain be an absolute minimizer for the total energy*".

As described in section 2, in cavitation we impose the boundary condition $u(x) = u_o(x)$ at $|x| = 1$, where $u_o(x) = \lambda x$. The displacement u_o is a homogeneous strain. Thus, *under quasiconvexity*, u_o

must be an absolute minimizer and *the phenomenon of cavitation should not occur.*

This is what happens under the present approach; while this fact contrasts sharply with the approach of Ball in [2], where the phenomenon of cavitation occurs with quasiconvex energy integrals.

In fact, under our approach the energy is defined by lower semicontinuity; thus, if inequality (5.2) holds for smooth test functions ϕ, then it holds for the extended energy too.

Therefore, if the stored-energy E in (1.1) is quasiconvex, then $u_o(x) = \lambda x$ is an absolute minimizer among displacements with the same boundary values. In terms of $v(r), v_o(r) = \lambda r$ is an absolute minimizer for E given by (2.4), and thus v_o is an absolute minimizer also for F_s, F_w in (3.5), (3.6).

Of interest is the Euler's first variation of the functional F_s in (4.3). We obtain (for a non formal derivation we can proceed like in theorem 7.3 of Ball [2]) that a minimum of F_s on $\bar{A}$, with the condition $v(1) = \lambda$, formally satisfies the Euler-Lagrange equation:

$$\frac{d}{dr}\left(r^{n-1}\phi_\xi\right) = (n-1)r^{n-2}\phi_\eta, \qquad \forall r \in (0,1), \tag{5.3}$$

and the boundary conditions

$$\text{at} \quad r = 1: \quad v(1) = \lambda, \tag{5.4}$$

$$\text{at} \quad r = 0: \quad \text{either} \quad v(0) = 0 \quad \text{or} \quad \lim_{r\to 0+}\left(\frac{r}{v(r)}\right)^{n-1}\phi_\xi = \tilde{h}. \tag{5.5}$$

The expression $(r/v)^{n-1}\phi_\xi$ that appears in (5.5) is called the *radial component of the Cauchy stress.* In principle we could imagine measuring the stress at $r = 0$ of an equilibrium solution with cavity. If $\tilde{h}$ is the value of this measure, then we have an indirect measure of the behaviour of the function $g(t)$ in (1.5) for large values of t; in fact $g(t)$ should behave like in figure 2, with $g(t)/t \to \tilde{h}$ as $t \to +\infty$.

Finally let us observe that the functional F_s in (4.3) can be represented in the form:

$$F_s(v) = \omega_n \int_o^1 r^{n-1}\left[\phi\left(v', \frac{v}{r}, \ldots, \frac{v}{r}\right) - \tilde{h}v'\left(\frac{v}{r}\right)^{n-1}\right]dr + \\ + \tilde{h}\frac{\omega_n}{n}[(v(1)]^n, \tag{5.6}$$

for every $v \in \bar{\mathcal{A}}$. Thus in the class of functions $v \in \bar{\mathcal{A}}$ such that $v(1) = \lambda$, F_s can be represented in integral form.

On the contrary, in the representation (4.3), the functional F_s is the sum of an integral and a measure concentrated at $r = 0$; notice that this measure is equal to the product of the constant $\tilde{h}$ by the volume of the cavity that forms around the origin. The measure $\tilde{h}(\omega_n/n)[v(0)]^n$ can be interpreted as the energy due to the cavity; this energy is proportional to the volume of the cavity and not to the surface area, like in some standard models.

6. The non radially symmetric case. Let us consider again the general situation without radial symmetry, and a stored-energy of the form

$$E(u) = \int_\Omega W_1(Du, adj\ Du)dx + \int_\Omega g(\det\ Du)dx, \tag{6.1}$$

where *adj* Du are the adjoints of the $n \times n$ matrix Du. Here W_1 and g are convex functions.

The reader interested in results on the existence of minima is referred to Ball [1], Ball and Murat [4] and Marcellini [14].

Similarly to (3.6), we define:

$$F_w(u) = \inf\{\liminf_k E(u_k) : u_k \in C^1,\ u_k \overset{w}{\to} u \quad \text{in} \quad H^{1,p}\}. \tag{6.2}$$

We use the following well known result from Ball, Currie and Olver [3], which we quote in loose form: *If* $u \in H^{1,p}(\Omega, \mathbf{R}^n)$ *with* $p > n^2/(n+1)$, *then* Det Du *is well defined as a distribution* (like in [1], [3] we use the notation Det Du, instead of det Du, to remember that the determinant is a distribution).

In the applications to nonlinear elasticity it is natural to impose the restriction that the determinant of the deformation gradient is positive. Since every positive distribution is a measure, *if* $u \in H^{1,p}(\Omega; \mathbf{R}^n)$ *with* $p > n^2/(n+1)$ *and if* Det $Du \geq 0$, *then the distribution* Det Du *is a positive measure.* We can operate on the measure Det Du by the *Lebesgue decomposition* (see for instance [21], theorem 6.9): we denote by $\text{Det}_R Du$ the *regular part* of Det Du, i.e. the part absolutely continuous with respect to the Lebesgue measure, and by $\text{Det}_S\ Du$ the *singular part*, i.e. the complement:

$$\text{Det}_S\ Du = \text{Det}\ Du - \text{Det}_R\ Du. \tag{6.3}$$

On the basis of the representation results of theorem 1, on the following theorem 5, and on the representation result obtained in section 5 of [13], we formulate the following:

Conjecture. *If* $p > n^2/(n+1)$ *and* Det $Du \geq 0$ *we have*

$$\begin{aligned} F_w(u) &= \int_\Omega W_1(Du, adj\ Du)dx + \\ &+ \int_\Omega g(\mathrm{Det}_R\ Du)dx + \tilde{g} \cdot \text{ Total Variation } (\mathrm{Det}_S Du) \end{aligned} \tag{6.4}$$

for every $u \in H^{1,p}(\Omega;\mathbf{R}^n)$, *where* $\tilde{g}$ *is the limit, as* $t \to +\infty$, *of* $g(t)/t$.

Up to now the above conjecture has been not proved even if the singular part $\mathrm{Det}_S\ Du$ is identically zero. It is known only the case $u \in C^1(\Omega;\mathbf{R}^n)$; in fact the following result has been obtained in [12], [13]:

Theorem 5. *Let* $p > n^2/(n+1)$; *for every* $u \in C^1(\Omega;\mathbf{R}^n)$ *we have* $F_w(u) = E(u)$; *thus formula (6.4) holds on* $C^1(\Omega;\mathbf{R}^n)$. *This means that*

$$E(u) \leq \liminf_k E(u_k), \tag{6.5}$$

for every $u, u_k \in C^1(\Omega;\mathbf{R}^n)$ *such that* u_k *converges to* u *in the weak topology of* $H^{1,p}(\Omega;\mathbf{R}^n)$, *with* $p > n^2/(n+1)$.

Acknowledgement. This research has been partially done when the author was visiting the Ecole Polytechnique Fédérale de Lausanne in January-February 1986. The author wishes to thank Bernard Dacorogna and Charles Stuart for their hospitality and for the discussions on this subject.

References

[1] J.M.Ball, *Convexity conditions and existence theorems in nonlinear elasticity*, Arch. Rat. Mech. Anal. **63** (1977), 337-403.

[2] J.M.Ball, *Discontinuous equilibrium solutions and cavitation in nonlinear elasticity*, Phil. Trans. R. Soc. London **306** (1982), 557-611.

[3] J.M.Ball, J.C.Currie, P.J.Olver, *Null lagrangians, weak continuity and variational problems of arbitrary order*, J. Funct. Anal. **41** (1981), 135-175.

[4] J.M.Ball, F.Murat, *$W^{1,p}$-quasiconvexity and variational problems for multiple integrals*, J. Funct. Anal. **58** (1984), 225-253.

[5] G.Buttazzo, G.Dal Maso, *Integral representation on $W^{1,\alpha}(\Omega)$ and $BV(\Omega)$ of limits of variational integrals*, Atti Accad. Naz. Lincei **66** (1979), 338-344.

[6] P.Charrier, B.Dacorogna, B.Hanouzet, P.Laborde, *An existence theorem for slightly compressible material in nonlinear elasticity*, S.I.A.M., Math. Anal. (1987).

[7] G.Dal Maso, *Integral representation on $BV(\Omega)$ of Γ limits of variational integrals*, Manuscripta Math. **30** (1980), 387-416.

[8] I.Ekeland, R.Temam, *Analyse convexe et problèmes variationnels*, Dunod & Gauthier-Villars (1974).

[9] M.Giaquinta, G.Modica, J.Soucek, *Functionals with linear growth in the calculus of variations I*, Commentationes Math. Univ. Caroline **20** (1979), 143-172.

[10] E.Giusti, *Non-parametric minimal surfaces with discontinuous and thin obstacles*, Arch. Rat. Mech. Anal. **49** (1972), 41-56.

[11] H.Lebesgue, *Intégrale, Longueur, aire*, Ann. Mat. Pura Appl. **7** (1902), 231-359.

[12] P.Marcellini, *Approximation of quasiconvex functions, and lower semicontinuity of multiple integrals*, Manuscripta Math. **51** (1985), 1-28.

[13] P.Marcellini, *On the definition and the lower semicontinuity of certain quasiconvex integrals*, Ann. Inst. H.Poincaré, Analyse Non Linéaire **3** (1986), 391-409.

[14] P.Marcellini, *Existence theorems in nonlinear elasticity*, Fermat Days 85, J.B. Hiriart-Urruty ed., North-Holland (1986), 241-247.

[15] P.Marcellini, C.Sbordone, *Semicontinuity problems in the calculus of variations*, Nonlinear Anal. Theory Methods Appl. **4** (1980), 241-257.

[16] M.Miranda, *Un teorema di esistenza e unicità per il problema dell'area minima in n variabili*, Ann. Scuola Norm. Sup. Pisa **19** (1965), 233-249.

[17] C.B.Morrey, *Multiple integrals in the calculus of variations*, Springer-Verlag (1966).

[18] R.W.Odgen, *Large deformation isotropic elasticity: on the correlation of theory and experiment for compressible rubberlike solids*, Proc. R. Soc. London A. **328** (1972), 567-583.

[19] P.Podio-Guidugli, G.Vergara Caffarelli, E.G.Virga, *Discontinuous energy minimizers in nonlinear elastostatics: an example of J.Ball revisited*, Journal of Elasticity **16** (1986), 75-96.

[20] P.Podio-Giudugli, G.Vergara Caffarelli, E.G.Virga, *Cavitation and phase transition of hyperelastic fluids*, Arch. Rat. Mech. Anal. **92** (1986), 121-136.

[21] W.Rudin, *Real and complex analysis*, McGraw-Hill (1966).

[22] J.Serrin, *On the definition and properties of certain variational integrals*, Trans. Amer. Math. Soc. **101** (1961), 139-167.

[23] J.Sivaloganathan, *Uniqueness of regular and singular equilibria for spherically symmetric problems of nonlinear elasticity*, Arch. Rat. Mech. Anal., to appear.

[24] J.Sivaloganathan, *A field theory approach to stability of radial equilibria in nonlinear elasticity*, Math. proc. Cambridge Philos., to appear.

[25] C.A.Stuart, *Radially symmetric cavitation for hyperelastic materials*, Ann. Inst. H.Poincaré, Analyse Non Linéaire **2** (1985), 33-66.

[26] C.A.Stuart, *Special problems involving uniqueness and multiplicity in hyperelasticity*, preprint.

[27] P.J.Blatz, W.L.Ko, *Application of finite elastic theory to the deformation of rubbery materials*. Trans. Soc. Rheology **6** (1962), 223-251.

[28] C.O.Horgan, R.Abeyaratne, *A bifurcation problem for a compressible nonlinear elastic medium: growth of a micro-void*, Journal of Elasticity **16** (1986), 189-200.

[29] J.K.Knowles, E.Sternberg, *On the ellipticity of the equations of nonlinear elastostatics for a special material*, Journal of Elastic-

ity **5** (1975), 341-361.

Istituto Matematico "U.Dini"
Viale Morgagni 67/A
I-50134 FIRENZE

THE CALCULUS OF VARIATIONS AND SOME SEMILINEAR VARIATIONAL INEQUALITIES OF ELLIPTIC AND PARABOLIC TYPE

ANTONIO MARINO

in collaboration with
M.Degiovanni - F.Giannoni - D.Passaseo - C.Saccon

Dedicated to Ennio De Giorgi on his sixtieth birthday

Introduction

We present here a survey on some techniques of nonsmooth calculus of variations, that we developed in these last years in collaboration with other authors, and some results concerning semilinear variational inequalities that can be deduced by means of such techniques.

All the problems presented are associated with a suitable lower semicontinuous functional f on a suitable constraint V. More precisely the problems of elliptic type treated in chapters I, II, IV can be seen as the search of "lower critical points for f"(in the sense of definition 1.2 of chapter I) while the parabolic one considered in Chapter III can be reduced to study the "strong evolution curves for

f"(according to definition 1.8 of Chapter III).

We remark that the constraint V involved in Chapter II is convex (but nonsmooth), while the one considered in the following chapters is non convex (and non smooth). In the latter, V is made up by the intersection of a convex set K and a hypersurface (a sphere) S. A capital point in many theorems is studying the "non-tancency"between K and S (see Theorem 2.2 of Chapter III).

The strong evolution curves are a necessary tool in all the problems studied. In Chapter II the theory of maximal monotone operators and their Lipschitz-continuous perturbation (see [10]) suffices for having such curves, since the constraint V is convex. On the contrary, in Chapters III and IV, where V is not convex, the use of the ϕ-convex functions theory has turned out to be very useful (see [22] and [31]).

As a premise, in Chapter I, we have felt worthwhile to point out the links between the concept of slope, some classical regularity results and the use of "supersolutions "as fictitious constraints: solutions of the elliptic variational inequality, with the condition $u \leq \varphi$, solve the corresponding equation, if φ is a supersolution.

Several indications on possible developments of the research and some open problems, are illustrated.

To conclude this introduction we wish to remark that the ideas and the techniques used in Chapters III and IV, for studying variational inequalities on non convex constraints, are based on the classical ones of the convex setting (see [8], [13], [14], [15], [35], [38], [40], [45], [46], [47], [75], [77]) and are a natural development of them. By means of these methods we could apply the standard techniques of the calculus of variations in the large to obtain the multiplicity results reported in Chapters II and IV.

Finally we wish to recall that some researches concerning strong evolution curves and some generalizations have been carried on recently, out of the ϕ-convex functions context. In these studies, which make use of some compactness arguments, some existence results (with no uniqueness) have been proved for larger classes of functionals (see [51],[70],[36]).

CHAPTER I

Some functionals. Preliminaries: slope, regularization, supersolutions.

In this chapter we want to point out two important facts concerning the notion of slope (see Definition 1.1): a) the relationship between the slope on a classical convex constraint and some regularity results (see [18]); b) the relationship between the slope and the use of supersolutions as fictitious constraints (see [50], [64], [66]). We remark that the latter technique can be seen as a non linear version of Perron's method for equations.

1. Slope and constrained slope. Let X be a metric space with metric d and $f : X \to \mathbf{R} \cup \{+\infty\}$ a given function. Set

$$\mathcal{D}(f) = \{u \in X \mid f(u) < +\infty\}.$$

The following definition was introduced in [22].

Definition 1.1. *Let $u \in X$ and $\rho \geq 0$. We set*

$$\chi_u(\rho) = \inf\{f(v) \mid d(u,v) \leq \rho\}.$$

We call "slope of f "the function $|\nabla f| : \mathcal{D}(f) \to \mathbf{R}^+ \cup \{+\infty\}$ defined by

$$|\nabla f|(u) = -\liminf_{\rho \to 0+} \frac{\chi_u(\rho) - \chi_u(0)}{\rho} \qquad (\chi_u(0) = f(u)).$$

A given point u in $\mathcal{D}(f)$ will be called a "lower critical point for f" if $|\nabla f|(u) = 0$.

Definition 1.2. *Let $V \subset X$. We define $I_V : X \to \mathbf{R} \cup \{+\infty\}$ by:*

$$I_V(u) = \begin{cases} 0 & \text{if } u \in V, \\ +\infty & \text{if } u \in X \setminus V. \end{cases}$$

Let $u \in V \cap \mathcal{D}(f)$ *(of course* $|\nabla(f+I_V)|(u) \le |\nabla f|(u)$*). We say that* u *is a "lower critical point for* f *on* V*" if* $|\nabla(f+I_V)|(u) = 0$.

The following statement is straightforward.

Lemma 1.3 (constrained slope). *Let* $V \subset X$ *and* $u \in V \cap \mathcal{D}(f)$*. Suppose that there exist a neighbourhood* U *of* u *and a map* $\psi : U \cap \{v \in X \mid f(v) \le f(u)\} \to V$ *with the properties:*

$$(1.4) \qquad \begin{cases} d(\psi(v),u) \le C_\psi d(v,u), \\ f(v) \ge f(\psi(v)) - C_f d(v,u) - \omega(d(u,v)) \end{cases}$$
for all v *in* U *with* $f(v) \le f(u)$

where C_f, C_ψ *are real numbers and* $\omega : \mathbf{R}^+ \to \mathbf{R}$ *is such that* $\lim_{t\to 0^+} \frac{\omega(t)}{t} = 0$.

Then we have:

a) $|\nabla f|(u) \le C_f + C_\psi |\nabla(f+I_V)|(u)$;

b) if $C_f = 0$*, then (as a consequence of a)) any lower critical point for* f *on* V *is a lower critical point for* f.

Remark 1.5. Consider the case in which X is an open subset of a Hilbert space H, f is Fréchet differentiable and V is a smooth manifold in H, contained in X.

If we assume, for instance, that for all u in V grad$f(u)$ is tangential to V, then for every u in V, (1.4) are verified in a suitable neighbourhood of u, with $C_f = 0$ and taking as ψ the function such that

$$\|\psi(v) - v\| = \inf\{\|w - v\| \mid w \in V\}.$$

2. A concrete functional, its slope and regularization. The functional f we are going to introduce is a well known one, which is strictly concerned with the problems we shall treat later.

In this section we wish to point out the links relating some regularization results which are known (see [13], [45]) and the slope of f on a suitable constraint (see theorem 2.5 in particular). The assumption that will be considered here are a simplified version of more general ones; for a detailed study of this topic see [17].

Let Ω be a bounded open subset of $\mathbf{R}^n$ and $g : \Omega \times \mathbf{R} \to \mathbf{R}$, $\varphi : \Omega \to \mathbf{R}$ be two functions. Set

$$K_\varphi = \{v \in \mathrm{L}^2(\Omega) \mid \varphi \le v \text{ a.e. in } \Omega\}.$$

(2.1) The following conditions will be considered.

$$(g) \quad \begin{cases} g \text{ is a Charatéodory function and there exist } a \text{ in} \\ \mathrm{L}^1(\Omega), b \text{ in } \mathbf{R} \text{ such that} \\ G(x,s) = \int_0^s g(x,\sigma)\,\mathrm{d}\sigma \le a(x) + bs^2 \quad \forall s, \text{a.a. } x; \\ \dfrac{g(x,s_2) - g(x,s_1)}{s_2 - s_1} \le C_g \quad \forall s_1, s_2, \text{ with } s_1 \ne s_2, \text{a.a. } x \\ g \text{ is integrable in } \Omega \times [\alpha,\beta] \quad \forall \alpha,\beta \text{ with } \alpha \le \beta; \end{cases}$$

$$(\varphi,r)_- \begin{cases} \varphi \in \mathrm{W}^{1,2}(\Omega) \text{ and, for every } v \text{ with } G(\cdot,v) \in \mathrm{L}^1(\Omega), \\ v - \varphi \in \mathrm{W}^{1,2}_\mathrm{o}(\Omega), \text{ if we set } w = v - \max(v,\varphi), \text{ we have} \\ \qquad g(\cdot,\varphi)w \in \mathrm{L}^1(\Omega), \\ \int_\Omega \mathrm{D}\varphi \mathrm{D}w\,\mathrm{dx} - \int_\Omega g(x,\varphi)w\,\mathrm{dx} \ge -r\left(\int_\Omega w^2\,\mathrm{dx}\right)^{1/2} \end{cases}$$

$$(\varphi,r)_+ \begin{cases} \varphi \in \mathrm{W}^{1,2}(\Omega) \text{ and, for every } v \text{ with } G(\cdot,v) \in \mathrm{L}^1(\Omega), \\ v - \varphi \in \mathrm{W}^{1,2}_\mathrm{o}(\Omega), \text{ if we set } w = v - \min(v,\varphi), \text{ we have} \\ \qquad g(\cdot,\varphi)w \in \mathrm{L}^1(\Omega), \\ \int_\Omega \mathrm{D}\varphi \mathrm{D}w\,\mathrm{dx} - \int_\Omega g(x,\varphi)w\,\mathrm{dx} \ge -r\left(\int_\Omega w^2\,\mathrm{dx}\right)^{1/2} \end{cases}$$

where $r \ge 0$.

The conditions above are verified if, for instance, $\Delta\varphi + g(\cdot,\varphi) \in \mathrm{L}^2(\Omega)$.

We introduce now the functional $f : \mathrm{L}^2(\Omega) \to \mathbf{R} \cup \{+\infty\}$ defined by

(2.2)

$$f(u) = \begin{cases} \int_\Omega \left(\frac{1}{2}|\mathrm{D}u|^2 - G(x,u)\right)\mathrm{dx} & \text{if } u \in \mathrm{W}^{1,2}_\mathrm{o}(\Omega), \\ +\infty & \text{if } u \in \mathrm{L}^2(\Omega) \setminus \mathrm{W}^{1,2}_\mathrm{o}(\Omega) \end{cases}$$

(here $Du = \left(\frac{\partial u}{\partial x_1}, \ldots, \frac{\partial u}{\partial x_n}\right)$). In the space $L^2(\Omega)$ we consider the usual metric $d(u,v) = \|v-u\| = \left(\int_\Omega (v-u)^2\,dx\right)^{1/2}$.

Proposition 2.3. *Under the assumptions (g) of (2.1) the following facts hold.*

a) $\mathcal{D}(f) = \{u \in W_0^{1,2}(\Omega) \mid G(\cdot,u) \in L^1(\Omega)\}$, *$\mathcal{D}(f)$ is convex; in addition if $u, v \in \mathcal{D}(f)$, then $\max(u,v)$, $\min(u,v)$ are in $\mathcal{D}(f)$ and for all w in $W_0^{1,2}(\Omega)$ with $u \le w \le v$ a.e., $w \in \mathcal{D}(f)$; moreover $W_0^{1,2}(\Omega) \cap L^\infty(\Omega) \subset \mathcal{D}(f)$.*

b) If $u, v \in \mathcal{D}(f)$, then

$$f(v) - f(u) \ge \int_\Omega (DuD(v-u) - g(x,u)(v-u))\,dx + \\ - \frac{C_g}{2}\int_\Omega (v-u)^2\,dx$$

(notice that $g(\cdot,u)(v-u)$ is lower integrable for all u,v in $\mathcal{D}(f)$).

The relationship between regularization, slope and constrained slope is cleared by the following statements (see [17], [18]).

Proposition 2.4. *Suppose (g) of (2.1) hold. If $u \in \mathcal{D}(f)$, then we have*

- $|\nabla f|(u) < +\infty$ *if and only if*
  $g(\cdot,u) \in L^1(\Omega), \Delta u + g(\cdot,u) \in L^2(\Omega)$ *in the sense of distributions;*

- $|\nabla f|(u) = \|\Delta u + g(\cdot,u)\|$, *if* $|\nabla f|(u) < +\infty$.

Theorem 2.5 (regularization and constrained slope).
Assume that (g) and $(\varphi, r)_-$ of (2.1) hold for some $r \ge 0$. Let $u \in \mathcal{D}(f) \cap K_\varphi$. If the variational inequality

$$\int_\Omega (DuD(v-u) - g(x,u)(v-u))\,dx \ge -C\|v-u\| \quad \forall v \text{ in } \mathcal{D}(f)\cap K_\varphi$$

holds (namely if $|\nabla(f+I_{K_\varphi})|(u)<+\infty$*: see b) of proposition 2.3), then* $g(\cdot,u)\in L^1(\Omega)$, $\Delta u+g(\cdot,u)\in L^2(\Omega)$ *in the sense of distributions (namely* $|\nabla f|(u)<+\infty$*: see proposition 2.4).*

It is worthwhile to remark that the proof is readily obtained by the constrained slope lemma 1.3, setting $\psi(v)=\max(v,\varphi)$ (see [18]).

3. Super-solutions as fictitious constraints. First consequences. We wish to present now a technique (see [50]) which is used for proving the results contained in chapter II.

In brief, we point out in theorem 3.3 that if $\bar\varphi$ is a super-solution for $\Delta+g$ (according to the definition 3.1), then any solution u of the variational inequality associated with $\Delta+g$ on the convex constraint $\{v\in\mathcal{D}(f)\mid v\le\bar\varphi\}$ is a solution of the equation $\Delta u+g(\cdot,u)=0$.

Definition 3.1. *Let* $\bar\varphi\in W^{1,2}(\Omega)$*. We say that* φ *is a super-solution [sub-solution] for the operator* $\Delta+g$*, if the condition* $(\varphi,0)_+$ *[*$(\varphi,0)_-$*] of (2.1) holds with* $\varphi=\bar\varphi$*.*

Theorem 3.2. *Suppose that the assumptions (g) of (2.1) hold. Then*

a) if $\bar\varphi_1,\bar\varphi_2$ *are super-solutions for* $\Delta+g$*, then* $\min(\varphi_1,\varphi_2)$ *is a super-solution for* $\Delta+g$*;*

b) if $\bar\varphi$ *is a super-solution for* $\Delta+g_1, g_2\le g_1$*, then* $\bar\varphi$ *is a super-solution for* $\Delta+g_2$*;*

c) if $\{s\mapsto g(x,s)\}$ *is convex a.e.in* Ω *and if* $\{x\mapsto G(x,u(x))\}$ $\left(=\int_0^{u(x)} g(x,\sigma)\,d\sigma\right)\in L^1(\Omega)$ *for every* u *in* $W^{1,2}_0(\Omega)$ *with* $u\ge 0$*, then every convex combination of super-solutions is a super-solution (the above conditions are verified if, for instance, the relation* $g(x,s)\ge\lambda s-c(x)$ *holds for all* $s\ge 0$*, with* $\lambda\in\mathbf{R}$*,* $c\in L^2(\Omega)$*);*

d) if $\varphi:\Omega\to\mathbf{R}$*,* $u\in W^{1,2}_0(\Omega)$ *are given,* u *is a solution of the problem* $P_\varphi(h)$ *stated in definition 1.1 of Chapter II, then* u *is* *a super-solution for* $\Delta + g$.

Theorem 3.3 (supersolutions as fictitious constraint).
Assume (g) of (2.1). Let $\bar{\varphi}$ *be a super-solution for* $\Delta + g$, $u \in \mathcal{D}(f)$ *and* $u \leq \bar{\varphi}$.
Then

a) *if* u *is a solution of the variational inequality*

$$\int_{\Omega} \mathrm{D}u\mathrm{D}(v-u)\,\mathrm{dx} - \int_{\Omega} g(x,u)(v-u)\,\mathrm{dx} \geq 0 \tag{3.4}$$

for all v *in* $\mathcal{D}(f)$ *with* $v \leq \bar{\varphi}$, *then*

$$g(\cdot,u) \in \mathrm{L}^1(\Omega),$$
$$\Delta u + g(\cdot,u) = 0 \qquad \textit{in the sense of distributions;}$$

b) *if* $\varphi : \Omega \to \mathbf{R}$ *is such that* $\varphi \leq u \leq \bar{\varphi}$ *a.a. in* Ω, u *verifies (3.4) for all* v *in* $\mathcal{D}(f)$ *with* $\varphi \leq v \leq \bar{\varphi}$, *then* u *verifies (3.4) for all* v *in* $\mathcal{D}(f)$ *with* $v \geq \varphi$.

We remark that, as in theorem 2.5, the proof is based on lemma 1.3, on costrained slope, taking $\psi(v) = \min(v, \bar{\varphi})$.

The previous results imply the following theorem which is concerned with a classical elliptic partial differential equations problem (for inequalities see theorem 2.1 in chapter II).

Theorem 3.5 (see [1] under different assumptions on g).
Assume (g) of (2.1). Let $\varphi_1, \varphi_2 : \Omega \to \mathbf{R}$ *be such that* $\varphi_1 \leq \varphi_2$ *a.e. in* Ω, φ_1 *is a sub-solution for* $\Delta + g$, φ_2 *is a super-solution (see Definition 3.1) and let* $\bar{u} \in \mathcal{D}(f)$, $\varphi_1 \leq \bar{u} \leq \varphi_2$ *a.e. in* Ω. *Then*

a) *there exists* u *in* $\mathcal{D}(f)$ *with* $\varphi_1 \leq u \leq \varphi_2$ *a.e. in* Ω, *such that*

$$g(\cdot,u) \in \mathrm{L}^1(\Omega) \quad , \quad \Delta u + g(\cdot,u) = 0;$$

b) *there exist* u_1, u_2 *in* $\mathcal{D}(f)$, *which solve the equation* $\Delta u + g(\cdot, u) = 0$, *with* $\varphi_1 \leq u_1 \leq u_2 \leq \varphi_2$ *a.e. in* Ω, *such that any other solution* u *of the equation, with* $\varphi_1 \leq u \leq \varphi_2$, *satisfies the inequality* $u_1 \leq u \leq u_2$.

The proof can be obtained by minimizing or maximizing the functional $\int_\Omega u \, dx$ (for instance), on the set of all solutions lying between φ_1 and φ_2.

CHAPTER II

Solvability results and jumping behaviour in some variational inequalities.

We present here some results which can be found in [66] for what concerns sections 1 and 2 and in [37] for what concerns section 3. The problems consist in proving existence and multiplicity of solutions of some variational inequalities, which are lower critical points for a suitable functional f on a certain convex constraint V.

Since these points are not, in general, minimizers for f, the use of the evolution equation associated with f on V (an equation of the type $\mathcal{U}' + \mathrm{grad}_V f(\mathcal{U}) = 0$), turns out to be a very important tool. This evolution equation, in the case treated in this chapter, is obtained essentially by using the theory of maximal monotone operators and their Lipschitz continuous perturbations, since the constraint V is convex.

We wish to remark, finally, the analogy between the situation described in theorems 2.3 and 2.4 and the one which takes place in the non constrained case, when a jumping nonlinearity occurs (see [5],[4]). This analogy is produced precisely by the presence of the obstacle.

1. Setting of the problem $P_\varphi(h)$ and its characterization. Let Ω be a bounded open subset of $\mathbf{R}^n$. Let $g : \Omega \times \mathbf{R} \to \mathbf{R}$, $\varphi : \Omega \to \mathbf{R}$, $h : \Omega \to \mathbf{R}$ be assigned functions. We assume that g is a Carathéodory function and set

$$G(x,s) = \int_0^s g(x,\sigma)\, \mathrm{d}\sigma,$$
$$D = \{u \in \mathrm{W}_o^{1,2}(\Omega) \mid G(\cdot,u) \in \mathrm{L}^1(\Omega)\},$$
$$K_\varphi = \{u \in \mathrm{L}^2(\Omega) \mid u \geq \varphi \text{ a.e. in } \Omega\}.$$

Definition 1.1. *We say that a function u in $\mathrm{W}_o^{1,2}(\Omega)$ is a solution of the problem $P_\varphi(h)$ if*

$$P_\varphi(h) \qquad \begin{cases} u \in D \cap K_\varphi \text{ and for all } v \text{ in } D \cap K_\varphi \\ \qquad (-g(\cdot,u) + h)\,(v-u) \in \mathrm{L}^1(\Omega) \\ \displaystyle\int_\Omega \mathrm{D}u\mathrm{D}(v-u)\,\mathrm{dx} + \int_\Omega (-g(\cdot,u)+h)\,(v-u)\,\mathrm{dx} \geq 0 \end{cases}$$

In what follows we shall use the conditions (g) introduced in (2.1) of chapter I and the functional f defined in (2.2) of chapter I.

Definition 1.2. *Let $h \in \mathrm{L}^2(\Omega)$ (for sake of simplicity) and let $f_h : \mathrm{L}^2(\Omega) \to \mathbf{R} \cup \{+\infty\}$ be defined by*

$$f_h(u) = f(u) + \int_\Omega hu\,\mathrm{dx} \qquad \forall u \text{ in } \mathrm{L}^2(\Omega).$$

It is apparent that

$$D = \mathcal{D}(f_h) \quad , \quad D \cap K_\varphi = \mathcal{D}(f_h + I_{K_\varphi}).$$

In the space $\mathrm{L}^2(\Omega)$ we consider the usual metric $d(u,v) = \|v-u\| = \left(\int_\Omega (v-u)^2\,\mathrm{dx}\right)^{1/2}$.

Using the notion of lower critical point introduced in definitions 1.1 and 1.2 of chapter I, we can characterize the solution of $P_\varphi(h)$ as follows.

Proposition 1.3. *Assume that (g) of (2.1) of chapter I hold and that $h \in L^2(\Omega)$. Let $u \in D \cap K_\varphi$.*

Then u is a solution of $P_\varphi(h)$ if and only if u is a lower critical point for f_h on K_φ, that is if $|\nabla(f + I_{K_\varphi})|(u) = 0$.

2. Solvability of the problem $P_\varphi(h)$ and properties of the set of solutions. The results reported in this section were presented in [66].

Theorem 2.1 (minimal solution). *Assume (g) of (2.1) of chapter I. If the set of all solutions of $P_\varphi(h)$ is non-empty, then there exists a minimal solution u_o, that is a solution u_o such that the relation*

$$u_o(x) \leq u(x) \qquad \textit{for a.a. } x \textit{ in } \Omega$$

holds for any u which solve the problem.

Furthermore the set of all solution of $P_\varphi(h)$ is closed with respect to the $L^2(\Omega)$-topology.

Theorem. 2.2 (first solvability theorem). *Assume (g) of (2.1) of chapter I. Denote by λ_1 the first eigenvalue of $-\Delta$ in $W_o^{1,2}(\Omega)$.*

Then

a) if $P_\varphi(h)$ has a solution, $\varphi_1 \leq \varphi$, $h_1 \geq h$ $(h, h_1 \in L^2(\Omega))$, then $P_{\varphi_1}(h_1)$ has a solution;

b) if $h_o, z \in L^2(\Omega)$, $z(x) > 0$ for all x in Ω, then there exists τ_o in $\mathbf{R}$ such that $P_\varphi(h_o + \tau z)$ has a solution for all $\tau \geq \tau_o$;

c) if $\int_0^s g(x,\sigma)\, d\sigma \leq \lambda' \frac{s^2}{2} + G_o(x)$ $\forall s \geq 0$ for suitable $\lambda' < \lambda_1$ and G_o in $L^1(\Omega)$, then $P_\varphi(h)$ has a solution for all h in $L^2(\Omega)$.

The following theorem shows that the presence of the obstacle φ produces a qualitative *change* in the nature of the problem. To see this consider, for instance

$$g(x,s) = \lambda s + g_o$$

where $g_o \in L^2(\Omega)$, $\lambda \in \mathbf{R}$, λ is not an eigenvalue of $-\Delta$. It is well known that the problem

$$\begin{cases} \Delta u + \lambda u + g_o = h & \text{in } \Omega \\ u = 0 & \text{on } \partial\Omega \end{cases}$$

has a unique solution for any h in $L^2(\Omega)$. On the contrary , looking at the statements which follow, one sees that if the obstacle is present, then the problem with the same g exhibits a "jumping" behaviour: things go as in the unconstrained case with a nonlinearity g such that

$$\lim_{s\to-\infty} \frac{g(x,s)}{s} < \lambda_1 < \lim_{s\to+\infty} \frac{g(x,s)}{s}$$

(λ_1 being the first eigenvalue of $-\Delta$).

Furthermore notice that in the following theorems the assumption $\lim\limits_{s\to+\infty} \frac{g(x,s)}{s} < \lambda_2$ (λ_2 is the second eigenvalue of $-\Delta$) is not required, while it is in [5], [78].

Theorem 2.3 (second solvability theorem, jumping behaviour). *Assume (g) of (2.1) of chapter I to hold. Let λ_1 be the first eigenvalue of $-\Delta$ and suppose that*

$$g(x,s) \geq \bar{\lambda}s - g_o(x) \quad \forall s \geq 0,\ \forall x \text{ in } \Omega$$

where $g_o \in L^2(\Omega)$ and $\bar{\lambda} > \lambda_1$. Then

a) if $P_\varphi(h)$ has a solution, $\varphi_1 \leq \varphi$, $h_1(x) > h(x)$ $\forall x$ in Ω, then $P_{\varphi_1}(h_1)$ has at least two *solutions;*

b) if $h_o, z \in L^2(\Omega)$, $z(x) > 0$ $\forall x$ in Ω, then there exists τ_o in $\mathbf{R}$ such that $P_\varphi(h_o + \tau z)$ has at least two *solutions for $\tau > \tau_o$,* at least one *solution for $\tau = \tau_o$ and* no *solutions for $\tau < \tau_o$;*

c) the set of all pairs (φ, h) such that $P_\varphi(h)$ is solvable has the following closure property:

if (φ_m, h_m) are such that $P_{\varphi_m}(h_m)$ is solvable, if $h_m \to h$ in $\mathrm{L}^2(\Omega)$ and $\varphi_m \to \varphi$ almost everywhere in Ω and φ is such that $K_\varphi \cap D \neq \emptyset$, then $P_\varphi(h)$ is solvable;
moreover if in addition $\varphi_m \leq \varphi$ $\forall m$, then any sequence $(u_m)_m$ of solutions of $P_{\varphi_m}(h_m)$ is bounded in $\mathrm{W}^{1,2}_o(\Omega)$ and, provided that $u_m \to u$ in $\mathrm{L}^2(\Omega)$, then u is a solution of $P_\varphi(h)$.

The previous theorem allows to prove a result similar to the one given in [5], in the unconstrained case with a jumping nonlinearity (with respect to the first eigenvalue). These results describe a situation in which the operator associated with the problem is not surjective, but produces a "fold"in the space.

Theorem 2.4 (of folding type). *Suppose that the assumptions of Theorem 2.3 hold.*

Let $z \in \mathrm{L}^2(\Omega)$ with $z(x) > 0$ $\forall x$ in Ω and $\int_\Omega z^2 \,\mathrm{dx} = 1$. Set $H_o = \{z\}^\perp$ and consider

$$\Phi = \{\varphi : \Omega \to \mathbf{R} \mid K_\varphi \cap D \neq \emptyset\} \qquad \textit{(see Section 1).}$$

Then there exists a lower semicontinuous function $\gamma : H_o \times \Phi \to \mathbf{R}$, such that for all (φ, h) in $\Phi \times \mathrm{L}^2(\Omega)$ one has

a) if $\int_\Omega hz \,\mathrm{dx} > \gamma\left(h - \left(\int_\Omega hz \,\mathrm{dx}\right) z, \varphi\right)$, then $P_\varphi(h)$ has at least two solutions;

b) if $\int_\Omega hz \,\mathrm{dx} = \gamma\left(h - \left(\int_\Omega hz \,\mathrm{dx}\right) z, \varphi\right)$, then $P_\varphi(h)$ has at least one solutions;

c) if $\int_\Omega hz \,\mathrm{dx} < \gamma\left(h - \left(\int_\Omega hz \,\mathrm{dx}\right) z, \varphi\right)$, then $P_\varphi(h)$ has no solutions.

In the case in which the function $s \mapsto g(x,s)$ is convex the previous theorem can be further precised.

Theorem 2.5 (convex case). *Assume that the function $s \mapsto g(x,s)$ is convex for all x in Ω and $G(\cdot, u) \in \mathrm{L}^1(\Omega)$ for all u in*

$W^{1,2}_o(\Omega)$ *with* $u \geq 0$ *(this is true if for instance* g *is linear with respect to* s*).*

a) *If the assumptions (g) of (2.1) of chapter I hold, then the set of all pairs* (φ, h)*, with* $h \in L^2(\Omega)$ *such that* $P_\varphi(h)$ *is solvable, is convex (in particular the function* γ *of Theorem 2.4 is convex).*

b) *If one has the additional assumptions*

$$\lim_{s \to +\infty} \frac{g(x,s)}{s} \leq \bar{\lambda} < \lambda_2 \quad \forall x \text{ in } \Omega,$$
$$g(x,s) \geq \underline{\lambda} s - g_o(x) \quad \forall x \text{ in } \Omega, \forall s \geq 0$$

with g_o *in* $L^2(\Omega)$ *and* $\underline{\lambda} > \lambda_1$*, then in the case a) of Theorem 2.4 the solutions are exactly two, in the case b) either there exists a unique solution or the set of all solutions is a segment of the form* $\{u_1 + \theta(u_2 - u_1) \mid 0 \leq \theta \leq 1\}$*, where* u_1, u_2 *are two solutions such that* $u_1(x) \leq u_2(x)$ *for almost all* x *in* Ω*.*

For other detailed informations concerning this problem we refer the reader to [50], [66].

Open Problems

a) If we assume that

$$\lim_{s \to +\infty} \frac{g(x,s)}{s} > \lambda_i,$$

where λ_i is the i-th eigenvalue of $-\Delta$ on Ω, do we have "multiple folds", in the sense that, for some φ and h there exist more than two solutions?

b) If we impose the obstacle condition not on all Ω, but just on an open (non empty) subset Ω' of Ω, does the number of solutions depende on Ω'? We conjecture that this dependence really takes place. How could it be expressed?

3. A parameter depending multiplicity result. Many typical results in smooth nonlinear analysis, concerning differential equation, can be extended to the constrained case, that is to the corresponding variational inequalities. In this section we present briefly one of such extensions (see [37]), which concerns one of the classical theorems of Rabinowitz. Unlike the situation which arises in the problem treated in the previous sections, in this case the results are quite similar to those proved in the unconstrained problem.

Also in this problem, as in the previous ones, the proof of the multiplicity result requires the use of the evolution curves associated with a suitable functional on a suitable constraint. Due to the convexity of the constraint, the existence of such curves is essentially obtained by means of the theory of maximal monotone operators and their Lipschitz continuous perturbations.

We remark also that the quadratic form $\int_\Omega |\mathrm{D}u|^2\,\mathrm{dx}$, associated with the Laplace operator, can be replaced by any form of the type $\sum_{ij}\int_\Omega a_{ij}\mathrm{D}_i u \mathrm{D}_j u\,\mathrm{d}x$ associated with a strictly elliptic operator of the second order with bounded, symmetric coefficents.

Let Ω be a bounded open subset of and let $\mathbf{R}^n$, $\varphi : \Omega \to \mathbf{R}$, $p : \Omega \times \mathbf{R} \to \mathbf{R}$ be two functions.

Theorem 3.1. *Assume that*

$$(p.1)\quad \begin{cases} p \text{ is a Carathéodory function and there exist } a, b, \gamma \\ \text{in } \mathbf{R} \text{ such that, for a.a. } x \text{ in } \mathbf{R} \\ |p(x,s)| \le a + b|s| \quad \forall s \text{ in } \mathbf{R}, \\ \dfrac{p(x,s_2) - p(x,s_1)}{s_2 - s_1} \ge -\gamma \quad \forall s_1, s_2 \text{ in } \Omega \text{ with } s_1 \ne s_2; \end{cases}$$

$$(p.2)\quad \lim_{s\to 0} \frac{p(x,s)}{s} = 0 \quad \text{uniformly with respect to } x;$$

$$(p.3)\quad (\text{symmetry})\ p(x,-s) = -p(x,s)\ \forall (x,s) \text{ in } \Omega \times \mathbf{R};$$

$$(\varphi)\quad \begin{cases} \varphi \in \mathrm{L}^2(\Omega), \varphi \ge 0 \text{ a.e. in } \Omega \text{ and there exists an open} \\ \text{subset } \Omega' \text{ of } \Omega \text{ such that } \operatorname{ess\,inf}_{\Omega'} \varphi > 0. \end{cases}$$

Then for any integer m there exist λ_m in $\mathbf{R}$ such that, for all $\lambda > \lambda_m$, *the problem*

$$\begin{cases} u \in W_o^{1,2}(\Omega) \quad , \quad -\varphi \le u \le \varphi \ a.e. \ in \ \Omega, \\ \displaystyle\int_\Omega DuD(v-u)\,dx - \lambda \int_\Omega (u - p(x,u))(v-u)\,dx \ge 0 \\ \qquad\qquad \forall v \ in \ W_o^{1,2}(\Omega) \ with \ -\varphi \le v \le \varphi \ a.e. \ in \ \Omega \end{cases}$$

admits at least m pairs of solutions $(u,-u)$ such that $u \neq 0$.

More precisely, if Ω' is a set satisfying (φ), then λ_m is precisely the m-th eigenvalue of the problem (in Ω')

$$\begin{cases} \Delta u + \lambda u = 0 \quad \textit{in the sense of distributions}, \\ u \in W_o^{1,2}(\Omega'). \end{cases}$$

The proof is carried out by introducing a suitable functional f (of the type considered in Chapter I), such that the lower critical points of f are solutions of the problem, and then by using an abstract multiplicity theorem similar to that stated in Chapter IV.

Chapter III

Φ-convex functions and some parabolic variational inequalities on non convex constraints.

In this chapter we shall consider an evolution equation on a non convex constraint. For this we premise, in section 1, some results on ϕ- convex functions (see [31]).

1. Subdifferential and ϕ-convex functions. Let H be a Hilbert space with inner product $\langle \cdot, \cdot \rangle$ and norm $\| \cdot \|$. Let $W \subset H$ and consider $f : W \to \mathbf{R} \cup \{+\infty\}$. Set

$$\mathcal{D}(f) = \{u \in W \mid f(u) < +\infty\}.$$

Definition 1.1. *Let $u \in \mathcal{D}(f)$. We define the "subdifferential of f at u" as the set $\partial^- f(u)$ of all α in H such that*

$$\liminf_{v \to u} \frac{f(v) - f(u) - \langle \alpha, v - u \rangle}{\|v - u\|} \geq 0$$

(it turns out trivially that $\partial^- f(u)$ is a closed and convex subset of H).

If $\partial^- f(u) \neq \emptyset$, then we define the "subgradient of f at u" as the element $\mathrm{grad}^- \mathrm{f(u)}$ *in $\partial^- f(u)$ which has minimal norm.*

If V is a subset of H, then we set

$$\begin{aligned} &\partial_V^- f(u) = \partial^- (f + I_V)(u), \\ &\mathrm{grad}_V^- f(u) = \mathrm{grad}^- (f + I_V)(u), \qquad \text{if } \partial_V^- f(u) \neq \emptyset, \end{aligned}$$

where $I_V(u) = 0$, if $u \in V$ and $I_V(u) = +\infty$, if $u \notin V$.

Remark 1.2.

a) An element u of $\mathcal{D}(f)$ is a lower critical point for f, according to the definition 1.2 of chapter I, if and only if $0 \in \partial^- f(u)$. An analogous remark holds for lower critical points on a constraint V.

b) If $u \in \mathcal{D}(f)$ and $\alpha \in \partial^- f(u)$, then $|\nabla f|(u) \leq \|\alpha\|$.

The previous definitions agree with the classical ones, if, for instance, f and V are smooth or if they are convex.

A remarkable property of $\partial^- f$ is the following one (see [31]) which is related to a theorem of Edelstein's (see [34]).

Theorem 1.3. *Let f be lower semicontinuous and let, for instance, $W = H$. Then the set $\{u \in \mathcal{D}(f) \mid \partial^- f(u) \neq \emptyset\}$ is dense in $\mathcal{D}(f)$.*

We introduce now a class of functions which allows to treat some non convex constraint.

Definition 1.4 (ϕ-convex functions). *Let $\phi : \mathcal{D}(f)^2 \times \mathbf{R}^3 \to \mathbf{R}$ be a continuous function. We say that f is ϕ-convex if*

$$f(v) \geq f(u) + \langle \alpha, v-u \rangle - \phi(u, v, f(u), f(v), \|\alpha\|)\|v-u\|^2$$

$$\forall u, v \text{ in } \mathcal{D}(f) \text{ with } \partial^- f(u) \neq \emptyset, \forall \alpha \text{ in } \partial^- f(u);$$

notice that we are not requiring explicitely $\partial^- f(u) \neq \emptyset$ at any u.

We say that f is ϕ-convex of order r ($r > 0$), if ϕ takes the special form $\phi(u, v, f(u), f(v), \|\alpha\|) = \phi_o(u, v, f(u), f(v))(1+\|\alpha\|^r)$.

Examples 1.5

a) Let $h_o : H \to \mathbf{R} \cup \{+\infty\}$ be a convex function and $h_1 : H \to \mathbf{R}$ be of class $C^{1,1}_{loc}$. Then $f = h_o + h_1$ is *ϕ-convex of order 0.*

b) Consider $f : H \to \mathbf{R} \cup \{+\infty\}$ defined as follows

$$f(u) = \begin{cases} 0, & \text{if } \|u\| \geq 1 \\ +\infty, & \text{if } \|u\| < 1. \end{cases}$$

Then f is *ϕ-convex of order 1.*

A first meaningful case of ϕ-convex function is produced by the following theorem (see [17] for a special case). More general statements hold, which can be found in [55]. To state the theorem we need a definition.

Definition 1.6. *Let $A, B \subset H$; we say that A and B are "externally non tangent", if*

$$(-\partial^- I_A(u)) \cap (\partial^- I_B(u)) = \{0\} \qquad \forall u \text{ in } A \cap B.$$

Theorem 1.7 (of Lagrange multipliers type). *Assume W to be open and f to be lower semicontinuous and ϕ-convex with $\phi = \phi_o(u, v)(1 + \|\alpha\|)$. Let M be a smooth manifold (possibly with* *boundary) of class $C^{1,1}_{loc}$ with finite codimension. Finally suppose M and $\mathcal{D}(f)$ to be externally non tangent.*

Then $f + I_M$ is ϕ-convex of order 1 and

$$\partial^-(f+I_M)(u) = \partial^- f(u) + \partial^- I_M(u) \qquad \forall u \text{ in } \mathcal{D}(f) \cap M.$$

For ϕ-convex functions a powerful evolution theorem holds (see [31]). First of all we introduce the notion of evolution curves.

Definition 1.8. *Let I be an interval with $\overset{\circ}{I} \neq \emptyset$ and let $\mathcal{U} : I \to H$ be a curve. We say that $\mathcal{U}$ is a "strong evolution curve for f", if*

$$\begin{cases} \mathcal{U} \text{ is absolutely continuous}, \mathcal{U}(t) \in \mathcal{D}(f) & \forall t \text{ in } I; \\ -\mathcal{U}'(t) \in \partial^- f(\mathcal{U}(t)) & \text{for almost all } t \text{ in } I; \\ f \circ \mathcal{U} \text{ is non increasing.} \end{cases}$$

Theorem 1.9 (evolution). *Assume W to be open and f to be a ϕ-convex, lower semicontinuous function. Then*

a) for any u_o in $\mathcal{D}(f)$ with $\partial^- f(u_o) \neq \emptyset$ there exists $T > 0$ and a unique strong evolution curve for f $\mathcal{U} : [0,T] \to W$ such that $\mathcal{U}(0) = u_o$; moreover $\mathcal{U}$ and $f \circ \mathcal{U}$ are Lipschitz-continuous and

$$(1.10) \qquad \begin{cases} \partial^- f(\mathcal{U}(t)) \neq \emptyset, \mathcal{U}'_+(t) = \mathrm{grad}^- \mathrm{f}(\mathcal{U}(\mathrm{t})) \\ (f \circ \mathcal{U})'_+(t) = -\|\mathrm{grad}^- \mathrm{f}(\mathcal{U}(\mathrm{t}))\|^2 \end{cases}$$

for every t in $[0,T[$;

b) if furthermore f is ϕ-convex of order less or equal than two, then for any u_o in $\mathcal{D}(f)$ there exists $T > 0$ and a unique strong evolution curve $\mathcal{U} : [0,T] \to W$ such that $\mathcal{U}(0) = u_o$; moreover $f \circ \mathcal{U}$ is absolutely continuous, (1.10) hold for every t in $]0,T[$ and the following continuous dependence on data holds:

if $(u_m)_m$ is a sequence in $\mathcal{D}(f)$ such that $u_m \to u_o \in \mathcal{D}(f)$ and $f(u_m) \to f(u_o)$, then the curves $\mathcal{U}_m$ starting from u_m and the associated functions $f \circ \mathcal{U}_m$ converge uniformly in any compact subset of $[0,T[$ to $\mathcal{U}$ and to $f \circ \mathcal{U}$ respectively.

For other properties of $\mathcal{U}$ see [31].

In the previous statement the following minimum theorem has great importance.

Theorem 1.11 (minimum). *Let W be open, f be a lower semicontinuous, bounded below, ϕ-convex function. Let $u \in \mathcal{D}(f)$ and $\partial^- f(u) \neq \emptyset$.*

Then for all $\rho > 0$ there exists $\lambda_o > 0$ such that the function

$$v \mapsto f(v) + \frac{1}{\lambda}\|v-u\|^2$$

has a unique minimum point u_λ in $B(u,\rho)$, for all λ in $]0,\lambda_o]$.

We remark that in [31] and in [17] some situations are also considered in which $\partial^- f(u) = \emptyset$.

2. Evolution equation associated with the eigenvalue problem for the Laplace operator with respect to an obstacle. The problem and the results treated in this section were given in [17]. It is clear that the Laplace operator can be replaced by any strictly elliptic operator of the second order with bounded and symmetric coefficents, obtaining the same results with the same methods.

Let Ω be a bounded open subset of $\mathbf{R}^n$, $g : \Omega \times \mathbf{R} \to \mathbf{R}$, $\varphi_1, \varphi_2 : \Omega \to \mathbf{R}$ measurable with $\varphi_1 \leq \varphi_2$ a.e. in Ω and $\rho > 0$. In the space $L^2(\Omega)$ we consider the usual norm $\|v\| = \int_\Omega v^2 \, dx$.

Set $G(x,s) = \int_0^s g(x,\sigma)\,\mathrm{d}\sigma$ and

$$\begin{aligned} K &= \{v \in \mathrm{L}^2(\Omega) \mid \varphi_1(x) \le v(x) \le \varphi_2(x) \text{ a.e. in } \Omega\}, \\ S_\rho &= \{v \in \mathrm{L}^2(\Omega) \mid \|v\| = \rho\}, \\ K_g &= \{v \in \mathrm{W}^{1,2}_o(\Omega) \cap K \mid G(\cdot, v) \in \mathrm{L}^1(\Omega)\}. \end{aligned}$$

We shall assume one or more of the following conditions.

(2.1)

- $(g.1)$ g is a Carathéodory function, there exist a in $\mathrm{L}^1(\Omega), b$ in $\mathbf{R}$ such that
$$G(x,s) \ge -a(x) - bs^2 \quad \forall s \text{ in } \mathbf{R}, \text{ a.a. } x \text{ in } \Omega;$$
- $(g.2)$ $\exists c$ in $\mathbf{R}$ such that if $s_1, s_2 \in \mathbf{R}$, $s_1 \ne s_2$
$$\frac{g(x,s_2) - g(x,s_1)}{s_2 - s_1} \ge -c \qquad \text{a.a. } x \text{ in } \Omega;$$
- $(g.3)$ g is integrable in $\Omega \times [a,b]$ for all a, b in $\mathbf{R}$;
- $(N.T.)$ K_g and S_ρ are externally non tangent, according to definition 1.6.

Theorem 2.2 (non tangency). *The assumption (N.T.) of (2.1) is verified for every* $\rho > \|\varphi_1 \vee 0 + \varphi_2 \wedge 0\|$ *(=*$\min\{\|v\| \mid v \in K\}$*), if (g.1),(g.2), (g.3) hold and if*

$\varphi_1, \varphi_2 \in \mathrm{W}^{1,2}(\Omega) \cap C(\Omega)$ *and there exist no open subset* Ω' *of* Ω *such that at least one of the following conditions hold:*

$\varphi_2 > 0$ *in* Ω' *and* $\varphi_2 \in \mathrm{W}^{1,2}_o(\Omega')$,

$\varphi_1 < 0$ *in* Ω' *and* $\varphi_1 \in \mathrm{W}^{1,2}_o(\Omega')$.

A complete characterization of (N.T.) is presented in [17].

Theorem 2.3 (evolution). *Let (g.1), (g.2), (N.T.) hold. Then for all u_o in $K_g \cap S_\rho$ there exist $\mathcal{U} : [0,+\infty[\to L^2(\Omega)$ absolutely continuous with $\mathcal{U}(0) = u_o$ and $\Lambda : [0,+\infty[\to \mathbf{R}$ such that:*

$$
(E.P.) \quad \begin{cases} \mathcal{U}(t) \in K_g \cap S_\rho \quad \forall t \geq 0 \quad \textit{and for almost all } t > 0 \\ \\ g(\cdot,\mathcal{U}(t))(v - \mathcal{U}(t)) \in L^1(\Omega) \qquad \forall v \textit{ in } K_g \\ \int_\Omega \mathcal{U}'(t)(v - \mathcal{U}(t))\,dx + \int_\Omega D\mathcal{U}(t)D(v - \mathcal{U}(t))\,dx + \\ \quad + \int_\Omega g(x,\mathcal{U}(t))(v - \mathcal{U}(t))\,dx \\ \quad + \Lambda(t)\int_\Omega \mathcal{U}(t)(v - \mathcal{U}(t))\,dx \geq 0 \quad \forall v \textit{ in } K_g \end{cases}
$$

(where $Du = (\frac{\partial u}{\partial x_1}, \ldots, \frac{\partial u}{\partial x_n})$). If moreover φ_1, φ_2 satisfy appropriate regularity conditions and (g.3) holds, then (E.P.) can be further specificated (see [17], theorem 4.5 part b)).

It is easy to see that the theorem still holds (with the same proof), if one replaces the form $\int_\Omega D\mathcal{U}(t)D(v - \mathcal{U}(t))\,dx$ by the form $\sum_{ij} a_{ij}D_i\mathcal{U}(t)D_j(v - \mathcal{U}(t))$, where $a_{ij} \in L^\infty(\Omega)$, $a_{ij} = a_{ji}$ and the relation $\sum_{ij} a_{ij}\xi_i\xi_j \geq a|\xi|^2$ holds with $a > 0$.

To get theorem 2.3 from theorem 1.9 the functionals f_o, f : $L^2(\Omega) \to \mathbf{R} \cup \{+\infty\}$ are introduced, defined by

$$
(2.4) \qquad f_o(u) = \begin{cases} \int_\Omega \left(\frac{1}{2}|Du|^2 + G(x,u)\right) dx, & \text{if } u \in W_o^{1,2} \cap K, \\ +\infty, & \text{elsewhere;} \end{cases}
$$

$$
f = f_o + I_{S_\rho}
$$

The proof is based on the following statement; here in the space $L^2(\Omega)$ we consider the usual inner product $\langle u, v\rangle = \int_\Omega uv\,dx$.

Theorem 2.5.

a) *Under the assumption (g.1)* $\mathcal{D}(f_o) = K_g, \mathcal{D}(f) = K_g \cap S_\rho$, f_o *and* f *are lower semicontinuous .*

b) *Under (g.1), (g.2)* f_o *is* ϕ*-convex with* $\phi = \phi_o(u,v)$.

c) *Under (g.1), (g.2) and (N.T.), by theorem 1.7, the functional* f *is* ϕ *- convex of order 1 and, if* $u \in K_g \cap S_\rho, \alpha \in \mathrm{L}^2(\Omega)$, *then* $\alpha \in \partial^- f(u)$ *if and only if*

$$g(\cdot,u)(v-u) \in \mathrm{L}^1(\Omega) \qquad \forall v \; in K_g;$$

$\exists \lambda$ *in* $\mathbf{R}$ *such that*

$$\int_\Omega \mathrm{D}u\mathrm{D}(v-u)\,\mathrm{dx} + \int_\Omega g(x,u)(v-u)\,\mathrm{dx} +$$

$$+\lambda \int_\Omega u(v-u)\,\mathrm{dx} \geq \int_\Omega \alpha(v-u)\,\mathrm{dx} \quad \forall v \; in \; K_g.$$

Now it is clear that Theorem 2.3 is a consequence of theorem 1.9.

Finally we wish to present two open problems

Problem 2.6. Can the non tangency condition (N.T.) in theorem 2.3 be weakened?

Problem 2.7. If Ω is a bounded open subset of $\mathbf{R}^n$ we introduce the functional $f : \mathrm{L}^2(\Omega; \mathbf{R}^\mathrm{N}) \to \mathbf{R} \cup \{+\infty\}$ defined by

$$f(u) = \begin{cases} \frac{1}{2}\int_\Omega |\mathrm{D}u|^2\,\mathrm{dx} & \text{if } u \in \mathrm{W}^{1,2}(\Omega,\mathbf{R}^\mathrm{N}),\ u \geq 1,\ u = \psi \text{ on } \partial\Omega \\ +\infty & \text{elsewhere} \end{cases}$$

ψ being an assigned boundary data, $\psi : \partial\Omega \to \mathbf{R}^\mathrm{N}$ with $\psi \geq 1$.

Is it possible to solve, in some suitable sense, the evolution equation associated with f?

Notice that in the case $n = 1$ (namely Ω is an interval) the problem 2.7 is that of geodesic with obstacle and their evolution, which were studied in [53].

Also notice that for $n = 3$ 2.7 is strictly related to the problem of "liquid cristals"(see [12]).

Chapter IV

Eigenvalue problems for some elliptic variational inequalities.

The problem described in section 1 can be seen as the search of the eigenvalues of a symmetric elliptic operator with respect to an obstacle. We wish to enphasize the fact that a very important role is played by the non tangency conditions, introduced in chapter III (see (N.T.) of (2.1)). The corresponding bifurcation problem is treated in section 2; we remark that in this problem non tangency is automatically fulfilled (due to the special nature of the "limit problem").

With the same methods the problem of Von Karman's plates equation with obstacle has been studied (see [19], and [26] for the bifurcation problem). In this case, up to now, the non tangency conditions are not completely characterized and it seems important to investigate for a better understanding of their meaning.

We recall also that, in the same framework, the geodesics with obstacle have been studied in [53] (existence and multiplicity).

1. Multiplicity of eigenvalues for the Laplace operator with respect to an obstacle. The problem and the results presented in this section can be found in [18]. It is easy to see that the Laplace operator can be replaced by any strictly elliptic, symmetric operator of the second order with bounded coefficents.

Let $\Omega, g, \varphi_1, \varphi_2, \rho$ be as in section 2 of chapter III. We shall also consider the sets K_g and S_ρ introduced there.

Problem 1.1. We shall be concerned with the following stationary problem: find (u, λ) such that

$$
(S.P.) \qquad \begin{cases} u \in K_g \cap S_\rho \quad , \quad \lambda \in \mathbf{R}, \\ g(\cdot, u)(v-u) \in \mathrm{L}^1(\Omega) \qquad \forall v \text{ in } K_g, \\ \displaystyle\int_\Omega \mathrm{D}u\mathrm{D}(v-u)\,\mathrm{dx} + \int_\Omega g(x,u)(v-u)\,\mathrm{dx} + \\ \qquad + \lambda \displaystyle\int_\Omega u(v-u)\,\mathrm{dx} \geq 0 \qquad \forall v \text{ in } K_g \end{cases}
$$

Theorem 1.2 (existence and multiplicity).

a) *Let $\rho > 0$ be such that $K_g \cap S_\rho \neq \emptyset$. Suppose assumptions (g.1),(g.2),(N.T.) of (2.1) in Chapter III hold.*

Then there exist u in $\mathrm{L}^2(\Omega)$ and λ in $\mathbf{R}$ which solve the stationary problem (S.P.).

b) *Suppose that the following symmmetry assumptions hold:*

$$
\begin{aligned} g(x,-s) &= -g(x,s) \qquad && \forall s \text{ in } \mathbf{R},\ \forall x \text{ in } \Omega, \\ \varphi_2(x) &= -\varphi_1(x) = \varphi(x) \qquad && \forall x \text{ in } \Omega. \end{aligned}
$$

Let $\varphi \in \mathrm{W}^{1,2}(\Omega)$ and $\rho \in [0, \int_\Omega \varphi^2\,\mathrm{dx}]$ and suppose that (2.1) of chapter III hold.

Then there are infinitely many (u, λ) such that (u, λ) and $(-u, \lambda)$ solve (S.P.). Moreover the set of λ's is bounded above and it infimum is equal to $-\infty$.

To get a) of Theorem 1.2 it suffices to consider the minimum of the funcional introduced in 2.4 and theorem 2.5, part c) in chapter III.

For what concerns b), in [18] the following abstract theorem is used, which we feel worth to point out.

Theorem 1.3 (of Lusternik-Schnirelman type). *Let H be a Hilbert space with norm $\|\cdot\|$ and let $h : H \to \mathbf{R} \cup \{+\infty\}$ be a function. If $u, v \in \mathcal{D}(h) = \{z \in H \mid h(z) < +\infty\}$, set*

$$
d^*(u,v) = \|v-u\| + |h(v) - h(u)|
$$

and denote by $\mathcal{D}(h)^*$ the set $\mathcal{D}(h)$ with the metric d^*.

We make the following assumptions.

a) *h is even: $h(-u) = h(u)$ $\forall u$ in H;*

b) *h is bounded below;*

c) *(regularity of $\mathcal{D}(h)$) for all u in $\mathcal{D}(h)$, lower critical point for h, there exist a neighbourhood V of u in $\mathcal{D}(h)^*$, such that V is contractible in $\mathcal{D}(h)^*$;*

d) *(Palais-Smale condition) if $c \in h(H)$ and if $(u_m)_m$ is a sequence such that*

$$h(u_m) \leq c \quad \forall m \quad , \quad \partial^- h(u_m) \neq \emptyset \quad \forall m,$$
$$\lim_{m \to \infty} \|\mathrm{grad}^- h(u_m)\| = 0,$$

then there exists a subsequence $(u_{m_j})_j$ which converges in $\mathcal{D}(h)^*$ to a point u in $\mathcal{D}(h)$ with the property $0 \in \partial^- h(u)$;

e) *for every u in $\mathcal{D}(h)$ there exists a unique strong evolution curve for h $t \mapsto \Psi(u,t)$, defined on $[0,+\infty[$, such that $\Psi : \mathcal{D}(h)^* \times [0,+\infty[\to \mathcal{D}(h)^*$ is continuous.*

Then h has at least k lower critical points, k being the Lusternik-Schnirelman category of the metric space $\tilde{\mathcal{D}}(h)^*$, obtained from $\mathcal{D}(h)^*$ by identifying the antipodal points.

If finally k=$+\infty$, then h has no maximum and

$$\sup\{h(v) \mid v \in \mathcal{D}(h), 0 \in \partial^- h(v)\} = \sup\{h(v) \mid v \in \mathcal{D}(h)\}.$$

Also for problem (S.P.) a question arises: is the non tangency assumption (N.T.) really necessary, or how can it be weakened?

Another very interesting problem is the study of the lower critical points of the functional introduced in problem 2.7 of chapter III, by means of the same ideas used in this section.

2. Bifurcation for elliptic variational inequalities. The problem and the results reported in this section were presented in [25].

Also in this section we shall consider an eigenvalue problem for an elliptic variational inequality of the form:

$$(2.1) \qquad \begin{cases} (\lambda, u) \in \mathbf{R} \times K \\ \displaystyle\int_\Omega (\mathrm{D}u\, \mathrm{D}(v-u) + p(x,u)(v-u))\, \mathrm{dx} \geq \\ \displaystyle\geq \lambda \int_\Omega u(v-u)\, \mathrm{dx} \qquad \forall v \in K. \end{cases}$$

More precisely, let Ω be a bounded open subset of $\mathbf{R}^n$ and let us assume that

(A1) $p : \Omega \times \mathbf{R} \to \mathbf{R}$ is a Carathéodory function such that for almost every x in Ω $p(x,0) = 0, \{s \mapsto p(x,s)\}$ is of class C^1 and there exists $c \in \mathbf{R}$ such that $|\mathrm{D}_s p(x,s)| \leq c$ in $\Omega \times \mathbf{R}$;

(A.2) K is a convex subset of $\mathrm{W}^{1,2}_\mathrm{o}(\Omega)$ of the form

$$\{u \in \mathrm{W}^{1,2}_\mathrm{o}(\Omega) \mid \varphi_1(x) \leq u(x) \leq \varphi_2(x) \text{ a.e. in } \Omega\},$$

where $\varphi_1 : \Omega \to [-\infty, 0]$ is an upper semicontinuous function and $\varphi_2 : [0, +\infty]$ is a lower semicontinuous one.

For a more general situation, including thin obstacles, we refer the reader to [25].

Our aim is to study the pairs (λ, u) satisfying (2.1). Because of (A.1) and (A.2), for every λ in $\mathbf{R}$, the pair $(\lambda, 0)$ verifies (2.1).

Definition 2.2. *A real number λ is said to be "of bifurcation" for (2.1), if there exists a sequence $((\lambda_h, u_h))_h$ of solutions of (2.1) with $u_h \neq 0$ and*

$$\lim_h \lambda_h = \lambda, \lim_h u_h = 0 \qquad \text{in } \mathrm{L}^2(\Omega).$$

As in the well known case of equations (see [9], [41], [48], [68]), we want to compare the bifurcation values with the eigenvalues of some "linearized"problem. To this aim, let us denote by K_o the closure in $W_o^{1,2}(\Omega)$ of the set

$$\{u \in C_o^\infty \mid u(x) \geq 0 \text{ in } \varphi_1^{-1}(0), u(x) \leq 0 \text{ in } \varphi_2^{-1}(0)\}.$$

It is readily seen that K_o is a closed convex cone. A reasonable "linarization"of (2.1) is given by the problem

$$\left\{\begin{aligned} & (\lambda, u) \in \mathbf{R} \times K_o \\ & \int_\Omega (\mathrm{D}u\ \mathrm{D}(v-u) + \mathrm{D}_s p(x,0) u(v-u))\,\mathrm{dx} \geq \\ & \geq \lambda \int_\Omega u(v-u)\,\mathrm{dx} \qquad \forall v \in K_o. \end{aligned}\right. \tag{2.3}$$

Definition 2.4. *A real number λ is said to be "an eigenvalue"of (2.3), if there exists a solution (λ, u) of (2.3) with $u \neq 0$.*

Theorem 2.5. *Let λ be a real number. If λ is of bifurcation for (2.1), then λ is an eigenvalue of (2.3).*

In the case of equations ($\varphi_1 = -\infty, \varphi_2 = +\infty$) also the converse is true (see [41]). For variational inequalities only partial results are known. We shall treat two particular situations in which Theorem 2.5 can be reversed.

The first one concerns the bifurcation from the first eigenvalue. If φ_1 takes only the values $-\infty, 0$ and φ_2 only the values $0, +\infty$, so that K is a cone, the problem was already treated in [69].

For the fourth order case, describing the buckling of an elastic plate, there are results in [33], again under the assumption that K is a cone and in [57], [59], under the hypothesis that the obstacles φ_1 and φ_2 are supersolution and subsolution of a suitable operator. The general case, with no further assumptions on φ_1 and φ_2, is treated in [26].

Theorem 2.6. *Let us assume that φ_1 and φ_2 are not both identically zero. Then*

$$\lambda = \inf\left\{ \int_\Omega \left(|\mathrm{D}u|^2 + \mathrm{D}_s p(x,0)u^2\right)\,\mathrm{dx} \mid u \in K_o,\ \int_\Omega u^2\,\mathrm{dx} = 1 \right\}$$

is achieved, λ is an eigenvalue of (2.3) (the first eigenvalue) and λ is of bifurcation for (2.1).

More precisely, there exist $\rho_o > 0$, $\{\lambda_\rho \mid 0 < \rho < \rho_o\} \subset \mathbf{R}$, $\{u_\rho \mid 0 < \rho < \rho_o\} \subset K$ such that (λ_ρ, u_ρ) satisfies (2.1), $\int_\Omega u^2\,\mathrm{dx} = \rho^2$ and

$$\lim_{\rho \to 0+} u_\rho = 0 \quad \textit{in } L^\infty(\Omega) \textit{ and in } \mathrm{W}_0^{1,2}(\Omega),$$

$$\lim_{\rho\to 0+} \lambda_\rho = \lim_{\rho \to 0+} \frac{1}{\rho^2} \int_\Omega \left(|\mathrm{D}u_\rho|^2 + \mathrm{D}_s p(x,0)u_\rho^2\right)\,\mathrm{dx} = \lambda.$$

In the second situation, concerning the case in which K_o is a linear space, we can give a bifurcation result also for higher eigenvalues.

Theorem 2.7. *Let us assume that*

$$\{x \in \Omega \mid \varphi_1(x) = 0\} = \{x \in \Omega \mid \varphi_2(x) = 0\}.$$

Then every eigenvalue λ of (2.3) is of bifurcation for (2.1). More precisely, there exist ρ_o, λ_ρ, and u_ρ as in the conlusion of Theorem 2.6.

The proof of the previous results is based on a variational technique. For this reason it is convenient to introduce the functionals:

$$f(u) = \begin{cases} \int_\Omega \left(\frac{1}{2}|\mathrm{D}u|^2 + P(x,u)\right)\,\mathrm{dx} & \text{if } u \in K \\ +\infty & \text{if } u \in \mathrm{L}^2(\Omega) \setminus K, \end{cases}$$

$$f_o(u) = \begin{cases} \frac{1}{2}\int_\Omega \left(|\mathrm{D}u|^2 + \mathrm{D}_s p(x,0)u^2\right)\,\mathrm{dx} & \text{if } u \in K_o \\ +\infty & \text{if } u \in \mathrm{L}^2(\Omega) \setminus K_o, \end{cases}$$

$$f_\rho(u) = \frac{1}{\rho^2} f(\rho u) \qquad \text{for } \rho > 0, u \in \mathrm{L}^2(\Omega),$$

where $P(x,s) = \int_0^s p(x,\sigma)\,d\sigma$.

Let us summarize the properties of the above functionals.

Proposition 2.8. *Let* $S_\rho = \{u \in L^2(\Omega) \mid \int_\Omega u^2\,dx = \rho^2\}$. *Then the following facts hold:*

a) $u \in K_o \cap S_1$ *is a lower critical point for* $f_o + I_{S_1}$ *if and only if for some* $\lambda \in \mathbf{R}$ *the pair* (λ, u) *satisfies (2.3);*

b) for every $b \in \mathbf{R}$ *there exists* $\rho_o > 0$ *such that for every* $u \in K_o \cap S_\rho$ *with* $0 < \rho < \rho_o$ *and* $f(u) \le b\rho^2$ *we have that* u *is a lower critical point for* $f + I_{S_\rho}$ *if and only if for some* $\lambda \in \mathbf{R}$ *the pair* (λ, u) *satisfies (2.1);*

c) for every sequence $(\rho_h)_h$ *in* $]0, +\infty[$ *converging to zero we have*

$$f_o = \Gamma^-(L^2(\Omega)) - \lim_h f_{\rho_h}$$

(we refer the reader to [7], [21] for the notion of variational convergence).

Finally, let us point out some open questions on this topic.

A first problem could be to remove or to weaken the assumption in theorem 2.7).

A second one could concern the study on the possible eigenvalues of (2.3). For studies in this direction see [42], [43], [44], [58], [67], [80].

References

[1] H.Amann, *Fixed point equations and nonlinear eigenvalue in ordered Banach spaces,* Siam Rew. **18** (1986), 620-709.

[2] H.Amann, *Saddle points and multiple solutions of differential equations,* Math.Z. **169** (1979).

[3] H.Amann, E.Zehnder, *Non trivial solutions for a class of non resonance problems and applications to nonlinear differential equations,* Ann. Sc. Norm. Sup. Pisa **7** (1980), 539-603.

[4] A.Ambrosetti, *Elliptic equations with jumping nonlinearities,* J. Math. and Phys. Sc. **18** (1984).

[5] A.Ambrosetti, G.Prodi, *On the inversion of some differentiable mappings with singularities between Banacyh spaces,* Ann. Mat. Pura e Appl. **93** (1972), 231-246.

[6] A.Ambrosetti, P.H.Rabinowitz, *Dual variational methods in critical point theory and applications,* J. Funct. Anal. **14** (1973), 349-381.

[7] H.Attouch, *Variational convergence for functions and operators,* Applicable Mathematics Series, Pitman (Advanced Pubblishing Program), Boston, Mass. - London, 1984.

[8] C.Baiocchi, A.Capelo, *Disequazioni variazionali e quasivariazionali,* Pitagora, 1978.

[9] R.Böhme, *Die Lösung der Verzweigunsgleichungen für nichtlineare Eigenwertprobleme,* Math.Z. **127** (1972), 105-126.

[10] H.Brezis, *Opérateurs maximaux monotone et semigroupes de contraction dans les espaces de Hilbert,* North Holland Mathematics Studies **5**, Notas de Matematica 50, Amsterdam, London 1973.

[11] H.Brezis, *Monotonicity methods in Hilbert spaces and some applications to nonlinear partial differential equations,* Contribution to nonlinear analysis, Zarantonello 1971, 101-156.

[12] H.Brezis, J.M.Coron, E.H.Lieb, *Harmonic maps with defects,* Comm. Math. Phys. **107** (1986), 649-705.

[13] H.Brezis, G.Stampacchia, *Sur la regularité de la solution d'inequations elliptiques,* Bull. Soc. Math. France **96** (1968), 153-180.

[14] F.E.Brodwer, *Non linear elliptic boundary value problems,* Bull. Am. Math. Soc. **69** (1963), 862-874.

[15] F.E.Brodwer, *Non linear monotone operators and convex subsets in Banach spaces,* Bull. Am. Math. Soc. **71** (1965), 780-785.

[16] A.Canino, *On p-convex sets and geodesics,* J. Diff. Eq., in press.

[17] G.Čobanov, A.Marino, D.Scolozzi, *Evolution equations for the eigenvalue problem for the Laplace operator with respect to an obstacle,* submitted to Ann. Scuola Normale Sup. Pisa.

[18] G.Čobanov, A.Marino, D.Scolozzi, *Multiplicity of eigenvalues of the Laplace operator with respect to an obstacle and non tancency conditions,* Nonlin. Anal. Th. Meth. Appl., to appear.

[19] G.Čobanov-D.Scolozzi, to appear.

[20] E.De Giorgi, M.Degiovanni, A.Marino, M.Tosques, *Evolution equations for a class of nonlinear operators,* Atti Accadem. Naz. Lincei Rend. Cl. Sci. Fis. Mat. Natur. **75** (1983), 1-8.

[21] E.De Giorgi, T.Franzoni, *Su un tipo di convergenza variazionale,* Rend. Sem. Mat. Brescia **3** (1979), 63-101.

[22] E.De Giorgi, A.Marino, M.Tosques, *Problemi di evoluzione in spazi metrici e curve di massima pendenza,* Atti Accadem. Naz. Lincei Rend. Cl. Sci. Fis. Mat. Natur. **68** (1980), 180-187.

[23] E.De Giorgi, A.Marino, M.Tosques, *Funzioni (p,q)-convesse* , Atti Accadem. Naz. Lincei Rend. Cl. Sci. Fis. Mat. Natur. **73** (1982), 6-14.

[24] M.Degiovanni, *Parabolic equations with time-dependent boundary conditions,* Ann. Mat. Pura Appl. **141** (1985), 223-263.

[25] M.Degiovanni, *Bifurcation problems for nonlinear elliptic variational inequalities,* Preprint Dip. Mat. Pisa **228** (1988).

[26] M.Degiovanni, *On the buckling of a thin elastic plate subjected to unilateral constraints,* Nonlinear variational problems (Isola d'Elba, 1986), Proceedings, in press.

[27] M.Degiovanni, *Bifurcation for odd nonlinear elliptic variational inequalities,* (preprint), Dip. Mat. Pisa **235**, Pisa, (1988).

[28] M.Degiovanni, A.Marino, *Nonsmooth variational bifurcation,* Atti Accad. Naz. Lincei Rend. Cl. Sci. Fis. Mat. Natur. **81** (1987), 259-269.

[29] M.Degiovanni, A.Marino, M.Tosques, *General properties of (p,q)-convex functions and (p,q)-monotone operators,* Ricerche Mat. **32** (1983), 285-319.

[30] M.Degiovanni, A.Marino, M.Tosques, *Evolution equations associated with (p,q)-convex functions and (p,q)-monotone operators,* Ricerche Mat. **33** (1984), 81-112.

[31] M.Degiovanni, A.Marino, M.Tosques, *Evolution equations with lack of convexity,* Nonlinear Anal. The. Meth. Appl. **12** (1985),1401-1443.

[32] M.Degiovanni, M.Tosques, *Evolution equations for (φ, f)-monotone operators,* Boll. Un. Mat. It. **B 5** (1986), 537-568.

[33] C.Do, *Bifurcation theory for elastic plates subjected to unilateral conditions,* J.Math. Anal. Appl. **60** (1977), 435-448.

[34] M.Edelstein, *On nearest points of sets in uniformly convex Banach spaces,* J. London Math. Soc. **43** (1968), 375-377.

[35] G.Fichera, *Problemi elastostatici con vincoli unilaterali : il problema di Signorini con ambigue condizioni al contorno,* Mem. Accad. Naz. Lincei **7** (1964), 91-140.

[36] M.Frigon, A.Marino, C.Saccon, to appear.

[37] F.Giannoni, to appear.

[38] P.Hartman, G.Stampacchia, *On some nonlinear elliptic differential-functional equations,* Acta Math. **115** (1966), 271-310.

[39] H.Hofer, *Variational and tological methods in partially ordered Hilbert spaces,* Math. Ann. **261** (1982), 493-514.

[40] D.Kinderleherer, G.Stampacchia, *An introduction to variational inequalities and their applications,* Pure and applied Mathematics, **88**, Academic Press, New York, London, Toronto, 1980.

[41] M.A.Krasnoselskii, *Topological Methods in the theory of nonlinear integral equations,* Gosudarstv. Izdat. Tehn.-Teor. Lit. , Moscow, 1956. The Macmillan Co. , New York, 1964.

[42] M.Kŭcera, *A new method for obtaining eigenvalues of variational inequalities: operators with multiple eigenvalues,* Czechoslovak Math. J. **32** (1982), 197-207.

[43] M.Kŭcera, *Bifurcation point of variational inequalities,* Czechoslovak Math. J. **32** (1982), 208-226.

[44] M.Kŭcera, *A global bifurcation theorem for obtaining eigenvalues and bifurcation points,* Czechoslovak Math. J. **38** (1988), 120-137.

[45] H.Lewy, G.Stampacchia, *On the regularity of the solutions of a variational inequality,* Comm. Pure Appl. Math. **22** (1969),153-188.

[46] J.L.Lions, *Quelques méthodes de résolutions des problèmes aux limites non linéaires,* Dunod Gauthier-Villars, 1969.

[47] J.L.Lions, G.Stampacchia, *Variational inequalities,* Comm. Pure and Appl. Math. **20** (1967), 493-519.

[48] A.Marino, *La biforcazione nel caso variazionale,* Confer. Sem. Mat. Univ. Bari **132** (1973).

[49] A.Marino, *Evolution equations and multiplicity of critical points with respect to an obstacle,* Contributions to modern calculus of variations (Bologna 1985), Res. Notes in Math. **148**, Pitman, London - New York, 1987, 123-144.

[50] A.Marino, D.Passaseo, to appear.

[51] A.Marino, C.Saccon, M.Tosques, *Curves of maximal slope and parabolic variational inequalities on non convex constraints,* Ann. Sc. Norm. Sup. Pisa, to appear.

[52] A.Marino, D.Scolozzi, *Punti inferiormente stazionari ed equazione di evoluzione con vincoli unilaterali non convessi,* Rend. Sem. Mat. e Fis. Milano **52** 1982, 393-414.

[53] A.Marino, D.Scolozzi, *Geodetiche con ostacolo,* Boll. Un. Mat. It. **B 2** (1983), 1-31.

[54] A.Marino, D.Scolozzi, *Autovalori dell'operatore di Laplace ed equazioni di evoluzione in presenza di ostacolo,* (Bari 1984) Problemi differenziali e teoria dei punti critici , Pitagora, Bologna, 1984, 137-155.

[55] A.Marino, D.Scolozzi, to appear.

[56] A.Marino, M.Tosques, *Curves of maximal slope for a certain class of non regular functions,* Boll. Un. Mat. It. **B 1** (1982), 143-170.

[57] E.Miersemann, *Eigenwertaufgaben für Variationsungleichungen,* Math. Nachr. **100** (1981), 221-228.

[58] E.Miersemann, *On higher eigenvalues of variational inequalities,* Comment. Math. New Carolin. **24** (1983), 657-665.

[59] E.Miersemann, *Eigenvalue problems in convex sets,* Mathematical Control Theory,401-408, Banach Center Pubbl. **14**, PWN, Warsaw, 1985.

[60] E.Mitidieri, M.Tosques, *Volterra integral equations associated with a class of nonlinear operators in Hilbert spaces,* Ann. Fac. Sci. Tolouse Math. , (5) **8** 2, (1986-87).

[61] E.Mitidieri, Tosques, *Nonlinear integrodifferential equations in Hilbert spaces: the variational case,* Proceedings of the congress "Volterra integral equations in Banach spaces and applications "(Trento, feb. 1987), to appear.

[62] L.Niremberg, *Variational and topological methods in nonlinear problems,* Bull. A.M.S. **4** (1981), 267-302.

[63] R.Palais, S.Smale, *A generalized Morse theory,* Bull. A.M.S. **70** (1964).

[64] D.Passaseo, *Molteplicitá di soluzioni per disequazioni variazionali,* Tesi di dottorato, Pisa 1988.

[65] D.Passaseo, *Molteplicitá di soluzioni per certe disequazioni variazionali di tipo ellittico,* Preprint Dip. Mat. Pisa **236**, Pisa, 1988.

[66] D.Passaseo, *Molteplicitá di soluzioni per disequazioni variazionali non lineari di tipo ellittico,* Preprint Dip. Mat. Pisa **248**, Pisa, 1988.

[67] P.Quittner, *Solvability and multiplicity results for variational inequalities,* (preprint), 1987.

[68] P.H.Rabinowitz, *Minimax methods in critical point theory with application to differential equations,* CBMS Regional Conference Series in Mathematics **65**, American Mathematical Society, Providence,R.I., 1986.

[69] R.C.Riddel, *Eigenvalue problems for nonlinear elliptic variational inequalities,* Nonlin. Anal. Th. Meth. Appl. **3** (1979), 1-33.

[70] C.Saccon, *Some parabolic equations on non convex constraints,* Boll. Un. Mat. It. , to appear.

[71] C.Saccon, *On a evolution problem with free boundary,* Houston J. Math., to appear.

[72] D.Scolozzi, *Esistenza e molteplicitá di geodetiche con ostacolo con estremi variabili,* Ricerche Mat. **33** (1984), 171-201.

[73] D.Scolozzi, *Un teorema di esistenza di una geodetica chiusa su varietá con bordo,* Boll. Un. Mat. It. **A 4** (1985), 451-457.

[74] M.Tosques, *Quasi autonomous evolution equations associated with (φ, f)-monotone operators,* (Trieste 1985), Rend. Circ. Mat. Palermo Suppl. **15** (1987), 163-180.

[75] G.Stampacchia, *Formes bilinéaires coercitives sur les ensembles convexes,* C. R. Acad. Sc. Paris **258** (1964), 4413-4416.

[76] G.Stampacchia, *Le problème de Dirichlet pour les équations elliptiques du second ordre a coefficents discontinus,* Ann. Inst. Fourier Grenoble **15** (1965),189-257.

[77] G.Stampacchia, *Regularity of solutions of some variational inequalities,* Proceeding of Symposia on Nonlinear Functional Analysis **18** I (1980).

[78] A.Szulkin, *On a class of variational inequalities involving gradient operators,* J. Math. Annal. Appl. **100** (1984).

[79] A.Szulkin, *On the solvability of a class of semilinear variational inequalities,* Rend. Mat. **4** (1984).

[80] A.Szulkin, *Positive solutions of variational inequalities: a degree theoretical approach,* J. Diff. Eq. **57** (1985), 90-111.

Dipartimento di Matematica
Università di Pisa
Via Buonarroti, 2
I-56127 PISA

SOME REMARKS ON THE DEPENDENCE DOMAIN FOR WEAKLY HYPERBOLIC EQUATIONS WITH CONSTANT MULTIPLICITY

SIGERU MIZOHATA

Dedicated to Ennio De Giorgi on his sixtieth birthday

1. Introduction. We are concerned with the Cauchy problem for the following weakly hyperbolic operator with constant multiplicity:

$$
(1.1)\qquad \begin{aligned} &P(x,t;D_x,D_t)u(x,t) = f(x,t), \quad (x,t) \in \mathbf{R}^n \times [t_0,T], \\ &\partial_t^j u\big|_{t=t_0} = u_j(x) \quad (0 \le j \le m-1), \quad t_0 \in [0,T], \end{aligned}
$$

where $D_x = i^{-1}\partial/\partial x$, $D_t = i^{-1}\partial/\partial t$. Hereafter we write D instead of D_x. We denote $\Omega = \mathbf{R}^n \times [0,T]$.

The principal symbol of P has the form

$$
(1.2)\qquad \prod_{j=1}^{k} (\tau - \lambda_j(x,t;\xi))^{m_j},
$$

where λ_j are *real and distinct*.

In order to consider the well-posedness in the space C^∞ or Gevrey class $\gamma^{(s)}$ $(s > 1)$, we proposed to use a *fine factorization* of operators, because it gives us a clear image to the well-posedness. A fine factorization means the factorization in the following form:

$$P = P_k \circ P_{k-1} \circ \ldots \circ P_1 + R, \tag{1.3}$$

where the principal symbol of P_j is $(\tau - \lambda_j)^{m_j}$. More precisely

$$\begin{aligned} P_j = {} & (D_t - \lambda_j(x,t;D))^{m_j} + a_{1,j}(x,t;D)(D_t - \lambda_j)^{m_j-1} + \\ & \ldots + a_{i,j}(D_t - \lambda_j)^{m_j-i} + \ldots + a_{m_j,j}\ , \quad \text{order } a_{i,j} \le i-1. \end{aligned} \tag{1.4}$$

$$R(x,t;D,D_t) = \sum_{j=0}^{m-1} r_j(x,t;D)D_t^j\ , \quad \text{order } r_j(x,t;\xi) \le -(m-1).$$

For simplicity, we assume that all the coefficients of P are C^∞ and with all their derivatives are bounded in Ω. Under the form (1.3) and (1.4), we state

(1) *Levi condition (C^∞ well-posedness).*

$$\text{(L)} \qquad \text{order } a_{i,j} \le 0\ , \quad \text{for } 1 \le i \le m_j,\ 1 \le j \le k.$$

(2) *Wellposed Gevrey index s.*

$$\begin{aligned} & \frac{1}{s} \ge \rho\ , \\ & \text{where} \quad \rho = \max_{1 \le j \le k} \rho_j\ , \rho_j = \max_{1 \le i \le m_j} \text{order } a_{ij}/i\ . \end{aligned} \tag{1.5}$$

Let us notice that, in the above factorization, the number ρ_j is invariant with respect to the order of factorization. Let us explain the above facts more precisely. We follow fairly classical definition.

Definition. *We say that the Cauchy problem (1.1) is C^∞ well-posed, with initial data at $t = t_0$, if for any $f(x,t) \in C^\infty([t_0,T] \times \mathbf{R}^n)$, and any initial data $u_j(x) \in C^\infty$, there exists a unique solution $u(x,t) \in C^\infty([t_0,T] \times \mathbf{R}^n)$. Further we say that (1.1) is uniformly C^∞ wellposed in Ω, when the above problem is wellposed for any $t_0 \in [0,T]$.*

The above definition is extended to the case of Gevrey classes, replacing the above conditions by $f(x,t) \in C^0([t_0,T];\gamma_x^{(s)})$, $u_j(x) \in$

$\gamma_x^{(s)}$, and $u(x,t) \in C^m([0,T];\gamma_x^{(s)})$ respectively, and we call (1.1) $\gamma_x^{(s)}$-wellposed and uniformly $\gamma_x^{(s)}$-wellposed respectively. Now we can state

Theorem 1.1. *In order that the Cauchy problem (1.1) be uniformly C^∞ wellposed in Ω, it is necessary and sufficient that P satisfies the Levi condition.*

Theorem 1.2. *In order that the Cauchy problem in Gevrey class $\gamma^{(s)}$ be uniformly wellposed in Ω, it is necessary that s satisfies $1/s > \rho$.*

Theorem 1.3. *If $1/s > \rho$, there exists a unique solution $u(x,t)$ for $t \in [t_0,T]$. If $1/s = \rho$, there exists only a local solution of (1.1).*

These theorems are new formulations obtained hitherto. We gave already their brief direct proofs ([12],[10]). Let us explain about the assumptions in Gevrey class. $f(x) \in \gamma_x^{(s)}$ means that

$$\sup_{x\in\mathbf{R}^n} |\partial^\alpha f(x)| \le M\alpha!^s C^{|\alpha|}, \quad \forall\alpha \ge 0, \quad (\exists M,\ \exists C > 0).$$

In Theorems 1.2 and 1.3, we assume that the coefficients of P satisfy the following type condition: $t \to a(x,t) \in \gamma_x^{(s)}$, $t \in [0,T]$, is continuous, with all its derivatives in t. Let us notice that we do not assume the Gevrey property in t. Incidentally, if we do not assume the Gevrey property of the coefficients in t of P, we cannot get the perfect factorization in Gevrey class. On the other hand, in order to obtain the standard result on the propagation of the wave front set WF_s in Gevrey class (see for instance [11]), it seems to the author, we need to require the Gevrey property of the coefficients in (x,t).

To illustrate the advantage of the above formulation, we explain the sufficient part of Theorem 1.1. For simplicity we assume

$$P = P_2 \circ P_1 + R.$$

Setting $(D_t - \lambda_1)^j u = u_j$ $(0 \le j \le m_1 - 1)$, $P_1 u = u_{m_1}$, $(D_t - \lambda_2)^j P_1 u = u_{m_1+j}$ $(0 \le j \le m_2 - 1)$, and denoting $U = {}^t(u_0, u_1, \ldots, u_{m-1})$, we get the (equivalent) system of $Pu = f$:

$$\partial_t U = iHU + B_0(x,t;D)U + F,$$

where

1) H is a diagonal matrix whose diagonal entries are $\lambda_i(x,t;D)$,
2) $B_0(x,t;\xi) \in S^0$, by Levi condition,
3) $F = {}^t(0,0,\ldots,0,f)$.

This shows that the problem is reduced essentially to consider the elementary hyperbolic operator

$$\partial_t u = i\lambda(x,t;D)u + b_0(x,t;D)u + f \qquad \text{with } b_0 \in S^0.$$

We must give some comments on the development of this subject. Since we explained it in C^∞ case in [13], we explain chiefly the case of Gevrey class. The systematic treatment of the Cauchy problem in Gevrey class began with Y.Ohya [14], and this was immediately extended by Leray-Ohya [8]. Let us notice that this latter gave precious subjects on (formally) hyperbolic operators with constant multiplicity. Leray-Ohya started from the following form:

(1.6)
$$P = q_1(x,t;D_x,D_t)^{\nu_1} \ldots q_\ell(x,t;D_x,D_t)^{\nu_\ell} + Q_{m-1}(x,t;D_x,D_t) =$$
$$= P_d(x,t;D_x,D_t) + Q_{m-1},$$

where q_j are all strictly hyperbolic differential operators, and Q_{m-1} is also differential operator of order $\leq m-1$. Here the suffix d of P_d signifies that P_d is a differential operator. S.Matsuura [9] proved that we can assume this always. J-C. De Paris has shown a kind of factorization by differential polynomials and defined "bien décomposable" operators which are known to be equivalent to satisfying Levi condition (see [1]). H.Komatsu [4], analyzing the above works, showed a general factorization theorem, defined the notion of irregularity through this factorization. See also [5]. On the other hand V.Ya.Ivrii has shown essentially the same results [3], and Theorems 1.2 and 1.3 have been obtained there. Both Komatsu and Ivrii, they make full use of De Paris decomposition.

Here we want to show the existence of a finite dependence domain. More precisely, we show that *the dependence domain for P is contained in the emission* for P_d, provided that we assume Levi condition (C^∞ case) or the index condition stated in Theorem 1.2 (Gevrey case). Recall that the emission with apex (x_0,t_0) is defined as the closure of all the time-like paths for P_d passing through (x_0,t_0). Notice that for P_d itself the property stated above is true. Concerning it, see [16].

As in [8], [3], [4], we take P_d as principal part, and we show that (1.1) is solved by the process of fairly simple successive approximations. Therefore our argument consists in essence of making bridge between the two decompositions (1.3) and (1.6). Let us notice that in C^∞ case this fact is shown fairly easily by following the argument which we shall show. Therefore we explain mainly the above fact in Gevrey case. Our result is not new. The purpose of this article is to show how to derive the existence of a finite dependence domain, starting from our formulation of well-posedness.

2. Preliminary. As Leray-Ohya [8], we take the expression of P as the starting point.

(2.1)
$$P = q_1(x,t;D_x,D_t)^{\nu_1} \ldots q_\ell(x,t;D_x,D_t)^{\nu_\ell} + Q_{m-1}(x,t;D_x,D_t),$$

where each $q_j(x,t;\xi,\tau)$ is a strictly hyperbolic differential polynomial, and its principal symbol has the form

$$q_j = \prod_{i=1}^{k_j}(\tau - \lambda_{j,i}(x,t;\xi)), \qquad \lambda_{j,i} \neq \lambda_{j,i'} \quad \text{for} \quad i \neq i'.$$

Set

$$P_d = q_1^{\nu_1} \ldots q_\ell^{\nu_\ell}. \tag{2.2}$$

Let

$$\begin{aligned} &Pu = f \in C^0([0,T];\gamma_x^{(s)}),\\ &\operatorname{supp}[f(.,t)] \subset K_0 \text{ (some compact set)}\\ &\partial_t^j u(x,0) = 0\ , \quad 0 \le j \le m-1. \end{aligned} \tag{2.3}$$

Since $P = P_d + Q_{m-1}$, defining $u_1, u_2, \ldots$ successively,

$$\begin{aligned} &P_d(u_1) = f, P_d(u_2) = -Q_{m-1}(u_1), \ldots,\\ &P_d(u_p) = -Q_{m-1}(u_{p-1}), \ldots \end{aligned} \tag{2.4}$$

with $\partial_t^j u(x,0) = 0$ $(0 \le j \le m-1)$ for all u_p, we want to show that the formal series

$$u(x,t) = u_1(x,t) + \ldots + u_p(x,t) + \ldots \tag{2.5}$$

converges in the topology of $\gamma_x^{(s)}$, with the derivatives in t up to order $m-1$. This is enough to show that the dependence, therefore the influence domain has the same property as in C^∞ case. Let us notice that the index s is restricted to $1/s > \rho$. For this purpose, we must know the structure of Q_{m-1}. The lemmas of next section show

Proposition 2.1. *We can express P_d and Q_{m-1} in the form*
(i) $P_d = \Pi_m + \sum_{|J|<m} b_J \Pi_J$ *with* $b_J(x,t;\xi) \in S^0 \cap \gamma_x^{(s)}$,
(ii) $Q_{m-1} = \sum_{|J|<m} c_J \Pi_J$ *with* $c_J(x,t;\xi) \in S^{(m-|J|)\rho} \cap \gamma_x^{(s)}$.

Let us explain the notations. Let $\partial = D_t - \lambda(x,t;D)$. We denote $\partial_1, \partial_2, \dots, \partial_{m_1}$ when $\lambda = \lambda_1$, and $\partial_{m_1+1}, \partial_{m_1+2}, \dots, \partial_{m_1+m_2}$ when $\lambda = \lambda_2$ and so on. Under this convention, denote

$$\Pi_m = \partial_1 \partial_2 \dots \partial_m.$$

Let $J = (j_1, j_2, \dots, j_k)$ with $J \subset \{1, 2, \dots, m\}$. We denote $\Pi_J = \partial_{j_1} \partial_{j_2} \dots \partial_{j_k}$ $|J| = k$. For convenience, we denote $\Pi_0 = I$ (identity).

Finally we call Π_J with $j_1 < j_2 < \dots < j_k$ and $k \le m-1$ (including Π_0) *standard basis.* As Lemma 2 of next section shows, we can assume that, in Proposition 2.1, all Π_J appearing there are standard basis.

Remark. It begins with A.Lax [7] to consider modules spanned by Π_J. This method was extended by M.Yamaguti [15]. A fairly detailed account of Π_J is also given in H.Kumano-go [6].

3. Some lemmas on the basis. The arguments in this section are fairly known. They can be carried out micro-locally. However, in view of our purpose, we consider the global version. Although we don't write explicitly, if we are concerned with the Gevrey symbols, all symbols belong to $\gamma_x^{(s)}$.

We often write ∂_i in the form $\tau - \lambda_i(x,t;\xi)$, λ_i being homogeneous of degree 1 in ξ. Moreover, we assume that, among λ_i, either $\lambda_i(x,t;\xi) = \lambda_j(x,t;\xi)$, or

$$(3.1) \qquad \left|\lambda_i(x,t;\xi) - \lambda_j(x,t;\xi)\right| \ge \delta|\xi|\,, \quad (\exists\delta > 0).$$

For convenience, we write

$$\Pi_J u = u_J. \tag{3.2}$$

We treat here only the symbols of class $S^m_{1,0}$. For simplicity we write it S^m. Remark that, in this section, we need not to assume that λ_i are real. It would be better to consider that λ_i are not directly related with those in the previous sections.

Let $a(x,t;\xi)$ and $b(x,t;\xi)$ be two symbols. We consider two products,

$$a(x,t;\xi)b(x,t;\xi)\ , \quad a(x,t;\xi)\circ b(x,t;\xi).$$

The former stands for the scalar product, and the latter stands for the operator product. So that

$$a\circ b \sim ab + \sum_{|\alpha|\geq 1}\frac{1}{\alpha!}a^{(\alpha)}b_{(\alpha)}.$$

Lemma 1. *Let $a(x,t;\xi)$ be any symbol of class S^1. Then $a(x,t;D)$ is spanned by ∂_i, ∂_j ($\lambda_i\neq\lambda_j$), and I with coefficients in S^0. In particular,*

$$[\partial_i,\partial_j] = a(x,t;D)\partial_i + b(x,t;D)\partial_j + c(x,t;D),$$

where $a,b,c\in S^0$.

PROOF. Assume $\lambda_i\neq\lambda_j$. Let $[\partial_i,\partial_j]=\tilde a(x,t;\xi)\in S^1$. Now

$$\tilde a(x,t;\xi) = \frac{\tilde a}{\lambda_i-\lambda_j}(\lambda_i-\lambda_j) = a(x,t;\xi)(\lambda_i-\lambda_j),$$

with $a\in S^0$. Hence, $\tilde a = a(x,t;\xi)\circ(\lambda_i-\lambda_j)+c$, with $c\in S^0$.

Definition of M_ρ. *Let us denote by M_ρ ($\rho\geq 0$), the module spanned by the basis Π_J ($J\subset\{1,2,\dots,m\}$, $|J|<m$) with coefficients in S^ρ. More explicitly, $N\in M_\rho$ means that*

$$N = \sum_{|J|\leq m-1} a^J(x,t;D)\Pi_J\ , \quad \textit{with } a^J(x,t;\xi)\in S^\rho.$$

Lemma 2. *Let $a(x,t;\xi) \in S^\rho$, then*

$$\Pi_m a(x,t;D) - a(x,t;D)\Pi_m \in M_\rho.$$

PROOF. Induction on m. For $m=1$, $(D_t-\lambda)\circ a - a\circ(D_t-\lambda) = D_t a - (\lambda \circ a - a \circ \lambda) \in S^\rho$. Now for general m (≥ 2),

$$\begin{aligned}\Pi_m a &= \partial_1(\partial_2 \ldots \partial_m)a = \partial_1 a(\partial_2 \ldots \partial_m) + \partial_1[\partial_2 \ldots \partial_m, a] = \\ &= a\Pi_m + [\partial_1, a](\partial_2 \ldots \partial_m) + \partial_1[\partial_2 \ldots \partial_m, a],\end{aligned}$$

so we can apply the induction.

Lemma 3. *Let $\Pi'_m = \partial_{j_1}\partial_{j_2}\ldots\partial_{j_m}$, where $(j_1, j_2, \ldots, j_m)$ is a permutation of $(1,2,\ldots,m)$. Then*

$$\Pi_m - \Pi'_m \in M_0.$$

PROOF. Induction on m. It will be enough to see that, in the case $\Pi'_m = \partial_2\partial_3\ldots\partial_m\partial_1$

$$\begin{aligned}\Pi'_m &= \partial_2 \ldots [\partial_m, \partial_1] + \partial_2 \ldots [\partial_{m-1}, \partial_1]\partial_m + \\ &+ \partial_2 \ldots [\partial_{m-2}, \partial_1]\partial_{m-1}\partial_m + \ldots + [\partial_2, \partial_1]\partial_3 \ldots \partial_m + \Pi_m.\end{aligned}$$

By applying Lemmas 1 and 2, we apply the hypothesis of the induction.

The following three lemmas are concerned with the relations between scalar products of symbols and operator products of symbols.

Lemma 4. *Let $\mathring{\Pi}_m = \prod_{j=1}^m(\tau - \lambda_j)$ (scalar product), where $\lambda_i \neq \lambda_j$ $(i \neq j)$. Let $a(x,t;\xi) \in S^\rho$ $(\rho \geq 0)$, it holds*

$$a\mathring{\Pi}_m = a \circ \mathring{\Pi}_m + \sum_{|J|<m} a^J \circ \mathring{\Pi}_J\ , \tag{3.3}$$

where $a^J(x,t;\xi) \in S^\rho$, and $J \subset \{1,2,\ldots,m\}$, and $\mathring{\Pi}_0 = I$.

PROOF. The case $m = 1$ is evident. We prove this by induction on m. For this purpose, let

$$\mathring{\Pi}_m = \tau^m + \sum_{j=1}^{m} b_j(x,t;\xi)\tau^{m-j}, \quad b_j \in S^j. \tag{3.4}$$

Hence

$$a \circ \mathring{\Pi}_m - a\mathring{\Pi}_m = \sum_{j=1}^{m} (a \circ b_j - ab_j)\tau^{m-j}.$$

Denote $a \circ b_j - ab_j = c_j(x,t;\xi)$ $(\in S^{j-1+\rho})$. On the other hand, by Lagrange,

$$c_j\tau^{m-j} = \sum_{k=1}^{m} b^{jk}(\tau - \lambda_1)\dots(\widehat{\tau - \lambda_k})\dots(\tau - \lambda_m),$$

where $b^{jk} \in S^\rho$. By applying the hypothesis of induction, we arrive at the conclusion.

Lemma 5. *Let $\lambda_1, \lambda_2, \dots, \lambda_m$ be distinct, then*

$$\Pi_m - \mathring{\Pi}_m \in M_0.$$

PROOF. In the case $m = 2$, the left hand side is written in the form of symbols,

$$(\tau - \lambda_1) \circ (\tau - \lambda_2) - (\tau - \lambda_1)(\tau - \lambda_2).$$

Denoting this symbol by $a_1(x,t;\xi) \in S^1$, and applying Lemma 1, we see the conclusion. In general m, we use the argument by induction on m.

$$\Pi_m = \Pi_{m-1} \circ (\tau - \lambda_m) = (\mathring{\Pi}_{m-1} + N) \circ (\tau - \lambda_m), \quad \text{where } N \in M_0.$$

It suffices to show that

$$\mathring{\Pi}_{m-1} \circ (\tau - \lambda_m) - \mathring{\Pi}_m \in M_0.$$

Using (3.4) replaced there m by $m-1$, the left hand side is expressed in the form

$$\sum_{j=0}^{m-1} b_j \tau^{m-1-j} \circ (\tau - \lambda_m) - \sum_{j=0}^{m-1} b_j \tau^{m-1-j}(\tau - \lambda_m)$$

$$b_j(x,t;\xi) \in S^j, \quad b_0 = 1,$$

$$= - \sum_{j=0}^{m-1} (b_j \circ \lambda_m - b_j \lambda_m)\tau^{m-1-j} - \sum_{j=0}^{m-1} b_j \circ [\tau^{m-1-j}, \lambda_m].$$

This expression has the form :

$$\sum_{j=0}^{m-1} c_j(x,t;\xi)\tau^{m-1-j}, \quad c_j \in S^j.$$

By using Lagrange's formula, this can be written

$$d_0^k(x,t;\xi)(\tau - \lambda_1)(\tau - \lambda_2)\dots\widehat{(\tau - \lambda_k)}\dots(\tau - \lambda_m), \quad d_0^k \in S^0.$$

Now each term has the form $a\mathring{\Pi}_{m-1}$ with $a \in S^0$. Thus after applying to it Lemma 4, we apply to this final form the induction on m.

We can extend Lemma 5 to a more general case. Let $\Pi_1, \dots, \Pi_\ell$ satisfy the assumption of Lemma 5. Let

$$\mathring{\Pi}_i = (\tau - \lambda_{i,1})\dots(\tau - \lambda_{i,n_i}).$$

For each i, we associate the basis of the form Π_{J_i} ($J_i \subset \{1, 2, \dots n_i\}$). Let us consider the basis of the form $\Pi_{J_1} \circ \Pi_{J_2} \circ \dots \circ \Pi_{J_\ell}$, where $|J| = |J_1| + \dots + |J_\ell| < n_1 + n_2 + \dots + n_\ell$ $(= m)$. We denote by M_0 the module spanned by these basis with coefficients in S^0. Then

Lemma 6. *Let* $\Pi'_m = \Pi_1 \circ \Pi_2 \circ \dots \circ \Pi_\ell$. *Then*

$$\mathring{\Pi}_1 \circ \mathring{\Pi}_2 \circ \dots \circ \mathring{\Pi}_\ell - \Pi'_m \in M_0.$$

PROOF. We prove this by induction on ℓ. The case $\ell = 1$ is Lemma 5. In the case $\ell = 2$, since $\Pi_j - \mathring{\Pi}_j = N_j \in M_0^{(j)}$ $(j = 1, 2)$ (Lemma 5),

$$\Pi_1 \circ \Pi_2 - \mathring{\Pi}_1 \circ \mathring{\Pi}_2 = N_1 \circ \Pi_2 + \Pi_1 \circ N_2 - N_1 \circ N_2.$$

In view of Lemma 2, we see the claim. Now, for general ℓ,

$$\begin{aligned}\Pi_1 \circ \ldots \circ \Pi_\ell - \mathring{\Pi}_1 \circ \ldots \circ \mathring{\Pi}_\ell = \\ = \Pi_1 \circ (\Pi_2 \circ \ldots \circ \Pi_\ell - \mathring{\Pi}_2 \circ \ldots \circ \mathring{\Pi}_\ell) + \\ + (\Pi_1 - \mathring{\Pi}_1) \circ (\mathring{\Pi}_2 \circ \ldots \circ \mathring{\Pi}_\ell),\end{aligned}$$

which shows that we can argue by induction and apply Lemma 2. This completes the proof.

Proof of Proposition 2.1.

Proof of (i). Remark that, according to our notation, an operator of *homogeneous* degree m of the form $D_t^m + \sum_{j=1}^m a_j(x,t;D)D_t^{m-j}$ is expressed by $\prod_{j=1}^m(\tau - \lambda_j(x,\mathring{t};\xi))$ (λ_j being the characteristic roots), and we expressed it by Π_m in Lemma 6. Thus Lemma 6 claims that

$$P_d - \Pi'_m = \sum_{|J|<m} \tilde{b}_J \Pi_J \ , \quad \text{with } \tilde{b}_J \in S^0.$$

By applying Lemma 3 to Π'_m, we get (i).

Proof of (ii). Recall that $Q_{m-1} = P - P_d$. We prove that P can be expressed in the form

$$P = \Pi_m + \sum_{|J|<m} a_J(x,t;D)\Pi_J \ , \quad \text{with } a_J \in S^{(m-|J|)\rho} \cap \gamma_x^{(s)}. \tag{3.5}$$

This is enough to prove (ii), taking account of (i).

First, the factorization (1.3) can be expressed in the form

$$P = \Pi'_m + \sum \left(a_{i_k,k}(x,t;D)\partial_k^{i_k}\right) \ldots \left(a_{i_j,j}\partial_j^{i_j}\right) \ldots \left(a_{i_1,1}\partial_1^{i_1}\right) + R \ , \tag{3.6}$$

where the sum is extended over $(i_1, \ldots, i_k)$ satisfying $i_1 + \ldots + i_k \leq m-1$. Recall that $a_{i_j,j} \in S^{(m_j - i_j)\rho} \cap \gamma_x^{(s)}$. By applying repeatedly Lemma 2, we see that each term in the sum is expressed in the form

$$\sum_{|J|<m} c^J(x,t;D)\Pi_J \ , \quad \text{where } c^J(x,t;\xi) \in S^{(m-|J|)\rho} \cap \gamma_x^{(s)}.$$

Finally we show that

$$(3.7) \qquad R = \sum_{j=0}^{m-1} r_j(x,t;D) D_t^j \in M_0.$$

Let $\Pi_j = \partial_j \partial_{j-1} \ldots \partial_1$ $(1 \leq j \leq m-1)$. We see easily, by induction on j, that for $1 \leq j \leq m-1$,

$$(3.8) \quad D_t^j = \Pi_j + \sum_{k=1}^{j} d_k^j(x,t;D) D_t^{j-k}, \quad \text{with } d_k^j(x,t;\xi) \in S^k \cap \gamma_x^{(s)}.$$

In fact $\Pi_j = D_t^j + \sum_{k=1}^j b_k^j(x,t;D) D_t^{j-k}$ with $b_k^j \in S^k$. Recalling that $r_j(x,t;\xi) \in S^{-(m-1)} \cap \gamma_x^{(s)}$, we verify (3.7). Finally, by applying Lemma 3 to Π'_m, (3.6) and (3.7) show (3.5).

4. Influence domain. As we mentioned in Section 2, we can assume that Π_J appearing in Proposition 2.1 are all standard basis. First, let us explain, by introducing the new unknown functions $u_J = \Pi_J u$ $(|J| < m)$, that the relation

$$(4.1) \qquad P_d(u) = f$$

can be expressed by a fairly simple first order system of pseudodifferential equations concerning them.

The conclusion is the following. Let

$$(4.2) \quad \tilde{u}_J = \langle \Lambda \rangle^{(m-1-|J|)\rho} u_J \quad (= \langle \Lambda \rangle^{(m-1-|J|)\rho} \Pi_J u), \quad (|J| < m),$$

where $\widehat{\langle \Lambda \rangle u}(\xi) = (1+|\xi|^2)^{1/2} \widehat{u}(\xi)$. Set

$$U = {}^t(\tilde{u}_J)_{|J|<m} \, .$$

Proposition 4.1. *The relation (4.1) is expressed in the following matrix form:*

$$(4.3) \qquad \partial_t U = iH(x,t;D)U + B_\rho(x,t;D)U + F,$$

where

i) *$H(x,t;D)$ is a diagonal matrix, whose diagonal entry is one of $\lambda_i(x,t;D)$,*

ii) Let $B_\rho(x,t;D) = (b_{ij}(x,t;D))_{1\le i,j\le N}$, then $b_{ij}(x,t;\xi) \in S^\rho \cap \gamma_x^{(s)}$,

iii) Let $F = {}^t(f_1,\dots,f_N)$, then either $f_i = f$ or $f_i = 0$.

Corollary.

$$P_d(u_p) = -Q_{m-1}(u_{p-1}) \tag{4.4}$$

can be expressed by

$$\partial_t U_p = iH(x,t;D)U_p + B_\rho(x,t;D)U_p + Q_\rho(x,t;D)U_{p-1}, \tag{4.5}$$

where $Q_\rho = (q_{ij}(x,t;D))_{1\le i,j\le N}$ satisfies $q_{ij}(x,t;\xi) \in S^\rho \cap \gamma_x^{(s)}$.

Proof of proposition 4.1. For J with $|J| = m-1$, there exists a unique j such that $\{j,J\} = \{1,2,\dots,m\}$. Hereafter we change the suffix of λ_j. We denote $\partial_j = \tau - \lambda_j$ $(1\le j\le m)$. After this convention,

$$\partial_j u_J = \Pi'_m u = \Pi_m u + N_J u\ ,$$

with $N_J \in M_0$ (see the definition of M_ρ in section 3). Hence Proposition 2.1 implies

$$\partial_j u_J = P_d u - \sum_{|J|<m} b_J \Pi_J u + N_J u = \sum_{|J|<m} b'_J u_J + f\ , \tag{4.6}$$

where $b'_J(x,t;\xi) \in S^0 \cap \gamma_x^{(s)}$.

Next, for J with $|J| \le m-2$, we denote by j the least integer in the complement of $J = (j_1,\dots,j_k)$ $(k\le m-2)$. Let J' be the sequence of $j, j_1,\dots,j_k$ arranged in increasing order (thus $\Pi_{J'}$ is a standard basis). Then denoting $(j,j_1,\dots,j_k)$ by $\tilde{J}$, we obtain

$$\partial_j u_J = u_{\tilde{J}} = u_{J'} + N_J u\ ,$$

where, by Lemma 3, N_J belongs to the module M_0, spanned by $\Pi_{J''}$ with $J'' \subset J'$, $|J''| < |J'| = |J|+1$. This implies, by applying $\langle\Lambda\rangle^{(m-1-|J|)\rho}$ on left, that

$$\partial_j \tilde{u}_J = b_0^J(x,t;D)\tilde{u}_J + \langle\Lambda\rangle^\rho \tilde{u}_{J'} + \sum_{J''\subset J'} b^J_{J''}(x,t;D)\tilde{u}_{J''}\ , \tag{4.7}$$

where $b_0^J(x,t;\xi)$, $b_{J''}^J(x,t;\xi) \in S^0 \cap \gamma_x^{(s)}$. In fact,

$$\left[\langle\Lambda\rangle^{(m-1-|J|)\rho}, \lambda_j\right] \langle\Lambda\rangle^{-(m-1-|J|)\rho} = b_0^J \in S^0 ,$$

and $\langle\Lambda\rangle^{(m-1-|J|)\rho} u_{J'} = \langle\Lambda\rangle^{\rho}\tilde{u}_{J'}$, moreover $\langle\Lambda\rangle^{(m-1-|J|)\rho} N_J$ is expressed in the form

$$\sum_{J''\subset J'} \langle\Lambda\rangle^{(m-1-|J|)\rho} a_{J''}\Pi_{J''} , \quad \text{with } a_{J''} \in S^0 \cap \gamma_x^{(s)},$$

and $|J''| \le |J|$. (4.6) and (4.7) prove Proposition.

Proof of Corollary. This is derived from Proposition 2.1, (ii).

In order to obtain fairly precise Gevrey estimates of U and U_p of (4.3) and (4.4), we use the following two basic propositions.

First, we are concerned with the solution u of

$$\text{(4.8)} \qquad \begin{cases} (\partial_t - i\lambda(x,t;D))\, u(x,t) = f(x,t), \\ u(x,0) = 0, \end{cases}$$

where we assume that $\lambda(x,t;\xi)$ is homogeneous of degree 1 in ξ, and real-valued. Putting $\lambda(x,t;\xi)/|\xi| = \lambda_0(x,t;\xi)$, we can assume (in view of the uniform Gevrey property of x of λ),

$$\text{(4.9)} \qquad \begin{aligned} &\|\partial_x^\alpha \lambda_0(x,t;D)\|_{\mathcal{L}(L^2;L^2)} \le \\ &\qquad \le c(\lambda_0)(|\alpha|-1)!^s C_0^{|\alpha|-1} , \quad \text{for all } |\alpha| \ge 1. \end{aligned}$$

The energy inequality for (4.8) has the form

$$\frac{d}{dt}\|u(.,t)\| \le \gamma_0\|u(.,t)\| + \|f(.,t)\|.$$

Then we put

$$\text{(4.10)} \qquad K_0(t) = e^{\gamma_0 t}, \quad K(t) = (1+\gamma t)e^{\gamma t},$$

where $\gamma = c(\lambda_0)\max(n^2, 2^n n)$. Denote

$$f_m(t) = \sum_{|\alpha|=m} \|\partial_x^\alpha f(.,t)\| , \quad u_m(t) = \sum_{|\alpha|=m} \|\partial_x^\alpha u(.,t)\| = \|u\|_m$$

Proposition 4.2 *Assume in (4.8)*

$$f_m(t) \le A_f K_0(t)(CK(t))^{m+k}(m+k)!^s \frac{t^j}{j!}\,, \qquad m = 0,1,2,\dots$$

then

$$u_m(t) \le 2A_f K_0(t)(CK(t))^{m+k}(m+k)!^s \frac{t^{j+1}}{(j+1)!}\,, \qquad m = 0,1,2,\dots$$

where k (≥ 0) is any real number, and $j \ge 0$ is an integer, and $C \ge 4C_0$, $C > 1$. (Note that C_0 appears in (4.9)).

The proof is elementary, but tedious. It is given in [11]. We must use another proposition, whose proof is also given in [11].

Proposition 4.3. *Let $a(x,\xi) \in S^{\rho}_{0,0}$, $0 < \rho < 1$. More precisely*

$$\left|a^{(\beta)}_{(\alpha)}\right| \le C_a |\alpha|!^s C_0^{|\alpha|}(|\xi|+1)^{\rho}\,, \quad \forall \alpha \ge 0,\ |\beta| \le n+2.$$

Assume

$$\sum_{|\alpha|=m} \|\partial^{\alpha} v\| \le M(m+k)!^s C^m\,, \qquad m = 0,1,2,\dots$$

Then

$$\sum_{|\alpha|=m} \|\partial^{\alpha}(a(x,D)v)\| \le K_a M(m+k+\rho)!^s C^{m+\rho}\,, \qquad m = 0,1,2,\dots$$

k being a non-negative number, and K_a is a positive constant, depending only on a, and $C > 4C_0$, $C \ge 1$.

We apply the above two propositions to (4.3) and (4.5). First, Proposition 4.2 implies the following estimates. For the solution U of

$$(\partial_t - iH)U = F\,, \quad U\big|_{t=0} = 0\,, \tag{4.11}$$

the estimate for F

$$\begin{cases} \|F(t)\|_m \le A_f K_0(t)(CK(t))^{m+k}(m+k)!^s \dfrac{t^j}{j!}, \ \forall m \\ \text{implies} \\ \|U(t)\|_m \le 2A_f K_0(t)(CK(t))^{m+k}(m+k)!^s \dfrac{t^{j+1}}{(j+1)!}, \ \forall m \end{cases} \tag{4.12}$$

Next, Proposition 4.3 gives:

$$(4.13)\quad \begin{cases} \|U\|_m \le MC^{m+k}(m+k)!^s \ , \ 0<\rho<1 \ , \\ \text{implies} \\ \|B_\rho U\|_m \le C_B M C^{m+k+\rho}(m+k+\rho)!^s \ , \\ \|Q_\rho U\|_m \le C_Q M C^{m+k+\rho}(m+k+\rho)1^s \ . \end{cases}$$

Now we consider U_1 corresponding to u_1 in (2.4) and (4.3) (replaced U by U_1). In order to estimate U_1, we solve (4.3) by successive approximations,

$$U_1 = U_{1,1} + U_{1,2} + \ldots + U_{1,j} + \cdots$$

where

$$(\partial_t - iH)(U_{1,1}) = F \ , \ (\partial_t - iH)(U_{1,2}) = B_\rho(U_{1,1}) \ , \ldots$$
$$\ldots(\partial_t - iH)(U_{1,j}) = B_\rho(U_{1,j-1}) \ , \ldots, \quad U_{1,j}\big|_{t=0} = 0 \ .$$

We apply here (4.12) and (4.13) alternately. First assume that $F(t)$ satisfies (4.12) with $j = k = 0$. Then

$$\|U_{1,1}\|_m \le 2A_f K_0(t)(CK(t))^m m!^s t.$$

Applying (4.13),

$$\|B_\rho(U_{1,1})\|_m \le 2A_f C_B K_0(t)(CK(t))^{m+\rho}(m+\rho)!^s t.$$

Applying (4.12), we get

$$\|U_{1,2}(t)\|_m \le 2A_f(2C_B)K_0(t)(CK(t))^{m+\rho}(m+\rho)!^s \frac{t^2}{2}.$$

We repeat this argument, which implies

$$(4.14)\quad \begin{aligned} \|U_{1,j}(t)\|_m \le\ & 2A_f(2C_B)^{j-1}K_0(t)\times \\ & \times (CK(t))^{m+(j-1)\rho}(m+(j-1)\rho)!^s \frac{t^j}{j!}. \end{aligned}$$

Next we consider the estimate of U_2:

$$(\partial_t - iH - B_\rho)(U_2) = Q_\rho(U_1) \ , \quad U_2\big|_{t=0} = 0 \ ,$$

in the form

$$U_2 = U_{2,1} + U_{2,2} + \ldots + U_{2,j} + \ldots,$$

where $U_{2,j}$ is the solution of

$$(\partial_t - iH - B_\rho)(U_{2,j}) = Q_\rho(U_{1,j}) \quad (j = 1, 2, 3, \ldots),$$
$$U_{2,j}\big|_{t=0} = 0 \ .$$

In view of (4.14),

$$\|Q_\rho(U_{1,j})\|_m \leq 2A_f(2C_B)^{j-1} C_Q K_0(t)(CK(t))^{m+j\rho}(m + j\rho)!^s \frac{t^j}{j!}.$$

We repeat the same argument as above taking $Q_\rho(U_{1,j})$ as a starting point. Then

$$\|U_{2,j}\|_m \ll 2A_f(2C_Q) \sum_{k=1}^{\infty} (2C_B)^{j+k-2} K_0(t) \times$$
$$\times (CK(t))^{m+(j+k-2)\rho}(m + (j + k - 2)\rho)!^s \frac{t^{j+k}}{(j+k)!} \ .$$

Hence

$$\|U_2\|_m \ll 2A_f(2C_Q) \sum_{j=1}^{\infty}\sum_{k=1}^{\infty} (2C_B)^{j+k-2} K_0(t) \times$$
$$\times (CK(t))^{m+(j+k-1)\rho}(m + (j + k - 1)\rho)!^s \frac{t^{j+k}}{(j+k)!} \ .$$

Now for general U_p (corresponding to u_p in (2.4)):

$$(\partial_t - iH - B_\rho)(U_p) = Q_\rho(U_{p-1}) \ , \quad U_p\big|_{t=0} = 0 \ , \tag{4.16}$$

we obtain

$$\|U_p\|_m \ll 2A_f(2C_Q)^{p-1} \sum_{j_1 \ldots j_p} (2C_B)^{j_1+\ldots+j_p-p} K_0(t) \times$$
$$\times (CK(t))^{m+(j_1+\ldots+j_p-1)\rho}(m + (j_1 + \ldots + j_p - 1)\rho)!^s \times \tag{4.17}$$
$$\times \frac{t^{j_1+\ldots+j_p}}{(j_1 + \ldots + j_p)!} \ ,$$

where the sum is taken over $(j_1, \ldots, j_p)$ satisfying $j_i \geq 1$. This estimate can be verified by induction on p.

Finally for $U = U_1 + U_2 + \ldots + U_p + \ldots$, we get (slightly rough estimate)

(4.18)

$$\|U\|_m \ll 2A_f \sum_{p=1}^{\infty} (2C_Q)^{p-1} \sum_{j_1 \cdots j_p} (2C_B)^{j_1+\ldots+j_p} K_0(t) \times$$
$$\times\, (CK(t))^{m+(j_1+..+j_p)\rho}(m + (j_1 + \ldots + j_p)\rho)!^s \frac{t^{j_1+..+j_p}}{(j_1 + .. + j_p)!} \cdot$$

Let us estimate the coefficient of $t^j/j!$ in the above right hand side. This is estimated by

$$A_f(2^3 C_B C_Q)^j K_0(t)(CK(t))^j (m + j\rho)!^s c(j),$$

where

$$c(j) = \sharp \quad \{(j_1, j_2, \ldots);\ j_1 + \ldots + j_p = j,\ j_i \geq 1\}.$$

One sees easily that $c(j) \leq 2^j$. In fact,

$$c(j) = \sum_{k=1}^{j} \binom{(j-k) + (k-1)}{k-1}.$$

This implies

$$\|U\|_m \ll A_f \sum_{j=1}^{\infty} K_0(t)(2^4 C_B C_Q)^j (CK(t))^{m+j\rho}(m + j\rho)!^s \frac{t^j}{j!}.$$

Using the majoration $(m + j\rho)! \leq 2^m 2^\rho m!(\rho j)!$, we reached the following desired result.

Let $U = U_1 + U_2 + \ldots + U_p + \ldots$

Then, for $m = 0, 1, 2, \ldots$

$$\|U(t)\|_m \ll A_f K_0(t)(CK(t))^m m!^s \Phi(t)\ , \tag{4.19}$$

where

$$\Phi(t) = \sum_{j=1}^{\infty} C'^j (CK(t))^{j\rho}(\rho j)!^s \frac{t^j}{j!}\ ,$$
$$C' = 2^{4+s\rho} C_B C_Q$$

(depending only on the operator P). In a form of theorem,

Theorem 4.1. *The formal series (2.5) defined by (2.4) converges uniformly (in t) in $\gamma_x^{(s)}$, including the derivatives in t up to order $m-1$. Hence, the influence, and dependence domains are the same as in the C^∞ case.*

Remark. (4.19) shows that $\Phi(t) < \infty$ if t (> 0) is small when $s\rho = 1$. Its radius of convergence depends on C defined by $f(x,t)$.

References

[1] J.Chazarain, *Opérateurs hyperboliques à caractéristiques de multiplicité constante*, Ann. Inst. Fourier **24** (1974),175-202.

[2] J-C.De Paris, *Problème de Cauchy oscillatoire pour un opérateur différentiel à caractéristique multiples: Lien avec l'hyperbolicité*, J. Math. Pure Appl. **51** (1972), 231-256.

[3] V.Ya.Ivrii, *Conditions for correctness in Gevrey classes of the Cauchy problem for weakly hyperbolic equations*, Siberian J. of Math. **17** (1976), 422-435.

[4] H.Komatsu, *Linear hyperbolic equations with Gevrey coefficients*, J. Math. Pure Appl. **59** (1980), 145-185.

[5] H.Komatsu, *Irregularity of hyperbolic operators*, Proc. Taniguchi Symposium on Hyperbolic Equations and Related Topics (1984), 193-233.

[6] H.Kumano-go, *Pseudo-differential operators*, MIT Press, Cambridge 1981.

[7] A.Lax, *On Cauchy's problem for partial differential equations with multiple characteristics*, Comm. Pure Appl. Math. **9** (1956), 135-169.

[8] J.Leray, Y.Ohya, *Systèmes linéaires, hyperboliques nonstricts*, Deuxième Colloque sur l'Analyse Fonctionnelle à Liège, 1964.

[9] S.Matsuura, *On non strict hyperbolicity*, Proc. Funct. Anal. Related Topics, 1969, Univ. Tokyo Press, 171-176.

[10] S.Mizohata, *Sur l'indice de Gevrey*, Séminaire sur propagation des singularités et opérateurs différentiels (J.Vaillant), Hermann, 1985.

[11] S.Mizohata, *On the Cauchy problem*, Academic Press and Science Press, Beijing 1986.

[12] S.Mizohata, *On the Cauchy problem for hyperbolic equations and related problems (micro-local energy method)*, Proc. Taniguchi Symposium on Hyperbolic Equations and Related Topics (1984), 193-233.
[13] S.Mizohata, *On weakly hyperbolic equations with constant multiplicities,* Patterns and Waves, 1-10, Studies in Math. and its Appl. **18** (1986), Kinokuniya/North-Holland.
[14] Y.Ohya, *Le problème de Cauchy pour les équations hyperboliques à caractéristique multiple*, J. Math. Soc. Japan **16** (1964), 268-286.
[15] M.Yamaguti, *Le problème de Cauchy et les opérateurs d'intégrale singulière*, Mem. College Sci. Univ. Kyoto, Ser. A **32** (1959), 121-151.
[16] J.Leray, *Hyperbolic differential equations*, Princeton Lecture Note, 1954.

Department of Mathematics
Faculty of Science
Kyoto University
KYOTO 606

MONOTONICITY OF THE ENERGY FOR ENTIRE SOLUTIONS OF SEMILINEAR ELLIPTIC EQUATIONS

LUCIANO MODICA

Dedicated to Ennio De Giorgi on his sixtieth birthday

1. Introduction and statements of the results. In the mathematical theory of phase transitions in Van der Waals fluids (see, for instance, Alikakos & Bates [AB], Caginalp [CA], Gurtin [GU], Hagan & Serrin [HS], Modica [MO3]) the following problem arises: to study the asymptotic behaviour as $\epsilon \to 0+$ of solutions to the semilinear elliptic equations

$$-\epsilon^2 \Delta u(x) + f(u(x)) = 0 \quad (x \in \Omega) \tag{1}$$

on an open subset Ω of $\mathbf{R}^n$. The model case is $f(u) = u^3 - u$.

Equation (1) is the Euler-Lagrange equation for the functional

$$\mathcal{E}_\epsilon(u;\Omega) = \int_\Omega \left(\frac{\epsilon}{2}|\nabla u|^2 + \frac{1}{\epsilon}F(u)\right)dx \tag{2}$$

where $F' = f$. A natural assumption on F within the gradient theory of phase transitions is that $F \geq 0$ and $F(t) = 0$ only for discrete values of t. The model case $f(u) = u^3 - u$ corresponds to $F(u) = \frac{1}{4}(1-u^2)^2$.Such general assumption on F turns out to be relevant even in other fields in which equation (1) arises, for instance in nuclear physics where (1) is considered with $f(u) = \pm \sin u$ and $F(u) = 1 \mp \cos u$.

By letting $v(x) = u(\epsilon x)$, equation (1) transforms into

$$-\Delta v + f(v) = 0 \tag{3}$$

on the open set $\epsilon^{-1}\Omega$, which becomes larger and larger as $\epsilon \to 0+$. Then it seems natural to consider an entire solution v (i.e. defined on the whole of $\mathbf{R}^n$) and to restate our problem as follows: to study the asymptotic behaviour as $\epsilon \to 0+$ of $v(x/\epsilon)$, or equivalently of $v_r(x) = v(rx)$ as $r \to +\infty$.

Existence, uniqueness, radial symmetry of strictly positive entire solutions of (3) going to zero at infinity (called ground states) have been recently studied in a series of very important papers by Atkinson & Peletier [AP], Berestycki & Lions [BL], Berestycki, Lions & Peletier [BLP], Coffman [CO], Gidas, Ni & Nirenberg [GNN1], [GNN2], McLeod & Serrin [MLS], Peletier & Serrin [PS1], [PS2], Strauss [STR].

The point of the present note is that we do not necessarily assume $v_r(x) \to 0+$ as $r \to +\infty$, as for the ground state solutions. To our aims a two-phases behaviour at infinity, like for instance that one of the solution

$$v(x_1, ..., x_n) = 4 \operatorname{arctg} \exp x_1$$

of the equation $-\Delta v + \sin v = 0$, must of course be allowed.

The main result we shall prove is the following.

1.1. Theorem. *Let F be a real function of class C^2 and u be a solution of class C^3 of the equation $-\Delta u + F'(u) = 0$ on the whole space $\mathbf{R}^n$. If $F(t) \geq 0$ for every $t \in \mathbf{R}$ and u is uniformly bounded on $\mathbf{R}^n$, then the function*

$$\phi(r) = \frac{1}{r^{n-1}} \int_{B_r} \left(\frac{1}{2}|\nabla u|^2 + F(u)\right) dx \quad , \quad B_r = \{x \in \mathbf{R}^n : |x| < r\}$$

is a non-decreasing function of $r > 0$.

Remark that, by taking in the previous result $r = a/\epsilon$ ($a > 0$, $\epsilon > 0$), $u_\epsilon(x) = u(\epsilon x)$, and by recalling (2), we easily obtain that

$$\phi(a/\epsilon) = \frac{1}{a^{n-1}} \mathcal{E}_\epsilon(u_\epsilon; B_a) \quad ;$$

hence the energy $\mathcal{E}_\epsilon$ associated with u_ϵ is monotone increasing as $\epsilon \downarrow 0$ on any ball B_a and there exists the limit of $\mathcal{E}_\epsilon(u_\epsilon; B_a)$ as $\epsilon \to 0+$. In a slightly different context, we have recently analyzed the geometrical and thermodynamical interpretation of this limit: see Modica [MO3], [MO4], Luckhaus & Modica [LM]. General results about the Γ-convergence of the functionals $\mathcal{E}_\epsilon$, all starting from a conjecture of De Giorgi (cf. [DGF]), are given in Modica & Mortola [MM1], [MM2], Modica [MO1], Sternberg [ST], Baldo [BA], Ambrosio [AM].

The following non-existence result of ground states can be deduced by Theorem 1.1.

1.2. Corollary. *Let F and u be as in Theorem 1.1. Assume in addition that $F \in C^3(\mathbf{R})$, $F(0) = 0$, $F''(0) > 0$, and*

$$\lim_{|x| \to +\infty} u(x) = 0.$$

Then we have $u \equiv 0$.

Note that this result does not contradict the existence theorems for strictly positive ground states quoted above, because in those theorems there is the assumption that $F(u) < 0$ for large values of u.

Finally, I want to mention that Theorem 1.1 is the crucial point for giving the proof of a rather general, positive answer to another interesting conjecture raised by De Giorgi [DG] in 1978, concerning the geometry of the level sets of bounded entire solutions of (3) (Modica [MO5]; cf. Modica & Mortola [MM3] for a positive answer in a particular case).

I am very glad to dedicate this paper to Ennio De Giorgi for his sixtieth birthday.

2. Proof of the results.

Proof of Theorem 1.1. We begin the proof by recalling the following pointwise gradient estimate.

2.1. Lemma (Modica [MO2]). *Let F and u be as in Theorem 1.1. Then*

$$|\nabla u(x)|^2 \leq 2\ F(u(x)) \quad \forall x \in \mathbf{R}^n.$$

The other tool in the proof of Theorem 1.1 is the well-known Pohozaev identity (cf. Pohozaev [PO], Berestycki & Lions [BL], Esteban & Lions [EL]), whose we give here an alternative proof in the following lemma.

2.2. Lemma (Pohozaev identity). *Let $u \in C^2(\mathbf{R}^n)$ be a solution on the whole of $\mathbf{R}^n$ of the equation $-\Delta u + f(u) = 0$, with $f \in C^0(\mathbf{R})$. Denote by F a primitive function of f. Let A be any ball in $\mathbf{R}^n$, centered at the origin. Then, for every $r > 0$,*

$$\begin{aligned}\int_{A_r} \Big((n-2)|\nabla u|^2 + 2n\ F(u)\Big)dx &= \\ = r \int_{\partial A_r} \Big(|\nabla u|^2 + 2F(u)\Big)d\mathcal{H}_{n-1} &+ \\ - 2 \int_{\partial A_r} (\nabla u \cdot \nu_r)(\nabla u \cdot x) d\mathcal{H}_{n-1}\end{aligned} \tag{4}$$

where $A_r = \{x \in \mathbf{R}^n : x/r \in A\}$, $\mathcal{H}_{n-1}$ denotes the Hausdorff $(n-1)$-dimensional measure, and $\nu_r(x)$ denotes the outer unit-normal vector in $x \in \partial A_r$.

Proof of Lemma 2.2. Consider

$$\psi(r) = r^{-n} \int_{A_r} \Big(|\nabla u|^2 + 2F(u)\Big)dx.$$

Then

$$\begin{aligned}\psi'(r) = &- nr^{-n-1} \int_{A_r} \Big(|\nabla u|^2 + 2F(u)\Big)dx + \\ &+ r^{-n} \int_{\partial A_r} \Big(|\nabla u|^2 + 2F(u)\Big)d\mathcal{H}_{n-1}.\end{aligned}$$

On the other hand, the equality

$$\psi(r) = \int_A \Big(|\nabla u|^2(ry) + 2F(u(ry)) \Big) dy$$

yields, after an integration by parts,

$$\begin{aligned}
&\psi'(r) = \\
&= 2\int_A \Big[\Big(\nabla(\nabla u)(ry)\nabla u(ry) \Big) \cdot y + f(u(ry))\nabla u(ry) \cdot y \Big] dy \\
&= 2r^{-n-1} \int_{A_r} \Big[\Big(\nabla(\nabla u)\nabla u \Big) \cdot x + \Delta u(\nabla u \cdot x) \Big] dx \\
&= 2r^{-n-1} \Big[- \int_{A_r} |\nabla u|^2 dx + \int_{\partial A_r} (\nabla u \cdot \nu_r)(\nabla u \cdot x) d\mathcal{H}_{n-1} \Big].
\end{aligned}$$

By equating the two formulae for $\psi'(r)$, the identity (4) follows.

Let us return to the proof of Theorem 1.1. By choosing A equal to the unit ball B in $\mathbf{R}^n$ so that $x = r\nu_r(x)$ on ∂B_r, and by using the Pohozaev identity (4) and Lemma 2.1, we obtain that

$$\begin{aligned}
2\phi'(r) = -(n-1)r^{-n} &\int_{B_r} (|\nabla u|^2 + 2F(u))dx + \\
&+ r^{-n+1} \int_{\partial B_r} (|\nabla u|^2 + 2F(u)) d\mathcal{H}_{n-1} = \\
= r^{-n} &\int_{B_r} (2F(u) - |\nabla u|^2) dx + \\
&+ 2r^{-n+1} \int_{\partial B_r} \left(\frac{\partial u}{\partial \nu_r} \right)^2 d\mathcal{H}_{n-1} \geq 0
\end{aligned}$$

and the proof of Theorem 1.1 is complete.

Proof of Corollary 1.2. Assume $n > 1$ (the proof for $n = 1$ is elementary). Assume by contradiction, that $u(0) > 0$. By applying Lemma 1.1 as in [M02], we infer that $u(x) > 0$ for every $x \in \mathbf{R}^n$. Since $f'(0) = F''(0) > 0$, and $f \in C^2(\mathbf{R})$, we can apply Proposition 4.1 of [GNN2] to obtain

$$u(x) = 0(e^{-|x|}|x|^{(1-n)/2}) \quad \text{as} \quad |x| \to +\infty;$$

hence $u \in L^2(\mathbf{R}^n)$. We now observe that

$$F(u) = 0(u^2) \quad \text{as} \quad u \to 0$$

because $F \geq 0$, $F(0) = 0$, and $F''(0) > 0$, so that

$$\int_{\mathbf{R}^n} F(u(x))\, dx \ < +\infty.$$

Then Lemma 2.1 yields that

$$\int_{\mathbf{R}^n} (|\nabla u|^2 + 2F(u(x)))\, dx \ < \ +\infty$$

and

$$\lim_{r \to +\infty} \phi(r) = 0$$

where ϕ is the function defined in Theorem 1.1. Since

$$\lim_{r \to 0+} \phi(r) = 0 \quad ,$$

$\phi \geq 0$, and ϕ is non-decreasing by Theorem 1.1, we conclude that $\phi \equiv 0$ and Corollary 1.2 immediately follows.

References

[AB] N.D.Alikakos, P.W.Bates, *On the Singular Limit in a Phase Field Model of Phase Transitions*, Ann. Inst.H.Poincaré, Anal. non lin., to appear.

[AM] L.Ambrosio, personal communication.

[AP] F.V.Atkinson, L.A.Peletier, *Ground States of* $-\Delta u = f(u)$ *and the Emden-Fowler Equation*, Arch. Rat. Mech. Anal. **93** (1986), 103-127.

[BA] S.Baldo, *Minimal Interface Criterion for Phase Transitions in Mixtures of Cahn-Hilliard Fluids*, to appear on Ann. Inst. H. Poincaré, Anal. non lin.

[BL] H.Berestycki, P.L.Lions, *Nonlinear Scalar Field Equations I, Existence of a Ground State; II, Existence of Infinitely Many Solutions*, Arch. Rat. Mech. Anal. **82** (1983), 313-375.

[BLP] H.Berestycki, P.L.Lions, L.A. Peletier, *An ODE Approach to the Existence of Positive Solutions for Semilinear Problems in* $\mathbf{R}^n$, Indiana Univ. Math. J. **30** (1981), 141-167.

[CA] G.Caginalp, *An Analysis of a Phase Field Model of a Free Boundary*, Arch. Rat. Mech. Anal. **92** (1986), 205-245.

[CO] C.V.Coffman, *Uniqueness of the Ground State Solution for* $\Delta u - u + u^3 = 0$ *and a Variational Characterization of Other Solutions,* Arch. Rat. Mech. Anal. **46** (1972), 12-95.

[DG] E.De Giorgi, *Convergence Problems for Functionals and Operators*, Proc. Int.Meet. on Recent Methods in Nonlinear Analysis, Rome, 1978, ed. by E.De Giorgi et al.,Pitagora,Bologna, 1979, 131-188.

[DGF] E.De Giorgi, T.Franzoni, *Su un tipo di convergenza variazionale*, Atti Accad.Naz.Lincei, Rend. Cl. Sc. Mat. Fis. Natur., **58** (1975), 842-850.

[EL] M.J.Esteban, P.L.Lions, *Existence and Non-existence Results for Semilinear Elliptic Problems in Unbounded Domains,* Proc. Royal Soc. Edinburgh **93A** (1982), 1-14.

[GNN1] B.Gidas, W.M.Ni, L.Nirenberg.*Symmetry and Related Properties via the Maximum Principles*, Comm. Math. Phys. **68** (1979), 209-243.

[GNN2] B.Gidas, W. M. Ni, L.Nirenberg, *Symmetry of Positive Solutions of Nonlinear Elliptic Equations in* $\mathbf{R}^n$, Mathematical Analysis and Applications, Part A, Advances in Mathematics, Suppl.Studies **7A**, ed. by L.Nachbin, Academic Press, New York, 1981, 369-402.

[GU] M.E.Gurtin, *Some Results and Conjectures in the Gradient Theory of Phase Transitions,* Institute for Mathematics and Its Applications, Univ. of Minnesota, preprint n.156 (1985).

[HS] R.Haga, J.Serrin, *Dynamic Changes of Phase in a Van der Waals Fluid.* New Perspectives in Thermodynamics, ed. by J.Serrin, Springer-Verlag, Berlin, 1986, 241-260.

[LM] S.Luckhaus, L.Modica, *The Gibbs-Thompson Relation within the Gradient Theory of Phase Transitions,* to appear on Arch. Rational Mech. Anal.

[MLS] K.McLeod, J.Serrin, *Uniqueness of Positive Radial Solutions of* $\Delta u = f(u)$ *in* $\mathbf{R}^n$. Arch. Rat. Mech. Anal. **99** (1987), 115-145.

[MO1] L.Modica, Γ-*Convergence to Minimal Surface Equation and Global Solutions to* $\Delta u = 2(u^3 - u)$, Proc. Int. Meet. on Recent

Methods in Nonlinear Analysis, Rome, 1978, ed. by E.De Giorgi et al., Pitagora, Bologna, 1979, 223-243.

[MO2] L.Modica, *A Gradient Bound and a Liouville Theorem for Nonlinear Poisson Equations*, Comm. Pure Appl. Math. **38** (1985), 679-684.

[MO3] L.Modica, *The Gradient Theory of Phase Transitions and the Minimal Interface Criterion*, Arch. Rat. Mech. Anal.**98** (1987), 123-142.

[MO4] L.Modica, *Gradient Theory of Phase Transitions with Boundary Contact Energy*, Ann. Inst. H.Poincaré, Anal. non lin., **4** (1987), 487-512.

[MO5] L.Modica, *Level Sets of Entire Solutions of Semilinear Elliptic Equations*, to appear.

[MM1] L.Modica, S.Mortola, *Un esempio di Γ^--convergenza,* Boll. Un. Mat. Ital. **14-B** (1977), 285-299.

[MM2] L.Modica, S.Mortola, *Il limite nella Γ-convergenza di una famiglia di funzionali ellittici*, Boll. Un. Mat. **14-A** (1977), 285-299.

[MM3] L.Modica, S.Mortola, *Some Entire Solutions in the Plane of Nonlinear Poisson Equations*, Boll. Un. Mat. Ital. **17-B** (1980), 614-622.

[PS1] L.A. Peletier, J.Serrin, *Uniqueness of Positive Solutions of Semilinear Equations in* $\mathbf{R}^n$, Arch. Rat. Mech. Anal. **81** (1983), 181-197.

[PS2] L.A.Peletier, J.Serrin, *Uniqueness of Non-negative Solutions of Semilinear Equations in* $\mathbf{R}^n$, J.Diff. Eq. **6** (1986), 380-397.

[PO] S.I.Pohozaev, *Eigenfunctions of the Equation* $\Delta u + \lambda f(u) = 0$, Soviet Math. Dokl. **5** (1965), 1408-1411.

[ST] P.Sternberg, *The Effect of a Singular Perturbation on Nonconvex Variational Problems*, Arch. Rat. Mech. Anal. **101** (1988), 209-260.

[STR] W.A.Strauss, *Existence of Solitary Waves in Higher Dimensions,* Comm. Math. Phys. **55** (1977), 149-162.

Dipartimento di Matematica
Università di Pisa
Via F.Buonarroti 2
I-56127 PISA

PSEUDO-DIFFERENTIAL OPERATORS OF VOLTERRA TYPE ON SPACES OF ULTRA DISTRIBUTIONS AND PARABOLIC MIXED PROBLEMS

M.K.Venkatesha Murthy

Dedicated to Ennio De Giorgi on his sixtieth birthday

1. Introduction. The calculus of classical pseudo-differential operators has been used in a fundamental way in the study of boudary value problems associated to (systems) of elliptic differential and pseudo-differential operators. This calculus was used by Hörmander in [7] to construct parametrices for the elliptic operators, using which the boundary value problem is reduced to a system of pseudo-differential operators on the boundary. In order to treat the boundary value problems for parabolic operators in bounded cylindrical domains Piriou introduced in [15] and [16] a class of operators which called pseudo-differential operators of Volterra type. The basic idea consists in developping a calculus of an appropriate class of anisotropic pseudo-differential operators as in an earlier note of Hunt and Piriou [8].

It is classical that, to a given parabolic operator, there is a naturally associated Gevrey class. De Giorgi showed in [3] and [4] the necessity of working in an appropriate Gevrey class, for instance, for the uniqueness in the Cauchy problem associated to a parabolic differential operator with smooth coefficients. Regularity results and isomorphism theorems for parabolic operators in classes of ultra-differentiable functions and in the corresponding spaces of ultra-distributions have been proved by Friedman in [6] and by Lions and Magenes in [9] and [10].

Boutet de Monvel and Krée introduced in [2] a class of pseudo-differential operators acting on spaces of functions and hyper-distributions of Gevrey type, which was extended by the author in [13] to an algebra of anisotropic pseudo-differential operators which act on spaces of ultra-distributions.

The object of this paper is to develop a class of pseudo-differential operators of Volterra type which act on suitable spaces of ultra-differentiable functions and ultra-distributions. We make use of the techniques of Piriou and those of Boutet de Monvel and Krée for this purpose. This calculus is used, in particular, to construct parametrices for parabolic systems in the sense of Petrowski (acting between sections of vector bundles). As a consequence, we deduce the existence of ultra-distributions solutions for right parabolic systems, and ultra-differentiable hypoellipticity for left parabolic systems on bounded cylindrical domains.

The existence of parametrices for parabolic systems (in the sense of Petrowski) as pseudo-differential operators of Volterra type together with the techniques of Hörmander [7] are used to reduce the study of mixed problems associated to Petrowski parabolic differential systems with coefficients in appropriate ultra-differentiable function spaces in bounded cylindrical domains to a system of pseudo-differential operators of Volterra type on the lateral boundary of the domain.

We shall use the following standard notation. Let U be an open set in $\mathbf{R}^{n+1}$, $X = (x_0, x) = (x_0, x_1, \ldots, x_n)$ a coordinate system at a point in U and $\Xi = (\xi_0, \xi) = (\xi_0, \xi_1, \ldots, \xi_n)$ the corresponding cotangential coordinates. We also use the standard notation x' to denote $(x_1, \ldots, x_{n-1})$ and write $x = (x', x_n)$, and similarly we write $\xi = (\xi', \xi_n)$ with $\xi' = (\xi_1, \ldots, \xi_{n-1})$. We set $\partial_j = \partial_{x_j} = \partial/\partial_{x_j}$, $D_j = D_{x_j} = -i\partial_j$ and similarly we use the notation ∂_{ξ_j} and D_{ξ_j}. If a_0 is an even integer we define a function $\rho : \mathbf{C}_{\zeta_0} \times \mathbf{C}^n_{\zeta} \to \mathbf{R}_+$ by

$$\rho(\zeta_0, \zeta) = \left[|\zeta_0|^{(2/a_0)} + |\zeta|^2 \right]^{1/2}, \quad \rho(0,0) = 0$$

and the parabolic distance function on $\mathbf{R}^{n+1}$ associated to the weight $a = (a_0, 1, \ldots, 1) \in \mathbf{N}^{n+1}$ by

$$\rho(\Xi) = \rho(i\xi_0, \xi).$$

Then the function ρ is continuous and C^∞ in $\mathbf{R}^{n+1}$. We have the associated polar type coordinates given by

$$\xi_0 = \rho^{a_0} \cos\omega_1 \ldots \cos\omega_n,\ \xi_1 = \rho\cos\omega_1 \ldots \sin\omega_n, \ldots, \xi_{n+1} = \rho \sin\omega_1.$$

For any multi-index $\alpha = (\alpha_0, \alpha_1, \ldots, \alpha_n) \in \mathbf{N}^{n+1}$ we define

$$(a, \alpha) = a_0\alpha_0 + \alpha_1 + \ldots + \alpha_n = a_0\alpha_0 + |\alpha|.$$

2. Anisotropic pseudo-differential operators. In this section we shall review the basic definitions and properties of anisotropic pseudo-differential operators of weight $a = (a_0, 1, \ldots, 1)$ introduced by Piriou in [15] and by the author in [13].

We shall say that a function $p \in C^\infty(U \times \mathbf{R}^{n+1})$ is *quasi-homogeneous* of degree m with respect to the weight $a = (a_0, 1, \ldots, 1)$ if

$$p\,(X, \lambda^{a_0}(i\xi_0), \lambda\xi) = \lambda^m p(X, i\xi_0, \xi) \qquad \text{for } \lambda > 0$$

and

$$(X, \Xi) \in U \times \mathbf{R}^{n+1} \simeq T^*U.$$

Symbol class $S^{m,a}(U)$. If m is a real number, we shall denote by $S^{m,a}(U)$ the set of all functions $p = p(X; i\xi_0, \xi) \in C^\infty(U \times \mathbf{R}^{n+1})$ such that $\exists p_j \in C^\infty(U \times \mathbf{R}^{n+1})$ with the following properties:

(a) p_j is quasi-homogeneous of degree $m - j$ in Ξ and

(b) $p \sim \Sigma p_j$ in the sense that, for any compact subset K of U, for any $\alpha, \beta \in \mathbf{N}^{n+1}$ and any integer $N \geq 1$, there exists a constant $c = c(K, \alpha, \beta, N) > 0$ such that

$$\left| D_\Xi^\alpha D_X^\beta \left[p(X; i\xi_0, \xi) - \sum_{j<N} p_j(X; i\xi_0, \xi) \right] \right| \leq \tag{2.1}$$
$$\leq c\,(1 + \rho(\Xi))^{m-N-(a,\alpha)}$$

in $K \times \mathbf{R}^{n+1}$.

Clearly we have $S^{m,a}(U) \subset S^{m,a}_{1,0}(U)$ (see [13]).

For $p \in S^{m,a}(U)$ we define the linear operator $p(X; \partial_{x_0}, D_x)$ on $\mathcal{D}(U)$ by

$$\begin{aligned} p(X; \partial_{x_0}, D_x)\phi(x) &= \\ &= (2\pi)^{-(n+1)} \int p(X; i\xi_0, \xi)\hat{\phi}(\Xi) \exp(i < X, \Xi >)\, d\Xi. \end{aligned} \tag{2.2}$$

Basic Properties

1) Continuity. $p(X; \partial_{x_0}, D_x)$ is a continuous linear operator from $\mathcal{D}(U)$ into $\mathcal{E}(U)$ and extends to a continuous linear mapping from $\mathcal{E}'(U)$ into $\mathcal{D}'(U)$ and has a very regular distribution kernel $\mathcal{K}_p \in \mathcal{D}'(U \times U)$ which is C^∞ outside the diagonal in $U \times U$ and C^r in $U \times U$ if $m/a_0 + (1 + n/a_0) + r < 0$, and sing.supp. $p(X; \partial_{x_0}, D_x)u \subseteq$ sing.supp. u for $u \in \mathcal{D}'(U)$.

2) Existence of asymptotic sums. If $p_j \in S^{m_j,a}(U)$ where m_j is a decreasing sequence of real numbers tending to $-\infty$ then there exists a $p \in S^{m_0,a}(U)$ satisfying

$$p - \sum_{j<N} p_j \in S^{m_N,a}(U) \tag{2.3}$$

and such a p is unique modulo $S^{-\infty,a}(U) = \bigcap_{r \in \mathbf{R}} S^{r,a}(U)$.

Anisotropic pseudo-differential operators. We now define the class of anisotropic pseudo-differential operators $\mathcal{L}^{m,a}(U)$ to be the set of all linear mappings $P : \mathcal{D}(U) \to \mathcal{E}(U)$ such that for any $f \in \mathcal{D}(U)$ there exists a $p_f \in S^{m,a}(U)$ such that

$$P(f\phi) = p_f(X; \partial_{x_0}, D_x)\phi, \quad \text{for all} \quad \phi \in \mathcal{D}(U). \tag{2.4}$$

We also have $\mathcal{L}^{m,a}(U) \subset \mathcal{L}^{m,a}_{1,0}$ (see [13]).

Clearly an anisotropic pseudo-differential operator has the regular properties as described in (1) above for $p_f(X; \partial_{x_0}, D_x)$ and has a symbol $p(X, \Xi)$ defined uniquely modulo $S^{-\infty,a}(U)$.

An anisotropic pseudo-differential operator P is said to be *properly supported* if the restrictions of the two canonical projections $U \times U \to U$ to the support of the distribution kernel $\mathcal{K}_P$ are proper maps.

We have the following standard calculus for the class of anisotropic pseudo-differential operators:

Composition. Let $P \in \mathcal{L}^{m,a}(U)$ and $Q \in \mathcal{L}^{\ell,a}(U)$ such that at least one of them is properly supported and let p and q respectively be their symbols, then $Q \circ P \in \mathcal{L}^{m+\ell,a}(U)$ and the symbol $q \circ p$ of $Q \circ P$ is given by the asymptotic sum

$$q \circ p \sim \Sigma(\alpha!)^{-1}(\partial_\Xi^\alpha\, p)(X,\Xi)(D_X^\alpha\, q)(X,\Xi). \tag{2.5}$$

Transposition. If $p \in \mathcal{L}^{m,a}(U)$ with symbol p then there exists a unique operator ${}^tP \in \mathcal{L}^{m,a}(U)$ such that $< Pu, v > = < u, {}^tPv >$ and its symbol ${}^tp$ is given by

$${}^tp(X,\Xi) \sim \Sigma(\alpha!)^{-1}(-D_X)^\alpha(\partial_\Xi)^\alpha\, p(X,-\Xi). \tag{2.6}$$

For $P \in \mathcal{L}^{m,a}(U)$, with symbol p, the L^2-adjoint P^* is also an operator in $\mathcal{L}^{m,a}(U)$ and its symbol p^* is given by

$$p^*(X,\Xi) \sim \Sigma(\alpha!)^{-1}(-D_X)^\alpha(\partial_\Xi)^\alpha\, \bar{p}(X,-\Xi). \tag{2.6$'$}$$

Globalization. As mentioned in the Introduction, the classical method of reducing a boundary value problem to a problem on the boundary necessarily leads to the introduction of systems of pseudo-differential operators on manifolds (namely, the boundary).

All the previous definitions and properties easily extend to systems in an obvious manner: if E and F are two finite dimensional vector spaces, the symbols of class $S^{m,a}(U; E, F)$ are defined using a norm in the space $L(E,F)$ in the estimates and the corresponding class of anisotropic pseudo-differential operators $P : \mathcal{D}(U,E) \to \mathcal{E}(U,F)$ are defined. Then the distribution kernel $\mathcal{K}_P$ belongs to $\mathcal{D}'(U \times U, L(E,F))$. The results on the composition and the adjoint can be appropriately modified using composition and transposition of linear maps respectively. Obviously, the operation of composition should be carried out taking into consideration the left and right compositions of the linear symbol-mappings p and q.

Change of variables. Suppose θ_0 is a diffeomorphism of an open interval $I_{x_0} \subset \mathbf{R}$ onto an interval $\tilde{I}$ and θ is a diffeomorphism of an open subset $V \subset \mathbf{R}^n_x$ onto $\widetilde{V}$ and $\Theta = (\theta_0, \theta)$ is the diffeomorphism of $U = I \times V$ onto $\widetilde{U} = \tilde{I} \times \widetilde{V}$.

If $P \in \mathcal{L}^{m,a}(U)$ with symbol p and if we define

$$P_\Theta(\phi) = P(\phi \circ \Theta) \circ \Theta^{-1} \quad \text{for} \quad \phi \in \mathcal{D}(U) \tag{2.7}$$

then $P_\Theta \in \mathcal{L}^{m,a}(\widetilde{U})$ and its symbol p_Θ is given by

$$p_\Theta(\Theta(X), \Xi) \sim \Sigma(\alpha!)^{-1}(\partial_\Xi)^\alpha \, p(X, {}^tJ_\Theta(X)\Xi) \cdot \\ \cdot \big[D_Y^\alpha \exp\big(i < \Theta(Y) - \Theta(X) - < d\Theta(X), Y - X >, \Xi > \big)\big]\Big|_{Y=X} \tag{2.8}$$

where J_Θ denotes the Jacobian matrix of the diffeomorphism Θ.

A completely analogous result holds also for diffeomorphisms of the form $\Theta(x_0, x) = \big(x_0, \theta(x_0, x)\big)$ and we shall omit the details.

Extension to sections of vector bundles on manifolds. Making use of this invariance property together with a locally finite partition of unity subordinated to an open covering by means of local coordinate patches of the above type, it is fairly standard to globalize the above considerations to define anisotropic pseudo-differential operators on cylindrical domains on manifolds.

Further, the globalization argument together with local trivialization frames for vector bundles and the remark on systems can be used to define anisotropic pseudo-differential operators acting between sections of vector bundles on cylindrical domains on manifolds.

Let $\mathcal{N}$ be a C^∞ manifold of dimension n and I be an interval. Consider two vector bundles E and F on $\mathcal{M} = I \times \mathcal{N}$.

We shall denote by $\mathcal{L}^{m,a}(\mathcal{M}; E, F)$ the class of linear operators $P : \mathcal{D}(\mathcal{M}, E) \to \mathcal{E}(\mathcal{M}, F)$ such that for every local coordinate chart (U, Θ) of $\mathcal{M}$, of the form $\Theta = (\theta_0, \theta)$ where θ_0 and θ are respectively local coordinate maps on I and $\mathcal{N}$, on which both E and F are trivial the restriction $P|_U$ is given by a matrix (with respect to the local frames of E and F) of anisotropic pseudo-differential operators $P_U = (P_{jk}) \in \mathcal{L}^{m,a}(U; \mathbf{C}^{\dim E}, \mathbf{C}^{\dim F})$.

3. Pseudo-differential operators of Volterra type

Definition 3.1 (Piriou). *An anisotropic pseudo-differential operator P is said to satisfy the* Volterra property with respect to the variable x_0 *if, when $\phi \in \mathcal{D}(U)$ vanishes in $x_0 < t$ for some $t \in \mathbf{R}$, then $P\phi$ also vanishes in $x_0 < t$.*

Spaces of operators $\mathcal{V}_{\pm}^{m,a}(U)$. We shall denote by $\mathcal{V}_{+}^{m,a}(U)$ the subclass of all anisotropic pseudo-differential operators $P \in \mathcal{L}^{m,a}(U)$ having the Volterra property with respect to the variable x_0 and by $\mathcal{V}_{-}^{m,a}(U)$ the subclass of those satisfying the condition of the above definition with $x_0 > t$ in place of $x_0 < t$.

$\mathcal{V}_{\pm,\mathrm{prop}}^{m,a}(U)$ stands for the subclass of properly supported operators in $\mathcal{V}_{\pm}^{m,a}(U)$.

The class of pseudo-differential operators of Volterra type have properties and a symbolic calculus, as a consequence of the corresponding properties of operators in $\mathcal{L}^{m,a}(U)$. We recall only the following basic properties:

Properties of operators in $\mathcal{V}_{+}^{m,a}(U)$

1) If $P \in \mathcal{V}_{\pm}^{m,a}(U)$ and $Q \in \mathcal{V}_{\pm}^{\ell,a}(U)$ and if, at least one of them is properly supported, then $Q \circ P \in \mathcal{V}_{\pm}^{m+\ell,a}(U)$.

2) The distribution kernel $\mathcal{K}_P$ of an operator $P \in \mathcal{V}_{\pm}^{m,a}(U)$ has the regularity properties as those of operators in $\mathcal{L}^{m,a}(U)$ and moreover $\mathcal{K}_P$ vanishes in the subset $\{(X,Y) \in U \times U;\ x_0 < y_0\}$.

3) If $\Theta = (\theta_0, \theta) : U = I \times V \to \widetilde{U} = \tilde{I} \times \widetilde{V}$ is a diffeomorphism of the kind considered before with the additional condition that if θ_0 is monotone increasing then $P \to P_\Theta$ maps $\mathcal{V}_{+}^{m,a}(U)$ bijectively onto $\mathcal{V}_{+}^{m,a}(\widetilde{U})$, while if θ_0 is monotone decreasing then $P \to P_\Theta$ maps $\mathcal{V}_{+}^{m,a}(U)$ bijectively onto $\mathcal{V}_{-}^{m,a}(\widetilde{U})$.

Symbols of operators in $\mathcal{V}_+^{m,a}(U)$: the space $\Sigma^{k,a}(U)$. Let $\Sigma^{k,a}(U)$ denote the space of all $p \in C^\infty(U \times \mathbf{R}^{n+1})$ such that

i) p is quasi-homogeneous of degree k in Ξ of weight a

ii) p has a continuous extension $p(X, \xi_0 + i\eta_0, \xi)$ to

$$\Omega = \{(\zeta_0, \xi) \in \mathbf{C} \times \mathbf{R}^n : Re\,\zeta_0 \geq 0 \text{ and } \rho(\zeta_0, \xi) > 0\} \tag{3.1}$$

which is holomorphic in ζ_0 in the subset $\{\zeta_0 \in \mathbf{C} : \ (\zeta_0, \xi) \in \Omega, Re\,\zeta_0 > 0\}$ and C^∞ with respect to (X, ξ) and all the derivatives $D_\xi^\alpha D_X^\beta\, p((X, \xi_0 + i\eta_0, \xi)$ are continuous in Ω, for $\alpha \in \mathbf{N}^n$ and $\beta \in \mathbf{N}^{n+1}$.

We set $\Sigma^{-\infty,a}(U) = \cap_k \Sigma^{k,a}(U)$.

Now we have the following

Theorem 3.2 (Piriou). *Suppose $p \sim \sum p_j$ is the symbol of an operator $P \in \mathcal{L}^{m,a}(U)$. Then $P \in \mathcal{V}_+^{m,a}(U)$ if and only if $p_j \in \Sigma^{m-j,a}(U)$, for all $j \geq 0$.*

For a proof we refer to the paper [15] of Piriou.

Definition 3.3. *If $P \in \mathcal{V}_+^{m,a}(U)$ then $p_0 \in \Sigma^{m,a}(U)$ (which is defined uniquely modulo $\Sigma^{-\infty,a}(U)$) is called the principal symbol of P.*

The above definitions of symbols and the Theorem 3.2 can be modified in an obvious manner to the operators in $\mathcal{V}_-^{m,a}(U)$.

Remark. The class of pseudo-differential operators of Volterra type acting between sections of vector bundles on manifolds (and on cylindrical domains on manifolds) of the type considered before are defined in the obvious way using local trivializing frames on local coordinate charts used earlier. We also write $\mathcal{V}_\pm^{m,a}(\mathcal{M}; E, F)$ and $\mathcal{V}_{\pm,\mathrm{prop}}^{m,a}(\mathcal{M}; E, F)$ to denote the class of pseudo-differential operators of Volterra type acting between sections of vector bundles E and F on the manifold $\mathcal{M}$.

4. Parabolic pseudo-differential operators .

Definition 4.1. *A pseudo-differential operator of Volterra type $P \in \mathcal{V}_{+}^{m,a}(\mathcal{M}; E, F)$ is said to be* right parabolic *(resp.* left parabolic*) of weight $a = (a_0, 1, \ldots, 1)$ if, for any $X \in \mathcal{M}$ there exists a local coordinate chart (U, Θ) of the type $\Theta = (\theta_0, \theta)$ and trivializations of E and F on U such that if $p_0 \in \Sigma^{m,a}(U; \mathbf{C}^{\dim E}, \mathbf{C}^{\dim F})$ is the principal symbol of $P|_U$ then there exists a $q_0 \in \Sigma^{-m,a}(U; \mathbf{C}^{\dim F}, \mathbf{C}^{\dim E})$ such that*

$$p_0(X, \zeta_0, \xi)\, q_0(X, \zeta_0, \xi) = Id_{F|_U} \quad in\ U \times \Omega \tag{4.1}$$

$$(resp. \quad q_0(X, \zeta_0, \xi)\, p_0(X, \zeta_0 \xi) = Id_{E|_U} \quad in\ U \times \Omega). \tag{4.1'}$$

Reduction to the local situation and making use of the composition formula of the symbolic calculus leads to the following existence result:

Theorem 4.2 (Piriou: Existence of parametrices). *If $P \in \mathcal{V}_{+}^{m,a}(\mathcal{M}; E, F)$ is right (resp. left) parabolic of weight $a = (a_0, 1, \ldots, 1)$ then there exists a right (resp. left) parametrix $Q \in \mathcal{V}_{+,\mathrm{prop}}^{-m,a}(\mathcal{M}; F, E)$ in the sense that*

$$PQ - Id \in \mathcal{V}_{+}^{-\infty,a}(\mathcal{M}; F, F) \tag{4.2}$$

$$(\text{resp.} \quad QP - Id \in \mathcal{V}_{+}^{-\infty,a}(\mathcal{M}; E, E)\ ). \tag{4.2'}$$

A class of pseudo-differential operators of parabolic type is provided by the class of a_0- parabolic differential operators introduced by Petrowski, whose definition we recall below.

We shall denote by $\mathcal{D}\mathrm{iff}(\mathcal{M}; E, F)$ the space of differential operators from sections of the vector bundle E into the sections of the vector bundle F.

If P is a differential operator of order m acting between sections of vector bundles E and F of the same rank h on a manifold $\mathcal{M}$ we identify P, with respect to a local coordinate chart (U, Θ) on which both E and F are trivialized, with a square system

$$(4.3) \qquad P = (P_{jk}) \text{ with } P_{jk} = P_{jk}(X; \partial_{x_0}, D_x), \quad 1 \le j, k \le h$$

of differential operators of order m with respect to the weight $a = (a_0, 1, \ldots, 1)$ in the sense that its symbol matrix is of the form

$$(4.4) \quad p = (p_{jk}) \text{ where } p_{jk} = p_{jk}(X, \Xi) = \sum_{(a,\alpha) \le m} p_{jk,\alpha}(X)\ \Xi^\alpha.$$

Definition. *P is said to be a_0-*parabolic *with respect to x_0 in the sense of Petrowski if, denoting by*

$$(4.5) \qquad p^0_{jk}(X, \Xi) = \sum_{(a,\alpha) = m} p_{jk,\alpha}(X)\ \Xi^\alpha$$

the principal part symbol matrix, then

$$(4.6) \qquad \det p^0_{jk}(X, \xi_0 + i\eta_0, \xi) \ne 0$$

for $\xi_0 \ge 0$, $(\eta_0, \xi) \in \mathbf{R}^{n+1}$ *and* $\Xi = (\xi_0, \xi) \in \mathbf{R}^{n+1} \backslash 0$.

That is, the zeros of the polynomial

$$(4.7) \qquad \zeta_0 \mapsto \det p^0_{jk}(X, \zeta_0, \xi), \quad \text{for} \quad \xi \in \mathbf{R}^n$$

satisfy

$$(4.8) \quad Re\, \eta_0 < 0, \quad (Im\, \zeta_0, \xi) \in \mathbf{R}^{n+1} \text{ and } (Re\, \zeta_0, \xi) \in \mathbf{R}^{n+1}.$$

Remark. The parabolic weight a_0 is necessarily an even integer and m is a multiple of a_0.

It is easily checked that a_0-parabolic differential operators with respect to x_0 in the sense of Petrowski are parabolic pseudo-differential operators belonging to the class $\mathcal{V}^{m,a}_{+,\text{prop}}(\mathcal{M}; E, F)$.

In particular, the heat operator $P = \partial_{x_0} - \Delta_x$ is 2-parabolic with respect to x_0 in the sense of Petrowski and belongs to $\mathcal{V}^{2,2}_{+}(I \times \mathbf{R}^n)$.

5. Parabolic mixed problems. Suppose $\mathcal{N}$ is C^∞ manifold of dimension n and $\mathcal{M} = I \times \mathcal{N}$. Let $\mathcal{U}$ be an open subset of $\mathcal{M}$ with a C^∞ boundary (of dimension n) in $\mathcal{M}$, which is nowhere tangential to $x_0 =$ const. That is, for any $X \in \partial\mathcal{U}$ the tangent space $T_x(\partial\mathcal{U}) \neq T_x\mathcal{N}$. Suppose that ν is a C^∞-vector field on $\mathcal{M}$ defined in a neighbourhood of $\partial\mathcal{U}$ transversal to $\partial\mathcal{U}$ and directed towards the interior of $\mathcal{U}$ such that $\nu(X) \in T_X\mathcal{N}$.

This can be described locally as follows: if (U, Θ) is a local coordinate chart for $\mathcal{M}$ at a point on $\partial\mathcal{U}$, namely,

$$\Theta : U \to \mathbf{R}^{n+1}, \quad \Theta(x_0, x) = \big(x_0, \theta(x_0, x)\big)$$

such that

$$\Theta(U \cap \mathcal{U}) \subset \{Y = (y_0, y', y_n);\ y_n > 0\},$$

$$\Theta(U \cap \partial\mathcal{U}) \subset \{Y = (y_0, y', y_n);\ y_n = 0\},$$

then ν is mapped onto the vector field $\partial/\partial y_n$.

For example, ν is the interior normal vector field with respect to a Riemannian metric.

We recall that a distribution section $u \in \mathcal{D}'(\mathcal{M}, E)$ is said to have *transversal trace of order* k $(\in \mathbf{N})$ on $\partial\mathcal{U}$ if locally, by means of local coordinate charts of the form (U, Θ) as above, u is mapped onto $u_\Theta = u \circ \Theta^{-1} \in C^k\big((\overline{\mathbf{R}}_{y_n})_+;\ \mathcal{D}'(\mathbf{R}^n_{(x_0,x')}),\ \mathbf{C}^{\dim E}\big)$.

We then define the *traces* $\gamma_j u \in \mathcal{D}'(\partial\mathcal{U}, E)$ locally in (U, Θ) by

$$(\gamma_j u) \circ \Theta^{-1} = (D_{y_n})^j \big(u \circ \Theta^{-1}\big)\big|_{y_n = 0}, \quad \text{for } 0 \leq j \leq k. \tag{5.1}$$

Now let $P \in \mathcal{D}\mathrm{iff}(\mathcal{M}; E, F)$ be a C^∞ differential operator which we assume is a_0-parabolic with respect to x_0 and of order m in the sense of Petrowski.

Consider vector bundles $G_1, \ldots, G_J$ on $\partial\mathcal{U}$ and pseudo-differential boundary operators B_{jr} ($j = 1, \ldots, J$ and $0 \leq r \leq m-1$). More precisely, we consider

$$B_{jr} \in \mathcal{V}^{m_j - r}(\partial\mathcal{U}; E|_{\partial\mathcal{U}}, G_j),\ B = (B_{jr}). \tag{5.2}$$

The spaces of distributions $\mathcal{D}'_+(\mathcal{M}, E)$. We denote by $\mathcal{D}'_+(\mathcal{M}, E)$ the space of distributions $u \in \mathcal{D}'(\mathcal{M}, E)$ with

$$\operatorname{supp} u \subset \{X = (x_0, x);\ x_0 \geq 0\} \tag{5.3}$$

i.e. $\mathcal{D}'_+(\mathcal{M}, E)$is the space of distributions vanishing in $x_0 < 0$.

Finally, for $u \in \mathcal{D}'_+(\mathcal{M}, E)$ having transversal trace of order $m-1$ we define

$$\gamma u = (\gamma_0 u, \gamma_1 u, \dots, \gamma_{m-1} u) \tag{5.4}$$

and

$$B(\gamma u) \in \mathcal{D}'_+(\partial \mathcal{U}, \oplus G_j)$$

by

$$[B(\gamma u)]_j = \sum_{0 \le r \le m-1} B_{jr}(\gamma u) \quad \text{for} \quad 1 \le j \le J. \tag{5.5}$$

We are now in a position to formulate the parabolic mixed problem. Suppose the differential operator P and the boundary pseudo-differential operators $B = (B_{jr})$ satisfy the above assumptions.

Let $\mathcal{U}_t$ denote the set $\{X = (x_0, x) \in \mathcal{U};\ x_0 < t\}$ and $(\partial\mathcal{U})_t$ denote the set $\{X = (x_0, x) \in \partial\mathcal{U};\ x_0 < t\}$.

Suppose given $f \in \mathcal{D}'_+(\mathcal{U}_t, F|\mathcal{U}_t)$ and $g \in \mathcal{D}'_+\left((\partial\mathcal{U})_t, \oplus G_j|(\partial\mathcal{U})_t\right)$, that is such that, $f = 0$ and $g = 0$ in $x_0 < 0$.

The *parabolic mixed boundary value problem* consists in finding a distribution section $u \in \mathcal{D}'_+(\mathcal{U}_t, E|\mathcal{U}_t)$ such that

$$Pu = f \ \text{ in } \mathcal{U}_t \quad \text{and} \quad B(\gamma u) = g \text{ on } (\partial\mathcal{U})_t. \tag{5.6}$$

6. Spaces of ultra-differentiable functions and ultra-distributions of class $[M]$. We consider a sequence $[M]$ of positive numbers M_k such that $M_0 = 1,\ k \le M_k$ for all integers $k \ge 0$ and satisfying the following assumptions:

1) M_k is increasing;
2) M_k is a logarithmically convex sequence in the sense that

$$M_k^2 \le M_{k-1} M_{k+1} \quad \text{for all} \quad k \ge 1;$$

3) there is a positive constant c_0 such that

$$\binom{k}{k'} M_{k'} M_{k-k'} \le c_0 M_k \quad \text{for all} \quad 0 \le k' \le k,\ k \in \mathbf{N};$$

4) there are two positive constants c_1, h_1 such that

$$M_{k+1} \leq c_1 h_1^k M_k \quad \text{for all} \quad k \geq 1;$$

5) there are two positive constants c_2, h_2 such that

$$M_{k+k'} \leq c_2 h_2^{k+k'} M_k M_{k'} \quad \text{for all} \quad k, k' \geq 1;$$

6) the sequence defines a non quasi-analytic class

$$\sum_{k=0}^{\infty} M_k / M_{k+1} < \infty.$$

Let U be an open set in $\mathbf{R}^n$. We shall denote by $\mathcal{E}(U, M_{k-r})$, for any $r \geq 0$, the space of all functions $f \in C^\infty(U)$ such that for any compact set K in U there exist constants $C = C(K, f) > 0$ and $h = h(K, f) > 0$ so that

$$\sum_{|\alpha| \leq k} \sup_{x \in K} |D_X^\alpha f(X)| \leq C h^k M_{k-r} \quad \text{for all} \quad k \in \mathbf{N}, \tag{6.1}$$

and by $\mathcal{D}(U, M_{k-r}) = \mathcal{D}(U) \cap \mathcal{E}(U, M_{k-r})$ the subspace of compactly supported functions. For any compact set K in U, $\mathcal{E}(K, M_{k-r})$ denotes the space of all C^∞ functions on K which belong to some $\mathcal{E}(V, M_{k-r})$, where V is a neighbourhood of K. The spaces $\mathcal{E}(U, M_{k-r})$, $\mathcal{D}(U, M_{k-r})$ and $\mathcal{E}(K, M_{k-r})$ are provided with the structure of locally convex topological vector spaces in a natural way (see for instance, the book of Lions and Magenes [10]).

Remark. It is well known, by a classical result of Mandelbrojt, that, under the assumption (5), $\mathcal{E}(U, M_{k-r})$ is non quasi-analytic in the sense that $\mathcal{D}(U, M_{k-r})$ is not trivial (that is, there exist sufficiently many non zero functions with compact support in $\mathcal{E}(U, M_{k-r})$). From now onwards we always assume the non quasi-analyticity condition (5).

$\mathcal{D}'(U, M_{k-r}), \mathcal{E}'(U, M_{k-r})$ and $\mathcal{E}'(K, M_{k-r})$ denote the dual spaces of $\mathcal{D}(U, M_{k-r})$, $\mathcal{E}(U, M_{k-r})$ and $\mathcal{E}(K, M_{k-r})$ respectively, and are provided with their strong dual topologies. The functionals in these spaces are respectively called *ultradistributions* of class M_{k-r} on U, *ultradistributions with compact support* of class M_{k-r} on U, *ultradistributions* of class M_{k-r} on K.

If K is a compact subset of U we shall denote by $\mathcal{E}'(U, M_{k-r}) \cap \mathcal{E}(K, M_{k-r})$ the space of ultra-distributions in $\mathcal{E}'(U, M_{k-r})$ whose restrictions to K belong to $\mathcal{E}(K, M_{k-r})$, which can again be provided with a locally convex topology and its strong dual is denoted by $\mathcal{D}'(U, M_{k-r}) \cap \mathcal{E}(U \setminus K, M_{k-r})$.

In order that the space of ultra-differentiable functions $\mathcal{E}(U, M_{k-r})$ contains the space of all real analytic functions we further assume the condition (A) (see Matsumoto [11]):

$$\text{(A)} \qquad \liminf_{k \to +\infty} (M_k/k!)^{1/k} > 0.$$

We have the following properties which are easily checked:

1) $\mathcal{E}(U, M_{k-r})$ and $\mathcal{D}(U, M_{k-r})$ are $\mathbf{C}$-algebras and $\mathcal{E}(U, M_{k-r})$ is a $\mathcal{D}(U, M_{k-r})$-module.

2) $\mathcal{E}(U, M_{k-r})$ and $\mathcal{D}(U, M_{k-r})$ are closed under derivation and composition by analytic mappings.

Proposition (Matsumoto).

(i) The space $\mathcal{E}(U, M_{k-r})$ is closed under division by non vanishing elements if there exists a constant $c \geq 1$ such that

$$(M_j/j!)^{1/j} \leq c\,(M_k/k!)^{1/k} \quad \textit{for all} \quad 1 \leq j \leq k,\ k \textit{ large}.$$

(ii) If there is a constant $c \geq 1$ such that

$$(M_j/j!)^{1/(j-1)} \leq c\,(M_k/k!)^{1/(k-1)} \quad \textit{for all} \quad 1 \leq j \leq k,\ k \textit{ large},$$

then the spaces $\mathcal{E}(U, M_{k-r})$ and $\mathcal{D}(U, M_{k-r})$ are closed under composition. Further, the implicite function theorem, existence and uniqueness theorems for ordinary differential equations hold.

Some classical examples of sequences M_k.

1) $M_k = (k!)^s$ with $s \geq 1$; we obtain the classical Gevrey spaces and Gevrey distributions of order s if $s > 1$; for $s = 1$ we obtain the class of all real analytic functions, which is well known to be

quasi-analytic, in the sense that there are no compactly supported functions and the "dual space" is the space of all real analytic functionals.

2) $M_k = (k!)^\nu \exp(be^{ak})$ with $\nu \in \mathbf{R}$ and a, b are positive real numbers.

We remark that there are spaces of ultra-differentiable functions larger than any Gevrey space. For an exhaustive discussion of these properties we refer to the paper of Matsumoto [11].

In view of these properties all the above notions can easily be extended to product spaces; the non quasi-analyticity assumption togheter with the appropriate assumption on the composition enable us to extend them, by localization and patching up using partitions of unity of class $[M]$, also to ultra-differentiable functions and ultra-distributions of class $[M]$ on real analytic manifolds, and under the assumption of (iv) in the above proposition, to manifolds of class $[M]$ and to sections of vector bundles with local trivializations of class $[M]$.

Finally, our considerations easily extend to function spaces and their dual spaces defined on cylindrical sets $U = I \times V$ in $\mathbf{R}^{n+1}$ where I is an interval in $\mathbf{R}$ and V is an open subset of $\mathbf{R}^n$ (and hence to manifolds $\mathcal{M} = I \times \mathcal{N}$). We shall mention only the following.

We can also introduce a somewhat more general class of functions and ultra-distributions better adapted to product sets as follows: suppose M_k and N_k are two sequences of positive numbers satisfying the previous assumptions and $U = I \times V$.

We shall denote by $\mathcal{E}(U, N_{k-s}, M_{k-r})$, for any $r \geq 0$ and $s \geq 0$ the space of all functions $f \in C^\infty(U)$ such that for any compact interval $J \subset I$ and any compact set K in V there exist constants $C > 0$ and $h > 0$ so that

$$\sum_{|\alpha| \leq k} \sup_{J \times K} |D^j_{x_0} D^\alpha_x f(X)| \leq C h^{j+k} N_{j-s} M_{k-r} \tag{6.2}$$

for all $j, k \in \mathbf{N}$.

In our proofs the following fundamental result of Paley-Wiener type, due to C. Roumieu, on Fourier transforms on ultra-differentiable functions in $\mathcal{D}(\mathbf{R}^n)$ and ultra-distributions in $\mathcal{E}'(\mathbf{R}^n)$, play an important role: given a sequence M_k satisfying the assumptions

(1)-(6) we define

$$M(\xi) = \sup_{\alpha \in \mathbf{N}^n} \left[\sum \alpha_j \log |\xi_j| - \log M_0^{-1} M_\alpha \right], \tag{6.3}$$

for $\xi \in \mathbf{R}^n$ with $\xi_j \neq 0$.

Theorem 6.1 (Roumieu). *Suppose the sequence M_k defines a non quasi-analytic class. Then we have the following.*

(a) An ultra-distribution $f \in \mathcal{D}'(\mathbf{R}^n, M_k)$ is the Fourier transform of a compactly supported function φ belonging to $\mathcal{D}(\mathbf{R}^n, M_k)$ if and only if f extends to an entire function of exponential type on $\mathbf{C}^n$ and for real $\xi \in \mathbf{R}^n$ satisfies an estimate of the form:

$$|f(\xi)| \leq B \exp[-M(h^{-1}\xi)] \quad \text{for all} \quad \xi \in \mathbf{R}^n \quad \text{with} \quad \xi_j \neq 0 \tag{6.4}$$

with a constant $B > 0$, where h is the constant in the definition of the class $\mathcal{D}(\mathbf{R}^n, M_k)$ and corresponding to the compact set, namely, the support of φ.

(b) An ultra-distribution $f \in \mathcal{D}'(\mathbf{R}^n, M_k)$ is the Fourier transform of a compactly supported ultra-distribution belonging to $\mathcal{E}'(\mathbf{R}^n, M_k)$ if and only if f extends to an entire function on $\mathbf{C}^n$ and satisfies the following two estimates:

(i) for any $\epsilon > 0$ there exists a constant $B_\epsilon > 0$ such that

$$|f(\zeta)| \leq B_\epsilon \exp\left[(b + \epsilon)(|\zeta_1| + \cdots + |\zeta_n|)\right] \quad \text{for } \zeta \in \mathbf{C}^n; \tag{6.5}$$

(ii) for any $h > 0$ there exists a constant $A_h > 0$ such that

$$|f(\xi)| \leq A_h \exp[M(h^{-1}\xi)] \quad \text{for } \xi \in \mathbf{R}^n \text{ with } \xi_j \neq 0. \tag{6.6}$$

7. Anisotropic pseudo-differential operators of ultra-differentiable class $[M]$. Symbol classes $S^{m,a}(U, [M])$. Let U be an open subset of $\mathbf{R}^{n+1}$ of the form $U = I \times V$ with I an interval in $\mathbf{R}_{x_0}$ and V an open subset of $\mathbf{R}^n_x$, and let $X = (x_0, x)$.

Definition 7.1. *A symbol* $p \sim \sum p_j$ *in* $S^{m,a}(U)$ *is said to be a* formal symbol of class $S^{m,a}(U,[M])$ *if for any compact set* K *in* U *there exist constants* $c = c(K)$ *and* $h = h(K) > 0$ *such that, for each* $j \in \mathbf{N}$ *and* $\tilde{\alpha} = (\alpha_0, \alpha)$, $\tilde{\beta} = (\beta_0, \beta) \in \mathbf{N}^{n+1}$, *we have*

$$
\begin{aligned}
(7.1) \qquad |D_{\xi_0}^{\alpha_0} D_{\xi}^{\alpha} D_{x_0}^{\beta_0} D_x^{\beta} p_j(X,\Xi)| &= |D_{\Xi}^{\tilde{\alpha}} D_X^{\tilde{\beta}} p_j(X,\Xi)| \le \\
&\le c h^{j+|\tilde{\alpha}|+|\tilde{\beta}|} M_{j+|\tilde{\beta}|}(\tilde{\alpha})! [1+\rho(\Xi)]^{m-j-(a,\alpha)}
\end{aligned}
$$

in $K_X \times \mathbf{R}_{\Xi}^{n+1}$.

Proposition 7.2. *The set* $\cup_{m\in\mathbf{R}} S^{m,a}(U,[M])$ *of all formal symbols of class* $[M]$ *is a* $\mathbf{C}$*-algebra which is closed under the operations of composition and trasposition (resp. taking adjoints) and the formulae for the composition of symbols (2.5) and of the transposed (resp. the adjoint) symbol (2.6) (resp. (2.6)') hold.*

Definition 7.3. *A formal symbol* $p \sim \sum p_j$ *of class* $S^{m,a}(U,[M])$ *is said to be a* true symbol of class $S^{m,a}(U,[M])$ *if for any compact subset* K *in* U *there exist constants* $c = c(K) > 0$ *and* $h = h(K) > 0$ *such that, for any integer* $\nu \ge 1$ *and* $\tilde{\alpha} = (\alpha_0, \alpha)$, $\tilde{\beta} = (\beta_0, \beta) \in \mathbf{N}^{n+1}$, *we have*

$$
\begin{aligned}
(7.2) \qquad \Big| D_{\Xi}^{\tilde{\alpha}} D_X^{\tilde{\beta}} \Big[p(X,\Xi) - \sum_{j<\nu} p_j(X,\Xi) \Big] \Big| &\le \\
&\le c h^{\nu+|\tilde{\alpha}|+|\tilde{\beta}|} M_{\nu+(a,\tilde{\beta})}(\tilde{\alpha})! [1+\rho(\Xi)]^{m-\nu-(a,\alpha)}
\end{aligned}
$$

in $K_X \times \mathbf{R}_{\Xi}^{n+1}$.

We shall write, when this assumption holds,

$$
(7.2)'. \qquad p \sim_{[M]} \sum p_j
$$

Definition 7.4 (Anisotropic pseudo-differential operator of weight a and of class $[M]$)**.** *An anisotropic pseudo-differential operator of weight* $a = (a_0, 1, \ldots, 1)$ *is said to be* strictly of class $[M]$ *if it is defined by a true symbol of class* $S^{m,a}(U,[M])$.

Following Boutet de Monvel and Krée we now introduce

Definition 7.4′. *A continuous linear operator* $P : \mathcal{D}(U, M_k) \to \mathcal{E}(U, M_{k-m})$ *is said to be an* anisotropic pseudo-differential operator of class $[M]$ *if its distribution kernel is of class* $[M]$ *outside the diagonal in* $U \times U$ *and the restriction of* P *to any relatively compact open subset* V *of* U *can be written as a sum of a pseudo-differential operator strictly of class* $[M]$ *on* V *and a linear operator having a kernel of class* $[M]$ *in* $V \times V$.

We denote the class of anisotropic pseudo-differential operators of weight $a = (a_0, 1, \dots, 1)$ and of class $[M]$ by $\mathcal{L}^{m,a}(U, [M])$.

Proposition 7.5.

(i) (Continuity). *If* $P \in \mathcal{L}^{m,a}(U, [M])$ *then* P *is a continuous linear map of* $\mathcal{D}(U, M_k)$ *into* $\mathcal{E}(U, M_{k-m})$.

(ii) $\cup_{m \in \mathbf{R}} \mathcal{L}^{m,a}(U, [M])$ *is closed under composition* $Q \circ P$, *if at least one of the two operators is properly supported, and under transposition, and the usual formulae for the symbols* $q \circ p$ *and* ${}^t p$ *hold for the true symbols.*

(ii)′ $\cup_{m \in \mathbf{R}} \mathcal{L}^{m,a}(U, [M])$, *modulo* $\mathcal{L}^{-\infty,a}(U, [M])$, *is a *-algebra.*

(iii) (Pseudo-local property of class $[M]$). *If* $P \in \mathcal{L}^{m,a}(U, [M])$ *then* P *extends to a continuous linear mapping of* $\mathcal{E}'(U, M_k)$ *into* $\mathcal{D}'(U, M_{k-m})$ *and the distribution kernel* $\mathcal{K}_P$ *of* P *belongs to the ultra-differentiable class* $\mathcal{E}(U \times U \setminus \Delta, M_{k-m})$ *outside the diagonal in* $U \times U$; *that is,* $\mathcal{K}_P|_{U \times U \setminus \Delta} \in \mathcal{E}(U \times U \setminus \Delta, M_{k-m})$.

(iv) If the symbol of P *belongs to* $S^{-\infty,a}(U, [M])$ *then* P *maps* $\mathcal{E}'(U, M_k)$ *into* $\mathcal{E}(U, M_k)$; *that is,* P *is a regularizing operator of class* $[M]$.

(v) If $\sum p_j$ *is a formal symbol of class* $S^{m,a}(U, [M])$ *then there exists a true symbol of class* $[M]$ *such that* $p \sim_{[M]} \sum p_j$.

Some remarks on the idea of proofs. For the proofs of the algebraic properties and the corresponding symbolic calculus we make use of the techniques of Boutet de Monvel and Krée [2]; namely, we introduce the formal norms in the space of symbols as follows: if $p \sim_{[M]} \sum p_j$ is a formal symbol of class $[M]$ then the formal norm of

p is defined by

$$(7.3) \qquad \sigma(p; h, [M]) =$$
$$= \sum (2n+2)^{-j} j! \left[M_{j+(a,\tilde{\beta})} \cdot (j + |\tilde{\alpha}|) \right]^{-1} h^{2j+|\tilde{\alpha}+\tilde{\beta}|} \left| D_{\Xi}^{\tilde{\alpha}} D_X^{\tilde{\beta}} p_j(X, \Xi) \right|$$

where the summation extends over all $j \in \mathbf{N}$, and all multi-indices $\tilde{\alpha} = (\alpha_0, \alpha)$, $\tilde{\beta} = (\beta_0, \beta) \in \mathbf{N}^{n+1}$ and $h > 0$. We note that the right hand side is well defined since the series clearly converges.

The family of these formal norms can be used to define the structure of a locally convex topological vector space on the space of formal symbols.

For the proofs of continuity and pseudo-local properties we use the nuclearity of the function spaces and their dual spaces involved together with the analytic properties of distributions of the type of p.f.(p_j) defined by quasi-homogeneous functions $p_j(\Xi)$.

We also have the following useful property on regularity of ultra-differentiable class as a consequence of the pseudo-local property of operators in $\mathcal{L}^{m,a}(U, [M])$:

Theorem 7.6. *Let $P \in \mathcal{L}^{m,a}(U, [M])$ be an anisotropic pseudo-differential operator of class $[M]$ and weight $a = (a_0, 1, \ldots, 1)$. If U' is any open subset of U such that $U \setminus U'$ is compact then P extends uniquely to a continuous linear map of*

$$\mathcal{E}'(U, M_k) \cap \mathcal{E}(U', M_k) \to \mathcal{D}'(U, M_{k-m}) \cap \mathcal{E}(U', M_{k-m}).$$

8. Pseudo-differential operators of Volterra type of class $[M]$. We define the class of pseudo-differential operators of Volterra type to be the subclass of pseudo-differential operators of weight $a = (a_0, 1, \ldots, 1)$ and of class $[M]$ by

$$(8.1) \qquad \mathcal{V}_{\pm}^{m,a}(U, [M]) = \mathcal{V}_{\pm}^{m,a}(U, [M]) \cap \mathcal{L}^{m,a}(U, [M])$$

and similarly the class of properly supported operators by

$$(8.2) \qquad \mathcal{V}_{\pm,\mathrm{prop}}^{m,a}(U, [M]) = \mathcal{V}_{\pm,\mathrm{prop}}^{m,a}(U) \cap \mathcal{L}^{m,a}(U, [M]).$$

We have the following analogues of Piriou's Theorem 3.2 for the symbols of operators in $\mathcal{V}_{\pm}^{m,a}(U,[M])$: let $\sum^{k,a}(U,[M])$ denote the space of all symbols $p \in C^{\infty}(U \times \mathbf{R}^{n+1})$ such that

(i) p is quasi-homogeneous of degree k in Ξ of weight $a = (a_0, 1, \ldots, 1)$

(ii) p has a continuous extension $p(X, \xi_0 + i\eta_0, \xi + i\eta) = p(X, \zeta_0, \zeta)$ to

$$\Omega_\epsilon = \{(\zeta_0, \zeta) \in \mathbf{C} \times \mathbf{C}^n : \quad Re\,\zeta_0 \geq 0,\ |Im\,\zeta| < \epsilon |Re\,\zeta| \text{ and } \rho(\zeta_0, \zeta) > 0\}$$

which is holomorphic in the subset $\{(\zeta_0, \zeta) \in \Omega : \quad Re\,\zeta_0 > 0\}$ and C^{∞} with respect to (X, ζ) such that for any multi-index $\tilde{\beta} = (\beta_0, \beta) \in \mathbf{N}^{n+1}$ we have

$$|D_X^{\tilde{\beta}} p(X, \zeta_0, \zeta)| \leq c h^{k+|\tilde{\beta}|} M_{k+(a,\tilde{\beta})} [1 + \rho(\zeta_0, \zeta)]^k. \tag{8.3}$$

We set $\sum^{-\infty,a}(U,[M]) = \bigcap_k \sum^{k,a}(U,[M])$. Then we have the

Proposition 8.1. *An operator $P \in \mathcal{L}^{m,a}(U,[M])$ with a true symbol $p \sim_{[M]} \sum p_j$ belongs to $\mathcal{V}_{+}^{m,a}(U,[M])$ if and only if $p_j \in \sum^{m-j,a}(U,[M])$.*

Remark 8.2. (Basic properties of operators in $\mathcal{V}_{\pm}^{m,a}(U,[M])$).

1) $\cup_{m \in \mathbf{R}} \mathcal{V}_{\pm}^{m,a}(U,[M])$ is closed under composition $Q \circ P$ if at least one of the two operators is properly supported, and for transposition (resp. for the operation of taking adjoints). Moreover, the formulae (2.5) for the true symbol $p \circ q$ of the composition and (2.6) for ${}^t p$ of the transpose (resp. (2.6)$'$ of the adjoint) hold.

2) The distribution kernel $\mathcal{K}_P$ of an operator P in $\mathcal{V}_{\pm}^{m,a}(U,[M])$ vanishes in the set $\{(X,Y) \in U \times U : x_0 < y_0\}$.

3) $\mathcal{V}_{\pm}^{m,a}(U,[M])$ is invariant under real analytic diffeomorphisms of the form $\Theta = (x_0, \theta(x_0, x))$ and the formula for the change of variables holds. It is also invariant under diffeomorphisms of class $[M]$ if the assumption (4) on the sequence M_k holds.

This last property (3) can be used to define the classes $\mathcal{V}_{\pm}^{m,a}(\mathcal{M}; E, F; [M])$ of pseudo-differential operators of Volterra type of class $[M]$ acting on the space of sections of a vector bundle E on the manifold $\mathcal{M}$ and mapping into the space of sections of a vector bundle F. We shall leave the details to the reader.

Definition 8.3. *A pseudo-differential operator P, belonging to $\mathcal{V}_+^{m,a}(U,[M])$, with principal symbol $p_0 \in \sum^{k,a}(U,[M])$, is said to be* right parabolic *(resp.* left*) of weight $a = (a_0, 1, \ldots, 1)$ if there exists a symbol $q_0 \in \sum^{-m,a}(U,[M])$ such that*

$$(8.4) \qquad p_0(X,\zeta_0,\zeta)\, q_0(X,\zeta_0,\zeta) = 1 \quad in \quad U \times \Omega_\epsilon$$

$$(8.5) \qquad (\text{resp. } q_0(X,\zeta_0,\zeta)\, p_0(X,\zeta_0,\zeta) = 1 \quad in \quad U \times \Omega_\epsilon).$$

Any differential operator $P(X, \partial_{x_0}, D_x)$ of parabolic type of weight $a = (a_0, 1, \ldots, 1)$ and of order m in the sense of Petrowski with coefficients in the space $\mathcal{E}(U, M_k)$ belongs to $\mathcal{V}_+^{m,a}(U,[M])$ and is parabolic (left as well as right).

We then have the following existence theorem:

Theorem 8.4 (Existence of parametrices). *If P, belonging to $\mathcal{V}_+^{m,a}(U,[M])$, is right parabolic (resp. left) of weight $a = (a_0, 1, \ldots, 1)$ then there exists a right (resp. left) parametrix $Q \in \mathcal{V}_+^{-m,a}(U,[M])$ in the sense that*

$$(8.6) \quad P \circ Q - I \in \mathcal{V}_+^{-\infty,a}(U,[M]) (resp.\ Q \circ P - I \in \mathcal{V}_+^{-\infty,a}(U,[M])).$$

Idea of the proof of Theorem 8.4. The construction of the parametrix reduces to seeking a true symbol $q \sim_{[M]} \sum q_j$ of a pseudo-differential operator of Volterra type

$$Q \in \mathcal{V}_+^{-m,a}(U,[M]) \text{ such that } p \circ q \sim_{[M]} 1 \text{ (resp. } q \circ p \sim_{[M]} 1).$$

In view of the Proposition 8.1 and the formula for the composition of true symbols (2.5) we are led to seek symbols $q_j \in \sum^{-m-j,a}(U,[M])$. The hypothesis of right (resp. left) parabolicity (8.4) (resp. (8.4)′) implies the existence of

$q_0 \in \sum^{-m,a}(U,[M])$ since $p_0 \in \sum^{m,a}(U,[M])$ and does not vanish in $U \times \Omega_\epsilon$. The q_j, for $j \geq 1$, are determined inductively by requiring that all the terms of quasi-homogeneous degree ≤ -1 in the asymptotic sum defining $p \circ q$ (resp. $q \circ p$) vanish; that is, we require

$$\sum_{r+s+(a,\tilde{\alpha})=j} [(\alpha)!]^{-1}(\partial_\Xi^{\tilde{\alpha}} p_r)(X,\Xi)(D_X^{\tilde{\alpha}} q_s)(X,\Xi) = 0, \quad \text{for all} \quad j \geq 1.$$

We then check inductively, using the fact that a derivation of the form $(\partial_\Xi^{\tilde{\alpha}} D_X^{\tilde{\beta}})$ maps the symbol class $\sum^{k,a}(U,[M])$ into the symbol class $\sum^{k-(a,\tilde{\alpha}),a}(U,[M])$, it follows that q_j thus defined belong to $\sum^{-m-j,a}(U,[M])$ thus determining a true symbol which defines the right (resp. left) parametrix Q.

Corollary 8.5. *If $P(X,\partial_{x_0},D_x)$ is a differential operator of parabolic type of weight $a = (a_0,1,\ldots,1)$ and of order m in the sense of Petrowski with coefficients in the space $\mathcal{E}(U,M_k)$ where $M_k = \sum_{|\alpha_0,\alpha)|\leq k}(\alpha_0!)^{a_0}\alpha!$ then, for any $f \in \mathcal{E}(U,M_k)$, any ultra-distribution solution $u \in \mathcal{D}'(U,M_k)$ of the parabolic equation*

$$P(X,\partial_{x_0},D_x)u = f \quad in \quad u \tag{8.8}$$

belongs to $\mathcal{E}(U,M_{k-m})$.

Idea of the proof of Corollary 8.5. It is enough to apply the left parametrix $Q \in \mathcal{V}^{-m,a}(U,[M])$ given by the Theorem 8.4 to both sides of the equation (8.8) and then use the pseudo-local property (7.iii) of class $[M]$ for $Q \in \mathcal{V}^{-m,a}(U,[M])$ together with the regularizing property (7.v) of ultra-differentiable class $[M]$ of operators in $\mathcal{V}^{-\infty,a}(U,[M])$.

9. Douglis-Nirenberg type parabolic systems. In this section we give a brief indication of a formal generalization of our preceeding considerations to more general systems of the type introduced by Douglis and Nirenberg (see [1]).

Let $\mathcal{N}$ be a real analytic manifold (resp. of class $[M]$ if the condition (iv) holds) of dimension n and $\mathcal{M} = I \times \mathcal{N}$, I being an interval. Let $E_1, \ldots, E_J$; $F_1, \ldots, F_K$ be real analytic (resp. of class $[M]$) vector bundles on $\mathcal{M}$.

We fix two systems of real numbers $s = (s_1, \ldots, s_J)$ and $t = (t_1, \ldots, t_K)$ and consider a system of pseudo-differential operators of Volterra type with weight $a = (a_0, 1, \ldots, 1)$ and of class $[M]$:

$$(9.1) \qquad P = (P_{kj}) \quad \text{where} \quad P_{kj} \in \mathcal{V}_+^{s_j - t_k, a}(\mathcal{M}; E_j, F_k; [M]).$$

If $p_{kj}^0 \in \sum^{s_j - t_k, a}(\mathcal{M}; E_j, F_k; [M])$ is the principal symbol of P_{kj} then the matrix $p^0 = (p_{kj}^0)$ is called the principal symbol of the system P.

Definition 9.1. *The system is said to be* right (*resp.* left) parabolic *of weight* $a = (a_0, 1, \ldots, 1)$ *if for every local coordinate chart* (U, Θ) *of* $\mathcal{M}$ *on which all the vector bundles are trivialized there exist*

$$(9.2) \quad q^0 = (q_{jk}^0) \quad \text{with} \quad q_{jk}^0 \in \sum^{t_k - s_j, a}(\mathcal{M}; \mathbf{C}^{\mathrm{rk}\, F_k}, \mathbf{C}^{\mathrm{rk}\, E_j}; [M])$$

such that

$$(9.3) \qquad p(X, \zeta_0, \zeta) q(X, \zeta_0, \zeta) = Id \;\; on \;\; \mathbf{C}^{\Sigma \mathrm{rk}\, F_k} \;\; in \;\; U \times \Omega_\epsilon$$

(resp.

$$(9.3)' \qquad p(X, \zeta_0, \zeta) q(X, \zeta_0, \zeta) = Id \;\; on \;\; \mathbf{C}^{\Sigma \mathrm{rk}\, E_j} \;\; in \;\; U \times \Omega_\epsilon).$$

As in the scalar case we have the following result on existence of parametrices for parabolic systems:

Theorem 9.2. *If* P *is a right (resp. left) parabolic system of weight* $a = (a_0, 1, \ldots, 1)$ *in the sense of the above definition then there exists a system of pseudo-differential operators*

$$(9.4) \qquad Q = (Q_{jk}) \;\; with \;\; Q_{jk} \in \mathcal{V}_+^{t_k - s_j, a}(\mathcal{M}; E_j, F_k; [M])$$

such that

$$(9.5) \qquad PQ - I \in \mathcal{V}_+^{-\infty, a}(\mathcal{M}; F, F; [M])$$

(resp.

$$(9.5)' \qquad PQ - I \in \mathcal{V}_{+}^{-\infty,a}(\mathcal{M}; E, E; [M]))$$

where $E = \oplus E_j$ *and* $F = \oplus F_k$.

As a consequence of the existence of parametrices we obtain, again as in the scalar case (Corollary 8.5), the regularity in the class of ultra-differentiable functions $\mathcal{E}(\mathcal{M}, [M])$ of solutions to Douglis-Nirenberg type systems of differential equations (with coefficients $\mathcal{E}(\mathcal{M}, [M])$ which are parabolic in the sense of Petrowski).

Theorem 9.3. *If* $P(X, \partial_{x_0}, D_x)$ *is a square system of differential operators from sections of* $E = \oplus E_j$ *into sections of* $F = \oplus F_k$ *(*$\operatorname{rk} E = \operatorname{rk} F$*) of parabolic type of weight* $a = (a_0, 1, \ldots, 1)$ *and of orders* $s_j - t_k$ *in the sense of Petrowski:*

$$(9.6) \qquad \begin{gathered} \det p_{jk}^{0}(X, \xi_0 + i\eta_0, \xi) \neq 0 \\ \text{for } \xi_0 \geq 0,\ (\eta_0, \xi) \in \mathbf{R}^{n+1} \text{ and } \Xi = (\xi_0, \xi) \in \mathbf{R}^{n+1} \setminus 0 \end{gathered}$$

with coefficients in $\mathcal{E}(U, M_\nu)$ *where* $M_\nu = \sum_{(a,\alpha) \leq \nu} (\alpha_0 !)^{a_0} \alpha!$ *then, for any* $f = (f_1, \ldots, f_J)$ *with* $f_j \in \mathcal{E}(U, M_\nu)$*, any ultra-distribution solution* $u = (u_1, \ldots, u_K)$ *with* $u_r \in \mathcal{D}'(U, M_\nu)$ *of the parabolic system of equations*

$$(9.7) \qquad \sum_{1 \leq j \leq J} P_{kj}(X, \partial_{x_0}, D_x) u_j = f_k \quad \text{in} \quad U$$

belongs to $\mathcal{E}(U, M_{\nu - s_j + t_k + (n+3)/2})$.

10. Parabolic mixed problems. In this section we shall apply the results on parabolic pseudo-differential operators of the previous sections to the mixed boundary value problems associated to parabolic systems of differential operators. These results in particular generalize the results on regularity in classes of ultra-differentiable functions for elliptic boundary problems due to the author [12] and those for solutions of parabolic equations due to Friedman [6].

All the manifolds and vector bundles considered in this section are assumed to be either real analytic or of class $[M]$ provided the sequence $\{M_\nu\}$ satisfies the assumption (iv). We shall say, for simplicity, that they are smooth.

Suppose $\mathcal{N}$ is a smooth manifold of dimension n and $\mathcal{M} = I \times \mathcal{N}$ be a cylinder on $\mathcal{N}$, with I a bounded interval in $\mathbf{R}_{x_0}$. Let $\mathcal{M}'$ be an open subset of $\mathcal{M}$ with smooth boundary $\partial\mathcal{M}'$ in $\mathcal{M}$ which is nowhere tangent to x_0 =const. Let $E_1, \dots, E_J$; $F_1, \dots, F_K$ be vector bundles on $\mathcal{M}'$ and further, let $G_1, \dots, G_H$ be vector bundles on the boundary manifold $\partial\mathcal{M}'$. Fix the system of real numbers $(s_1, \dots, s_J)$ and $(t_1, \dots, t_K)$ and assume rk $(\oplus E_j) =$ rk $(\oplus F_k)$.

Let

$$P = (P_{kj}) \in \mathcal{D}\text{iff}(\mathcal{M}; \oplus E_j, \oplus F_k; [M]), \quad \text{with ord } P_{kj} = s_j - t_k \tag{10.1}$$

be a Douglis-Nirenberg (square) system of differential operators on $\mathcal{M}$ which is assumed to be parabolic in the sense of Petrowski of weight $a = (a_0, 1, \dots, 1)$ and with coefficients in $\mathcal{E}(\mathcal{M}, [M])$.

Then, by Theorem 9.3, there exist a right parametrix

$$Q = (Q_{jk}) \text{ with } Q_{jk} \in \mathcal{V}_+^{t_k - s_j, a}(\mathcal{M}'; F_k, E_j; [M]) \tag{10.2}$$

and a left parametrix

$$Q' = (Q'_{jk}) \text{ with } Q'_{jk} \in \mathcal{V}_+^{t_k - s_j, a}(\mathcal{M}'; F_k, E_j; [M]) \tag{10.2$'$}$$

for P.

Let us fix another set of real numbers $(r_1, \dots, r_L)$ and consider pseudo-differential operators of Volterra type

$$B_{hj} \in \mathcal{V}_+^{t_j - r_h, a}(\mathcal{M}'; E_j|\partial\mathcal{M}', G_h; [M]), \quad B = (B_{hj}) \tag{10.3}$$

(boundary operators) acting between the spaces of sections of ultradistributions on $\partial\mathcal{M}'$ such that the composed Volterra pseudo-differential operators satisfy the following hypothesis:

$$B \circ Q \text{ is right and } (I - Q') \oplus B \text{ is left parabolic of weight } a = (a_0, 1, \dots, 1) \text{ on } \partial\mathcal{M}'. \tag{10.4}$$

Remark 10.1. We remark that the assumption on the boundary operators holds when rk $(\oplus_{1\leq h\leq H} G_h) = (m/2)\text{rk}\,(\oplus_{1\leq j\leq J} E_j)$ where $m = \sum s_j - \sum t_k$ and the well known complementing condition of Agmon, Douglis and Nirenberg (Lopatinski-Shapiro condition) is satisfied by principal symbols (b_{hj}) of the boundary operators with respect to the given system P of differential operators.

For a real number τ we shall write

$$\mathcal{M}'_\tau = \{X = (x_0, x) \in \mathcal{M}';\ x_0 < \tau\}$$

and

$$(\partial\mathcal{M}')_\tau = \{X = (x_0, x) \in \partial\mathcal{M}';\ x_0 < \tau\}.$$

Then we have the following theorem:

Theorem 10.2. *Suppose given* $f = (f_1, \ldots, f_K)$ *and* $g = (g_1, \ldots, g_H)$ *with*

$$f_k \in \mathcal{D}'(\mathcal{M}'_\tau, \oplus F_k | \mathcal{M}'_\tau, M_{\nu - s_k}), \tag{10.5}$$

$$g_h \in \mathcal{D}'((\partial\mathcal{M}')_\tau, G_h | (\partial\mathcal{M}')_\tau, M_{\nu - r_h}). \tag{10.6}$$

Under the above assumptions on the system of differential operators P *and on the boundary operators* B *the parabolic mixed boundary value problem*

$$Pu = f \ \text{ in } \ \mathcal{M}'_\tau \ \text{ and } \ \sum B_{hj}(\gamma u_j) = g_h \ \text{ on } \ (\partial\mathcal{M}')_\tau \tag{10.7}$$

has a unique solution $u = (u_1, \ldots, u_J)$ *with*

$$u_j \in \mathcal{D}'(\mathcal{M}'_\tau, \oplus E_j | \mathcal{M}'_\tau, M_{\nu - t_j}). \tag{10.8}$$

Further, we have the following hypoellipticity of class $[M]$: *if, for an open subset* V *of* $\mathcal{M}'_\tau$ *such that* $\mathcal{M}'_\tau \setminus V$ *is compact, we have*

$$f_k \in \mathcal{D}'(\mathcal{M}'_\tau, \oplus F_k | \mathcal{M}'_\tau, M_{\nu - s_k}) \cap \mathcal{E}(V, \oplus F_k | \mathcal{M}'_\tau \cap V, M_{\nu - s_k}) \tag{10.5$'$}$$

and

$$g_h \in \mathcal{D}'((\partial\mathcal{M}')_\tau, G_h|(\partial\mathcal{M}')_\tau, M_{\nu-r_h}) \cap \\ \cap\, \mathcal{E}(V, G_h|(\partial\mathcal{M}')_\tau \cap V, M_{\nu-r_h}) \tag{10.6$'$}$$

then

$$u_j \in \mathcal{E}(V, E_j|V, M_{\nu-t_j+(n+3)/2}).$$

The proof makes use of the method of Hörmander which consists in reducing (locally in every coordinate chart on which all the bundles involved are trivialized) to the construction of projectors (using the parabolicity condition together with the trace operation) and then showing that these projectors are systems of pseudo-differential operators of Volterra type acting on traces γu on $(\partial\mathcal{M}')_\tau$. Finally the calculus and the pseudo-local properties of these operators developed earlier will yield the result.

References

[1] S.Agmon, A.Douglis, L.Nirenberg, *Estimates near the boundary for elliptic partial differential equations satisfying general boundary conditions II,* Comm. Pure Appl. Math. **17** (1964), 35-92.

[2] L.Boutet de Monvel, P.Krée, *Pseudo-differential operators and Gevrey classes,* Ann. Inst. Fourier **17** (1967), 295-323.

[3] E.De Giorgi, *Un esempio di non unicità della soluzione del problema di Cauchy relativo ad una equazione differenziale lineare a derivate parziali di tipo parabolico,* Rend. Mat. **14** (1955), 382-387.

[4] E.De Giorgi, *Un teorema di unicità per il problema di Cauchy relativo ad equazioni differenziali lineari a derivate parziali di tipo parabolico,* Ann. Mat. Pura Appl. **40** (1955), 371-377.

[5] A.Friedman, *Classes of solutions of systems of partial differential equations of parabolic type,* Duke Math. J. **24** (1957), 433-442.

[6] A.Friedman, *On the regularity of the solutions of non-linear elliptic and parabolic systems of partial differential equations,* J. Math. and Mech. **7** (1958), 43-59.

[7] L.Hörmander, *Pseudo-differential operators and non-elliptic boundary problems,* Ann. Math . **83** (1966), 129-209.

[8] C.Hunt, A.Piriou, *Opérateurs pseudo-différentiels anisotropes d'ordre variable,* C.R. Acad. Sc. Paris, Sér. A **268** (1969), 214-217.

[9] J.L.Lions, E.Magenes, *Sur certains aspects des problèmes aux limites non-homogènes pour des opérateurs paraboliques,* Ann. Sc. Norm. Sup. Pisa **18** (1964), 303-344.

[10] J.L.Lions, E.Magenes, *Problèmes aux limites non-homogènes et applications,* Vol.3, Dunod, Paris (1970).

[11] W.Matsumoto, *Theory of pseudo-differential operators of ultradifferentiable class,* J. Math. Kyoto Univ. **27** (1987), 453-500.

[12] M.K.V.Murthy, *A remark on the regularity at the boundary for solutions of elliptic equations,* Ann. Sc. Norm. Sup. Pisa **15** (1961), 353-368.

[13] M.K.V.Murthy, *Some remarks in the theory of pseudo-differential operators,* Proc. Int. Conf. on Functional Analysis and Related Topics, Tokyo (1969), 122-130.

[14] M.K.V.Murthy, *Pseudo-differential operators on Gevrey spaces and regularity for elliptic boundary problems,* Bull. Soc. Royale des Sciences de Liège **40** (1971), 537-539.

[15] A.Piriou, *Une classe d'opérateurs pseudo-différentiels du type de Volterra,* Ann. Inst. Fourier **20** (1970), 77-94.

[16] A.Piriou, *Problèmes aux limites généraux pour des opérateurs différentiels paraboliques dans un domaine borné,* Ann. Inst. Fourier **21** (1970), 59-78.

Dipartimento di Matematica
Università di Pisa
Via Buonarroti, 2
I-56127 PISA

THE NEUMANN PROBLEM FOR SECOND ORDER ELLIPTIC EQUATIONS WITH RAPIDLY OSCILLATING PERIODIC COEFFICIENTS IN A PERFORATED DOMAIN

OLGA A.OLEINIK A.S.SHAMAEV G.A.YOSIFIAN

Dedicated to Ennio De Giorgi on his sixtieth birthday

During the last decades various problems in mechanics and physics have stimulated the appearence and intensive development of a new branch of the theory of partial differential equations-homogeneization and G-convergence of differential operators. The fundamental contribution in this field has been made by E.De Giorgi and his school. In papers [1],[2] the G-convergence was considered and its important properties were studied, in particular theorems on the homogenization of elliptic operators with rapidly oscillating and periodic coefficients and the convergence of energy integrals were proved. Results on the G-convergence of higher order differential equations with rapidly oscillating and random coefficients were obtained in [3].

In this paper we consider the Neumann problem for a second order elliptic equation with ϵ-periodic coefficients in a perforated domain with an ϵ-periodic structure. We prove estimates with respect to the small parametr ϵ for the difference between the solution of this problem and the solution of the related homogenized problem. We also obtain some estimates which characterize the convergence of the energy integrals and generalized gradients (flows) of the Neumann problem in a perforated domain. Similar results have been earlier established for mixed boundary value problems of linear elasticity with the Dirichlet data on the outer part of the boundary and the Neumann conditions on the surface of the cavities (see [4]-[6]).

In the case of non-homogeneous Neumann conditions the perforated structure of the domain brings about some new effects in the boundary conditions of the homogenized problem.

On the basis of the estimates obtained here it is possible using the methods of [7] to study the behaviour of eigenvalues of the considered problem.

Similar results can be obtained by the method suggested here, for the Neumann problem for the elasticity system as well as for other boundary value problems for second order elliptic equations.

Let Ω be a smooth bounded domain in $\mathbf{R}^n$ with boundary $\partial\Omega$. We consider a perforated domain $\Omega^\epsilon \subset \Omega$ whose boundary $\partial\Omega^\epsilon$ consists of $\partial\Omega$ and the surface of cavities $S_\epsilon \subset \Omega$. We define Ω^ϵ and S_ϵ as follows.

Let ω be a smooth domain in $\mathbf{R}^n$ with a 1-periodic structure, i.e. ω is invariant with respect to the shifts by the vector $z = (z_1, ..., z_n)$ with integer components ($z \in \mathbf{Z}^n$). Denote by T_ϵ the set of all $z \in \mathbf{Z}^n$ such that

$$\epsilon(z + \bar{Q}) \subset \Omega, \quad \rho(\epsilon(z + \bar{Q}), \partial\Omega) \geq \epsilon$$

where

$$Q = \{x : 0 < x_j < 1, \ j = 1, ..., n\}, \quad \epsilon G = \{x : \epsilon^{-1} x \in G\},$$

$\bar{G}$ is the clousure of G in $\mathbf{R}^n$, $\rho(A, B)$ is the distance in $\mathbf{R}^n$ between the sets A and B. Set

$$(1) \quad \bar{\Omega}_1 = \bigcup_{z \in T_\epsilon} \epsilon(z + \bar{Q}), \ \bar{\Omega}_1^\epsilon = \bar{\Omega}_1 \cap \epsilon\bar{\omega}, \ \bar{\Omega}^\epsilon = \bar{\Omega}_1^\epsilon \cup (\bar{\Omega} \backslash \Omega_1),$$

where ϵ is a small parameter. We assume that $\Omega_1, \Omega_1^\epsilon, \Omega^\epsilon$ (the sets of interior points of $\bar{\Omega}_1, \bar{\Omega}_1^\epsilon, \bar{\Omega}^\epsilon$ respectively) are domains with Lipschitz boundaries.

Obviously the boundary of Ω^ϵ consists of $\partial\Omega$ and $S_\epsilon = \partial\Omega^\epsilon \cap \Omega$. We also assume that the cell of periodicity $Q \cap \omega$ is a Lipschitz domain and $Q \backslash \bar{\omega}$ is a union of a finite number of Lipschitz domains separated from one another and from the edges of cube Q by a positive distance.

By $H^m(G)$ (for integer $m \geq 0$) we denote the Hilbert space consisting of functions which have weak derivatives up to the order m belonging to $L^2(G)$. The norm in $H^m(G)$ is given by the formula

$$\|u\|_m = \sum_{|\alpha|\leq m} \Big(\int_G |D^\alpha u|^2 dx\Big)^{1/2},$$

$$D^\alpha u = \frac{\partial^{|\alpha|} u}{\partial x_1^{\alpha_1} \dots \partial x_n^{\alpha_n}}, \quad |\alpha| = \alpha_1 + \dots + \alpha_n.$$

By $H^{m+\frac{1}{2}}(\partial G)$ we denote the space of traces on ∂G of functions $v \in H^{m+1}(G)$ (see [8]).

In Ω^ϵ we consider the following Neumann problem

$$(2) \qquad \begin{cases} \mathcal{L}_\epsilon u^\epsilon = \dfrac{\partial}{\partial x_h}\left(C^{hk}\left(\dfrac{x}{\epsilon}\right)\dfrac{\partial u^\epsilon}{\partial x_k}\right) = f^\epsilon(x) & \text{in } \Omega^\epsilon, \\ \dfrac{\partial u^\epsilon}{\partial \nu_\epsilon} = 0 \quad \text{on } S_\epsilon, \quad \dfrac{\partial u^\epsilon}{\partial \nu_\epsilon} = \psi^\epsilon & \text{on } \partial\Omega, \\ (u^\epsilon, 1)_{L^2(\Omega^\epsilon)} = 0. \end{cases}$$

Here $f^\epsilon \in L^2(\Omega^\epsilon)$, $\psi^\epsilon \in L^2(\partial\Omega)$,

$$\frac{\partial u}{\partial \nu_\epsilon} \equiv C^{hk}\left(\frac{x}{\epsilon}\right)\nu_h \frac{\partial u}{\partial x_k}, \quad (u,v)_{L^2(G)} = \int_G uv \, dx,$$

$(\nu_1, \dots, \nu_n)$ is the unit outward normal to $\partial\Omega^\epsilon, C^{hk}(\xi)$ are piece-wise smooth functions in $\mathbf{R}^n$, 1-periodic in $\xi = (\xi_1, \dots, \xi_n)$ and satisfying the conditions

$$(3) \qquad \begin{aligned} &\lambda_1 |\eta|^2 \leq C^{hk}(\xi)\eta_h\eta_k \leq \lambda_2|\eta|^2, \ \forall \eta \in \mathbf{R}^n, \\ &\lambda_1, \lambda_2 = \text{const.} > 0, \ C^{hk} = C^{kh}. \end{aligned}$$

Here and in what follows we adopt the convention of summation over repeated indices from 1 to n unless pointed otherwise.

We also assume that the surfaces across which $C^{hk}(\xi)$ may be discontinuous are smooth and do not intersect $\partial\omega$.

Consider the Neumann problem for the homogenized operator

$$(4) \qquad \begin{cases} \hat{\mathcal{L}} u^o = \dfrac{\partial}{\partial x_h}(\hat{C}^{hk}\dfrac{\partial u^0}{\partial x_k}) = f^o & \text{in } \Omega, \\ \text{meas}(Q \cap \omega)\dfrac{\partial u^0}{\partial \nu} = \psi^o & \text{on } \partial\Omega, \ (u^o, 1)_{L^2(\Omega)} = 0, \end{cases}$$

where

$$\phi^o \in L^2(\partial\Omega),\ f^o \in L^2(\Omega),\ \frac{\partial u}{\partial \nu} \equiv \hat{C}^{hk}\nu_h \frac{\partial u}{\partial x_k}.$$

It is assumed throughout the paper that the solvability conditions for problems (2) and (4) are satisfied, i.e.

$$(f^\epsilon, 1)_{L^2(\Omega^\epsilon)} = (\psi^\epsilon, 1)_{L^2(\partial\Omega)}(f^0, 1)_{L^2(\Omega)} = \\ = \big(\operatorname{meas}(Q \cap \omega)\big)^{-1}(\psi^o, 1)_{L^2(\partial\Omega)}.$$

The homogenized coefficients $\hat{C}^{hk}$ are defined by the formulas

$$\hat{C}^{pq} = \big(\operatorname{meas}(Q \cap \omega)\big)^{-1} \int_{Q\cap\omega} \left(C^{pq}(\xi) + C^{pj} \frac{\partial N^q}{\partial \xi_j} \right) d\xi, \tag{5}$$

where the functions $N^q(\xi), q = 1, ..., n$, are solutions of the problems

$$\begin{cases} \dfrac{\partial}{\partial \xi_k}\left(C^{k\ell}(\xi) \dfrac{\partial N^q}{\partial \xi_\ell} \right) = -\dfrac{\partial}{\partial \xi_k} C^{kq}(\xi) & \text{in } \omega, \\ \nu_k C^{k\ell}(\xi) \dfrac{\partial N^q}{\partial \xi_\ell} = -\nu_k C^{kq}(\xi) & \text{on } \partial\omega, \\ N^q(\xi) \ \text{ is } 1-\text{periodic in } \xi, \ (N^q, 1)_{L^2(Q\cap\omega)} = 0. \end{cases} \tag{6}$$

Existence of N^q can be easily proved by the Riesz theorem (see e.g.[4]).

It is interesting to note that in the case of a perforated domain Ω^ϵ, when $Q \cap \omega \neq Q$, there appears the coefficient mes$(Q \cap \omega)$ by the conormal derivative in the Neumann boundary conditions on $\partial\Omega$ of the homogenized problem (4).

In order to prove the estimate for solutions of problems (2) and (4) we shall have to consider a more general elliptic boundary value problem of Neumann type

$$\frac{\partial}{\partial x_i}(A^{ij} \frac{\partial u}{\partial x_j}) = f^o + \frac{\partial f^i}{\partial x_i} \ \text{ in } \ G,\ \frac{\partial u}{\partial \nu_A} = \psi + f^i \nu_i \ \text{ on } \ \partial G, \tag{7}$$

where G is a bounded domain in $\mathbf{R}^n$,

$$f^i \in L^2(G), i = 0, ..., n, \psi \in L^2(\partial G), \frac{\partial u}{\partial \nu_A} \equiv A^{ij} \frac{\partial u}{\partial x_j} \nu_i,$$

the coefficients A^{ij} satisfy conditions similar to (3).

We say that a function u is a solution of problem (7) if $u \in H^1(G)$ and for any $v \in H^1(G)$ the following integral identity is satisfied

$$\int_G A^{ij} \frac{\partial u}{\partial x_j} \frac{\partial v}{\partial x_i} dx = \int_G (f^i \frac{\partial v}{\partial x_i} - f^o v) dx + \int_{\partial G} \psi v dS.$$

Let $f \in L^2(\Omega^\epsilon)$, $\psi \in L^2(\partial\Omega)$. The scalar products

$$(f, v)_{L^2(\Omega^\epsilon)}, \quad (\psi, v)_{L^2(\partial\Omega)}$$

define continuous linear functionals belonging to $(H^1(\Omega^\epsilon))^*$ which is the dual space to $H^1(\Omega^\epsilon)$. The norms of f, ψ as elements of $(H^1(\Omega^\epsilon))^*$ are denoted by $\|f\|_*, \|\psi\|_*$ respectively.

Consider a truncating function $\varphi(x)$ which has the following properties:

$$\begin{aligned} &\varphi \in C_0^\infty(\Omega), |\nabla\varphi| \le c\epsilon^{-1}, \varphi(x) \equiv 0 \ \text{ for } \ x \in \Omega \backslash \Omega_1, \\ &\varphi(x) \equiv 1 \ \text{ for such } \ x \in \Omega_1 \ \text{ that } \ \rho(x, \partial\Omega_1) > c_1\epsilon, \end{aligned} \tag{8}$$

where c, c_1 are constants independent of ϵ.

Theorem 1. *Let*

$$f^\epsilon \in L^2(\Omega^\epsilon), \ f^o \in H^1(\Omega), \ \psi^\epsilon \in L^2(\partial\Omega), \ \psi^o \in H^{3/2}(\partial\Omega)$$

and u^ϵ, u^o be solutions of problems (2),(4) respectively. Then

$$\begin{aligned} &\|u^\epsilon - u^o - \epsilon\varphi N^s(\frac{x}{\epsilon})\frac{\partial u^o}{\partial x_s}\|_{H^1(\Omega^\epsilon)} \le \\ &c(\epsilon^{1/2}\|f^o\|_{H^1(\Omega)} + \epsilon^{1/2}\|\psi^o\|_{H^{3/2}(\partial\Omega)} + \|f^o - f^\epsilon\|_* + \|\psi^o - \psi^\epsilon\|_*), \end{aligned} \tag{9}$$

where c is a constant independent of $\epsilon, \psi^o, \psi^\epsilon, f^o, f^\epsilon$.

In order to prove this theorem we first establish some auxiliary results.

Let G be a bounded domain in $\mathbf{R}^n$ with a Lipschitz boundary. Consider a non-empty set $\gamma \subset \partial G$ such that in neighbourhood of each of its points γ coincides with ∂G. By $H^1(G, \gamma)$ we denote the completion in the norm of $H^1(G)$ of the space of functions belonging to $C^\infty(\bar{G})$ and vanishing in a neighbourhood of γ. Since ∂G is a Lipschitz surface, $v \in H^1(G)$ has a trace on γ (see [14]). For $v, w \in H^1(G)$ we say that $v = w$ on γ, if $v - w \in H^1(G,\gamma)$.

The space of traces on γ of functions from $H^1(G)$ denote by $H^{1/2}(\gamma)$. In $H^{1/2}(\gamma) = H^1(G)/H^1(G,\gamma)$ the norm is given by the formula

$$\|\varphi\|_{H^{1/2}(\gamma)} = \inf_w\{\|w\|_{H^1(G)}, w = \varphi \text{ on } \gamma\}.$$

Lemma 1. *Let $D, G, D\backslash G$ be bounded domains with Lipschitz boundaries, $G \subset D$, and let the set $\gamma = (\partial G) \cap D$ coincide with ∂G in a neighbourhood of each of its points. Then there exists a linear extension operator $P : H^1(D\backslash G) \to H^1(D)$ such that*

$$Pv = v \qquad \textit{for any} \quad v \equiv \text{ const.},$$

$$\|Pv\|_{H^1(D)} \leq c_o\|v\|_{H^1(D\backslash G)}, \tag{10}$$

$$\|\nabla Pv\|_{L^2(D)} \leq c_1\|\nabla v\|_{L^2(D\backslash G)}. \tag{11}$$

PROOF. For each $v \in H^1(D\backslash G)$ there exits an extension $V \in H^1(D)$ to the set D, since the boundary of $D\backslash\bar{G}$ is a Lipschitz surface (see e.g. [14]). Moreover V can be chosen such as to satisfy the inequality

$$\|V\|_{H^1(D)} \leq c\|v\|_{H^1(D\backslash G)},$$

where c is a constant independent of v (see [14]). Consider a function W which is weak solution of the following boundary value problem

$$\Delta W = 0 \text{ in } G; W = V \text{ on } \gamma; \ \frac{\partial W}{\partial x_i}\nu_i = 0 \text{ on } (\partial G)\backslash\gamma.$$

This means that W belongs to $H^1(G)$ and satisfies the integral identity

$$\int_G \frac{\partial W}{\partial x_i}\frac{\partial w}{\partial x_i}dx = 0 \tag{12}$$

for any $w \in H^1(G,\gamma); W - V \in H^1(G,\gamma)$. Taking into account the Friedrichs inequality for functions from $H^1(G,\gamma)$ (see [10]) and setting $Pv = W$ in G, $Pv = v$ in $D\backslash G$ we deduce from (12) that

$$\|Pv\|_{H^1(D)} \leq c\|v\|_{H^1(D\backslash G)}.$$

Obviously $Pv = v$ if $v \equiv$ const. Therefore the inequality (10) is valid. Let us prove the inequality (11). Since $P(v+c) = Pv + c, c =$const., the inequality (10) yields

$$\begin{aligned} \|\nabla(Pv)\|_{L^2(D)} = \|\nabla P(v+c)\|_{L^2(D)} \leq \\ \leq c_o(\|\nabla v\|_{L^2(D\backslash G)} + \|v+c\|_{L^2(D\backslash G)}). \end{aligned} \tag{13}$$

Let us choose c such that $(v+c,1)_{L^2(D\backslash G)} = 0$. Therefore using (13) and the Poincaré inequality

$$\|v+c\|_{L^2(D\backslash G)} \leq c_1\|\nabla v\|_{L^2(D\backslash G)}$$

we obtain (11). The lemma is proved.

Lemma 2. *There exists a linear extension operator $P : H^1(\Omega^\epsilon) \to H^1(\Omega)$ such that*

$$Pv = v \qquad \text{for any} \quad v \equiv \text{const},$$

$$\|\nabla Pv\|_{L^2(\Omega)} \leq c_0\|\nabla v\|_{L^2(\Omega^\epsilon)}, \tag{14}$$

$$\|Pv\|_{H^1(\Omega)} \leq c_1\|v\|_{H^1(\Omega^\epsilon)} \tag{15}$$

for any $v \in H^1(\Omega^\epsilon)$, where constants c_o, c_1 do not depend on ϵ, v.

Proof. Let $v \in H^1(\Omega^\epsilon)$ and set $V(\xi) = v(\epsilon\xi)$. Fix $z \in T_\epsilon$ where $T_\epsilon \subset \mathbf{Z}^n$ and is the same as in the formula (1). Consider $V(\xi)$ in $\omega \cap (z+Q)$. By virtue of Lemma 1 we can extend $V(\xi)$ to a function $P_1V \in H^1(z+Q)$ such that

$$\begin{aligned} \|P_1V\|_{H^1(z+Q)} \leq c_o\|V\|_{H^1(\omega\cap(z+Q))}, \\ \|\nabla P_1V\|_{L^2(z+Q)} \leq c_1\|\nabla V\|_{L^2(\omega\cap(z+Q))}. \end{aligned} \tag{16}$$

Extending $V(\xi)$ in this way for each $z \in T_\epsilon$ we obtain a function $P_1 V$ satisfying inequalities (16) for all $z \in T_\epsilon$ with constant c_o, c_1 independent of z.

If the set $Q \backslash \bar{\omega}$ lies strictly inside the cube Q then the function $(P_1 V)(x/\epsilon)$ furnishes the required extension. However if $\bar{Q} \backslash \bar{\omega}$ has a non-empty intersection with ∂Q the function $P_1 V$ may not belong to $H^1(\epsilon^{-1}\Omega)$ since its traces on some adjacent faces of different cubes $\epsilon(z+Q), z \in T_\epsilon$, may differ. In this case we can alter $P_1 V$ near these sets as follows.

We have assumed above that $Q \backslash \bar{\omega}$ consists of a finite number of Lipschitz domains separated from each other by a positve distance, say δ. Let $\gamma_1, ..., \gamma_m$ be mutually non-intersecting $(n-1)$-dimensional domains such that $\bar{\gamma}_1 \cup ... \cup \bar{\gamma}_m = (\partial Q) \backslash \omega$. Due the conditions on $\partial \omega$ there exist Lipschitz domains $g_1, ..., g_m$ such that $g_j \subset \mathbf{R}^n \backslash \bar{\omega}$, $\gamma_j \subset g_j$, $g_j = g_i + z$ if $\gamma_j = \gamma_i + z, (z \in \mathbf{Z}^n)$, $g_j \cap Q$, $g_j \backslash Q$ are non-empty Lipschitz domains, g_j belongs to a $\delta/4$ neighbourhood of γ_j. Denote by $G_1, ..., G_N$ all mutually non-intersecting domains of the form

$$G_j = (g_s + z) \cap \epsilon^{-1}\Omega_1, \ z \in T_\epsilon, \ j = 1, ..., N, \ s = 1, ..., m.$$

Set $G_0 = G_1 \cup ... \cup G_N$. Denote by $\tilde{G}_j$ the $\delta/2$ neighbourhood of G_j. It is easy to see that $P_1 V \in H^1(\epsilon^{-1}\Omega \backslash G_0)$. Using Lemma 1 we can extend $P_1 V$ to each G_j as a function $P_2 V$ such that

$$(17) \qquad \begin{aligned} \|\nabla_\xi P_2 V\|_{L^2(G_j)} &\leq c_o \|\nabla_\xi P_1 V\|_{L^2(\tilde{G}_j \backslash G_j)}, \\ \|P_2 V\|_{H^1(G_j)} &\leq c_1 \|P_1 V\|_{H^1(\tilde{G}_j \backslash G_j)}. \end{aligned}$$

Set $U(\xi) = (P_1 V)(\xi)$ for $\xi \in (\epsilon^{-1}\Omega) \backslash G_0$,

$$U(\xi) = (P_2 V)(\xi) \ \text{ for } \xi \in G_0.$$

Setting $(PV)(x) = U(x/\epsilon)$ and using estimates (16),(17) we obtain the required extension.

Lemma 3. *Suppose that the set γ belongs to $\partial\Omega$ and that, in a neighbourhood of each of its points, γ coincides with $\partial\Omega$. Then, for any $v \in H^1(\Omega^\epsilon)$ the following inequalities are satisfied*

$$(18) \qquad \|v\|_{L^2(\gamma)} \leq c_1 \|v\|_{H^1(\Omega^\epsilon)},$$

$$\|v-\mu_\epsilon\|_{L^2(\Omega^\epsilon)} \le c_2\|\nabla v\|_{L^2(\Omega^\epsilon)}, \tag{19}$$

where the constants c_1, c_2 *do not depend on* ϵ *and*

$$\mu_\epsilon = \big(\operatorname{meas}(\Omega^\epsilon)\big)^{-1}(v,1)_{L^2(\Omega^\epsilon)}.$$

PROOF. Inequality (18) is a direct consequence of a similar inequality for the domain Ω and lemma 2.

It is easy to show that

$$\|v-\mu_\epsilon\|^2_{L^2(\Omega^\epsilon)} \le \|v-\mu\|^2_{L^2(\Omega^\epsilon)} \tag{20}$$

for any $\mu \in \mathbf{R}$. Let Pv be the extension of v to the domain Ω, constructed in lemma 2, and let

$$\mu = (\operatorname{meas}\Omega)^{-1}(Pv,1)_{L^2(\Omega)}.$$

It follows from (20),(14) and the Poincaré inequality for Ω that

$$\begin{aligned}\|v-\mu_\epsilon\|^2_{L^2(\Omega^\epsilon)} \le \|v-\mu\|^2_{L^2(\Omega^\epsilon)} \le \|Pv-\mu\|^2_{L^2(\Omega)} \le \\ \le c\|\nabla Pv\|^2_{L^2(\Omega)} \le c_1\|\nabla v\|^2_{L^2(\Omega^\epsilon)}.\end{aligned}$$

The lemma is proved.

We introduce the following notation

$$\begin{aligned}&\alpha^{is}(\xi) = \hat{C}^{is} - C^{is}(\xi) - C^{ij}(\xi)\frac{\partial N^s}{\partial \xi_j}, \quad i,s=1,\dots,n,\\ &\hat{\xi}_j = (\xi_1,\dots,\xi_{j-1},\xi_{j+1},\dots,\xi_n) \in \mathbf{R}^{n-1},\\ &S^j_t = \{\xi : \xi_j = t,\ 0<\xi_\ell<1,\ \ell\ne j\}\cap\omega,\\ &Q^j_{t_1t_2} = \{\xi : t_1<\xi_j<t_2,\ 0<\xi_\ell<1,\ \ell\ne j\}.\end{aligned} \tag{21}$$

Due to our assumtpions on C^{hk}, it follows from [9] that α^{ij} are piecewise continuous functions in $\overline{Q\cap\omega}$

Lemma 4. *The functions* α^{is} *satisfy the following relations*

$$\int_{Q\cap\omega} \alpha^{is}(\xi)d\xi = 0, \quad i,s=1,\dots,n, \tag{22}$$

(23) $$\frac{\partial}{\partial \xi_i}\alpha^{is}(\xi) = 0 \quad in \quad \omega, \; s = 1, ..., n,$$

(24) $$\nu_i \alpha^{is} = \nu_i \hat{C}^{is} \quad on \quad \partial\omega, \; s = 1, ..., n,$$

(25) $$\int_{S_t^j} \alpha^{js}(\xi) d\hat{\xi}_j = \big(\text{meas}(S_t^j) - \text{meas}(Q \cap \omega)\big) \hat{C}^{js}$$

(there is no summation over j).

PROOF. Equalities (22), (23), (24) follows directly from (5), (6). Multiplying the equation (6) by $\xi_j - t + 1$ integrating it over the set $Q^j_{t-1,t} \cap \omega$ and using the boundary conditions in (6) we get

$$-\int_{Q^j_{t-1,t}\cap\omega} C^{j\ell}\frac{\partial N^q}{\partial \xi_\ell} d\xi =$$
$$= \int_{Q^j_{t-1,t}\cap\omega} C^{jq}\, d\xi - \int_{\omega\cap\partial Q^j_{t-1,t}} C^{jl}\frac{\partial N^q}{\partial \xi_\ell}\nu_j(\xi_j - (t-1))dS +$$
$$- \int_{\omega\cap\partial Q^j_{t-1,t}} C^{jq}\nu_j(\xi_j - (t-1))dS.$$

Due to (22) and the periodicity of $N^q, C^{j\ell}$ this equality yields (25). The lemma is proved.

Remark 1. If Ω^ϵ is a non-perforated domain, i.e. $\omega = \mathbf{R}^n$, $\Omega^\epsilon = \Omega$, then $\int_{S_t^j} \alpha^{js}(\xi) d\hat{\xi}_j = 0$, since $\text{meas}(S_t^j) = \text{meas}(Q \cap \omega) = 1$.

Lemma 5. *Let Ω be a smooth bounded domain and*

$$B_\delta = \{x \in \Omega, \rho(x, \partial\Omega) < \delta\}.$$

Then there exists a $\delta_0 > 0$ such that for any $\delta \in (0, \delta_0)$ and any $v \in H^1(\Omega)$ the estimate

(26) $$\|v\|^2_{L^2(B_\delta)} \le c\delta \|v\|^2_{H^1(\Omega)}$$

holds with a constant c independent of δ, v.

PROOF. Let $\Gamma \subset \Omega$ be a surface which is sufficiently close in C^2 to $\partial\Omega$. Then it follows from the imbedding theorem that for any $v \in H^1(\Omega)$ we have

$$\|v\|_{L^2(\Gamma)} \leq c\|v\|_{H^1(\Omega)}$$

with a constant c independent of Γ. Therefore one can easily verify that (26) is valid. The lemma is proved.

Lemma 6. *Let $\sigma_1, ..., \sigma_{2n}$ be $(n-1)$-dimensional faces of the cube*

$$\epsilon Q = \{x : 0 < x_j < \epsilon, j = 1, ..., n\}.$$

Then

$$\int_{\sigma_i} |u|^2 dS - \int_{\sigma_j} |u|^2 dS \ \leq \ c\|u\|^2_{L^2(\epsilon Q)} \tag{27}$$

for any $u \in H^1(\epsilon Q)$ where c is a constant independent of ϵ and u.

PROOF. Let

$$\sigma_1 = \{x_1 = 0\} \cap \epsilon\bar{Q}, \ \sigma_2 = \{x_2 = 0\} \cap \epsilon\bar{Q},$$
$$S_1 = \epsilon^{-1}\sigma_1, \quad S_2 = \epsilon^{-1}\sigma_2, \ y = \frac{x}{\epsilon}.$$

Consider the points

$$\hat{y}^1 = (0, y_2, ..., y_n), \ \hat{y}^2 = (y_2, 0, y_3, ..., y_n)$$

belonging to the faces S_1, S_2 of the cube Q respectively. The segment

$$y(t, y_2, y_3, ..., y_n) = t\hat{y}^1 + (1-t)\hat{y}^2$$

for $t \in [0, 1]$ belongs to $\bar{Q}$. Therefore for any $u \in H^1(Q)$ we have

$$\begin{aligned} u^2(y(1, y_2, ..., y_n)) - u^2(y(0, t_2, ...y_n)) = \\ = \int_0^1 \frac{\partial u^2}{\partial t}(y(t, y_2, ..., y_n))dt \\ = 2\int_0^1 u(y(t, y_2, ..., y_n))\frac{\partial u}{\partial \xi_j}(y(t, y_2, ...y_n))\frac{\partial y_j}{\partial t}dt, \end{aligned}$$

where $\xi = x/\epsilon$.

Integrating this equality with respect to $y_2, ..., y_n$ from 0 to 1 we get

$$\int_{S_1} |u|^2 dS - \int_{S_2} |u|^2 dS \leq c \int_Q |u||\nabla_\xi u| d\xi.$$

Changing the variables $\xi = \epsilon^{-1}x$ in this inequality we obtain (27) for σ_1, σ_2. Estimate (27) for other σ_i, σ_j can be obtained in a similar way. The lemma is proved.

Lemma 7. *Let $u \in H^1(\Omega)$ and Ω_1 be the subdomain of Ω, defined in (1). Then*

$$\|u\|_{L^2(\partial\Omega_1)} \leq c\|u\|_{H^1(\Omega)}, \tag{28}$$

where c is a constant independent of ϵ, u.

PROOF. The domain Ω_1 depends on ϵ and its boundary $\partial\Omega_1$ consists of the $(n-1)$-dimensional faces of the cubes $\epsilon(z+Q)$. Denote by $\sigma_j, ..., \sigma_j^{\ell j}$ the faces of the cubes $\epsilon(z+\bar{Q}), z \in T_\epsilon$, parallel to the hyperplanes $x_j = 0, j = 1, ..., n$, and lying on $\partial\Omega_1$. Then

$$\partial\Omega_1 = \bigcup_{s=1}^{n} \bigcup_{s=1}^{\ell j} \bar{\sigma}_j^s. \tag{29}$$

The cube $\epsilon(z+Q)$, $z \in T_\epsilon$, on whose boundary lies the set σ_j^s is denoted by q_j^s. It is easy to see that among $q_j^s, s = 1, ..., \ell_j, j = 1, ..., n$, the number of cubes identical to q_m^t is not greater than $2n$.

The boundary $\partial\Omega$ is a smooth surface, therefore each cube q_j^s possesses a face $\sigma_{s,j}$ such that $\sigma_{s,j}$ is parallel to the hyperplane $x_{m(j,s)} = 0$ and $\sigma_{s,j}$ is a projection along the axis $0x_{m(j,s)}$ of a surface $S_{s,j} \subset \partial\Omega$ which is given by the equation

$$x_m = \psi_m(\hat{x}_m), \ \hat{x}_m \in \sigma_{s,j}, \ \|\psi_m\|_{C^2} \leq M, \ m = m(j,s),$$

and $c_1\epsilon \leq |x-y| \leq c_2\epsilon$ for any $x \in \sigma_{s,j}, y \in S_{s,j}$, where constants c_1, c_2, M do not depend on ϵ, s, j.

Denote by $Q_{s,j}$ the set formed by the segments orthogonal to $\sigma_{s,j}$ and connecting the points of $\sigma_{s,j}$ and $S_{s,j}$. Then using a suitable diffeomorphism mapping $Q_{s,j}$ to ϵQ and taking into account Lemma 6 we find that

$$\|u\|^2_{L^2(\sigma_{s,j})} \leq c(\|u\|^2_{L^2(S_{s,j})} + \|u\|^2_{H^1(Q_{s,j})}).$$

Therefore by Lemma 6 we get

$$\|u\|^2_{L^2(\sigma_j^s)} \leq c_1(\|u\|^2_{L^2(S_{s,j})} + \|u\|^2_{H^1(Q_{s,j})}).$$

Summing up these inequalities with respect to j, s we obtain estimate (28), since due to the smoothness of $\partial\Omega$ there is an integer K independent of ϵ and such that each $\bar{Q}_{s,j}$ can have a non-empty intesection only with a finite number of $\bar{Q}_{\ell,t}$ which is not greater than K. The lemma is proved.

Lemma 8. *Let*

$$\gamma^h(x) \in L^\infty(\partial\Omega_1), \ h = 1, ..., n,$$

be such that for any $h = 1, \ldots, n$ and any $m = 1, \ldots, \ell_h$,

$$\int_{\sigma_h^m} \gamma^h(x) dS = 0, \quad \sup_{\partial\Omega_1} |\gamma^h| \leq \gamma, \tag{30}$$

where γ =const., and σ_h^m are the same as in (29). Then for any $v \in H^2(\Omega), w \in H^1(\Omega)$ the inequality

$$\left| \int_{\partial\Omega_1} \gamma_h v w \ dS \right| \leq c\epsilon^{1/2} \gamma \|v\|_{H^2(\Omega)} \|w\|_{H^1(\Omega)} \tag{31}$$

holds with a constant c independent of γ, v, w, ϵ.

Proof. Consider a function $\Gamma(x)$ defined almost everywhere on $\partial\Omega_1$ by the formula

$$\Gamma(x) = \left(\text{meas}(\sigma_\ell^m)\right)^{-1} \int_{\sigma_\ell^m} \nu_h \gamma^h v \ dS \qquad \text{for } \ x \in \sigma_\ell^m.$$

Obviously $\Gamma(x)$ is constant on each σ_ℓ^m. Therefore setting

$$\zeta(x) = \left(\text{meas}(\sigma_\ell^m)\right)^{-1} \int_{\sigma_\ell^m} v dS \qquad \text{for } \ x \in \sigma_\ell^m$$

we obtain from (30) and the Poincaré inequality for σ_ℓ^m

$$\int_{\partial\Omega_1} |\Gamma|^2 dS = \sum_{j=1}^{n}\sum_{s=1}^{\ell j} \int_{\sigma_j^s} |\Gamma|^2 dS =$$

$$= \sum_{j=1}^{n}\sum_{s=1}^{\ell j} \left(\operatorname{meas}(\sigma_j^s)\right)^{-1} \left(\int_{\sigma_j^s} \gamma^j (v-\zeta) dS \right)^2 \leq$$

$$\leq \sum_{j=1}^{n}\sum_{s=1}^{\ell j} \left(\operatorname{meas}(\sigma_j^s)\right)^{-1} \int_{\sigma_j^s} |\gamma^j|^2 dS \int_{\sigma_j^s} (v-\zeta)^2 dS$$

$$\leq c\epsilon^2\gamma^2 \int_{\partial\Omega_1} |\nabla v|^2 dS.$$

It follows, by virtue of Lemma 7, that

$$\int_{\partial\Omega_1} |\Gamma|^2 dS \leq c_1 \epsilon^2 \gamma^2 \|v\|^2_{H^2(\Omega)}, \tag{32}$$

where c_1 is a constant independent of ϵ, v.

It is easy to see that

$$\int_{\partial\Omega_1} \nu_h \gamma^h v w dS = \int_{\partial\Omega_1} \Gamma w dS + \int_{\partial\Omega_1} (\nu_h \gamma^h v - \Gamma) w dS. \tag{33}$$

Due to (32) and Lemma 7 we have

$$\begin{aligned} \left| \int_{\partial\Omega_1} \Gamma w \, dS \right| &\leq \|\Gamma\|_{L^2(\partial\Omega_1)} \|w\|_{L^2(\partial\Omega_1)} \leq \\ &\leq c_2 \epsilon\gamma \|v\|_{H^2(\Omega)} \|w\|_{H^1(\Omega)}. \end{aligned} \tag{34}$$

Let us estimate the second integral in the right-hand side of (33). Define a function $\eta(x)$ on $\partial\Omega_1$ by the formula

$$\eta(x) = \left(\operatorname{meas}(q_m^\ell)\right)^{-1} \int_{q_m^\ell} w dx \qquad \text{for } x \in \sigma_m^\ell.$$

Therefore

$$\begin{aligned} \|w-\eta\|^2_{L^2(\partial\Omega_1)} &= \sum_{j=1}^{n}\sum_{s=1}^{\ell j} \|w-\eta\|^2_{L^2(\sigma_j^s)} \leq \\ &\leq c_3\epsilon \sum_{j=1}^{n}\sum_{s=1}^{\ell j} \|\nabla w\|^2_{L^2(q_j^s)} \leq c_4 \epsilon \|\nabla w\|^2_{L^2(\Omega)}. \end{aligned} \tag{35}$$

Here we have used the fact that among the $q_j^s, s = 1, ..., \ell_j, j = 1, ..., n$, the number of cubes identical to q_m^t is not greater than $2n$; and we have applied the estimate

$$\|w - \eta\|^2_{L^2(\sigma_j^s)} \leq c\|\nabla w\|^2_{L^2(q_j^s)}$$

which holds thanks to Poincaré inequality and the continuity of the imbedding

$$H^1(\epsilon^{-1}q_j^s) \subset L^2(\epsilon^{-1}\sigma_j^s).$$

By the definition of $\Gamma(x)$ we have

$$\int_{\sigma^\ell} (\nu_h \gamma^h v - \Gamma) dS = 0.$$

Therefore, since η is constant on each σ_ℓ^m, we find by virtue of (35),(32) and Lemma 7 that

$$\begin{aligned}\left|\int_{\partial\Omega_1} (\nu_h\gamma^h v - \Gamma) w dS\right| &\leq \|\nu_h\gamma^h v - \Gamma\|_{L^2(\partial\Omega_1)}\|w - \eta\|_{L^2(\partial\Omega_1)} \leq \\ &\leq c_5 \left(\gamma\|v\|_{L^2(\partial\Omega_1)} + \|\Gamma\|_{L^2(\partial\Omega_1)}\right) \epsilon^{1/2}\|\nabla w\|_{L^2(\Omega)} \\ &\leq c_6\gamma\epsilon^{1/2}\|v\|_{H^2(\Omega)}\|w\|_{H^1(\Omega)}.\end{aligned}$$

This inequality together with (33),(34) yields (31). The lemma is proved.

Proof of the Theorem 1. Set

$$\tilde{u} = u^o + \epsilon\varphi N^s(\frac{x}{\epsilon})\frac{\partial u^o}{\partial x_s}.$$

Let us apply the operator $\mathcal{L}_\epsilon$ to $u^\epsilon - \tilde{u}$. Taking into account the equations in (2),(4),(6) we have

$$\begin{aligned}\mathcal{L}_\epsilon(u^\epsilon - \tilde{u}) = f^\epsilon - f^o + \frac{\partial}{\partial x_h}\left[\left(\hat{C}^{hk} - C^{hk} - \epsilon C^{hj}\frac{\partial(\varphi N^k)}{\partial x_j}\right)\frac{\partial u^o}{\partial x_k}\right] + \\ - \epsilon\frac{\partial}{\partial x_h}\left(C^{hk}\varphi N^s\frac{\partial^2 u^o}{\partial x_k \partial x_s}\right) =\end{aligned}$$

$$
\begin{aligned}
&= f^\epsilon - f^o + \frac{\partial}{\partial x_h}\left[(1-\varphi)(\hat{C}^{hk} - C^{hk})\frac{\partial u^o}{\partial x_k}\right] + \\
&+ \left[\hat{C}^{hk} - C^{hk} - \epsilon C^{hj}\frac{\partial N^k}{\partial x_j} - \epsilon\frac{\partial (C^{sk}N^h)}{\partial x_s}\right]\varphi\frac{\partial^2 u^o}{\partial x_k \partial x_h} + \\
&+ \frac{\partial \varphi}{\partial x_h}\alpha^{hk}\frac{\partial u^o}{\partial x_k} - \epsilon\frac{\partial}{\partial x_h}\left[\frac{\partial \varphi}{\partial x_j}C^{hj}N^k\frac{\partial u^o}{\partial x_k}\right] + \\
&- \epsilon\frac{\partial \varphi}{\partial x_h}C^{hk}N^s\frac{\partial^2 u^0}{\partial x_k \partial x_s} - \epsilon\varphi C^{hk}N^s\frac{\partial^3 u^o}{\partial x_k \partial x_h \partial x_s},
\end{aligned}
\tag{36}
$$

where α^{hk} are defined by (21). The equality (36) is satisfied in the sense of distributions.

Let us define the functions $N^{hk}(\xi)$ as the solutions of the boundary value problems

$$
\begin{cases}
\dfrac{\partial}{\partial \xi_j}(C^{j\ell}(\xi)\dfrac{\partial N^{hk}}{\partial \xi_\ell}) = \\
= -\dfrac{\partial}{\partial \xi_s}(C^{sh}N^k) - C^{hj}\dfrac{\partial N^k}{\partial \xi_j} - C^{hk} + \hat{C}^{hk} & \text{in } \omega, \\
C^{j\ell}\dfrac{\partial N^{hk}}{\partial \xi_\ell}\nu_j = -\nu_s C^{sh}N^k & \text{on } \partial\omega, \\
N^{hk}(\xi) \quad \text{is } 1-\text{periodic in } \xi, \quad (N^{hk}, 1)_{L^2(Q\cap\omega)} = 0.
\end{cases}
\tag{37}
$$

It follows from (36),(37) that

$$
\mathcal{L}_\epsilon(u^\epsilon - \tilde{u}) = f^\epsilon - f^o + \frac{\partial F_h^1}{\partial x_h} + \frac{\partial F_j^2}{\partial x_j} + F_1^o + F_2^o + F_3^o,
\tag{38}
$$

where

$$
\begin{cases}
F_h^1 = (1-\varphi)(\hat{C}^{hk} - C^{hk})\dfrac{\partial u^o}{\partial x_k} - \epsilon\dfrac{\partial \varphi}{\partial x_j}C^{hj}N^k\dfrac{\partial u^o}{\partial x_k} \\
F_j^2 = \epsilon\varphi C^{j\ell}\dfrac{\partial N^{hk}}{\partial \xi_\ell}\dfrac{\partial^2 u^o}{\partial x_k \partial x_h} \\
F_1^o = -\epsilon\dfrac{\partial \varphi}{\partial x_j}C^{j\ell}\dfrac{\partial N^{hk}}{\partial \xi_\ell}\dfrac{\partial^2 u^o}{\partial x_h \partial x_k} - \epsilon\dfrac{\partial \varphi}{\partial x_h}C^{hk}N^s\dfrac{\partial^2 u^o}{\partial x_k \partial x_s} \\
F_2^o = -\epsilon\varphi C^{j\ell}\dfrac{\partial N^{hk}}{\partial \xi_\ell}\dfrac{\partial^3 u^o}{\partial x_k \partial x_h \partial x_j} - \epsilon\varphi C^{hk}N^s\dfrac{\partial^3 u^o}{\partial x_k \partial x_h \partial x_s} \\
F_3^o = \dfrac{\partial \varphi}{\partial x_h}\alpha^{hk}\dfrac{\partial u^o}{\partial x_k}.
\end{cases}
\tag{39}
$$

Consider the boundary conditions for $u^\epsilon - \tilde{u}$. By virtue of the boundary conditions for N^s, N^{js} on $\partial\omega$ and (8) we have

$$\varphi(\nu_i C^{ij} + \epsilon \nu_i C^{i\ell} \frac{\partial N^j}{\partial x_\ell}) = 0 \quad \text{on} \quad \partial\Omega^\epsilon,$$

$$- \epsilon \nu_i \varphi C^{ih} N^k = \epsilon \varphi \nu_j C^{j\ell} \frac{\partial N^{hk}}{\partial \xi_\ell} \quad \text{on} \quad \partial\Omega^\epsilon.$$

Therefore performing appropriate computations we get

$$\frac{\partial(u^\epsilon - \tilde{u})}{\partial \nu_\epsilon} = (1-\varphi)\nu_i C^{ij} \frac{\partial u^\epsilon}{\partial x_j} - (1-\varphi)\hat{C}^{ij}\nu_i \frac{\partial u^o}{\partial x_j} + F_h^1 \nu_h + F_h^2 \nu_h. \tag{40}$$

Set $w = u^\epsilon - \tilde{u} - \mu_\epsilon$ where μ_ϵ is such that $(w,1)_{L^2(\Omega^\epsilon)} = 0$. Due to the boundary conditions for u^ϵ, u^o and the fact that $\varphi \equiv 0$ in $\Omega\backslash\Omega_1$, it follows from (40),(38) that

$$\begin{aligned} &- \int_{\Omega^\epsilon} C^{hk} \frac{\partial w}{\partial x_k} \frac{\partial w}{\partial x_h} dx = \int_{\Omega^\epsilon} (f^\epsilon - f^o) w dx - \int_{\Omega^\epsilon} F_h^1 \frac{\partial w}{\partial x_h} dx + \\ &- \int_{\Omega^\epsilon} F_h^2 \frac{\partial w}{\partial x_h} dx + \int_{\Omega^\epsilon} (F_1^o w + F_2^o w + F_3^o w) dx + \\ &+ \int_{S_\epsilon} \nu_i \hat{C}^{ij} (1-\varphi) \frac{\partial u^0}{\partial x_j} w dS + \int_{\partial\Omega} ((\text{meas}(Q \cap \omega))^{-1} \psi^o - \psi^\epsilon) w dS. \end{aligned} \tag{41}$$

Let us estimate the integrals in the right-hand side of this equality. Note that owing to (8) the functions $\frac{\partial \varphi}{\partial x_j}$ and $(1-\varphi)$ vanish at points $x \in \Omega_1$ such that $\rho(x, \partial\Omega_1) \geq c_1 \epsilon$, and $|\epsilon \nabla \varphi| \leq c$, where c, c_1 are constants independent of ϵ. Therefore by Lemma 5 we obtain

(42)
$$\begin{aligned} \left| \int_{\Omega^\epsilon} F_h^1 \frac{\partial w}{\partial x_h} dx \right| + \left| \int_{\Omega^\epsilon} F_1^o w dx \right| &\leq c_2 \|w\|_{H^1(\Omega^\epsilon\backslash\Omega_2)} \|u^o\|_{H^2(\Omega\backslash\Omega_2)} \\ &\leq c_3 \|w\|_{H^1(\Omega^\epsilon)} \epsilon^{1/2} \|u^o\|_{H^3(\Omega)} \end{aligned}$$

where c_3 is constant and does not depend on ϵ, and

$$\Omega_2 = \{x \in \Omega_1, \rho(x, \partial\Omega_1) > c_1 \epsilon\}.$$

It follows from (39) that

$$\left|\int_{\Omega^\epsilon} F_h^2 \frac{\partial w}{\partial x_h} dx\right| + \left|\int_{\Omega^\epsilon} F_2^o w dx\right| \leq c_4 \epsilon \|w\|_{H^1(\Omega^\epsilon)} \|u^o\|_{H^3(\Omega)}, \tag{43}$$

where c_4 is a constant independent of ϵ. Taking (23) and (24) into account we get

(44)

$$\begin{aligned}
\mathcal{I}_1 = &\int_{\Omega^\epsilon} F_3^o w dx + \int_{S_\epsilon} \hat{C}^{ij} \nu_i (1-\varphi) \frac{\partial u^o}{\partial x_j} w dS + \\
&+ \int_{\partial\Omega} \left(\left(\text{meas}(Q \cap \omega)\right)^{-1} \psi^0 - \psi^\epsilon\right) w dS = \\
= &\int_{\Omega^\epsilon} \left(\frac{\partial \varphi}{\partial x_h} \alpha^{hk} \frac{\partial u^o}{\partial x_k} w\right) dx + \int_{S_\epsilon} \hat{C}^{ij} \nu_i (1-\varphi) \frac{\partial u^o}{\partial x_j} w dS + \\
&+ \int_{\partial\Omega} \left(\left(\text{meas}(Q \cap \omega)\right)^{-1} \psi^o - \psi^\epsilon\right) w dS = \\
= &\int_{\Omega_1^\epsilon} \frac{\partial(\varphi - 1)}{\partial x_h} \alpha^{hk} \frac{\partial u^o}{\partial x_k} w dx - \int_{S_\epsilon} (\varphi - 1) \hat{C}^{ij} \nu_i \frac{\partial u^o}{\partial x_j} w dS + \\
&+ \int_{\partial\Omega} \left(\left(\text{meas}(Q \cap \omega)\right)^{-1} \psi^o - \psi^\epsilon\right) w dS = \\
= -&\int_{(\partial\Omega_1)\setminus S_\epsilon} \nu_h \alpha^{hk} \frac{\partial u^o}{\partial x_k} w dS + \int_{(\partial\Omega_1)\cap S_\epsilon} \hat{C}^{ij} \nu_i \frac{\partial u^o}{\partial x_j} w dS + \\
&+ \int_{\partial\Omega} \left(\left(\text{meas}(Q \cap \omega)\right)^{-1} \psi^o - \psi^\epsilon\right) w dS + \\
&- \int_{\Omega_1^\epsilon} (\varphi - 1) \alpha^{hk} \frac{\partial^2 u^o}{\partial x_k \partial x_h} w dx - \int_{\Omega_1^\epsilon} (\varphi - 1) \alpha^{hk} \frac{\partial u^o}{\partial x_k} \frac{\partial w}{\partial x_h} dx.
\end{aligned}$$

It should be noted here, that in the integral over $(\partial\Omega_1)\setminus S_\epsilon$ the normal ν is exterior to $\partial\Omega_1$, whereas in the integral over $(\partial\Omega_1) \cap S_\epsilon$ the normal ν is exterior to $\partial\Omega^\epsilon$. The last two integrals in the right-hand side of (44) can be estimated by

$$c\epsilon^{1/2} \|u^o\|_{H^3(\Omega)} \|w\|_{H^1(\Omega^\epsilon)}, \quad c = \text{const.}$$

similarly to (42). Set

$$\beta^{hk}(\xi) = \begin{cases} \alpha^{hk}(\xi) & \text{in } \omega, \\ \hat{C}^{hk} & \text{in } \mathbf{R}^n \setminus \omega. \end{cases}$$

Then

$$\beta^{hk}\left(\frac{x}{\epsilon}\right) = \begin{cases} \alpha^{hk}(x/\epsilon) & \text{in } (\partial\Omega_1)\backslash S_\epsilon, \\ \hat{C}^{hk} & \text{in } (\partial\Omega_1)\cap S_\epsilon. \end{cases} \tag{45}$$

It follows from (44),(45) that

$$|\mathcal{I}_1| \le |\mathcal{I}_2| + c\epsilon^{1/2}\|u^o\|_{H^3(\Omega)}\|w\|_{H^1(\Omega^\epsilon)}, \tag{46}$$

where

$$\begin{aligned} \mathcal{I}_2 = &- \int_{\partial\Omega_1}\left(\beta^{hk}\left(\frac{x}{\epsilon}\right)\nu_h\frac{\partial u^o}{\partial x_h}w\right)dS+ \\ &+ \frac{\text{meas}(Q\backslash\omega)}{\text{meas}(Q\cap\omega)}\int_{\partial\Omega}\psi^o w dS + \int_{\partial\Omega}(\psi^o - \psi^\epsilon)wdS. \end{aligned} \tag{47}$$

The integral identity for u^o yields

$$\begin{aligned} -\int_{\Omega\backslash\Omega_1} C^{hk}\frac{\partial u^o}{\partial x_k}\frac{\partial w}{\partial x_h}dx + \int_{\partial\Omega}\frac{\psi^o w}{\text{meas}(Q\cap\omega)}dS+ \\ -\int_{\partial\Omega_1}\hat{C}^{hk}\frac{\partial u^o}{\partial x_k}\nu_h w dS = \int_{\Omega\backslash\Omega_1} f^o w dx. \end{aligned}$$

Therefore by virtue of Lemma 5

$$\begin{aligned} &\frac{\text{meas}(Q\backslash\omega)}{\text{meas}(Q\cap\omega)}\int_{\partial\Omega}\psi^o w dS = \\ &\qquad = \text{meas}(Q\backslash\omega)\int_{\partial\Omega_1} C^{hk}\nu_h\frac{\partial u^o}{\partial x_k}wdS + \mathcal{I}_3, \end{aligned} \tag{48}$$

where

$$|\mathcal{I}_3| \le c\epsilon^{1/2}(\|u^o\|_{H^2(\Omega)}\|v\|_{H^1(\Omega^\epsilon)} + \|f^o\|_{L^2(\Omega)}\|w\|_{H^1(\Omega^\epsilon)}). \tag{49}$$

We thus obtain from (47),(48) that

$$\begin{aligned} \mathcal{I}_2 = &\int_{\partial\Omega_1}(\text{meas}(Q\backslash\omega)\nu_h\hat{C}^{hk} - \beta^{hk}\nu_h)\frac{\partial u^o}{\partial x_k}wdS+ \\ &+ \int_{\partial\Omega}(\psi^o - \psi^\epsilon)wdS + \mathcal{I}_3. \end{aligned} \tag{50}$$

Set

$$v = \frac{\partial u^o}{\partial x_k}, \ \gamma^h = \text{meas}(Q\backslash\omega)\hat{C}^{hk} - \beta^{hk}$$

in Lemma 8. Using (45) and (25) it is easy to verify that conditions (30) are satisfied. We conclude from (48),(49) and Lemma 8 that

$$(51) \quad \begin{aligned} |\mathcal{I}_2| \leq c(\epsilon^{1/2}\|u^o\|_{H^3(\Omega)}\|w\|_{H^1(\Omega^\epsilon)} + \epsilon^{1/2}\|f^o\|_{L^2(\Omega)}\|w\|_{H^1(\Omega^\epsilon)} + \\ + \|\psi^o - \psi^\epsilon\|_* \ \|w\|_{H^1(\Omega^\epsilon)}). \end{aligned}$$

It follows from (41),(42),(43),(46),(51) that

$$(52) \quad \begin{aligned} \int_{\Omega^\epsilon} |\nabla w|^2 dx \leq c_1(\epsilon^{1/2}\|u^o\|_{H^3(\Omega)} + \epsilon^{1/2}\|f^o\|_{L^2(\Omega)}\|w\|_{H^1(\Omega^\epsilon)} + \\ + \|f^\epsilon - f^o\|_* \ \|w\|_{H^1(\Omega^\epsilon)} + \|\psi^\epsilon - \psi^o\|_* \ \|w\|_{H^1(\Omega^\epsilon)}). \end{aligned}$$

From the well known results on the smothness of solutions of elliptic equations [8] we have

$$(53) \qquad \|u^o\|_{H^3(\Omega)} \leq c(\|f^0\|_{H^1(\Omega)} + \|\psi^o\|_{H^{3/2}(\partial\Omega)}).$$

Therefore inequalities (52), (53), (19), yield (9). The theorem is proved.

Let us consider now the convergence of energy integrals related to solutions of problem (2). Set

$$(54) \quad \begin{aligned} E_\epsilon(u^\epsilon) &\equiv \int_{\Omega^\epsilon \cap \Omega'} C^{jk}(\frac{x}{\epsilon})\frac{\partial u^\epsilon}{\partial x_k}\frac{\partial u^\epsilon}{\partial x_j}dx, \\ E_0(u^o) &\equiv \int_{\Omega'} \hat{C}^{jk}\frac{\partial u^o}{\partial x_k}\frac{\partial u^0}{\partial x_j}, \end{aligned}$$

where Ω' is a smooth subdomain of $\Omega, \bar{\Omega}' \subset \Omega$.

Theorem 2 (On the convergence of energy integrals). *Suppose that all conditions of Theorem 1 are satisfied. Then*

$$(55) \quad \begin{aligned} &|E_\epsilon(u^\epsilon) - \text{meas}(Q \cap \omega)E_o(u^o)| \leq \\ &\leq c[\epsilon^{1/2}(\|f^o\|_{H^1(\Omega)} + \|\psi^o\|_{H^{3/2}(\partial\Omega)}) + \|f^o - f^\epsilon\|_*^2 + \|\psi^o - \psi^\epsilon\|_*^2 + \\ &+ (\|f^o\|_{L^2(\Omega)} + \|\psi^o\|_{H^{1/2}(\partial\Omega)})(\|f^o - f^\epsilon\|_* + \|\psi^o - \psi^\epsilon\|_*)], \end{aligned}$$

where c is a constant independent of ϵ.

PROOF. By virtue of Theorem 1 we have

$$\frac{\partial u^\epsilon}{\partial x_i} = \frac{\partial u^o}{\partial x_i} + \frac{\partial N^p(\epsilon^{-1}x)}{\partial \xi_i}\frac{\partial u^o}{\partial x_p} + q_i^\epsilon(x), \quad x \in \Omega^\epsilon, \tag{56}$$

where

$$\begin{aligned}\|q_i^\epsilon\|_{L^2(\Omega^\epsilon\cap\Omega')} \le c\Big(&\epsilon^{1/2}\|f^o\|_{H^1(\Omega)} + \epsilon^{1/2}\|\psi^o\|_{H^{3/2}(\partial\Omega)} + \\ &+ \|f^o - f^\epsilon\|_* + \|\psi^o - \psi^\epsilon\|_*\Big)\end{aligned} \tag{57}$$

and the constant c does not depend on ϵ. Therefore for small ϵ

$$\begin{aligned}E_\epsilon(u_\epsilon) = \int_{\Omega^\epsilon\cap\Omega'} C^{ij}&\left(\frac{\partial u^o}{\partial x_i} + \frac{\partial u^o}{\partial x_s}\frac{\partial N^s}{\partial \xi_i}\right)\cdot \\ &\cdot\left(\frac{\partial u^o}{\partial x_j} + \frac{\partial u^o}{\partial x_t}\frac{\partial N^t}{\partial \xi_j}\right)dx + p^\epsilon,\end{aligned} \tag{58}$$

where

$$p^\epsilon = 2\int_\Omega C^{ij}\Big(\frac{\partial u^0}{\partial x_i} + \frac{\partial u^o}{\partial x_s}\frac{\partial N^s}{\partial \xi_i}\Big)q_j^\epsilon dx + \int_{\Omega^\epsilon\cap\Omega'} q_i^\epsilon C^{ij} q_j^\epsilon dx.$$

Since C^{ij}, $\frac{\partial N^s}{\partial \xi_i}$ are bounded functions, it follows from (57) that

$$|p^\epsilon| \le \left[\int_{\Omega^\epsilon\cap\Omega'} q_i^\epsilon q_i^\epsilon dx + \|u^o\|_{H^1(\Omega^\epsilon\cap\Omega')}\left(\int_{\Omega^\epsilon\cap\Omega'} q_i^\epsilon q_i^\epsilon dx\right)^{1/2}\right].$$

Therefore

$$\begin{aligned}|p^\epsilon| \le c_1\Big[&\epsilon^{1/2}\|f^o\|^2_{H^1(\Omega)} + \epsilon^{1/2}\|\psi^o\|^2_{H^{3/2}(\partial\Omega)} + \\ &+ \|f^o - f^\epsilon\|_*^2 + \|\psi^o - \psi^\epsilon\|_*^2 + \\ &+ (\|f^o\|_{L^2(\Omega)} + \|\psi^o\|_{H^{1/2}(\partial\Omega)})(\|f^o - f^\epsilon\|_* + \|\psi^o - \psi^\epsilon\|_*)\Big].\end{aligned} \tag{59}$$

Let us introduce functions $H^{st}(\xi)$ by the formula

$$(60)\quad H^{st} \equiv C^{ij}(\xi)\frac{\partial}{\partial \xi_j}(N^t + \xi_t)\frac{\partial}{\partial \xi_i}(N^s + \xi_s) - \operatorname{meas}(Q \cap \omega)\hat{C}^{st}.$$

Henceforth we shall assume that $C^{ij}, N^s(\xi), \frac{\partial N^s(\xi)}{\partial \xi_i}$ are extended to $Q \backslash \omega$ as zeroes. Therefore $H^{st}(\xi)$ are defined in Q and one can easily verify using (5) that

$$(61)\qquad \int_Q H^{st}(\xi)d\xi = 0.$$

Thus we can replace the domain $\Omega^\epsilon \cap \Omega'$ by Ω' in (58) and rewrite (58) in the form

$$(62)\quad E_\epsilon(u^\epsilon) = \int_{\Omega'} \frac{\partial}{\partial \xi_i}(N^s + \xi_s)C^{ij}\frac{\partial}{\partial \xi_j}(N^t + \xi_t)\frac{\partial u^o}{\partial x_s}\frac{\partial u^o}{\partial x_t}dx + p^\epsilon.$$

It follows from (62), (60) that

$$(63)\quad E_\epsilon(u^\epsilon) - \operatorname{meas}(Q \cap \omega)E_o(u^o) = \int_{\Omega'} H^{st}\left(\frac{x}{\epsilon}\right)\frac{\partial u^o}{\partial x_s}\frac{\partial u^0}{\partial x_t}dx + p^\epsilon.$$

Let us denote by $\mathcal{I}_\epsilon$ the set of $z \in \mathbf{Z}^n$ such that $\epsilon(Q + z) \subset \Omega'$. By $\mathcal{I}'_\epsilon$ we denote the set of $z \in \mathbf{Z}^n$ such that $\epsilon(Q + z) \cap \partial\Omega' \neq \emptyset$. We then have

$$(64)\quad \begin{aligned} &E_\epsilon(u^\epsilon) - \operatorname{meas}(Q \cap \omega)E_o(u^o) = \\ &\quad = \sum_{z \in \mathcal{I}'_\epsilon} \int_{\epsilon(z+Q)\cap\Omega'} H^{st}\left(\frac{x}{\epsilon}\right)\frac{\partial u^o}{\partial x_s}\frac{\partial u^o}{\partial x_t}dx + \\ &\qquad + \sum_{z \in \mathcal{I}_\epsilon} \int_{\epsilon(z+Q)} H^{st}\left(\frac{x}{\epsilon}\right)\frac{\partial u^o}{\partial x_s}\frac{\partial u^o}{\partial x_t}dx + p^\epsilon = \\ &\quad = k_1 + k_2 + p^\epsilon. \end{aligned}$$

It is easy to see that the first sum in the right-hand side of (64) can be written in the form

$$k_1 = \int_{G_\epsilon} H^{st}\frac{\partial u^o}{\partial x_s}\frac{\partial u^o}{\partial x_t}dx,$$

where $G_\epsilon \subset \Omega'$ and belongs to a neighbourhood of $\partial\Omega'$ of order ϵ. Therefore using Lemma 5 we get

$$|k_1| \le c\,\epsilon\, \|u^o\|^2_{H^2(\Omega')}. \tag{65}$$

Let us estimate the second sum in the right hand side of (64). Denote by Ω'' a subdomain formed by the cubes $\epsilon(z+Q)$ when $z \in \mathcal{I}_\epsilon$. Set

$$\gamma_t(x) = -\big(\operatorname{meas}(\epsilon Q)\big)^{-1} \int_{\epsilon(z+Q)} \frac{\partial u^0}{\partial x_t} dx, \;\; x \in \epsilon(z+Q).$$

The functions $\gamma_t(x)$ are constant on each of the cubes $\epsilon(z+Q)$. We have

$$\begin{aligned} k_2 &= \int_{\Omega'} H^{st} \frac{\partial u^o}{\partial x_s}\left(\frac{\partial u^o}{\partial x_t} + \gamma_t\right) dx - \int_{\Omega''} H^{st}\gamma_t dx = \\ &= \int_{\Omega''} H^{st} \frac{\partial u^o}{\partial x_s}\left(\frac{\partial u^o}{\partial x_t} + \gamma_t\right) dx + \\ &\quad - \int_{\Omega''} H^{st}\left(\frac{\partial u^o}{\partial x_s} + \gamma_s\right)\gamma_t dx + \int_{\Omega''} H^{st}\gamma_s\gamma_t dx. \end{aligned} \tag{66}$$

It is easy to see that

$$\begin{aligned} \int_{\Omega''} |\gamma_t|^2 dx &= \sum_{z\in\mathcal{I}_\epsilon} \operatorname{meas}(\epsilon Q)\cdot \frac{1}{\big(\operatorname{meas}(\epsilon Q)\big)^2}\left(\int_{\epsilon(z+Q)} \frac{\partial u^o}{\partial x_t} dx\right)^2 \le \\ &\le \sum_{z\in\mathcal{I}_\epsilon} \int_{\epsilon(z+Q)} \left|\frac{\partial u^o}{\partial x_t}\right|^2 dx = \int_{\Omega''} \left|\frac{\partial u^o}{\partial x_t}\right|^2 dx. \end{aligned} \tag{67}$$

By virtue of the Poincaré inequality we have

$$\int_{\Omega''} \left|\frac{\partial u^o}{\partial x_t} + \gamma_t\right|^2 dx \le c\epsilon^2 \int_{\Omega''} \left|\nabla \frac{\partial u^o}{\partial x_t}\right|^2 dx. \tag{68}$$

Due to (61) the last integral in (66) is equal to zero, since γ_t are constant on $\epsilon(z+Q), z \in \mathcal{I}_\epsilon$. Therefore

$$\begin{aligned} |k_2| &\le c_o \left(\int_{\Omega''} |\nabla u^o|^2 dx\right)^{1/2} \left(\sum_{t=1}^n \int_{\Omega''} \left(\frac{\partial u^o}{\partial x_t} + \gamma_t\right)^2 dx\right)^{1/2} + \\ &\quad + c_1 \left(\sum_{t=1}^n \int_{\Omega''} |\gamma_t|^2 dx\right)^{1/2} \left(\sum_{s=1}^n \int_{\Omega''} \left|\frac{\partial u^o}{\partial x_s} + \gamma_s\right|^2\right)^{1/2}, \end{aligned}$$

and taking into account (67), (68) we obtain

$$|k_2| \leq c_3 \epsilon \|u^o\|_{H^2(\Omega'')}. \tag{69}$$

Estimate (55) follows from (69),(65),(59),(64), since

$$\|u^o\|_{H^3(\Omega)} \leq c\|f^o\|_{H^1(\Omega)} + \|\psi^o\|_{H^{3/2}(\partial\Omega)}).$$

The theorem is proved.

Now we consider the convergence of generalized gradients (flows) related to solutions of problem (2).

Set

$$\gamma_\epsilon^p \equiv C^{pk}\left(\frac{x}{\epsilon}\right), \quad \gamma_o^p(x) \equiv \hat{C}^{pk}\frac{\partial u^o}{\partial x_k}, \quad p = 1, ..., n, \tag{70}$$

where u^ϵ, u^o are solutions of problems (2), (4) respectively. We assume that $\gamma_\epsilon^p(x)$ is equal to zero in $\Omega \backslash \Omega^\epsilon$ and

$$C^{jk}(\xi) \equiv \frac{\partial N^s}{\partial \xi_\ell}(\xi) \equiv N^p(\xi) \equiv 0 \quad \text{in } Q\backslash\omega.$$

Theorem 3. *Under the assumptions of Theorem 1 the following estimate is valid*

$$\begin{aligned} \|\gamma_\epsilon^p - \ \text{meas}(Q \cap \omega)\gamma_o^p - A^{pq}\left(\frac{x}{\epsilon}\right)\frac{\partial u^o}{\partial x_q}\|_{L^2(\Omega)} \leq \\ \leq c(\epsilon^{1/2}\|f^o\|_{H^1(\Omega)} + \epsilon^{1/2}\|\psi^o\|_{H^{3/2}(\partial\Omega)} + \\ + \|f^o - f^\epsilon\|_* + \|\psi^o - \psi^\epsilon\|_*), \end{aligned} \tag{71}$$

where c is a constant independent of ϵ, the functions $A^{pq}(\xi)$ are defined by the formulas

(72)

$$A^{ps}(\xi) = C^{ps}(\xi) + C^{pi}\frac{\partial N^s}{\partial \xi_i} - \ \text{meas}(Q \cap \omega)\hat{C}^{ps} \quad \textit{for} \ \ \xi \in Q \cap \omega,$$

$$A^{ps}(\xi) = - \ \text{meas}(Q \cap \omega)C^{ps} \qquad \textit{for} \ \ \xi \in Q\backslash\omega.$$

Moreover $\gamma_\epsilon^p(x) \to$ meas$(Q \cap \omega)\gamma_o^p(x)$ *weakly in* $L^2(\Omega)$ *as* $\epsilon \to 0$.

Proof. It follows from (56),(57),(70),(72) that

$$\gamma_\epsilon^p(x) \equiv C^{pi}\frac{\partial u_0}{\partial x_i} = C^{pi}\frac{\partial u_o}{\partial x_i} + C^{pi}\frac{\partial N^s}{\partial \xi_i}\frac{\partial u_o}{\partial x_s} + C^{pi}q_i^\epsilon(x) =$$
$$= (C^{ps} + C^{pi}\frac{\partial N^s}{\partial \xi_i})\frac{\partial u_o}{\partial x_s} + C^{pi}q_i^\epsilon(x).$$

Hence by (70),(72) we obtain

$$\gamma_\epsilon^p(x) - \text{meas}(Q \cap \omega)\gamma_o^p(x) = A^{ps}\left(\frac{x}{\epsilon}\right)\frac{\partial u^o}{\partial x_s} + C^{pi}q_i^\epsilon(x). \tag{73}$$

Estimate (71) follows from (73) and (57).

Weak convergence of $\gamma_\epsilon^p(x)$ to meas$(Q \cap \omega)\gamma_o^p(x)$ is due to (73), (57) and the fact that $\int_Q A^{ps}(\xi)d\xi = 0$ and therefore $A^{ps}(x/\epsilon)(\partial/\partial x_s)u^o \to 0$ weakly in $L^2(\Omega)$ as $\epsilon \to 0$.

The theorem is proved.

References

[1] E.De Giorgi, *G-operators and G-convergence,* Proc. Intern. Congr. of Mathematicians. August 16-24, Warszawa, 1983 Vol.II, North-Holland, 1175-1191.

[2] E.De Giorgi, S.Spagnolo, *Sulla convergenza degli integrali dell'energia per operatori ellittici del secondo ordine,* Boll. Un. Mat. It. **8** (1973), 391-411.

[3] V.V.Zhikov, S.M.Kozlov, O.A.Oleinik, T'en Ngoan Kha, *Averaging and G-convergence of differential operators,* Russian Math. Surv. **34** (1979), 69-147.

[4] O.A.Oleinik, G.A.Yosifian, *On homogenization of the linear elasticity system with rapidly oscillating coefficients in perforated domains,* in: N.E.Kochin, Advances in mechanics, Nauka, Moscow, 1984, 237-249.

[5] O.A.Oleinik, A.S.Shamaev, G.A.Yosifian, *On homogenization problems for the elasticity system with non-uniformly oscillating*

coefficients, Teubner Texte zur Mathematik **79**, Mathematical Analysis, 192-202.

[6] O.A.Oleinik, A.S.Shamaev. G.A.Yosifian, *On the convergence of the energy, stress tensors and eigenvalues in homogenization problems of elasticity,* ZAMM **65** (1985), 13-17.

[7] O.A.Oleinik, A.S.Shamaev, G.A.Yosifian, *Homogenization of eigenvalues and eigenfunctions of the boundary value problems in perforated domains for elliptic equations with non-uniformly oscillating coefficients,* in: Current Topics in Partial Differential equations. Papers dedicated to Prof. Mizohata, ed. by Y.Ohya, K.Kasahara, N.Shimakura, Kinokuniya Co. Tokyo, 1986, 187-216.

[8] J.L.Lions, E.Magenes, *Problèmes aux limites non homogènes et applications,* Vol.1, Dunod, Paris, 1968.

[9] O.A.Oleinik, *Boundary value problems for linear elliptic and parabolic equations with discontinuous coefficients,* Izvestiya AN SSSR, Ser. Mat. **25** (1961), 3-20.

[10] V.A.Kondratiev, *On the solvability of the first boundary problem for strongly elliptic equations,* Trans. Mosc. Math. Soc. **16** (1967), 293-318.

[11] O.A.Oleinik, *On homogenization problems,* in: Trends and Applications of Pure Mathematics to Mechanics, Lect. Notes in Phys. **195**, Springer Verlag, 1984, 248-272.

[12] O.A.Oleinik, A.S.Shamaev, G.A.Yosifian, *On asymptotic expansion of solutions of the Dirichlet problem for elliptic equations and the elasticity system in perforated domains,* Dokl. AN SSSR **284** (1985), 1062-1066.

[13] O.A.Oleinik, A.S.Shamaev, G.A.Yosifian, *On eigenvalues of boundary value problems for the elasticity system with rapidly oscillating coefficients in perforated domains,* Matem. Sbornik **132** (174), n.4 (1987), 517-531.

[14] O.A.Ladyzhenskaya, N.N.Uraltseva, *Linear and Quasi-linear Equations of Elliptic Type,* Nauka, Moscow, 1973.

Institute for Problems in Mechanics
Academy of Sciences of the USSR
Prospekt Vernadskogo 101
MOSCOW

DISCRETE EXTERIOR MEASURES AND THEIR MEANING IN APPLICATIONS

LIVIO CLEMENTE PICCININI

Dedicated to Ennio De Giorgi on his sixtieth birthday

1. Introduction. E.De Giorgi uses to deliver very original lectures during his courses at Scuola Normale Superiore in Pisa. Among the many subjects he dealt with, we recall that the course of the academic year 1971-72 was mostly devoted to the theory of exterior measures. Some of it was collected in the "Quaderno della Scuola Normale" [2]; namely Chapter 3 contains the fundamentals of exterior measures (following mostly Carathéodory's approach [1]), while chapter 4 contains the definition of an original geometric measure, used to solve problems of minimal surfaces with obstacles.

The course of the following year, unfortunately not available but as handwritten notes, dealt with a general theory of exterior measures. Particular stress was laid upon the theory of exterior integral for functions whatsoever. In particular it could be found there a subtle analysis of the possible definitions for the measurability of functions, and a clear partition between those properties of the integral which still hold in absence of measurability and those that get lost.

Some years later, during a course that I held at ISAS in Triest,

I happened to repeat some parts of De Giorgi's course of the year 1972-73; so I had the problem of finding examples both suitable and convincing.

Everybody that teaches measure theory knows that, speaking of Lebesgue-Stiltjes measure, it is very hard to exhibit convincing counterexamples arising from the non measurability of the functions. As a matter of fact one should use an exterior measure (hence endowed with countable sub-additivity), where there are "few" measurable sets, but not "too few", and at the same time this measure should not be too abstract, in order that computations could be easily performed and not only believed to exist.

An exterior measure that seems to enjoy all these properties is at hand, but it appears to be so trivial that it be good only for the examples given to the students during a lecture, not even to be written on a text-book.

A fortiori it should not even be mentioned in a scientific paper. It is only thanks to computer science that it achieves its right to an official life out of the darkness of triviality. Actually in the last two years, while working to applied problems of pattern recognition and of optical character recognition (OCR), I remarked that this measure, so poor at first sight, appears to have really deep meanings.

What is important, as we shall see later, is just its enormous richness of (constructable) non measurable sets. This fact reverts to some extent the traditional way we look at measure theory, where non measurable sets are considered rather a regrettable accident. Even more in most theories we start quite from a σ-algebra of subsets and we define the measure only on this σ-algebra, defining the exterior measure only as a generalization of the "good" measure.

The computer science arguments that I shall introduce later were based not only on the theoretical computer science but also on the study of perception psychology. This is a field in which discrete exterior measures should be meaningful. They seem actually to have the power of explaining many optical illusion phenomena. I have formulated some hypothesis, specially concerning Poggendorf's illusion and similar, but as it is obvious this is a research field where wide experimentation is needed in order to get meaningful evidence; so it is too early to present undue theories now. I wish to thank Prof. G.Vicario of Padova, who introduced me to the study of perception psychology in many conversations we had at our department. I express too the hope of experimenting with his research team these theories as soon as possible.

2. Discrete exterior measures. First of all we recall the definition of exterior measure.

Let X be a set and let P be the family of all its subsets. A function $\Phi : P \to [0, +\infty]$ satisfying the following properties

i) $\phi(\emptyset) = 0$

ii) monotonicity: $A \subset B \Longrightarrow \Phi(A) \leq \Phi(B)$

iii) countable subadditivity:

$$F \subset \bigcup_i F_i \quad \Longrightarrow \quad \Phi(F) \leq \Sigma\Phi(F_i)$$

is called exterior measure on the set X.

Remark that monotonicity is actually an automatic consequence of properties i) and iii).

The definition of Carathéodory's measurability stresses the fact that measurable sets are "consistent tools for measuring" rather then their additivity property. Actually we give the following definition

Let Φ be an exterior measure on the space X. A subset M is said to be C-measurable for Φ if for any subset $E \subset X$ it holds

$$\Phi(E) = \Phi(E \cap M) + \Phi(E - M).$$

The usual properties still hold without restriction: the family of C-measurable subsets is a σ-algebra and the exterior measure restricted to this σ-algebra is still completely additive on disjoint subsets.

In order to build exterior measures there are two main constructions due to Carathéodory. We shall use the first of them, while the second is usually required, together with the first, in almost all geometric measures, with the only apparent exception of De Giorgi's measure ([2], chap. 4).

The first Carathéodory's construction is as follows:

Let **F** *be a family of parts of* X *containing* $\emptyset$; *let* α *be a set function (not necessarily subadditive) such that*

$$\alpha : \mathbf{F} \to [0, +\infty] \quad \textit{and} \quad \alpha(\emptyset) = 0.$$

We set, for each $E \subset X$,

$$\alpha^*(E) = \inf\{\Sigma\alpha(F_i), F_i \in \mathbf{F}, \cup F_i \supset E\}$$

if there exists at least one sequence $\{F_i\}$ *that covers* $E, \alpha^*(E) = +\infty$ *otherwise.*

It holds automatically that α^* is an exterior measure; in general it is $\alpha^* \leq \alpha$, but whenever α is countably subadditive on its definition domain, then α^* is an extension of α, namely $\alpha^* = \alpha$ on **F**. Furthermore the sets that are C-measurable for α (remark that in this case they do not necessarily belong to **F**) are C-measurable also for the generated exterior measure α^*.

Carathéodory's construction allows, as it is well known, a very simple definition of Lebesgue-Stiltjes measures. Actually it is enough

i) set as family **F** the set of the intervals of **R** of the form $[a, b[$, (resp. $]a, b]$)

ii) Assign a monotone non-decreasing function ϕ left semicontinuous (resp. right)

iii) define $\alpha([a, b[) = \phi(b) - \phi(a)$, $\alpha(\emptyset) = 0$.

Then, using Heine-Borel compactness theorem, it follows that α is countably subadditive, hence α^* is one of its extensions. As for C-measurable sets we note that the half lines $]-\infty, b[$ are C-measurable for the set function α, hence the σ-algebra generated by them is C-measurable for the exterior measure α^*. In particular the Borel sets are C-measurable.

In the construction of the discrete external measures one proceeds essentially in the same way, only the family **F** is made poorer.

Definition 1. *Let* **A** *be a subset of* **R** *with no accumulation point. We choose for* **F** *the family of the intervals* $[a, b[$ *with* $b \geq a$, $a \in \mathbf{A}, b \in \mathbf{A}$. *Let now* ϕ *be a non-decreasing funtion: we let, for any* $E \in \mathbf{F}$, $\alpha(E) = \phi(b) - \phi(a)$.

We say that α^*, *obtained from* α, *using first Carathéodory's construction, is the discrete exterior measure associated to* A *and* ϕ.

The simplest example is given by the *indicator measure,* which is obtained choosing $\mathbf{A} = \mathbf{Z}, \phi = x$. This measure simply counts the number of unit intervals (left-closed and right open) which the generic set E intersects. As we shall see in section 3 this is of fundamental importance in applications to computer science.

For sake of convenience we give a notation also for the homothetic of this measure:

Definition 2. *For each $p \in \mathbf{R}^+$ we denote by $MI(p)$ the indicator measure obtained letting in def.1 $\mathbf{A} = \{x : x = kp, k \in \mathbf{Z}\}$ and $\phi = x/p$.*

The generalization of choosing the left ends of the intervals from one set and the right ends from another one does not seem interesting, since the family of C-measurable sets would be too much decreased.

We remark that the indicator measure, that yet has some resemblance with the Lebesgue-Stiltjes measure associated to the step-function ($\phi = x - 1$ if $x \in \mathbf{Z}, \phi = [x]$ otherwise), coincides with it only on the sets which are unions of intervals starting on an integer. As a matter of fact this Lebesgue-Stiltjes measure works as a test of the characteristic function performed only at integer values, while the indicator measure is non-zero (except trivial cases) as soon as the intersection with the interval is non-empty. Of course this fact deeply reduces the family of measurable sets.

Theorem 1. *Let associate to $\mathbf{A}$ and to ϕ a discrete exterior measure, then the family of C-measurable sets contains the σ-algebra generated by the half-lines of the form $[-\infty, a[$ with $a \in \mathbf{A}$. In particular if ϕ is strictly monotone then the C-measurable sets are exactly those formed by (at most) countable unions of elements of $\mathbf{F}$ and half-lines of the form $[-\infty, a[$ or $[a, +\infty[$, where $a \in \mathbf{A}$.*

PROOF. For each interval of the form $[a, b[$, non trivial, namely decomposable into $[a, c[\cup [c, b[$, where $a, b, c \in A$, it holds obviously

$$\alpha([a,b[) = \alpha(]-\infty, c[\cap[a,b[) + \alpha(]-\infty, c[-[a,b[) = \alpha([a,c[) + \alpha([c,b[)$$

As we recalled before the sets which are measurable for the function α are also C-measurable for α^*.

If ϕ is not strictly monotone there can be non-empty sets of measure 0, which therefore happen to be C-measurable. On the contrary if ϕ is strictly monotone it follows that any non-empty set has positive exterior measure. If there were a measurable set M not belonging to those we listed in the statement, there should exist at least one interval that could not be further decomposed, say $E = [a, b[$, belonging to $\mathbf{F}$ such that $M \cap E \neq \emptyset$ and $M - E \neq \emptyset$, for which it may hold

$$\alpha^*(E) = \alpha^*(M \cap E) + \alpha^*(E - M).$$

But this is a contradiction inasmuch on E there exists no covering smaller then E itself; it should then follow $\alpha^*(E) = \alpha(E) = 2\alpha(E)$, hence $\alpha(E) = 0$, thus contradicting the strict monotonicity hypothesis which implies $\alpha(E) > 0$ as soon as $E \neq \emptyset$.

QED

The (exterior) integral with respect to any esterior measure μ is directly defined for any function, not depending on its measurability. Nonetheless for sake of simplicity in the notations we shall suppose that an integral is already defined for real monotone functions.

Definition 3. *Let μ be an exterior measure on the space X. For each*

$$f : X \to [0, +\infty[$$

we let

$$\int f d\mu = \int_0^{+\infty} \mu(\{x : f > \lambda\}) d\lambda.$$

We recall that a function f is said to be C-measurable for the exterior measure μ if all its level sets $\{x : f > \lambda, \ \lambda \in \mathbf{R}\}$ are C-measurable.

From the general theory of integration we get that the integral is a monotone functional on all the functions, while additivity in general holds only for measurable functions.

A restricted form of additivity holds for all the functions, namely additivity by truncation:

let $f_1 = (f - k) \vee 0$ *and* $f_2 = f \wedge k$ *for* $k > 0$
then it holds $\int f d\mu = \int f_1 d\mu + \int f_2 d\mu$.

Then usual theorems on the limit in the integral hold without restriction on the form of the convergence when we consider sequences of C-measurable functions; if on the contrary we wish to consider sequences of functions whatsoever it is necessary to introduce stronger hypothesis on the convergence, the simplest of which is given by the uniform convergence.

The indicator measure provides elementary counterexamples: actually the C-measurable functions are only those made up by "steps" with integer and left closed and right open. In particular non-trivial continuous functions are not C-measurable.

In order to have an example of non-additivity for continuous functions we let

$$f_1(x) = \begin{cases} x & \text{if } 0 \le x \le 1 \\ 2 - x & \text{if } 1 \le x \le 2 \\ 0 & \text{elsewhere} \end{cases}$$

and respectively

$$f_2(x) = \begin{cases} x - 1 & \text{if } 1 \le x \le 2 \\ 3 - x & \text{if } 2 \le x \le 3 \\ 0 & \text{elsewhere.} \end{cases}$$

Then it holds

$$\int f_1 dMI(1) = 2, \int f_2 dMI(1) = 2, \quad \text{while} \quad \int (f_1 + f_2) dMI(1) = 3.$$

Remark anyhow that subadditivity of the integrals still holds. In our case it is related in an obvious way to the subadditivity of maximum (or upper bound).

In order to have a counterexample to Lebesgue theorem of dominated convergence we may consider the sequence of functions

$$f_n = x^n \ \text{ if } \ 0 \le x \le 1, \qquad f_n = 0 \text{ elsewhere.}$$

that are not C-measurable for the measure $MI(1)$, and converge non-unifomly to the function

$$f = 1 \ \text{ if } \ x = 1, \qquad f = 0 \text{ elsewhere.}$$

Therefore we have, for each n, $\int f_n dMI(1) = 2$, while $\int f dMI(1) = 1$. For the exterior measure there exists for each set E a C-measurable set M such that $\mu(M) = \mu(E)$, where $M \supset E$. In particular for the indicator measures the set M which enjoys this property is unique. Since in what follows it will play a particular task we shall use a special symbol for it:

Definition 4. *Let $MI(p)$ be an indicator measure. For any set E we shall denote by the symbol $M(E;p)$ the unique C-measurable set for the measure $MI(p)$ which contains E and for which it holds*

$$MI(p)(E) = MI(p)(M(E,p)).$$

We shall denote by $H(E;p)$ the homothetic of $M(E;p)$ defined as

$$H(E;p) = \{x : px \in M(E;p)\}.$$

We shall need also the corresponding measures and operations in the two-dimensional space. We shall therefore use the notations $MI(p,q), M(E;p,q), H(E;p,q)$, where the first parameter is referred to the first coordinate and the second to the second coordinate.

3. Use of indicator measures in computer science. The most interesting cases in which indicator measure may be explicitely used belong to the theory of pattern recognition.

Obviously there the first step of grabbing the image is solved at the hardware level, very often using a scanner. This phase corresponds to the physiological activity of retina. Since each single cell averages the signal, it would be improper to think that grabbing the image could be described by an indicator function. Actually monotonicity is preserved, but subadditivity can fail, since the average analogic signal is that one that approximates the nearest digital level.

It could happen that two signals of intensity 1.3, taken separately, are digitalized as signals of intensity 1, while joined together in the same cell they would reach the level 2.6, digitalized as 3, hence a signal greater than the sum of two elementary signals.

This non-additivity explains some apparent effects of bad scanning of an image which are due to the resolution capability of the scanners, which is inferior to the corresponding capability of the human eye.

Anyhow when scanning written texts the grey scale is usually eliminated, and the perception threshold is taken to be equal to the light intensity of the paper, so the result is very near to the transformation $M(E; 1, 1)$.

Anyhow we are more interested to transformations of a digitalized image that follow its grabbing. The reason why indicator measures are of use seems to lie in the fact that they are performed in a natural and fast way on the processors used in the personal computers. In particular it is not necessary to encode the raster of the image, since operations are rather performed on the bytes of the raster themselves.

Of course the same measures and transformations can be obtained in different ways, which, according to the processor, may be faster or slower. For example we have the equality

$$MI(E, 16) = MI(H(E, 4), 4).$$

On the Macintosh SE, that we used in the tests, the sequence on the right is twice faster then the left one. Therefore, even when we indicate the operation in the simplest way, it must be intended that it means the best of its actual implementations.

The first picture (see fig.1) shows the effect of the trasformation $H(E; ., .)$. In the first example the set E is a geometric figure. Parameters are not exact power of two, so we get an overlapping effect due to the pattern of the rectangle on the left side. The second example shows a section of a text grabbed by a 300 dots/inch scanner (hence enlarged some 4 times) which undergoes the transformation $H(E; 16, 1)$. We can remark, especially in the second example, the effect of compactification of the image, which allows us to appreciate the different behaviour of a discrete exterior measure versus averaging.

The second example is of fundamental importance in Optical Character Recognition (OCR), since not only it supplies a useful tool for dividing the text into single lines, but also allows to evaluate the body of the letters, their ascent and their descent. This thinner analysis, according both to the experience of the author and to the existing literature is of great use (compare [4] for many references),

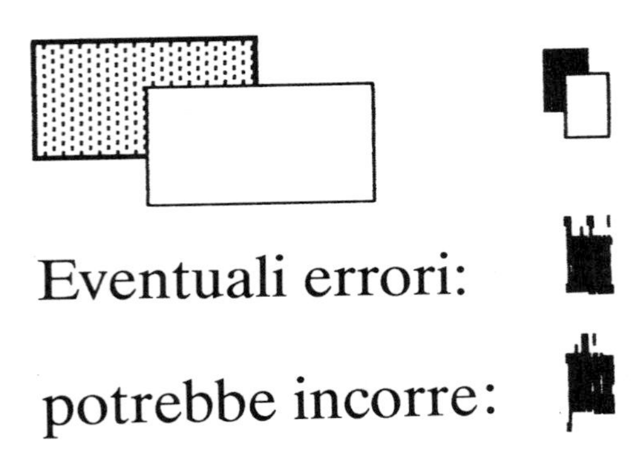

Figure 1

but apparently has not been used in the commercial OCR's. Personally, on the contrary, in the OCR we are preparing together with Dr. Brainik, of the Università di Udine, we used systematically this sharper analysis; recalling also the experiments in perception psychology, it seems really to be a powerful tool for solving ambiguous strings and for validating results with greater precision.

In this analysis one starts from each horizontal line of pixels, say $E(n)$, then evaluates the number

$$A(n) = MI(E(n), 16) \ (= MI(H(E(n), 16), 1)).$$

Starting from the top of a line of text we find at first $A(n) = 0$, what amounts to interspace between lines, then we find a level at which small positive values appear (the ascent), a second level at which we approach the maximum (the threshold usually is 75% of the maximum); at this level we have reached the top of the body of the letters. In the exit we find a first level where we go suddenly under the 50% of the maximum, and this corresponds to the base line and finally we find the last step at which $A(n)$ becomes 0 again (the low of the descent). Of course image can be noisy, so in the practice some step values must be corrected, both filtering the original image, and, more efficiently, filtering $H(E; 16, 1)$.

The reader can remark that the same analysis performed on the original image, namely using $B(n) = MI(E(n), 1)$, would lead to results much less sure. Actually in this case the lack of C-measurability of the sets allows some form of regularization. Similar techniques are used to determine the global structure of the page, where the search of columns uses transformations of the type $H(E; 4, 64)$. Possible horizontal lines are found in a similar way (after suitable changes to the original image) using transformations of the type $H(NOT(E); 4, 64)$.

We come now to describe a case where the indicator measure really operates as a binary indicator of presence/absence of the phenomenon. The situation seems to emulate the psychological process by which we can establish if a sheet of paper is perfectly clean or if it has spots. At first glance we have only global perception that neither estimates the exact number of existing spots nor estimates the average. In fact a small spot is perceived if the average of the sheet would be under the threshold of perception. Perhaps this fact means that also in the human brain there is a phase of image processing that is neither statistical nor analytic, but rather ressembles discrete exterior measures.

In computer science these techniques are well suited for detecting rare events, say a few black pixels on the screen, one word out of a vocabolary, or the search according to multiple keys in a data base. A tecnique that goes in this direction can be found in [3], where its high efficiency is proved both theoretically and experimentally.

In view of the way how indicator measures work, it will be necessary a preprocessing that reduces each event to a single bit of information, that is to be localized, in order to state the original position of the event in a straightforward way (otherwise the procedure would lose much of its efficiency): as we shall see in the simple example that follows, the function H allows the simplest transformation. In pattern recognition of course more complicated procedures are required in order to reduce phaenomena to single bits of information.

Such procedures appear to be strongly influenced by the a-priori hypothesis that are done about the phaenomena that are to be investigated and distinguished between each other. Such problem is by no means new, and its theoretical analysis was begun much before the existence of computer science as a standing alone field of research. We quote for all Kant, in particular notes I and II to section 13 of

Prolegomena [5].

Coming back to our example we illustrate the reduction of the information to a single bit. Suppose for sake of simplicity to have a vocabulary in which each word has the same lenght (in the figures, for lack of space, the lenght is three words of three letters, but the real case would use a vocabulary of many thousands of words, with a greater number of letters: each letter is represented by one byte). The vocabulary is represented by the set T.

Suppose now a defective string S is assigned (in the figure the first and the third letter are assigned); our goal is to find (that is to associate a bit) to those words of the vocabulary T which coincide with the string S at the places where it is defined. This problem is better solved using indexes when the vocabulary is ordered and *the defective string contains the initial letters.* In the general case it seems to be needed a comparison with all the elementes of the vocabulary, or particulary sophisticated indexes are required (which easily grow to dimensions much greater than the vocabulary itself).

The figure shows the method of indicator funtions: it consists of sending a copy of the defective string *at the same time* to the whole vocabulary, namely $V = H(S;1;1/L)$, where L is the cardinality of the vocabulary. The symmetric difference operation leaves white bytes where the match is found, for those positions that are defined. Using the mask M the neutralized positions are eliminated: the result is the set R where only solutions to the problem appear to be white. The reduction to a single bit is performed by $H(R;b^*p,1)$, where $b = 8$ is the lenght of any letter, while p is the lenght of the single words. Since such image appears in the negative an inversion takes place, which associates bit 1 to each match that has been found.

As one sees the result is a set of horizontal width 1 and length L. For the next analysis this dimension is not very functional, hence in practice we shall use line of 4, 8 or 16 words, according to the processor, and change the structure of V and M correspondingly.

The following step is the localization of rare events of one single bit. Let $p << 1$ be the probability of the event. We fix a sequence $p_i, q_i, i = 1, 2, ..., n$, such that, denoting by $k_i = p_i q_i$, it holds $\Pi k_i = 1/p$ and we create by recursion the sets

$$E_0 = E; \quad E_i = H(E_{i-1}; p_i, q_i))$$

For the next analysis the optimal values of k_i would be 2 or 4 according to the machine but problems of memory allocation and of

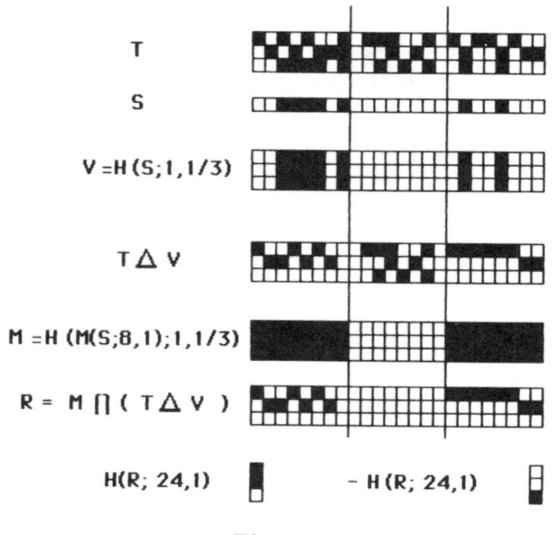

Figure 2

time elapsed in constructing the sequence E_i suggest to actually use the values 8 or 16.

The sequence $E_i, i = n, n-1, ..., 0$ may be considered a sequence of images of greater and greater detail. Hence the analysis can be performed by a depth first search. That is we find recursively the squares from which the event was drawn.

Find the $Q(n, s_n) \subset E_n$ such that $MI(Q(n, s_n)) = 1$.
Find the $Q(m-1, s_m) \subset H^{-1}(Q(m, s_m); p_m, q_m)$ such that $MI(Q(m-1, s_m)) = 1$.

So we arrive in optimal time to localize the event. Remark that the use of non-measurable sets gives a signal that in its smallest measurable overset the event *has happened one or more times.*

References

[1] C.Carathéodory, *Vorlesungen über reelle Funktionen,* Teubner, 1918.

[2] E.De Giorgi, L.C.Piccinini, F.Colombini, *Frontiere orientate di misura minima e questioni collegate,* Quaderni della Scuola

Normale di Pisa, 1972.

[3] M.Edahiro, I.Kokubo, T.Asano, *A new Point-Location Algorithm and its Practical Efficiency-Comparison with existing Algorithms,* ACM Transactions on Graphics **3** (1984), 86-109.

[4] S.Kahan, T.Pavlidis, H.Baird, *On the recognition of Printed Characters of any Font and Size,* IEEE Trans. on Pattern Analysis and Machine Intelligence **9** (1987), 247-288.

[5] I.Kant, *Prolegomena zu einer jeden künftigen Metaphysik, die als Wissenschaft wird auftreten können,* 1783.

Dipartimento di Matematica e Informatica
Università di Udine
Via Zanon 6
I-33100 UDINE

AN EMBEDDING THEOREM

Giorgio Talenti

Dedicated to Ennio De Giorgi on his sixtieth birthday

Suppose A is a *Young function;* i.e. A maps $[0,\infty[$ into $[0,\infty[$, is convex and vanishes at 0. Let u be real-valued, locally integrable in euclidean n-dimensional space $\mathbf{R}^n$; assume the (distributional) derivatives of u satisfy

$$\int_{\mathbf{R}^n} A(|\text{grad } u(x)|)dx < \infty \tag{1}$$

and the support of u is bounded.

Theorem. *If A grows so fast that*

$$\int^{+\infty} \left[\frac{r}{A(r)}\right]^{\frac{1}{n-1}} dr < \infty, \tag{2}$$

then

$$\text{ess.}\sup |u| < \infty, \tag{3}$$

i.e. u is (essentially) bounded.

PROOF. Recall that

$$\operatorname{div}\,(x|x|^{-n}) = nC_n\delta(x),$$

where

$$C_n = \frac{\pi^{\frac{n}{2}}}{\Gamma(1+\frac{n}{2})},$$

the measure of the unit n-dimensional ball, and

$$\delta = \text{Dirac mass}.$$

Thus

$$nC_n u(x) = \int_{\mathbf{R}^n} \left(Du(y), \frac{x-y}{|x-y|^n} \right) dy \tag{4}$$

by the very definition of distributional derivatives, since u has compact support. Here we denote gradients dy D, and scalar products between vectors by (,). Therefore

$$nC_n|u(x)| \leq \int_{\mathbf{R}^n} |Du(y)|\; |x-y|^{1-n} dy. \tag{5}$$

A theorem by Hardy & Littlewood says that, if ϕ and ψ are nonnegative and measurable, then

$$\int_{\mathbf{R}^n} \phi(x)\psi(x)dx \leq \int_0^{+\infty} \phi^*(s)\psi^*(s)ds,$$

where stars denote *decreasing rearrangements,* i.e.

$$\phi^*(s) = \sup\left\{t \geq 0 : \text{meas}\{x \in \mathbf{R}^n : \phi(x) > t\} > s\right\}.$$

See [2], for instance. Notice the pair

$$\phi(x) = |x|^{1-n}, \quad \phi^*(s) = \left(\frac{C_n}{s}\right)^{1-\frac{1}{n}}.$$

Thus inequality (5) implies

$$nC_n^{\frac{1}{n}} \operatorname{ess.sup} |u| \leq \int_0^V s^{-1+\frac{1}{n}} |Du|^*(s) ds, \tag{6}$$

where

$$V = \text{meas (support of } u) \tag{7}$$

and $|Du|^*$ is *the decreasing rearrangement of the lenght of* Du. Inequality (6) is sharp: as is easy to see, equality holds in (6) if $u(x) = f(|x|)$, where f is nonnegative decreasing convex and compactly supported.

For the sake of completeness, let us incidentally quote the following inequality:

$$nC_n^{\frac{1}{n}} u^*(s) \leq \int_0^{V-s} (t+s)^{-1+\frac{1}{n}} |Du|^*(t)dt,$$

which holds for $0 \leq s \leq V$ and generalizes (6) considerably.

Our next task is to estimate the right-hand side of (6). To this end we introduce

$$\tilde{A}(s) = \sup\{rs - A(r) : 0 \leq r < \infty\},$$

the *Young conjugate* of A, i.e. the smallest Young function such that

$$rs \leq A(r) + \tilde{A}(s)$$

for all $r \geq 0$ and $s \geq 0$.

Let n' stand for $n/(n-1)$ and let B denote the Young function whose conjugate, $\tilde{B}$, is given by

$$\tilde{B}(s) = n' \int_1^{+\infty} \tilde{A}(st) \frac{dt}{t^{n'+1}}. \tag{8}$$

Hypothesis (2) and lemma below guarantee that the right-hand side of (8) is *finite* for every $s > 0$: thus B is actually *non-zero*.

Incidentally,

$$\tilde{B}(s) = n' \int_s^{+\infty} \tilde{A}(t) \left(\frac{s}{t}\right)^{n'} \frac{dt}{t},$$

a convolution on the multiplicative topological group of positive real numbers - where dt/t is the Haar measure;

$$\left(\frac{s}{n'}\frac{d}{ds} - 1\right) \tilde{B}(s) + \tilde{A}(s) = 0,$$

a differential equation;

$$\tilde{B}(s) \geq \tilde{A}(ns),$$

as an easy application of Jensen inequality shows.

The right-hand side of (6)

$$\begin{aligned}
&= \frac{1}{sV^{1-\frac{1}{n}}} \int_0^V s\frac{V^{1-\frac{1}{n}}}{t^{1-\frac{1}{n}}} |Du|^*(t)dt \\
&\leq \frac{1}{sV^{1-\frac{1}{n}}} \left(\int_0^V A(|Du|^*(t))dt + \int_0^V \tilde{A}\left(s\frac{V^{1-\frac{1}{n}}}{t^{1-\frac{1}{n}}} \right) dt \right) \\
&= \frac{1}{sV^{1-\frac{1}{n}}} \int_{\mathbf{R}^n} A(|Du|)dx + \frac{V^{\frac{1}{n}}}{s} \tilde{B}(s),
\end{aligned}$$

since

$$\int_0^V A(|Du|^*(t))dt = \int_{\mathbf{R}^n} A(|Du|)dx$$

by the equimeasurability of rearrangements. Here s is any positive number. Hence (6) gives

$$s(nC_n^{\frac{1}{n}} V^{\frac{1}{n}} \text{ess. sup} |\text{u}|) \leq \frac{1}{\text{V}} \int_{\mathbf{R}}^{\text{n}} A(|Du|)dx + \tilde{B}(s)$$

for every $s > 0$. In other words, we must have

$$B(nC_n^{\frac{1}{n}} V^{\frac{1}{n}} \text{ess. sup} |u|) \leq \frac{1}{V} \int_{\mathbf{R}^n} A(|Du|)dx. \tag{9}$$

Inequality (9) yields an estimate of ess.sup. $|u|$ in terms of $\int_{\mathbf{R}^n} A(|Du|)dx$ and meas(support of u). The theorem follows.

Lemma. *If (2) holds, then the Young conjugate, $\tilde{A}$, of A satisfies*

$$\int^{+\infty} \tilde{A}(s)s^{-1-n/(n-1)}ds < \infty. \tag{10}$$

PROOF. Recall the alternative formula (see e.g. [3])

$$\tilde{A}(s) = \int_0^s a^{-1}(t)dt,$$

where a is the *derivative* of A, i.e.

$$A(r) = \int_0^r a(t)dt,$$

and a^{-1} is a *generalized inverse* of a, i.e.

$$a^{-1}(t) = \inf\{t' \geq 0 : a(t') \geq t\}.$$

Fix any positive number k. An integration by parts shows

$$\int_k^{+\infty} \tilde{A}(s)s^{-1-n/(n-1)}ds \leq \left(1 - \frac{1}{n}\right)\tilde{A}(k)k^{-n/(n-1)} + \\ +(n-1)\left(1 - \frac{1}{n}\right)\int_k^{+\infty} a^{-1}(s)d[-s^{-1/(n-1)}].$$

We have

$$\int_k^{+\infty} a^{-1}(s)d[-s^{-1/(n-1)}] \leq \int_h^{+\infty} rd[-a(r)^{-1/(n-1)}],$$

since $a^{-1}(a(r)) \leq r$. Here h stands for $a^{-1}(k)$. Another integration by parts gives

$$\int_h^{+\infty} rd[-a(r)^{-1/(n-1)}] \leq ha(h)^{-1/(n-1)} + \int_h^{+\infty} a(r)^{-1/(n-1)}dr.$$

Obviously

$$a(r) \geq \frac{A(r)}{r}.$$

The lemma follows.

Remarks. Our theorem includes a well-known embedding theorem for Sobolev spaces, see [1, chap.5] for instance.

Also, it improves [1, Thm 8.35], where (3) is derived from (1) under the following hypothesis:

$$\int^{+\infty} A^{-1}(t)t^{-1-1/n}dt < \infty.$$

Note that such a condition is equivalent to the following

$$\int^{+\infty} A(r)^{-1/n}dr < \infty,$$

and the latter is more stringent than (2). Indeed

$$\int_h^{+\infty} \left(\frac{r}{A(r)}\right)^{\frac{1}{n-1}} dr \leq \left(hA(h)^{-1/n} + (1-\frac{1}{n}) \int_h^{+\infty} A(r)^{-1/n}dr\right)^{\frac{n}{n-1}}$$

for any $h > 0$, because $A(r)/r$ increases and

$$\left(\int_0^{+\infty} \phi(r)^{n/(n-1)}dr\right)^{1-\frac{1}{n}} \leq (1-\frac{1}{n}) \int_o^{+\infty} r^{-1/n}\phi(r)dr$$

for any *nonnegative decreasing* function ϕ.

References

[1] R.A.Adams, *Sobolev spaces,* Academic Press (1975).
[2] Hardy, Littlewood, Polya, *Inequalities,* Cambridge University Press (1964).
[3] A.W.Roberts, D.E.Varberg, *Convex functions,* Academic Press (1973).

Istituto Matematico "U.Dini"
Università di Firenze
Viale Morgagni 67/A
I-50134 FIRENZE

NONLOCAL EFFECTS INDUCED BY HOMOGENIZATION

LUC TARTAR

Dedicated to Ennio De Giorgi on his sixtieth birthday

It is indeed surprising that a weak limit of inverses of second order elliptic operators is also the inverse of a second order elliptic operator. Of course in order to obtain this result one has to consider a class larger than the class of operators of the form

$$-\sum_{i=1}^{N} \frac{\partial}{\partial x_i}\left[a(x)\frac{\partial}{\partial x_i}\right] \tag{1}$$

with a positive function a, and include the operators of the form

$$-\sum_{i,j=1}^{N} \frac{\partial}{\partial x_i}\left[a_{ij}(x)\frac{\partial}{\partial x_j}\right] \tag{2}$$

where the functions a_{ij} are the entries of a symmetric positive matrix $A(x)$.

If one chooses to consider the physical interpretation of (stationary) diffusion of heat or electricity, (1) corresponds to a material which is locally isotropic, while (2) corresponds to a locally anisotropic material. Although there was no mathematical definition of what they meant by an effective coefficient, physicists were aware of this fact long before mathematicians; their knowledge was certainly based on the fact that they knew from experience that anisotropic materials existed (crystals for example).

After the first mathematical results in this direction obtained by S.Spagnolo, an Italian school grew influenced by pioneering ideas of E.De Giorgi. G-convergence and then Γ-convergence were then the usual mathematical tools involved to treat those problems of physics which were connected to minimizing functionals.

While I was visiting De Giorgi in Pisa, he mentioned to me some rather different questions that he had been investigating. His motivation seemed quite different from mine, as he was looking at interesting mathematical questions that were challenging from the point of view of weak convergences, and his remarks were connected to linear hyperbolic equations; I was myself motivated by the list of questions of mechanics and physics that I knew and I was becoming more and more convinced that some of the strange rules invented by physicists could be related to homogenization results for hyperbolic equations, and there was no real mathematical understanding of this question.

De Giorgi's remark was related to the equation

$$(3)\qquad \begin{cases} \dfrac{\partial u_\epsilon(x,y,t)}{\partial t} + \sin\dfrac{y}{\epsilon}\cdot\dfrac{\partial u_\epsilon(x,y,t)}{\partial x} = 0 \\ u_\epsilon(x,y,0) = v(x,y) \end{cases}$$

whose solution u_ϵ converges only weakly as ϵ tends to 0; from the formula

$$(4)\qquad u_\epsilon(x,y,t) = v\left(x - t\sin\frac{y}{\epsilon}, y\right)$$

one finds that u_ϵ converges weakly to u_o given by

$$(5)\qquad \begin{aligned} u_o(x,y,t) &= \frac{1}{2\pi}\int_o^{2\pi} v(x - t\sin z, y)dz = \\ &= \int_{-1}^{+1} v(x - t\lambda, y)\frac{d\lambda}{\pi\sqrt{1-\lambda^2}}. \end{aligned}$$

Equation (3) describes a phenomenon of propagation in the x direction and the remark shows that if the coefficients oscillate in a different direction y, it induces for the limiting equation (i.e. at a macroscopic level) some kind of diffusion in the propagating direction, but not of the usual kind because here we have a finite propagation speed effect.

It was not clear to me what would be a natural class of partial differential equations that would contain equations like (3) and be stable by homogenization; actually as I was aware of more general questions for fluids and of the attempts of D. McLaughlin, G.Papanicolaou and O. Pironneau in that direction, I had a wider goal that included semilinear hyperbolic systems.

The apparition of nonlocal effects by homogenization was not new to me. In realistic situations of physics, the earliest reference that I knew (in a language that I could understand) was some work of E. Sanchez-Palencia who had treated some examples related to acoustics, viscoelasticity and electromagnetism in a periodic setting. At that time, I did not understand these results from the continuum mechanics/physics point of view, but I knew what was the mathematical tool to attack that kind of questions because J.L.Lions had made a similar remark in one of his lectures at Collège de France: in order to show that homogenization could sometime lead to pseudo-differential operators, he had considered the following academic example

$$\frac{d^2}{dt^2}(A_\epsilon u_\epsilon) + B_\epsilon u_\epsilon = f; \quad u_\epsilon(0) = v; \quad \frac{du_\epsilon}{dt}(0) = w \tag{6}$$

where A_ϵ and B_ϵ were second order elliptic operators with periodically oscillating coefficients; he had applied a Laplace transform in t which gave him a one parameter homogenization problem

$$(p^2 A_\epsilon + B_\epsilon) L u_\epsilon(p) = Lf(p) + w + pv \tag{7}$$

for the Laplace transform Lu_ϵ of u_ϵ and so the limit u_o satisfied an equation of the type

$$C(p) L u_o(p) = Lf(p) + w + pv \tag{8}$$

where $C(p)$ denoted the homogenization limit of $(p^2 A_\epsilon + B_\epsilon)$. The main remark was that, as $C(p)$ would generally not be a polynomial in p, (8) would give a pseudo-differential operator after applying the inverse Laplace transform.

In this example talking about pseudo-differential operators was a fancy way to say that the limiting equation was showing a memory effect (or relaxation effect for physicists). The causality principle, saying that only the present and the past were necessary to deduce the future, was an easy consequence of the homogenization procedure (which is also a fancy name for expressing the fact of taking many weak limits), and I had started in 1980 investigating which memory kernels could appear by this homogenization process.

1. A simpler problem. I had decided to work as a starting point with a much simpler equation

$$\frac{\partial u_\epsilon(x,t)}{\partial t} + a_\epsilon(x)u_\epsilon(x,t) = 0; \quad u_\epsilon(x,0) = v(x) \tag{9}$$

which I could consider as a model describing a mixture of products having different rates of decay (for example radioactive materials). Considering the oscillatory nature of the solution, I then realized that having an intial data without oscillations was not really physically relevant, and that a way to have a problem that would not emphasize a special creation time like $t = 0$, could be to consider a variant

$$\begin{aligned} &\frac{\partial u_\epsilon(x,t)}{\partial t} + a_\epsilon(x)u_\epsilon(x,t) = f(x,t) \\ &u_\epsilon(x,\tau) = 0 \quad \text{for some negative time} \quad \tau. \end{aligned} \tag{10}$$

As working with (10) seemed only to be a slight technical application after understanding (9), I decided to continue to work with (9). From the explicit solution of (9)

$$u_\epsilon(x,t) \equiv v(x)e^{-ta_\epsilon(x)} \tag{11}$$

we deduce that the weak limit u_o of u_ϵ (after extracting a subsequence, for which we keep the index ϵ) will be of the form

$$\begin{aligned} &u_o(x,t) \equiv v(x)B(x,t) \\ &\text{where} \quad B(x,t) \quad \text{is the weak limit of} \quad e^{-ta_\epsilon(x)} \end{aligned} \tag{12}$$

and so the memory effect would be a mathematical consequence of the fact that

(13) $B(x,t)$ is not of the form $e^{-tb(x)}$ except if $a_\epsilon \to b$ strongly.

(This remark seemed to me to be the simplest example where a limit of semi-groups was not a semi-group. Although I had heard of different theorems about limits of semi-groups, I had never heard anyone mention a counter-example like this one).

If one looks at an equation for u_o with a memory kernel,

$$\begin{aligned} &\frac{\partial u_o(x,t)}{\partial t} + b(x)u_0(x,t) + \int_o^t K(x,t-s)u_o(x,s)ds = 0 \\ &u_o(x,0) = v(x) \end{aligned} \tag{14}$$

one sees that the Laplace transform of K is easy to obtain, as (12) and (14) give

$$\begin{cases} Lu_o(x,p) \equiv v(x)LB(x,p) \\ (p + b(x) + LK(x,p))Lu_o(x,p) = v(x) \end{cases} \tag{15}$$

so that

$$p + b(x) + LK(x,p) \equiv \frac{1}{LB(x,p)}. \tag{16}$$

The problem was then to find K, and its properties, through an inverse Laplace transform. As I was visiting Madison, I asked J. Nohel about this special delay equation, and was advised to check the analyticity and behaviour at infinity that were the key to classical theorems for inverse Laplace transform. As I expected the kernel K to be ≤ 0 in order to preserve nonnegativity of the solution, it reminded me of using Bernstein's theorem which states that a function $g(p)$ defined on $]0,\infty[$ is the Laplace transform of a nonnegative measure μ if and only if

$$(-1)^m \frac{d^m g}{dp^m} \geq 0 \quad \text{in} \quad]0,+\infty[\quad \text{for} \quad m = 0,1,\ldots \quad . \tag{17}$$

(I had heard of this result, mixed with questions on Padé approximants, in talks of theoretical physicists, the first time from D.Bessis).

The Laplace transform of B can be studied easily if one introduces the Young's measures associated to the sequence a_ϵ. Assuming that a_ϵ is bounded in L^∞, for example

$$0 < \alpha \leq a_\epsilon(x) \leq \beta \quad \text{a.e.,} \tag{18}$$

there exists a family of probability measures $d\nu_x$ with support in the interval $[\alpha, \beta]$ such that, after extracting a subsequence, for which we keep the index ϵ, for every real continuous function G one has

$$\begin{aligned} &G(a_\epsilon) \to g \quad \text{in} \quad L^\infty(R) \text{ weak}^*, \\ &\text{with} \quad g(x) = \int G(\lambda) d\nu_x(\lambda) \text{ a.e.} \end{aligned} \tag{19}$$

This shows that

$$\begin{aligned} B(x,t) &\equiv \int e^{-t\lambda} d\nu_x(\lambda), \\ LB(x,p) &= \int \frac{1}{p+\lambda} d\nu_x(\lambda) \quad \text{a.e.} \end{aligned} \tag{20}$$

Because of (18), $LB(x,p)$ is defined for $p \geq 0$, and from (16) the Laplace transform of our unknown kernel K is

$$LK(x,p) \equiv \left(\int \frac{1}{p+\lambda} d\nu_x(\lambda) \right)^{-1} - p - b(x). \tag{21}$$

When p goes to ∞, $LB(x,p)$ has the asymptotic expansion

$$LB(x,p) \equiv \frac{1}{p} - \frac{1}{p^2} \int \lambda d\nu_x(\lambda) + \frac{1}{p^3} \int \lambda^2 d\nu_x(\lambda) - \ldots \tag{22}$$

so that, if we denote by A_m the weak* limit of $(a_\epsilon)^m$, i.e.

$$\begin{aligned} &[a_\epsilon]^m \to A_m \quad \text{in} \quad L^\infty(R) \text{ weak}^*, \\ &\text{with} \quad A_m(x) = \int \lambda^m d\nu_x(\lambda) \quad \text{a.e. ,} \end{aligned} \tag{23}$$

we obtain

$$LK(x,p) \equiv \left\{ p + A_1(x) - \frac{A_2(x) - A_1^2(x)}{p} \ldots \right\} - p - b(x). \tag{24}$$

From this, one sees that the natural choice for b is A_1 and that one can hope that Bernstein's theorem could be applied in order to prove that K exists and is ≤ 0. I indeed carried out the details in 1980, and I used the following lemma: if ϕ and ψ satisfy the hypotheses of Bernstein's theorem and $0 \leq \phi < 1$ on $]0, +\infty[$, then $\psi/(1-\phi)$ satisfies also the same hypotheses. I later asked Y.Meyer if this result was known and he remarked that it could be seen directly as a result about convolution equation: it suffices to notice that if $d\mu$ is a nonnegative measure on $[0, +\infty[$ satisfying $d\mu(\{0\}) < 1$, then the series of powers of $d\mu$ (for the convolution product) converges to a nonnegative measure which is the inverse of $(\delta_o - d\mu)$.

I quoted the above result to Luisa Mascarenhas (with the method using convolution which was quicker) and asked her to study the case where the coefficients a_ϵ could depend on t. She solved that question, and it involved a lot of technical estimates, as part of her thesis [1].

Around the same time that I had done these computations, I learned a different tool, also connected to homogenization. In his approach to questions of bounds for effective coefficients for two component mixtures D.Bergman had extended some functions that were naturally defined on the positive real axis to the complex plane cut on the real negative axis and he had noticed that this extended function was mapping the upper half plane into itself (independently G.Milton had a similar idea). The mathematical properties were simple consequences of Cauchy's representation theorem for this class of functions, and one idea was to use the Taylor expansion at $z = 1$ in order to obtain better estimates of the value at another point (I think that this idea had been used in previous work by S.Prager) and I understood later (from G.Milton's talk at a meeting in New-York in June 1981) that there was some connection with Padé approximants.

When encountering an holomorphic function $f(z)$ defined in the complement of a compact set J of the real axis, real on $R \backslash J$ and satisfying $Im f(z) > 0$ for $Im\ z > 0$, one may use a representation formula

$$f(z) \equiv \alpha z + \beta + \int_J \frac{d\mu(\lambda)}{\lambda - z} \tag{25}$$

where α and β are real numbers with $\alpha \geq 0$ and $d\mu$ is a nonnegative measure with support in J (in our examples we would have obtained easily an expansion near $z = \infty$, and thus know the values of α and β).

From equation (24) we have decided to choose $b(x) \equiv A_1(x)$, so (21) is replaced by

$$p + A_1(x) + LK(x,p) = \left(\int \frac{d\nu_x(\lambda)}{p+\lambda} \right)^{-1}. \tag{26}$$

The new idea consists in noticing that the function F_1 defined by

$$F_1(z) \equiv \int_{J_{1x}} \frac{d\nu_x(\lambda)}{z+\lambda} \quad \text{for } z \text{ outside } J_{1x} \equiv -\text{support } \nu_x \tag{27}$$

is holomorphic outside J_{1x}, and satisfies $ImF_1(z) < 0$ for $Imz > 0$ because for λ in J_{1x}, $1/(x + iy + \lambda)$ satisfies this property. Near $z = \infty$, F_1 admits the expansion $F_1(z) = 1/z - A_1(x)/z^2 + A_2(x)/z^3 + O(1/z^4)$ and we notice that if we denote by I_{1x} the smallest interval containing the support of ν_x (i.e. the convex hull of $-J_{1x}$), then $F_1(z) \neq 0$ outside $-I_{1x}$. Then we define $F_2 = 1/F_1$ which is holomorphic outside $-I_{1x}$, satisfies $ImF_2(z) > 0$ for $Imz > 0$ and admits near $z = \infty$ the expansion $F_2(z) = z + A_1(x) - [A_2(x) - A_1(x)^2]/z + O(1/z^2)$. We deduce from the representation theorem that there exists a nonnegative measure $d\mu_x$ with support in I_{1x} such that

$$\left(\int_{I_{1x}} \frac{d\nu_x(\lambda)}{p+\lambda} \right)^{-1} \equiv p + A_1(x) + \int_{I_{1x}} \frac{d\mu_x(\lambda)}{-\lambda - p} \quad \text{for } p \text{ outside } -I_{1x}. \tag{28}$$

This gives the Laplace transform of K by

$$LK(x,p) \equiv -\int_{I_{1x}} \frac{d\mu_x(\lambda)}{\lambda + p} \qquad \text{for } p \text{ outside } -I_{1x} \tag{29}$$

which can be inverted immediately, giving

$$K(x,t) \equiv -\int_{I_{1x}} e^{-\lambda t} d\mu_x(\lambda). \tag{30}$$

Notice that we have obtained $K \leq 0$, but also the unexpected fact that the derivatives of K alternate sign.

Let us remark that I_{1x} corresponds to the smallest interval containing the support of $d\nu_x$ and so can be much smaller than the interval where $a_\epsilon(x)$ takes its values. The different moments of the measure $d\mu_x$ can be computed from information on the oscillations of the sequence a_ϵ; for example, we have already obtained from the expansion at $z = \infty$ for F_2 that the mass of $d\mu_x$ is $A_2(x) - A_1(x)^2$ (so that $d\mu_x \equiv 0$ if and only if the sequence a_ϵ converges strongly). If we denote by $B_1(x)^{-1}$ the weak limit of $1/a_\epsilon(x)$, then the integral in time of $K(x,t)$ can be found by taking $p = 0$ in (26) and is equal to $B_1(x) - A_1(x)$.

We can now attack the interesting question of characterizing all possible kernels $K(x,t)$ when one knows that $a_\epsilon(x)$ satisfies the inequalities

$$a_-(x) \le a_\epsilon(x) \le a_+(x) \quad \text{a.e.} \tag{31}$$

i.e. accepting $I_{1x} = [a_-(x), a_+(x)]$ as a maximal interval for the support of the measures $d\mu_x$ or $d\nu_x$. So let $K(x,t)$ be defined by a formula like (30) with a nonnegative measure $d\mu_x$ with support in $[a_-(x), a_+(x)]$; we then define a function $G(p)$ for p outside $[-a_+(x), -a_-(x)]$, by the formula

$$G(p) \equiv \int_{I_{1x}} \frac{d\mu_x(\lambda)}{-\lambda - p} \quad \text{for} \quad p \text{ outside } -I_{1x} \tag{32}$$

and we want to find a function $A_1(x)$ such that $G(p)+p+A_1(x)$ map the upper half plane into itself (which is automatically true) and be $\neq 0$ for p outside $[-a_+(x), -a_-(x)]$. The inverse of this function will then be defined for p outside $[-a_+(x), -a_-(x)]$ and by using the representation theorem, we will deduce the existence of a measure $d\nu_x$ satisfying (28), and the behaviour near $z = \infty$ will imply that $d\nu_x$ is a probability.

Setting $p = \xi + i\eta$, this gives the following condition

$$\xi + A_1(x) - \int_{I_{1x}} \frac{d\mu_x(\lambda)}{(\lambda + \xi)} \neq 0 \quad \text{for } \xi \text{ outside } -I_{1x} \equiv [-a_+(x), -a_-(x)]. \tag{33}$$

As this function of ξ is increasing for $\xi < -a_+(x)$ and for

$-a_-(x) < \xi$, this is equivalent to

$$(34)\qquad \begin{aligned} A_1(x) - a_+(x) + \int_{I_{1x}} \frac{d\mu_x(\lambda)}{a_+(x) - \lambda} \leq \ 0 \ \leq \\ \leq A_1(x) - a_-(x) - \int_{I_{1x}} \frac{d\mu_x(\lambda)}{\lambda - a_-(x)} \end{aligned}$$

i.e.

$$(35)\qquad \begin{aligned} a_-(x) + \int_{I_{1x}} \frac{d\mu_x(\lambda)}{\lambda - a_-(x)} \leq \ A_1(x) \ \leq \\ \leq a_+(x) - \int_{I_{1x}} \frac{d\mu_x(\lambda)}{a_+(x) - \lambda} \end{aligned}$$

which gives the following necessary and sufficient condition for the nonnegative measure $d\mu_x$ having support in the interval $[a_-(x), a_+(x)]$ (assuming that it is not restricted to a point): eliminating A_1 in (35) and dividing by $(a_+ - a_-)$ we obtain

$$(36)\qquad \int_{I_{1x}} \frac{d\mu_x(\lambda)}{(a_+(x) - \lambda)(\lambda - a_-(x))} \leq 1.$$

We have incidently answered another natural question which is to characterize the nonnegative measures $d\mu_x$ that can arise from a sequence a_ϵ satisfying (31) and converging weakly to A_1 (which satisfies also (31)): the answer is given by (35).

2. Generalization. The same idea can be used with Fourier transform instead of Laplace transform, and I had studied with François Murat an example of this type [2]. We will now apply the same idea to De Giorgi's remark and this will involve both Fourier and Laplace transforms: this has been done recently, with motivation coming from flows in porous media, by Amirat, Hamdache and Ziani [3].

We will consider the following generalization of (3)

$$(37)\qquad \begin{aligned} &\frac{\partial u_\epsilon(x,y,t)}{\partial t} + a_\epsilon(y)\frac{\partial u_\epsilon(x,y,t)}{\partial x} = 0, \\ &u_\epsilon(x,y,0) = v(x,y), \end{aligned}$$

where the oscillating coefficient a_ϵ satisfies

$$a_-(y) \leq a_\epsilon(y) \leq a_+(y) \tag{38}$$

for two bounded functions a_- and a_+. Assuming, for example, that v has compact support we apply Fourier transform in x and then Laplace transform in t and this gives

$$LF_x u_\epsilon(\xi, y, p) = \frac{F_x v(\xi, y)}{p + 2i\pi\xi a_\epsilon(y)} \tag{39}$$

which is valid for $Re\, p > 0$. If we extract a subsequence (for which we keep the index ϵ) in order to obtain a family of Young measures $d\nu_y$ associated with the sequence a_ϵ (they will be probability measures with support in the interval $[a_-(y), a_+(y)]$ we find that the weak limit u_o of the subsequence u_ϵ satisfies

$$LF_x u_o(\xi, y, p) = F_x v(\xi, y) \int \frac{d\nu_y(\lambda)}{p + 2i\pi\xi\lambda}. \tag{40}$$

If we define the function $H(\xi, p)$ by

$$H(\xi, p) = \int \frac{d\nu_y(\lambda)}{p + 2i\pi\xi\lambda} \tag{41}$$

we obtain a function which is holomorphic in a strip $|Im\xi| < \eta_o$ if $2\pi\eta_o a_\pm(y) < Re\ p$. We will then be able to apply some of the results of our preceding analysis, like the formula (28), by analytic continuation. The analogous of formula (28) is

$$\left(\int \frac{d\nu_y(\lambda)}{q + \lambda}\right)^{-1} \equiv q + A_1(y) - \int \frac{d\mu_y(\lambda)}{q + \lambda} \tag{42}$$

for q complex outside the real segment $[-a_+(y), -a_-(y)]$,

where A_1 is the weak limit of a_ϵ and the nonnegative measures $d\mu_y$ have support in the interval $[a_-(y), a_+(y)]$. By choosing $q = p/2i\pi\xi$ we obtain

$$\left(\int \frac{d\nu_y(\lambda)}{p + 2i\pi\xi\lambda}\right)^{-1} = p + 2i\pi\xi A_1(y) - (2i\pi\xi)^2 \int \frac{d\mu_y(\lambda)}{p + 2i\pi\xi\lambda} \tag{43}$$

if $\quad Re\, p > 0$.

Applying the inverse Laplace transform we obtain a delay equation for the Fourier transform in x of u_o:

$$(44)\quad \begin{aligned} &\frac{\partial}{\partial t}F_x u_o(\xi,y,t) + 2i\pi\xi A_1(y) F_x u_o(\xi,y,t) + \\ &\qquad - \int_o^t K(\xi,y,t-s) F_x u_o(\xi,y,s)\,ds = 0 \\ &F_x u_o(\xi,y,0) = F_x v(\xi,y) \\ &\text{with } \ K(\xi,y,\tau) = (2i\pi\xi)^2 \int e^{-2i\pi\xi\lambda\tau} d\mu_y(\lambda) \end{aligned}$$

which, after applying the inverse Fourier transform, gives

$$(45)\quad \begin{aligned} &\frac{\partial}{\partial t}u_o(x,y,t) + A_1(y)\frac{\partial}{\partial x}u_o(x,y,t) + \\ &\qquad - \int_o^t \int \frac{\partial^2}{\partial x^2} u_o(x-\lambda(t-s),y,s)\; d\mu_y(\lambda)\,ds = 0 \\ &u_o(x,y,0) = v(x,y). \end{aligned}$$

3. Conclusion. Did we gain anything by expressing that the weak limit of

$$(46)\qquad u_\epsilon(x,y,t) = v(x - a_\epsilon(y)t, y)$$

is solution of (45) as we knew it by a formula analog to (5)

$$(47)\qquad u_o(x,y,t) = \int v(x - t\lambda, y)\,d\nu_y(\lambda),$$

or did we lure ourselves into playing the strange game (for a mathematician) of having a solution and looking for an equation?

We were indeed looking for a convolution equation because our problem was invariant by translation in x or t, and we know that linear operators commuting with translations have to be given by convolutions.

If we had been looking at a nonlinear equation like

$$\frac{\partial u_\epsilon(x,t)}{\partial t} + a_\epsilon(x)\big(u_\epsilon(x,t)\big)^2 = 0, \quad u_\epsilon(x,0) = v(x) \tag{48}$$

whose solution is

$$u_\epsilon(x,t) = \frac{v(x)}{1 + tv(x)a_\epsilon(x)} \tag{49}$$

(assuming $a_\epsilon \geq 0$ and choosing $v \geq 0$ in order to avoid blow up), we would have found that (for the right subsequence) u_ϵ converges weakly to u_o given by

$$u_o(x,t) = v(x) \int \frac{d\nu_x(\lambda)}{1 + tv(x)\lambda}. \tag{50}$$

What would then be a reasonable class of equations to consider for that problem?

The above examples and methods which I have described are well suited to some linear problems and my goal was to attack the general question of understanding what kind of memory kernels can appear by homogenization. This seems of importance for applications because memory effects do occur in models of continuum mechanics, an example being viscoelasticity for which I refer to the recent book of Renardy, Hrusa and Nohel [4] which gives an up to date description of both the models of continuum mechanics and the corresponding mathematical results.

References

[1] L.Mascarenhas, *A linear homogenization problem with time dependent coefficient*, Trans. A.M.S. **281** (1984), 179-195. (Part of thesis, Lisbon, July 1983).

[2] L.Tartar, *Remarks on homogenization*, Homogenization and Effective Moduli of Materials and Media, 228-246, the IMA Vol. in Math. and its Appl., Vol.1., Springer, New York, 1986.

[3] Y.Amirat, K.Hamdache, A.Ziani, *Homogénéisation d'équations hyperboliques du premier ordre - Application aux milieux poreux*, preprint.

[4] M.Renardy, W.J.Hrusa, J.A.Nohel, *Mathematical Problems in Viscoelasticity*, Longman, 1987.

Carnegie Mellon University
Department of Mathematics
PITTSBURGH, PA 15213

ON REGULARITY AND EXISTENCE OF VISCOSITY SOLUTIONS OF NONLINEAR SECOND ORDER, ELLIPTIC EQUATIONS

NEIL S. TRUDINGER

Dedicated to Ennio De Giorgi on his sixtieth birthday

1. Introduction. In 1957 there appeared De Giorgi's famous paper [2] on Hölder estimates for weak solutions of linear elliptic equations in divergence form,

$$Lu = D_i(a^{ij}D_j u) = 0. \tag{1.1}$$

As is well known, the results and techniques of this paper brilliantly opened the study of second order, quasilinear elliptic equations in more than two variables, including the regularity theory of extremals of multiple integrals of the form,

$$\int_\Omega F(x,u,Du)\,, \tag{1.2}$$

and weak solutions of quasilinear, elliptic equations in divergence form,

$$D_i A^i(x,u,Du) + B(x,u,Du) = 0. \tag{1.3}$$

The regularity theory was further developed by Ladyzhenskaya and Ural'tseva and Morrey in their respective books [16], [22] and the relationship between De Giorgi's estimate and the general existence theory is also described for example in the book [4].

In this paper we shall consider a regularity theory for weak solutions of fully nonlinear, elliptic equations in the general form

$$F(u) = F(x, u, Du, D^2u) = 0. \tag{1.4}$$

Here the role of the De Giorgi estimates is played by the Hölder and pointwise estimates of Krylov-Safonov [14] for solutions of linear, elliptic equations in non-divergence form,

$$Lu = a^{ij} D_{ij} u = 0. \tag{1.5}$$

The notion of weak solution we employ is implicit in the classical Perron process but has been resurrected recently under the name "viscosity" solution by Crandall and Lions [1] in the context of first order equations, with an extension to degenerate elliptic, second order equations being formulated by Lions [20]. Their terminology arises from the stability of the notion under elliptic regularization and the analogy with the vanishing viscosity method in fluid mechanics. The relation between viscosity solutions of second order equations and stochastic control and game theory is treated in the articles [20], [21]. For the existence theory, the notion of viscosity solution turns out to be important as classical existence theorems for equation (4) are only known, in more than two dimensions, when the function F is convex or concave with respect to D^2u.

We proceed now to some definitions. Let Ω be a domain in Euclidean n space, $\mathbf{R}^n$ and set $\Gamma = \Omega \times \mathbf{R} \times \mathbf{R}^n \times \mathbf{S}^n$, where $\mathbf{S}^n$ denotes the linear space of real, $n \times n$ symmetric matrices. The function F in (1.4) is assumed to be defined and continuous on Γ and the associated operator degenerate elliptic so that F is non-decreasing with respect to D^2u, that is

$$F(x, z, p, r) \leq F(x, z, p, r + \eta) \tag{1.6}$$

for all $x, z, p, r \in \Gamma$ and non-negative matrices $\eta \in \mathbf{S}^n$. A function $u \in C^o(\Omega)$ is then called a *viscosity subsolution* of equation (1.4) if for any function $\varphi \in C^2(\Omega)$ and local maximum x_o of $u - \varphi$, we have

$$F(x_o, u(x_o), D\varphi(x_o), D^2\varphi(x_o)) \geq 0. \tag{1.7}$$

The function u is called a *viscosity supersolution* of (1.4) if for any $\varphi \in C^2(\Omega)$ and local minimum x_o of $u - \varphi$, we have

$$F(x_o, u(x_o), D\varphi(x_o), D^2\varphi(x_o)) \leq 0. \tag{1.8}$$

A function u is a *viscosity (*or *Perron) solution* of (1.4) if it is both a viscosity subsolution and supersolution. Classical solutions (subsolutions, supersolutions) are easily seen to be viscosity solutions (subsolutions, supersolutions) by virtue of the degenerate ellipticity of F. Some basic properties of viscosity solutions of second order equations are treated in [20]. In particular, a function u is a viscosity subsolution (supersolution) of (1.4) if and only if

$$\pm F(x, u(x), p, r) \geq 0 \quad \text{for all} \quad (p, r) \in D^{1,2}_{\pm} u(x), \quad x \in \Omega, \tag{1.9}$$

where $D^{1,2}_{\pm} u(x)$ denotes respectively the second superdifferential, subdifferential of the function u at the point x. The viscosity notion is also stable with respect to uniform convergence in the sense that if $\{u_m\} \subset C^o(\Omega)$, $\{F_m\} \subset C^0(\Gamma)$ converge locally uniformly to u, F respectively and $F_m[u_m] \geq 0$, (≤ 0), in the viscosity sense, then also $F[u] \geq 0$, (≤ 0), in the viscosity sense, [20].

Unless otherwise stated, all notation in this paper follows the book [4].

2. Regularity. Our initial goal in this article is to show that continuous viscosity solutions of uniformly elliptic equations are continuously differentiable with Hölder continuous first derivatives. This extends our work in [28] where only Lipschitz continuous solutions are considered. Utilizing the Perron approach of Ishii [6,7,8], we are then able to infer existence theorems for continuously differentiable solutions of the Dirichlet problem. The uniqueness of such solutions is treated in the papers [8], [9], [11], [29]. The function F will initially be subject to the following structure conditions:

F1. *(Uniform ellipticity)*
$\lambda \cdot \operatorname{trace} \eta \leq F(x, z, p, r + \eta) - F(x, z, p, r) \leq \Lambda \cdot \operatorname{trace} \eta$

F2. $|F(x, z, p, 0)| \leq \mu_o + \mu_1 |p|$

F3. $|F(x, z, p, r) - F(y, t, q, r)| \leq \mu_o + \mu_1(|p| + |q|) + \omega(|x - y| + |z - t|)|r|$

for all $x, y, \in \Omega$, $|z|, |t| \leq K_o$, $p, q \in \mathbf{R}^n$, $r \in \mathbf{S}^n$, $\eta \geq 0$, $\eta \in \mathbf{S}^n$, $K_o > 0$, where $\lambda, \Lambda, \mu_o, \mu_1$, are positive constants (depending possibly on K_o) and ω is a non-decreasing function on $\mathbf{R}^+$ with $\omega(a) \to 0$ as $a \to 0$. For the regularity assertions we shall require that ω be a power function, that is

$$\omega(a) = \mu_2 a^\tau \tag{2.1}$$

for positive constants μ_2 and τ, but this can be dispensed with for our later existence theorems. We observe that condition F1 implies the Lipschitz continuity of F with respect to r and hence may be expressed equivalently as

$$\lambda I \leq F_r \leq \Lambda I. \tag{2.2}$$

The structure conditions F1, F2, F3 should be viewed in conjunction with their prime motivating example, the Isaac's equation from stochastic game theory. Let $\{L_{\alpha\beta}\}$ be a family of linear operators, indexed by two parameters $\alpha \in A$, $\beta \in B$ and given by

$$L_{\alpha\beta} u = a^{ij}_{\alpha\beta} D_{ij} u + b^i_{\alpha\beta} D_i u + c_{\alpha\beta} u \tag{2.3}$$

where $a^{ij}_{\alpha\beta}$, $b^i_{\alpha\beta}$, $c_{\alpha\beta}$, $f_{\alpha\beta} \in C^o(\Omega)$. The corresponding Isaac's equation,

$$F[u] = \inf_{\alpha \in A} \sup_{\beta \in B} (L_{\alpha\beta} u - f_{\alpha\beta}) = 0 \tag{2.4}$$

will satisfy conditions F1 to F3 if the operators are *uniformly elliptic* with respect to α and β, that is,

$$\lambda I \leq \left[a^{ij}_{\alpha\beta}\right] \leq \Lambda I \tag{2.5}$$

for all $\alpha \in A$, $\beta \in B$ for fixed positive constants λ, Λ, and the coefficients $a^{ij}_{\alpha\beta}$ are uniformly continuous (Hölder continuous to also satisfy (2.1)) while the coefficients $b^i_{\alpha\beta}$, $c_{\alpha\beta}$, $f_{\alpha\beta}$ are uniformly bounded, with respect to α and β. When either of the sets A or B are singletons, we obtain the Bellman equations of stochastic control theory.

We can now state our regularity assertion.

Theorem 2.1. *Let u be a continuous viscosity solution of equation (1.4) in the domain Ω, where F satisfies the structure conditions* F1, F2 *and* F3, *together with (2.1). Then u is continuously differentiable in Ω, with first derivatives locally Hölder continuous with exponent α depending only on n, Λ/λ and τ.*

PROOF. In our previous paper, we utilized the key idea of Jensen [9] of approximating the function u by functions whose graphs have fixed distance from the graph of u. While this technique still permits us to treat non Lipschitz solutions, it will be simpler to follow the approximations invoked by Jensen, Lions and Souganidis [11] in their extension of Jensen's comparison principle [9] to continuous solutions. Accordingly let us define for positive ϵ, the functions,

$$
\begin{aligned}
u_\epsilon^+(x) &= \sup_{y\in\Omega}\left(u(y) - \frac{|x-y|^2}{\epsilon^2}\right), \\
u_\epsilon^-(x) &= \inf_{y\in\Omega}\left(u(y) + \frac{|x-y|^2}{\epsilon^2}\right).
\end{aligned} \tag{2.6}
$$

We first note that the supremum and infimum will be achieved in (2.6) at points $x^\pm$ satisfying

$$
|x - x^\pm|^2 \le \epsilon^2 \operatorname{osc}_\Omega u \equiv \epsilon^2 \omega_o \tag{2.7}
$$

provided $x \in \Omega(\epsilon\sqrt{\omega_o})$ where, for any $\delta > 0$, $\Omega(\delta)$ is defined as $\{x \in \Omega |\ \operatorname{dist}(x, \partial\Omega) > \delta\}$. It then follows that

$$
|Du_\epsilon^\pm| \le \frac{2\sqrt{\omega_o}}{\epsilon}, \quad \pm D^2 u_\epsilon^\pm \ge -\frac{2}{\epsilon^2} \tag{2.8}
$$

in the sense of distributions. Accordingly the functions $u_\epsilon^\pm$ possess second differentials almost everywhere in $\Omega(\epsilon\sqrt{\omega_o})$ and moreover [11] at any point x of twice differentiability,

$$
\pm F(x^\pm, u(x^\pm), Du_\epsilon^\pm(x), D^2u_\epsilon^\pm(x)) \ge 0. \tag{2.9}
$$

As indicated in [29], the approximating differential inequalities (2.9) are sufficient to ensure a Hölder estimate for u, by virtue of the structure conditions F1 and F2. To see this, we write the inequality (2.9) for u_ϵ^+ in the form,

$$
\text{(2.10)} \qquad \begin{aligned} a_0^{ij} D_{ij} u_\epsilon^+ &\geq -F\left(x^+, u(x^+), Du_\epsilon^+, 0\right) \\ &\geq -\mu_0 - \mu_1 |Du_\epsilon^+| \end{aligned}
$$

by F2, where the coefficient matrix $a_o = [a_o^{ij}]$, given by

$$
\text{(2.11)} \qquad a_o(x) = \int_o^1 F_r(x^+, u(x^+), Du_\epsilon^+, tD^2 u_\epsilon^+) dt,
$$

satisfies by (2.2) the condition of uniform ellipticity,

$$
\text{(2.12)} \qquad \lambda I \leq a_o \leq \Lambda I.
$$

Because of the second derivative bounds (2.8), the Krylov-Safonov theory is applicable to the differential inequality (2.10), (see [29]). In particular, if the ball $B_2 = B_{2R}(y) \subset \Omega(\epsilon\sqrt{\omega}_o)$ and $B_1 = B_R(y)$,

$$
M_2 = \sup_{B_2} u_\epsilon^+, \; M_1 = \sup_{B_1} u_\epsilon^+,
$$

we have the weak Harnack inequality ([25],[4] Theorem 9.22),

$$
\text{(2.13)} \qquad \left(R^{-n} \int_{B_1} (M_2 - u_\epsilon^+)^p\right)^{1/p} \leq C\left(M_2 - M_1 + \frac{\mu_o}{\lambda} R^2\right),
$$

where p and C are positive constants depending on n, Λ/λ and μ_1 diam Ω. But now, we can let ϵ tend to zero in (2.13), to obtain the same inequality for the function u itself. Similarly, we get the same inequality for the function $-u$, from the approximation u_ϵ^-, and in the usual way, (for example see [4], Corollary 9.23), we deduce $u \in C^\alpha(\Omega)$ for some $\alpha > 0$, depending only on n, Λ/λ and μ_1 diam Ω, together with the Hölder estimation,

$$
\text{(2.14)} \qquad \operatorname{osc}_{B_{\sigma R}} u \leq C\sigma^\alpha \left(\operatorname{osc}_{B_R} u + \frac{\mu_o}{\lambda} R^2\right),
$$

for any ball $B_R \subset \Omega$, $0 < \sigma < 1$, where also C depends on n, Λ/λ and $\mu_1 \cdot \operatorname{diam} \Omega$. Henceforth there will be no loss of generality in assuming the function F is independent of z.

The strategy of our ensuing proof is to successively improve the Hölder exponent of the solution u. Following our technique in [28], we shall consider the differences v_ϵ, defined by

$$v_\epsilon(x,\xi) = u_\epsilon^+(x+h\xi) - u_\epsilon^-(x). \tag{2.15}$$

as functions of the $2n$ variables (x,ξ) in the domain

$$\tilde{\Omega} = \Omega(3h) \times B_2(0), \qquad \text{where} \quad 2\epsilon\sqrt{\omega_o} \leq h.$$

By using the inequalities (2.9), we obtain for almost all $x \in \Omega(3h)$

$$\begin{aligned} F(x^-, &Du_\epsilon^-(x), D^2u_\epsilon^+(x+h\xi)) - F(x^-, Du_\epsilon^-(x), D^2u_\epsilon^-(x)) \geq \\ &\geq F(x^-, Du_\epsilon^-(x), D^2u_\epsilon^+(x+h\xi)) + \\ &\quad - F((x+h\xi)^+, Du_\epsilon^+(x+h\xi), D^2u_\epsilon^+(x+h\xi)) \end{aligned} \tag{2.16}$$

so that, writing

$$\begin{aligned} a(x) &= [a^{ij}(x)] = \\ &= \int_o^1 F_r(x^-, Du_\epsilon^-(x), tD^2u_\epsilon^+(x+h\xi) + (1-t)D^2u_\epsilon^-(x))dt, \end{aligned} \tag{2.17}$$

and using the structural condition F3, we have

$$\begin{aligned} a^{ij}D_{ij}v_\epsilon &\geq -\omega(3h)\left|D^2u_\epsilon^+(x+h\xi)\right| + \\ &\qquad - \mu_1\left(|Du_\epsilon^+(x+h\xi)| + |Du_\epsilon^-(x)|\right) - \mu_o \\ &\geq -\frac{\omega(3h)}{h}\sigma^{ij}D_{i\xi_j}v_\epsilon - \mu_1\left(|Dv_\epsilon| + \frac{2}{h}|D_\xi v_\epsilon|\right) - \mu_o, \end{aligned} \tag{2.18}$$

where

$$\sigma^{ij}(x) = \frac{D_{ij}u_\epsilon^+(x+h\xi)}{|D^2u_\epsilon^+(x+h\xi)|} \quad \text{if} \quad D^2u_\epsilon^+(x+h\xi) \neq 0, \quad 0 \quad \text{otherwise}.$$

Let us now fix a vector ξ_o with $|\xi_o| = 1$, and simplify (2.18) through a coordinate transformation

$$\xi \mapsto \xi_o + \frac{\lambda h}{\omega}(\xi - \xi_o), \tag{2.19}$$

to give

$$(2.20)\qquad a^{ij}D_{ij}v_\epsilon + \lambda\sigma^{ij}D_{i\xi_j}v_\epsilon \geq -\mu_1\left(|Dv_\epsilon| + \frac{2\lambda}{\omega}|D_\xi v_\epsilon|\right) - \mu_o.$$

Using (2.10), we then get a *uniformly elliptic* differential inequality in the $2n$ variables (x,ξ), namely

$$(2.21)\qquad \begin{aligned} \tilde{L}v_\epsilon \equiv a^{ij}D_{ij}v_\epsilon + \lambda\sigma^{ij}D_{i\xi_j}v_\epsilon + na_o^{ij}D_{\xi_i\xi_j}v_\epsilon \geq \\ \geq -\mu_1\left[\left(1+\frac{n\omega^2}{\lambda^2}\right)|Dv_\epsilon| + \frac{2\lambda}{\omega}|D_\xi v_\epsilon|\right] - \mu_o\left(1+\frac{n\omega^2}{\lambda^2}\right) \end{aligned}$$

which has coefficient matrix $\tilde{a}$ satisfying

$$(2.22)\qquad \frac{\lambda}{2}I \leq \tilde{a} \leq n\Lambda I.$$

We are now in a position to apply the weak Harnack inequality again. To do this, we fix a point $y \in \Omega$ and $R > 0$ such that $\mathrm{dist}(y,\partial\Omega) > 2R + 3h$, $2R < \frac{\lambda h}{\omega}$ and denote

$$(2.23)\qquad \begin{aligned} &\tilde{B}_R = B_R(y) \times B_R(\xi_o), \\ &\tilde{M}_2 = \sup_{\tilde{B}_{2R}} v_\epsilon, \quad \tilde{M}_1 = \sup_{\tilde{B}_R} v_\epsilon. \end{aligned}$$

Applying the weak Harnack inequality, ([4], Theorem 9.22), to the function $w = \tilde{M}_2 - v_\epsilon$ in the domain $\tilde{B}_{2R}$, we then obtain

$$(2.24)\qquad \left(R^{-2n}\int_{\tilde{B}_R}(\tilde{M}_2 - v_\epsilon)^p\right)^{1/p} \leq C\left(\tilde{M}_2 - \tilde{M}_1 + \frac{\mu_o R^2}{\lambda}(1 + n\omega^2/\lambda^2)\right),$$

where p and C are positive constants depending only on n, Λ/λ and $\frac{\mu_1 R}{\lambda}(1 + \frac{\omega^2}{\lambda^2} + \frac{\lambda}{\omega})$. At this stage, we can let $\epsilon \to 0$, thereby obtaining (2.24) for the difference,

$$(2.25)\qquad v(x,\xi) = u(x + h\xi) - u(x).$$

By replacing u by $-u$, we see that (2.24) also holds for the function $-v$, and we subsequently obtain the Hölder estimate,

$$\operatorname{osc}_{\tilde{B}_{\sigma R}} v \leq C\sigma^\alpha \left(\operatorname{osc}_{\tilde{B}_R} v + \frac{\mu_o R^2}{\lambda}(1+\frac{n\omega^2}{\lambda^2})\right) \tag{2.26}$$

for any ball $B_3 = B_{3R} \subset \Omega$, $h < R \leq \frac{\lambda h}{\omega}$, $0<\sigma<1$, where α and C are positive constants depending only on n, Λ/λ and $\frac{\mu_1 R}{\lambda}(1+\frac{\omega^2}{\lambda^2}+\frac{\lambda}{\omega})$. Hence, if $u \in C^{o,\beta}(\Omega)$ for some $\beta \leq 1$, we can reduce (2.26) to an estimate in the x variables only, namely

$$\operatorname{osc}_{\tilde{B}_{\sigma R}} v(x,\xi_o) \leq C\sigma^\alpha \left(h^\beta [u]_{\beta;B_3} + \frac{\mu_o R^2}{\lambda}(1+\frac{n\omega^2}{\lambda^2})\right) \tag{2.27}$$

and choosing $\sigma = h/R$, we thus have, for any $|\xi|=1$,

$$\begin{aligned} &\left|u(y+h\xi)+u(y-h\xi)-2u(y)\right| \leq \\ &\qquad \leq Ch^{\alpha+\beta}R^{-\alpha}\left[[u]_\beta + \frac{\mu_o R^2}{\lambda h^\beta}\left(1+\frac{n\omega^2}{\lambda^2}\right)\right]. \end{aligned} \tag{2.28}$$

If ω is a Hölder modulus (2.1), we choose

$$R = h^{1-\tau/2}(\frac{d}{3})^{\tau/2} \quad , \quad d = d_y = \operatorname{dist}(y,\partial\Omega),$$

so that for any $h < d/3$, $h^{\tau/2} < \lambda/3\mu_2$, we have

$$\begin{aligned} &\left|u(y+h\xi)+u(y-h\xi)-2u(y)\right|h^{-(\beta+\alpha\tau/2)} \leq \\ &\qquad \leq Cd^{-\alpha\tau/2}\left([u]_\beta + \frac{\mu_o}{\lambda}d^{2-\beta}\left(1+\frac{\mu_2^2 d^{2\tau}}{\lambda^2}\right)\right), \end{aligned} \tag{2.29}$$

where α and C are positive constants depending on n, Λ/λ, μ_1/λ, μ_2/λ and τ. Consequently (see [24]) we obtain $u \in C^{\beta'}(\Omega)$ if $\beta' = \beta + \alpha\tau/2 \leq 1$ and $u \in C^{1,\beta'-1}$ if $\beta' > 1$ and hence by iteration we conclude $u \in C^{1,\beta}(\Omega)$ for some $\beta > 0$. Finally we observe that once we have established $u \in C^{o,1}(\Omega)$, we can assume $\mu_1 = 0$ in the structure conditions F2, F3 with the result that the constants α and C in (2.29) will only depend on n, Λ/λ and τ. Theorem 1 is thus proved and moreover we obtain from (2.29) for any subdomain $\Omega' \subset\subset \Omega$, the estimate (cf. [28], Theorem 2.1)

$$[Du]_{\alpha;\Omega'} \leq C(1 + \delta^{-(1+\alpha)}|u|_{o;\Omega}), \tag{2.30}$$

where α is a positive constant depending only on n, Λ/λ and τ, while C depends also on μ_o/λ, μ_1/λ, μ_2/λ, diam Ω and $|u|_{o;\Omega}$ and $\delta =$ dist$(\Omega', \partial\Omega)$.

Global regularity. As in [28], we may readily deduce, from the interior estimate (2.30), a corresponding global estimate for the Dirichlet problem. This is accomplished through the following boundary estimate which arises from the differential inequality (2.10) and extends Krylov's boundary Hölder estimate [13] to viscosity solutions; (see also [14,19]).

Lemma 2.2. *Let $u \in C^o(\bar{\Omega})$ be a viscosity solution of equation (1.4) in the domain Ω, where F satisfies the structure conditions* F1 *and* F2. *If T is an open $C^{1,\tau}$ boundary portion of $\partial\Omega$, with $u = g$ on T for some $g \in C^{1,\tau}(\Omega \cup T)$, it follows that u is differentiable on T, $Du \in C^\alpha(T)$ and moreover for any point $y \in T$, $R <$ dist$(y, \partial\Omega - T)$, $B^+ = B_R(y) \cap \Omega$, we have the estimates,*

$$\begin{aligned} \operatorname{osc}_{\Omega \cap B_{\sigma R}} u &\leq C\sigma \left(\operatorname{osc}_{B^+} u + R^{1+\tau}[Dg]_{\tau;B^+} + \frac{\mu_o}{\lambda} R^2 \right), \\ \operatorname{osc}_{\Omega \cap B_{\sigma R}} v &\leq C\sigma^\alpha \left(\frac{1}{R} \operatorname{osc}_{B^+} u + R^\tau [Dg]_{\tau;B^+} + \frac{\mu_o}{\lambda} R \right), \end{aligned} \tag{2.31}$$

where $\alpha > 0$ depends only on n, Λ/λ, τ and $\mu_1 \cdot$ diam Ω, *C depends also on T and the function v is given by*

$$v(x) = \frac{u(x) - g(x)}{\operatorname{dist}(x, T)}. \tag{2.32}$$

By combining the estimates (2.30), (2.31) we now obtain a global regularity result.

Theorem 2.3. *Let $u \in C^o(\bar{\Omega})$ be a viscosity solution of the Dirichlet problem*

$$F[u] = 0 \quad in \quad \Omega, \; u = g \quad in \quad \partial\Omega, \tag{2.33}$$

where F satisfies the structure conditions F1, F2, F3, $\partial\Omega \in C^{1,\tau}$, $g \in C^{1,\tau}(\bar{\Omega})$ *for some $\tau > 0$. Then $u \in C^{1,\alpha}(\bar{\Omega})$ for some positive constant α depending only on n, Λ/λ and τ and we have the estimate*

$$|u|_{1,\alpha;\Omega} \leq C \tag{2.34}$$

where C depends on n, Λ/λ, τ, μ_o/λ, μ_1/λ, μ_2/λ, $|g|_{1,\tau}$, $|u|_{0;\Omega}$ and Ω.

As noted above, once we infer the solution $u \in C^{0,1}(\bar{\Omega})$ we are in the situation of our previous paper [28] and there is no loss of generality in assuming $\mu_1 = 0$, (with consequent adjustments to μ_o). Furthermore weaker conditions on the boundary data $\partial\Omega, g$ suffice for the first inequality in (2.31) and the consequent global boundedness of Du, [19].

A priori estimates. If we know in advance that a viscosity solution u of equation (1.4) belongs to a Hölder space $C^{1,\tau}(\Omega)$ for some $\tau > 0$, we can dispense with the hypothesis (2.1) in the estimates (2.30), (2.34). To demonstrate this assertion, we return to the estimates (2.27), (2.28) with $\mu_1 = 0$ (as above) and $\beta = 1$. Normalizing u so that $Du(y) = 0$, we obtain from (2.28)

$$\begin{aligned} &|u(y+h\xi) + u(y-h\xi) - 2u(y)| \leq \\ &\leq C\sigma^{\alpha-\tau}h^{1+\tau}\left([Du]_{\tau;B_3} + \frac{\mu_0 R^2}{\lambda h^{1+\tau}}(1+\frac{n\omega^2}{\lambda^2})\right), \end{aligned} \tag{2.35}$$

provided $\omega \leq \lambda\sigma$. Hence if $\tau < \alpha$, by choosing σ sufficiently small, we can deduce from (2.35) a weighted interior estimate, for any $\Omega' \subset\subset \Omega$,

$$[Du]_{\tau;\Omega'} \leq C\delta^{-\tau}\left(|Du|_o + \frac{\delta}{\lambda}(\mu_o + \mu_1|Du|_o)\right) \tag{2.36}$$

where C depends on n, Λ/λ and ω/λ. An estimate for Du follows immediately from (2.36) by standard interpolation [4], but by invoking the more general interpolation inequality of [26], we can permit more general structure conditions than F2, F3, namely,

F2*. $|F(x,z,p,0)| \leq \mu_o + \mu_2|p|^2$

F3*. $$|F(x,z,p,r) - F(y,t,q,r)| \leq \\ \leq \mu_o + \mu_2(|p|^2 + |q|^2) + \omega(|x-y| + |z-t|)|r|.$$

Since the Hölder estimate (2.14) and the boundary estimates (2.31) extend to the conditions F1, F2* (with the constant dependence on $\mu_1 \cdot \text{diam}\, \Omega$ replaced by $\mu_2|u|_0$), we conclude the following a priori estimate.

Theorem 2.4. *Let u be a continuous viscosity solution of equation (1.4) in the domain Ω, where F satisfies the structure conditions F1, F2*, F3*. Then there exists a constant α depending only on n, Λ/λ such that if $u \in C^{1,\alpha}(\Omega)$ we have for any $0 \leq \tau \leq \alpha$, $\Omega' \subset\subset \Omega$, the estimate*

$$[Du]_{\tau;\Omega'} \leq C(1 + \delta^{-1-\tau}|u|_0) \tag{2.37}$$

where C depends only on n, Λ/λ, μ_o/λ, $\mu_2|u|_0/\lambda$, ω/λ and diam Ω. *Furthermore, if $\partial\Omega \in C^{1,\tau}$, $g \in C^{1,\tau}(\bar{\Omega})$, $0 < \tau < \alpha$, and $u = g$ continuously on $\partial\Omega$, then $u \in C^{1,\tau}(\bar{\Omega})$ and*

$$|u|_{1,\tau;\Omega} \leq C \tag{2.38}$$

where C depends on n, Λ/λ, μ_o/λ, $\mu_2|u|_o/\lambda, \omega/\lambda$, $|g|_{1,\tau;\Omega}$, $|u|_{0,\Omega}$ and Ω.

We remark also that Theorem 2.1 could have been proved with $F2, F3$ replaced by F2*, F3*.

Further regularity. We comment briefly on the twice differentiability of viscosity solutions. First, if only conditions F1 and F2 hold and u is a continuous viscosity solution of equation (1.4) in the domain Ω, it follows that u is twice differentiable almost everywhere in Ω, with equation (1.4) then holding almost everywhere, [30]. Next, if in addition to F1, F2, F3, the function F is either concave or convex with respect to the r variables, then it follows that $u \in C^{1,\alpha}(\Omega)$ for all $\alpha < 1$. Finally, if we also assume:

$$F4. \ |F(x,z,p,r) - F(y,t,q,r)| \leq \omega(|x-y| + |z-t| + |p-q|)(1+|r|),$$

for all $x, y \in \Omega$, $|z|$, $|t|$, $|p|$, $|q| \leq K_o$, $r \in \mathbf{S}^n$, where ω satisfies (2.1), then $u \in C^{2,\alpha}(\Omega)$ for some $\alpha > 0$ depending only on n, $\Lambda/\lambda, \tau$. If $\partial\Omega \in C^{2,\tau}$, $g \in C^{2,\tau}(\bar{\Omega})$ and u is a solution of the Dirichlet problem, (2.33), then $u \in C^{2,\alpha}(\bar{\Omega})$; ($C^{1,\alpha}(\bar{\Omega})$ $\forall$ $\alpha < 1$ if F4 is not assumed). These last assertions follow by consideration of the second order difference quotients; (cf. [23], [27]).

Alternative techniques. The proofs of interior regularity, Theorem 2.1, and the interior estimate Theorem 2.4, may also be effected by varying the approximation parameter ϵ as a function of h or by perturbation from the case $\omega = 0$. These techniques avoid the addition of the new variables ξ but the latter requires an auxiliary existence theorem for its execution. In fact our approach to regularity originally arose from a previous observation that the continuous differentiability of weak solutions of divergence structure equations (1.3), when A is only Hölder continuous in x, could be deduced directly by application of De Giorgi's estimate to the *differenced* equation rather than by the perturbation argument of Giaquinta and Giusti [6].

L^p spaces. When the solution u lies in the Sobolev space, $W^{2,n}(\Omega)$, there is no need for the approximation (2.6) as the weak Harnack inequality (2.24) and, indeed the oscillation estimate (2.26), are directly applicable to the differences (2.25). In this case the constants μ_0, μ_1, in the structure conditions F2, F3 can be replaced by functions in certain L^p spaces, as indicated in [28].

3. Existence. There are various approaches to demonstrating existence of solutions to boundary value problems, particularly the Dirichlet problem, for fully nonlinear elliptic equations of the form (1.4). These comprise: (i) the classical method of continuity which can be employed when the function F is concave (or convex) in the r

variables, (see for example [4]); (ii) utilization of the stochastic game or control theory representations, (see for example [12], [20], [10]); (iii) approximation by nonlinear Poisson equations as suggested by Evans and Jensen, based on Evans [3]; (iv) the Perron process as developed by Ishii [6,7,8] and (v) a discretization method which we have developed recently in collaboration with H. Kuo. While the last mentioned technique has the advantage of yielding continuous viscosity solutions as limits of iterative schemes, it will be simplest for our present exposition to adopt the Ishii approach, which under very general hypotheses, provides viscosity solutions which are not necessarily continuous. Their continuity becomes a separate consideration but follows immediately from appropriate comparison principles. Further regularity may then be deduced from our results in the preceding section. Under the structural hypotheses F1, F2, F3 various existence theorems may be formulated depending on the type of boundary behaviour encompassed and a priori bounds for the solutions. A standard condition ensuring the latter is, for example:

$$\begin{aligned} &\lambda \cdot \operatorname{trace} \eta \leq F(x,z,p,r+\eta) - F(x,z,p,r), \\ &(\operatorname{sign} z) F(x,z,p,0) \leq \mu_o + \mu_1 |p| \end{aligned} \tag{3.1}$$

for all $x,z,p,r \in \Gamma$, $\eta \geq 0$, $\in \mathbf{S}^n$, where now λ, μ_o, μ_1 are fixed constants, (see [4], chapter 17). For smooth boundary data, we have the following result.

Theorem 3.1. *Let $F \in C^o(\Gamma)$ satisfy the structure conditions* F1, F2, F3 *together with (3.1). Then, if $\partial\Omega \in C^{1,\tau}$, $g \in C^{1,\tau}(\bar{\Omega})$ for some $\tau > 0$, there exists a viscosity solution $u \in C^{1,\alpha}(\bar{\Omega})$ of the Dirichlet problem (2.33), for some $\alpha > 0$, depending only on $n, \Lambda/\lambda$ and τ.*

The proof of Theorem 3.1 combines the existence of continuous solutions from Ishii [8] with the regularity assertions and estimates of the preceding section. The work of Ishii involves two stages, namely a Perron approach to get generalized solutions and a comparison principle to infer their continuity. As indicated above the first stage requires minimal hypotheses. Following Ishii, a function u on Ω is called a *viscosity subsolution (supersolution)* of equation (1.4) in

Ω if its upper (lower) semicontinuous envelope, $u^*(u_*)$, fulfills the definition in Section 1, and as before is a viscosity solution if it is both subsolution and supersolution. The existence of such generalized viscosity solutions is guaranteed by the following lemma.

Lemma 3.2. *Let $F \in C^o(\Gamma)$ be degenerate elliptic and suppose there exists a viscosity supersolution u_1 and subsolution u_2 of (1.4) in Ω with $u_1, u_2 \in C^o(\bar{\Omega})$ and $u_1 \leq u_2$. Then there exists a viscosity solution u of equation (1.4) in Ω satisfying $u_1 \leq u \leq u_2$.*

For the comparison principle, we replace condition F3 by

$$F5. \quad \begin{cases} F \quad \text{is monotone increasing with respect to } z, \\ |F(x,z,p,r) - F(y,z,q,r)| \leq \\ \qquad \leq \omega(|x-y|) + \mu(|x-y||r| + |q-p|), \end{cases}$$

for $x, y \in \Omega$, $|z| \leq K_o$, $p, q \in \mathbf{R}^n$, $r \in \mathbf{S}^n$, where ω is as in F3 and μ is a positive constant. Then we have, from Ishii [8], (see also [29]),

Lemma 3.3. *Let $u, v \in L^\infty(\Omega)$ be respectively viscosity subsolution and supersolution of equation (1.4) in Ω with $u^* \leq v_*$ on $\partial\Omega$, and suppose F satisfies* F1 *(or at least the first inequality) and* F5. *Then $u^* \leq v_*$ in Ω.*

Combining Lemmas 3.2 and 3.3 we deduce the existence of a unique continuous viscosity solution of the Dirichlet problem (2.33) provided (3.1) and F5 hold, $g \in C^o(\bar{\Omega})$ and $\partial\Omega$ satisfies a barrier condition, for example an exterior cone condition. If we adjoin conditions F1, F2, F3, we obtain a unique solution $u \in C^{1,\alpha}(\Omega) \cap C^0(\bar{\Omega})$, by virtue of Theorem 3.1. The removal of condition F5 may be accomplished by approximation and use of the Leray-Schauder theorem, ([4], Chapter 11).

We remark here that the solution of the Dirichlet problem in Theorem 3.1 will be unique if F5 holds with $|x-y|$ replaced by $|x-y|^{1/2}$, as the comparison principle, Lemma 3.3, can be extended accordingly if the functions $u, v \in C^{0,1}(\Omega)$, [29]. For the Isaac's equation (2.4) with countable index sets A, B, we can ensure condition (3.1) by assuming $c_{\alpha\beta} \leq 0 \quad \forall \alpha \in A, \quad \beta \in B$, in addition to the uniform ellipticity and coefficient continuity as formulated in Section 2. Under these conditions we thus have the existence result.

Corollary 3.4. *For $\partial\Omega \in C^{1,\tau}$, $g \in C^{1,\tau}(\bar\Omega)$ there exists a viscosity solution $u \in C^{1,\alpha}(\bar\Omega)$ for some $\alpha > 0$, of the Dirichlet problem (2.33) for the Isaac's equation (2.4) which is unique if the coefficents $a^{ij}_{\alpha\beta} \in C^{1/2}(\Omega)$, uniformly with respect to A, B.*

Through Lemma 3.2, barrier considerations for the Isaac's equation reduce to those for linear equations, so that if, for example, the coefficients $a^{ij}_{\alpha\beta} \in C^{\tau}(\bar\Omega)$, $\tau > 0$, the $C^o(\bar\Omega)$ solvability of the Dirichlet Problem (2.33) for arbitrary $g \in C^o(\bar\Omega)$ follows from that for Laplace's equation.

Obstacle problems. Because the structure conditions F1, F2 and F3 are preserved when the operator F is perturbed by a bounded function, the estimates of Theorem 2.4 are applicable to obstacle problems. When we consider a bilateral obstacle equation of the form,

$$F[u;\psi_1,\psi_2] = \inf\{F[u], u-\psi_1, \psi_2 - u\} = 0, \tag{3.2}$$

in Ω, we can dispense with conditions such as (3.1), implying a priori solution bounds. We then have the following extension of Theorem 3.1.

Theorem 3.5. *Let $F \in C^o(\Gamma)$ satisfy conditions* F1, F2*, F3*. *Then if $\partial\Omega \in C^{1,\tau}$, $g \in C^{1,\tau}(\bar\Omega), \psi_1, \psi_2 \in C^{1,1}(\bar\Omega)$, $\psi_1 \le g \le \psi_2$ in Ω, there exists a viscosity solution $u \in C^{1,\alpha}(\bar\Omega)$, for some $\alpha > 0$, of the Dirichlet obstacle problem,*

$$F[u;\psi_1,\psi_2] = 0 \quad in \quad \Omega, \quad u = g \quad on \quad \partial\Omega. \tag{3.3}$$

By choosing ψ_1, ψ_2 as sufficiently large constants, we see that Theorem 3.5 includes Theorem 3.1. To prove Theorem 3.5 we solve the approximating problems,

$$\begin{aligned} F[u_\eta] &= \beta_\eta(u_\eta - \psi_2) - \beta_\eta(\psi_1 - u_\eta) \quad \text{in} \quad \Omega, \\ u_\eta &= g \quad \text{on} \quad \partial\Omega, \end{aligned} \tag{3.4}$$

where $\eta > 0$, $\beta_\eta(t) = (t^+)^2/\eta^2$ is a typical penalty function. The boundedness of the penalty terms on the right hand side, independently of η, follows from application of the usual maximum principle arguments to the approximation (2.6). Theorem 3.5 includes the uniformly elliptic Isaac's equation as a special case. For earlier work on the Bellman equation see [17].

References

[1] M.G.Crandall, P.-L.Lions, *Viscosity solutions of Hamilton-Jacobi equations*, Trans. Amer. Math. Soc. **277** (1983), 1-42.

[2] E.De Giorgi, *Sulla differenziabilità e l'analicità delle estremali degli integrali multipli*, Mem. Accad. Sci. Torino Cl. Sci. Fis. Mat. Natur. (3) **3** (1957), 25-43.

[3] L.C.Evans, *A convergence theorem for solutions of non-linear second order elliptic equations*, Indiana University Math. J. **27** (1978), 875-887.

[4] D.Gilbarg, N.S.Trudinger, *Elliptic partial differential equations of second order*, 2nd edition, Springer-Verlag, Berlin-Heidelberg-New York-Tokyo (1983).

[5] M.Giaquinta, E.Giusti, *Boundary $C^{1,\alpha}$ regularity for second-order elliptic equations in divergence form*, J. Reine Angewandte Mathematik **351** (1984), 55-65.

[6] H.Ishii, *Perron's method for Hamilton-Jacobi equations*, Duke Math. J. **55** (1987), 369-384.

[7] E.Ishii, *A boundary value problem of the Dirichlet type for Hamilton Jacobi equations*, to appear.

[8] E.Ishii, *On uniqueness and existence of viscosity solutions of fully nonlinear second order elliptic PDE's*, to appear.

[9] R.Jensen, *The maximum principle for viscosity solutions of fully nonlinear second order partial differential equations*, Arch. Rat. Mech. Anal., to appear.

[10] R.Jensen, P.-L.Lions, *Some asymptotic problems in fully nonlinear equations and stochastic control*, Ann. Sc. Norm. Pisa **11** (1984), 129-176.

[11] R.Jensen, P.-L.Lions, P.E.Souganidis, *A uniqueness result for viscosity solutions of fully nonlinear second order partial differential equations*, Proc. Amer. Math. Soc., to appear.

[12] N.V.Krylov, *Controlled diffusion processes*, Springer-Verlag, Berlin-Heidelberg-New York (1980).

[13] N.V.Krylov, *Boundedly nonhomogeneous elliptic and parabolic equations in a domain*, Izv. Akad. Nauk SSSR Ser. Mat. **47** (1983), 75-108; Mat. USSR Izv. **21** (1984), 67-98.

[14] N.V.Krylov, *Nonlinear elliptic and parabolic equations of the second order*, Reidel, Dordrecht (1987).

[15] N.V.Krylov, M.V.Safonov, *Certain properties of parabolic equations with measurable coefficients*, Izv. Akad. Nauk SSSR Ser. Mat. **40** (1980), 161-175; Math. USSR Izv. **16** (1981), 151-164.

[16] O.A.Ladyzhenskaya, N.N.Ural'tseva, *Linear and quasilinear elliptic equations*, Academic Press, New York (1968).

[17] S.Lenhardt, *Bellman equation for optimal time stopping problems*, Indiana Univ. Math. J. **32** (1983), 363-375.

[18] G.M.Lieberman, *The quasilinear Dirichlet problem with decreased regularity at the boundary*, Comm. Part. Diff. Eq. **6** (1981), 437-497.

[19] G.M.Lieberman, *The Dirichlet problem for quasilinear elliptic equations with continuously differentiable boundary values*, Comm. Part. Diff. Eq. **11** (1986), 167-229.

[20] P.-L.Lions, *Optimal control of diffusion processes and Hamilton-Jacobi-Bellman equations II*, Comm. Part. Diff. Eq. **8** (1983), 1229-1276.

[21] P.-L.Lions P.E.Souganidis, *Viscosity solutions of second order equations, stochastic control and stochastic differential games*, in IMA Vol.10, Stochastic Differential Systems - Stochastic Control Theory and Applications, Springer-Verlag, Berlin (1988).

[22] C.B.Morrey Jr., *Multiple integrals in the calculus of variations*, Springer-Verlag, Berlin-Heidelberg-New York (1966).

[23] M.V.Safonov, *On the classical Bellman's elliptic equation*, Soviet Math. Dokl. **30** (1984), 482-485.

[24] E.Stein, *Singular integrals and differentiability properties of functions*, Princeton University Press, Princeton (1970).

[25] N.S.Trudinger, *Local estimates for subsolutions and supersolutions of general second order elliptic quasilinear equations*, Invent. Math. **61** (1980), 67-79.

[26] N.S.Trudinger, *On an interpolation inequality and its application to nonlinear elliptic equations*, Proc. Amer. Math. Soc. **95** (1985), 75-78.

[27] N.S.Trudinger, *Lectures on nonlinear second order elliptic equations*, Nankai Institute of Mathematics, Tianjin, China (1985).

[28] N.S.Trudinger, *Hölder gradient estimates for fully nonlinear elliptic equations*, Proc. Royal Soc. Edinburgh **108A** (1988), 57-65.

[29] N.S.Trudinger, *Comparison principles and pointwise estimates for viscosity solutions of second order, elliptic equations*, Centre for Math. Anal., Aust. Mat. Univ., Research Report **R45** (1987).

[30] N.S.Trudinger, *On the twice differentiability of viscosity solutions of nonlinear elliptic equations*, Centre for Math. Anal., Aust. Mat. Univ., Research Report **R20** (1988).

Centre for Mathematical Analysis
Australian National University
GPO Box 2601
CANBERRA ACT 2601

ETUDE D'UN SYSTÈME EN MULTIPLICITÉ 4, LORSQUE LE DEGRÉE DU POLYNÔME MINIMAL EST PETIT

JEAN VAILLANT

Dédié à Ennio De Giorgi pour son soixantième anniversaire

0. Introduction. Nous avons dans des publications précédentes [8], [9], [10] donné de nouvelles conditions de Levi, pour les systèmes à multiplicité constante, qui nous semblent permettre d'envisager la caractérisation générale de l'hyperbolicité de ces systèmes.

On remarque que les cas où le degré p du polynône minimal correspondant à une racine caractéristique de multiplicité m satisfait à $p \leq \frac{m}{2}$ présentent le plus de difficulté ; en outre, ces cas ne sont possibles (en dehors du cas fortement hyperbolique) que si $m \geq 4$. Nous traitons ici l'un des deux cas possibles si $m = 4$, $p = 2$; nous avons choisi le plus difficile: la condition nécessaire s'obtient moins directement que d'habitude, la condition suffisante, aisée, si le rang généralisé est constant, présente plus de difficulté que dans les autre cas. Nous donnons aussi des indications permettant d'obtenir l'invariance des conditions de Levi introduites.

Ce travail a été rédigé en partie à Pise, nous exprimons ici nos vifs remerciements.

1. Définitions et position du problème.

1.a. $x = (x_o, x_1) \in \Omega$; Ω est un voisinage ouvert de 0 dans $\mathbf{R}^2$. $\alpha(x)$ et $b(x)$ sont des matrices 4×4 fonctions analytiques de x dans Ω; on considère l'opérateur différentiel matriciel:

$$h(x, D) = a(x, D) + b(x) = ID_0 + \alpha(x)D_1 + b(x)$$

où : $D_o = \frac{\partial}{\partial x_o}$, $D_1 = \frac{\partial}{\partial x_1}$. On note: $\xi = (\xi_o, \xi_1) \in \mathbf{R}^2$ la variable duale de x et on considère le déterminant caractéristique:

$$\text{dét}[\xi_o\ I + \alpha(x)\xi_1];$$

on le suppose hyperbolique par rapport à la direction $(1, 0)$ et ayant une seule racine caractéristique, soit:

$$\text{dét}[\xi_o\ I + \alpha(x)\xi_1] = [\xi_o - \lambda(x)\xi_1]^4.$$

Remarque. L'étude générale d'un opérateur hyperbolique à multiplicité constante permet, par une microlocalisation convenable [2], [10], de se ramener à une seule racine caractéristique, l'opérateur devient pseudo-différentiel, mais l'étude est essentiellement analogue à celle que nous allons faire.

Par un changement de coordonnées évident, les bicaractéristiques deviennent droites et verticales et:

$$\text{dét}[\xi_o\ I + \alpha(x)\xi_1] = \xi_o^4.$$

1.b. On considère la matrice: $a(x, \chi, 1) = \chi I + \alpha(x)$ comme une matrice dont les éléments appartiennent à l'anneau des polynôme en χ, ayant pour coefficients des germes de fonctions analytiques de x en 0, par rapport à l'idéal engendré par χ [6]; c'est l'anneau des fractions construites à partir de ces polynômes et dont le dénominateur n'est pas divisible par χ; il est principal et ses idéaux sont engendrés par les puissances de χ.

Dans cet anneaux $a(x, \chi, 1)$ est équivalente à une matrice diagonale, dont les éléments sont les facteurs invariants de $a(x, \chi, 1)$.

Nous ferons l'hypothèse suivante sur ces facteurs invariants:

$$a(x, \chi, 1) = P(x, \chi) \begin{pmatrix} \chi^2 & & & 0 \\ & \chi^2 & & \\ & & 1 & \\ 0 & & & 1 \end{pmatrix} Q(x, \chi),$$

dét $P(x,0) \not\equiv 0$, dét $Q(x,0) \not\equiv 0$. On remarque que:

$$\text{dét}\, P(x,\chi) \cdot \text{dét}\, Q(x,\chi) \equiv 1;$$

les calculs de [6] permettent des précisions utiles: $Q(x,\chi)$ peut être choisie comme une matrice de polynômes en χ que l'on peut calculer explicitement et son déterminant s'exprime à l'aide des mineurs de $a(x,\chi,1)$; on peut aussi calculer P explicitement et les dénominateurs qui apparaissent sont les facteurs de dét $Q(x,\chi)$. On note $A(x,\xi)$ la matrice des cofacteurs de a, de sorte que:

$$a\,A \;=\; A\,a \;= \xi_o^4\, I.$$

Dans un voisinage de 0, il résulte de l'hypothèse sur les facteurs invariants que:

$$A = \mathcal{A}\xi_o^2,$$

où $\mathcal{A}(x,\xi)$ est une matrice de polynômes en ξ à coefficients analytiques et:

$$\mathcal{A}(x,0,1) \not\equiv 0;$$

de plus, il existe un mineur d'ordre 2 de a soit $A^{i_1 i_2}_{j_1 j_2}(x,\xi)$ tel que: $A^{i_1 i_2}_{j_1 j_2}(x,0,1) \not\equiv 0$. On supposera que Ω est inclus dans un tel voisinage.

On sait aussi que:

$$\alpha^2 \equiv 0$$

dans Ω.

1.c. On dira que le rang généralisé de la matrice caractéristique (pour la racine caractéristique $\xi_o = 0$) est constant si et seulement si:

$$\exists Q \text{ tel que } \quad \text{dét} Q(x,0) \neq 0 \quad , \quad \text{pour tout} \quad x \in \Omega.$$

Ceci équivant à dire que:

$$\mathcal{A}(x,0,1) \neq 0 \ , \ \forall x \quad \text{et} \quad A^{i_1 i_2}_{j_1 j_2}(x,0,1) \neq 0, \quad \forall x.$$

Si le rang généralisé est constant, il existe une matrice 4×4, $\Delta_o(x)$ de germes analytiques en 0, de déterminant jamais nul telle que:

$$\tilde{\alpha} = \Delta_o^{-1}(x)\alpha(x)\Delta_o(x) = \begin{pmatrix} 0 & 1 & 0 & 0 \\ 0 & 0 & 0 & 0 \\ 0 & 0 & 0 & 1 \\ 0 & 0 & 0 & 0 \end{pmatrix} = \begin{pmatrix} J & 0 \\ 0 & J \end{pmatrix}$$

(on suppose Ω choisi convenablement); on a aussi:

$$\Delta_o^{-1}(x)a(x,\xi)\Delta_o(x) = I\xi_o + \begin{pmatrix} J & 0 \\ 0 & J \end{pmatrix}\xi_1.$$

1.d. La proposition suivante montre que, par le choix d'un opérateur pseudodifférentiel elliptique Δ de symbole principal Δ_0 , on peut aussi simplifier les coefficients b, grâce à un résultat d'Arnold appliqué par Petkov [4] au problème considéré :

Proposition. *Si le rang généralisé est constant il existe un opérateur* $\Delta(x,D_1) = \Delta_o(x) + \Delta_1(x)D_1^{-1} + ... + \Delta_j(x)D_1^{-j} = ...$, *où les* $\Delta_j(x)$ *sont analytiques, tel que:*

$$h(x,D)\Delta(x,D_1) = \Delta(x,D_1)\tilde{h}(x,D),$$

modulo un opérateur d'ordre $-\infty$ *en* ξ_1, *où :*

$$\tilde{h}(x,D) = I\ D_o + \tilde{\alpha}D_1 + \tilde{b}(x,D_1)$$

$$\tilde{b}(x,D_1) = \tilde{b}_o(x) + \tilde{b}_1(x)D_1^{-1} + ... + \tilde{b}_j(x)D_1^{-j} + ...,$$

$$\tilde{b}(x,\xi_1) = \begin{pmatrix} 0 & 0 & 0 & 0 \\ \tilde{b}_1^2 & 0 & \tilde{b}_3^2 & \tilde{b}_4^2 \\ \tilde{b}_1^3 & 0 & 0 & 0 \\ \tilde{b}_1^4 & 0 & \tilde{b}_3^4 & 0 \end{pmatrix},$$

les coefficients matriciels $\tilde{b}_j(x)$ *sont analytiques.*

1.e. On considère le problème de Cauchy:

$$h(x,D)u = f,$$

$$u|_{x_o=t} \ = \ g_t(x_1),$$

f est le second membre donnée, g_t la donnée de Cauchy sur $x_o = t$ et u l'inconnue.

Définitions. *Le problème de Cauchy est bien posé en* $(\underline{t},\underline{x}_1) \in \Omega$ *si et seulement si il existe un voisinage ouvert* ω *de* $(\underline{t},\underline{x}_1)$, $\omega \subset \Omega$, *tel que:* $\forall f \in C^\infty(\omega)$, $\forall g_{\underline{t}} \in C^\infty(\omega_{\underline{t}})$, *où :* $\omega_{\underline{t}} = \omega \cap \{x_o = \underline{t}\}$, *il existe une solution* C^∞, *unique, du problème de Cauchy dans* ω*; le problème de Cauchy est bien posé dans* Ω, *si il est bien posé en tout point de* Ω.

Le problème de Cauchy est bien posé au voisinage de 0, *ou brièvement localement bien posé , si il existe un voisinage ouvert de* 0, Ω, *où le problème de Cauchy est bien posé .*

2. Conditions de Levi.

2.a. On écrira ces conditions pour un opérateur:

$$h = I\ D_o + a(x)D_1 + b_o(x) + b_1(x)D_1^{-1} + .. + b_j(x)D_1^{-j} + ..$$

Notation. A un symbole homogène $\Lambda(x,\xi)$, polynômial en ξ_o, on associera un opérateur pseudodifferentiel $\Lambda'(x,D)$, différentiel en D_o, de symbole développable en symboles homogènes dont Λ est le symbole principal. On note σ l'opération prendre le symbole principal.

Exemple. A $\mathcal{A}(x,\xi)$ on associe $\mathcal{A}'(x,D)$; $\sigma(\mathcal{A}') = \mathcal{A}$; à $\xi_o I$, on associe D'_o; $\sigma(D'_o) = \xi_o I$.

Définition.

$$\mathcal{L}_o^\# = \sigma\big(\mathcal{A}'[h\mathcal{A}'(h\mathcal{A}' - D_o'^2) - (h\mathcal{A}' - D_o'^2)D_o'^2]\big).$$

Définition. *La condition* $L_1^\#$ *s'énonce: il existe un opérateur* $\mathcal{A}'$ *et un opérateur* D'_o *tel que* $\mathcal{L}_o^\#$ *soit divisible par* ξ_o.

Définition. *On pose alors:*

$$\mathcal{L}_o^\# = \Lambda_1^\# \xi_o$$

et:

$$\mathcal{L}_1^\# = \sigma\big(\mathcal{A}'\big[h{\Lambda^\#}'_1 - h\mathcal{A}'(h\mathcal{A}' - D_o'^2)D'_o + \\ + (h\mathcal{A}' - D_o'^2)D_o'^3\big]\big).$$

Définition. *La condition $L_2^{\#}$ s'énonce alors: il existe un opérateur ${\Lambda^{\#}}'_1$ tel que $\mathcal{L}_1^{\#}$ soit divisible par ξ_o.*

Définition. *On pose alors*

$$\mathcal{L}_1^{\#} = \Lambda_2^{\#}\xi_o$$

et

$$\mathcal{L}_2^{\#} = \sigma\big(\mathcal{A}'[h{\Lambda^{\#}}'_2 - h{\Lambda^{\#}}'_1 D'_o + h\mathcal{A}'(h\mathcal{A}' - D'^2_o)D'^2_o - \\ - (h\mathcal{A}' - D'^2_o)D'^4_o]\big).$$

Définition. *La condition $L_3^{\#}$ s'énonce alors: il existe un opérateur ${\Lambda^{\#}}'_2$ tel que $\mathcal{L}_2^{\#}$ soit divisible par ξ_o.*

Définitions. *On définit en coordonnées locales [10], les opérateurs sur l'espace cotangent, privé de $\xi_1 = 0$:*

$$\mathcal{L}_o(x,\xi,D_x) = \mathcal{A}(x,\xi)[a(x,D_x) + b_o(x)]$$
$$\mathcal{L}_1(x,\xi) = \mathcal{A}(x,\xi)b_1(x)\xi_1^{-1}.$$

On définit des conditions en coordonnées locales; nous montrerons ensuite qu'elles sont équivalentes aux conditions $L^{\#}$ précédentes.

Définitions. *La condition L_1 s'énonce:*

$$\mathcal{L}_o\mathcal{L}_o(\mathcal{A}) \quad \textit{est divisible par} \quad \xi_o;$$

on pose alors:

$$\mathcal{L}_o\mathcal{L}_o(\mathcal{A}) = \Lambda_1\xi_o.$$

La condition L_2 s'énonce:

$$\mathcal{L}_o\Lambda_1 \quad \textit{est divisible par} \quad \xi_o;$$

on pose alors:

$$\mathcal{L}_o\Lambda_1 = \Lambda_2\xi_o.$$

La condition L_3 s'énonce:

$$\mathcal{L}_o(\Lambda_2) - \mathcal{L}_o\mathcal{L}_1\mathcal{L}_o(\mathcal{A}) \quad \textit{est divisible par} \quad \xi_o.$$

Proposition. *$L_1^{\#}$ et L_1 sont équivalentes; si elles sont réalisées, $L_2^{\#}$ et L_2 sont équivalentes et de même si $L_1^{\#}$ et $L_2^{\#}$ sont réalisées $L_3^{\#}$ et L_3 sont équivalentes.*

PREUVE. On calcule $\mathcal{L}_o^{\#}$ explicitement; on trouve, si $D_o' = ID_o + H^*(x, D)...$, où H^* est d'ordre 0:

$$\mathcal{L}_o^{\#} = \mathcal{L}_o\mathcal{L}_o(\mathcal{A}) - 4\xi_o\mathcal{L}_o(\mathcal{A})H^* + \xi_o^2 N(x,\xi) + \xi_o^3 N_1(x,\xi)$$

le symbole homogène $N(x,\xi)$ doit être explicité , mais nous ne le détaillerons pas ici.

On a donc: $L_1^{\#}$ équivaut à L_1 et:

$$\Lambda_1^{\#} = \Lambda_1 - 4\mathcal{L}_o(\mathcal{A})H^* + \xi_o N(x,\xi) + \xi_o^2 N_1(x,\xi).$$

On a ensuite:

$$\mathcal{L}_1^{\#} = \mathcal{L}_o(\Lambda_1^{\#}) + \xi_o N_2(x,\xi) + \xi_o^2 N_3(x,\xi);$$

et $L_2^{\#}$ équivaut à L_2; on doit encore expliciter $N_2(x,\xi)$ et on a:

$$\Lambda_2^{\#} = \Lambda_2 + N_4(x,\xi) + \xi_o N_5(x,\xi);$$

le symbole homogène N_4 doit être explicité .

On a alors:

$$\mathcal{L}_2^{\#} = \mathcal{L}_o(\Lambda_2^{\#}) + N_6(x,\xi) + \xi_o N_7(x,\xi);$$

on explicite N_6 et on obtient:

$$\mathcal{L}_2^{\#} = \mathcal{L}_o(\Lambda_2) - \mathcal{L}_o\mathcal{L}_1\mathcal{L}_o(\mathcal{A}) + \xi_o N_8(x,\xi)$$

d'où l'équivalence de $L_3^{\#}$ et L_3.

Conséquences.

1) Les condition $L^{\#}$ ne dépendent pas du choix des opérateurs D_o', $\mathcal{A}', {\Lambda^{\#}}_1', {\Lambda^{\#}}_2'$ et ne dépendent donc que de h;

2) L'opérateur D_o' est l'expression dans les coordonnées choisies en 1.a d'un opérateur dont le symbole principal est $(\xi_o - \lambda(x)\xi_1)I$;

on en déduit facilement l'invariance des conditions de Levi, par changement de coordonnées;

3) En vue d'utiliser la réduction vue au 1.d, on montrera l'invariance des conditions de Levi par l'action de l'opérateur elliptique $\Delta(x, D_1)$ du 1.d. On notera $\Delta^{(i)}(x, D_1)$ un inverse de $\Delta(x, D_1)$ modulo un opérateur d'ordre $-\infty$ en ξ_1. On obtient la:

Proposition. *Si $h(x,D)$ vérifie la condition de Levi $L_1^{\#}$, son transformé $\tilde{h}(x,D)$ satisfait la condition de Levi correspondante $\tilde{L}_1^{\#}$; si h vérifie $L_1^{\#}$ et $L_2^{\#}$, $\tilde{h}$ vérifie $\tilde{L}_1^{\#}$ et $\tilde{L}_2^{\#}$; si h vérifie $L_1^{\#}, L_2^{\#}, L_3^{\#}$, $\tilde{h}$ vérifie $\tilde{L}_1^{\#}, \tilde{L}_2^{\#}$ et $\tilde{L}_3^{\#}$.*

PREUVE. On note $\tilde{\mathcal{A}}' = \Delta^i \mathcal{A}' \Delta$, $\tilde{D}_o = \Delta^i D_o \Delta$; $\sigma(\tilde{D}_o) = \xi_o\, I$

$$\sigma(\tilde{\mathcal{A}}') = \Delta_o^{-1} \mathcal{A} \Delta_o \quad \text{et} \quad \Delta_o^{-1} \mathcal{A} \Delta_o \Delta_o^{-1} a \Delta_o = \xi_o^2 I,$$

donc: $\tilde{\mathcal{A}}' \tilde{D}_o^2$ a pour symbole, la matrice des cofacteurs de $\tilde{a}(x,\xi) = \xi_o I + \tilde{\alpha}(x)\xi_1$; on pose:

$$\tilde{\mathcal{L}}_o^{\#} = \sigma\{\tilde{\mathcal{A}}'[\tilde{h}\tilde{\mathcal{A}}'(\tilde{h}\tilde{\mathcal{A}} - \tilde{D}'^2_o) - (\tilde{h}\tilde{\mathcal{A}}' - \tilde{D}'^2_o)\tilde{D}'^2_0]\}$$

et:

$$\tilde{\mathcal{L}}_o^{\#} = \Delta_o^{-1} \mathcal{L}_o^{\#} \Delta_o; \quad \tilde{L}_1^{\#} \quad \text{equivaut a } L_1^{\#}$$

$$\text{et} \quad \tilde{\Lambda}_1^{\#} = \Delta_o^{-1} \Lambda_1^{\#} \Delta_o.$$

La fin de la preuve utilise les mêmes remarques.

2.b. Nous allons expliciter les conditions L_1, L_2, L_3 pour l'opérateur:

$$h(x,D) = I\ D_o + \alpha(x) D_1 + b(x),$$

où :

$$\alpha(x) = \begin{pmatrix} 0 & \mu(x) & 0 & 0 \\ 0 & 0 & 0 & 0 \\ 0 & 0 & 0 & \nu(x) \\ 0 & 0 & 0 & 0 \end{pmatrix},$$

$\mu(x)$ et $\nu(x)$ sont des germes analytiques $\mu \not\equiv 0$, $\nu \not\equiv 0$; on a donc: $b(x) = b_o(x)$, $b_j(x) = 0$, $\mathcal{L}_1 = 0$; on peut prendre les $b_i^i(x) \equiv 0$. On est amené à distinguer les cas suivantes:

i) $b_3^2\, b_1^4 \not\equiv 0$. L_1 s' écrit:

$$\mu b_1^2 + \nu b_3^4 \equiv 0 \quad , \quad b_1^2 b_3^4 - b_3^2 b_1^4 \equiv 0$$

L_2 s'écrit:

$$\mu\left[b_1^3 - b_3^1\left(\frac{b_1^2}{b_3^2}\right)^2 - D_o\left(\frac{b_1^2}{b_3^2}\right)\right] + \nu\left[b_2^4 - b_4^2\left(\frac{b_3^4}{b_3^2}\right)^2 + D_o\left(\frac{b_3^4}{b_3^2}\right)\right] \equiv 0$$

L_3 s'écrit:

$$\frac{(b_3^4 b_4^2 - b_3^1 b_1^2)}{\nu b_3^2}\left[b_1^3 - \left(\frac{b_1^2}{b_3^2}\right)^2 b_3^1 - D_o\left(\frac{b_1^2}{b_3^2}\right)\right] +$$
$$- D_o\left\{\frac{1}{\nu}\left[b_1^3 - \left(\frac{b_1^2}{b_3^2}\right)^2 b_3^1 - D_o\left(\frac{b_1^2}{b_3^2}\right)\right]\right\} \equiv 0$$

ii) Si $b_3^2 \not\equiv 0,\ b_1^4 \equiv 0$:

$$L_1 : b_1^2 \equiv b_3^4 \equiv 0$$
$$L_2 : \mu b_1^3 + \nu b_2^4 \equiv 0$$
$$L_3 : D_o\left(\frac{b_1^3}{\nu}\right) \equiv 0$$

iii) $b_3^2 \equiv 0\ ,\ b_1^4 \not\equiv 0$:

$$L_1 : b_1^2 \equiv b_3^4 \equiv 0$$
$$L_2 : \mu b_4^2 + \nu b_3^1 \equiv 0$$
$$L_3 : D_o\left(\frac{b_3^1}{\mu}\right) \equiv 0$$

iv) $b_3^2 \equiv 0\ ,\ b_1^4 \equiv 0$

$L_1 : b_1^2 \equiv b_3^4 \equiv 0$

L_2 et L_3 sont vérifiées sans condition supplémentaire.

Lors de l'étude la condition nécessaire, nous utiliserons pour l'opérateur:

$$h(x,D) = I\ D_o + \alpha(x)D_1 + b(x),$$

du 1.b, la réduction du 1.c, 1.d et 2.a, Conséquence 3.

L'opérateur transformé $\tilde{h}$ est de la forme donnée au 1.d; les conditions de Levi s'écrivent pour cet opérateur:

i) si $(\tilde{b}_o)_3^2(\tilde{b}_o)_1^4 \not\equiv 0$,

$$L_1 : (\tilde{b}_o)_1^2 + (\tilde{b}_o)_3^4 \equiv 0, \quad (\tilde{b}_o)_1^2(\tilde{b}_o)_3^4 - (\tilde{b}_o)_3^2(\tilde{b}_o)_1^4 \equiv 0$$

$$L_2 : (\tilde{b}_o)_1^3 - (\tilde{b}_o)_4^2 \left[\frac{(\tilde{b}_o)_1^2}{(\tilde{b}_o)_3^2}\right]^2 - 2D_o\frac{(\tilde{b}_o)_1^2}{(\tilde{b}_o)_3^2} \equiv 0$$

$$L_3 : \frac{(\tilde{b}_o)_1^2(\tilde{b}_o)_4^2}{(\tilde{b}_o)_3^2}\left[(\tilde{b}_o)_1^3 - D_o\frac{(\tilde{b}_o)_1^2}{(\tilde{b}_o)_3^2}\right] + D_o\left[(\tilde{b}_o)_1^3 - D_o\frac{(\tilde{b}_o)_1^2}{(\tilde{b}_o)_3^2}\right] +$$

$$- \left[\frac{(\tilde{b}_o)_1^2}{(\tilde{b}_o)_3^2}(\tilde{b}_1)_1^2 + (\tilde{b}_1)_1^4\right] - (\tilde{b}_o)_1^4\left[\frac{(\tilde{b}_1)_3^2}{(\tilde{b}_o)_3^2} + \frac{(\tilde{b}_1)_3^4}{(\tilde{b}_o)_1^2}\right] \equiv 0$$

ii) si $(\tilde{b}_o)_3^2 \not\equiv 0$ et $(\tilde{b}_o)_1^4 \equiv 0$:

$$L_1 : (\tilde{b}_o)_1^2 \equiv (\tilde{b}_o)_3^4 \equiv 0$$

$$L_2 : (\tilde{b}_o)_1^3 \equiv 0$$

iii) si $(\tilde{b}_o)_1^4 \not\equiv 0$ et $(\tilde{b}_o)_3^2 \equiv 0$:

$$L_1 : (\tilde{b}_o)_1^2 \equiv (\tilde{b}_o)_3^4 \equiv 0$$

$$L_2 : (\tilde{b}_o)_4^2 \equiv 0$$

iv) si $(\tilde{b}_o)_3^2 \equiv (\tilde{b}_o)_1^4 \equiv 0$:

$$L_1 : (\tilde{b}_o)_1^2 \equiv (\tilde{b}_o)_3^4 \equiv 0.$$

3. Condition nécessaire pour que le problème de Cauchy soit localement bien posé . On considère l'opérateur différentiel h du 1.a, 1.b.

Théorème. *Pour que le problème de Cauchy soit localement bien posé pour l'opérateur h, il faut que les conditions de Levi L_1, L_2, L_3 soient satisfaites.*

PREUVE.

3.a. Il suffit de démontrer le résultat pour l'ensemble des points $(\underline{t}, \underline{x}')$ où le rang généralisé est constant; en effet cet ouvert est partout dense, car son complémentaire est l'ensemble analytique des x où le rang varie, considéré au 1.c et, d'autre part, les conditions de Levi s'expriment par l'annulation de fonctions continues.

3.b. Proposition L_1. *La condition: "$\mathcal{L}_o\mathcal{L}_o\mathcal{A}$ est divisible par ξ_o" est nécessaire pour que le problème de Cauchy soit localement bien posé.*

PREUVE. La condition: $\mathcal{L}_o\mathcal{L}_o\mathcal{A}$ est divisible par ξ_o s'exprime par l'annulation d'une fonction matricielle analytique $\mathcal{M}_1(x)$. On va démontrer que si, en un point $\underline{x}, \mathcal{M}_1(\underline{x}) \neq 0$, le problème de Cauchy n'est pas bien posé pour h dans un voisinage de ce point.

On utilise alors l'invariance de la condition de Levi L_1 par l'action de l'opérateur Δ. En explicitant la condition pour l'opérateur transformé $\tilde{h}$, on obtient que:

$\mathcal{M}_1(x) = 0$ équivaut aux conditions sur $\tilde{b}_o(x)$ données à la fin du 2.b.

On a donc à démontrer d'abord le

Lemme 1.

i) *Si* $(\tilde{b}_o)^2_3(b_o)^4_1 \not\equiv 0$, *si* $[(\tilde{b}_o)^2_1 + (\tilde{b}_o)^4_3](\underline{x}) \neq 0$ *ou si:*

$$[(\tilde{b}_o)^2_1(\tilde{b}_o)^4_3 - (\tilde{b}_o)^2_3(\tilde{b}_o)^4_1](\underline{x}) \neq 0,$$

il existe un voisinage de $\underline{x}$ *où le problème de Cauchy n'est pas bien posé pour* h.

ii) *Si* $(\tilde{b}_o)^2_3 \not\equiv 0$ *et* $(\tilde{b}_o)^4_1 \equiv 0$, *si* $(\tilde{b}_o)^2_1(\underline{x}) \neq 0$ *ou si* $(\tilde{b}_o)^4_3(\underline{x}) \neq 0$, *il existe un voisinage de* $\underline{x}$, *où le problème de Cauchy n'est pas bien posé pour* h.

iii) *Si* $(\tilde{b}_o)^2_3 \equiv 0$ *et* $(\tilde{b}_o)^4_1 \not\equiv 0$, *si* $(\tilde{b}_o)^2_1(\underline{x}) \neq 0$, $(\tilde{b}_o)^4_3(\underline{x}) \neq 0$, *il existe un voisinage de* $\underline{x}$, *où le problème de Cauchy n'est pas bien posé.*

iv) *Si* $(\tilde{b}_o)^2_3 \equiv (\tilde{b}_o)^4_1 \equiv 0$, *si* $(\tilde{b}_o)^2_1(\underline{x}) \neq 0$ *ou* $(\tilde{b}_o)^4_3(\underline{x}) \neq 0$, *il existe un voisinage de* $\underline{x}$ *où le problème de Cauchy n'est pas bien posé pour* h.

Preuve. Nous ne donnerons la démonstration que dans le cas i), les cas particuliers ii), iii), iv) ayant des démonstrations analogues.

On peut supposer que $(\tilde{b}_o)_3^2(\tilde{b}_o)_1^4(\underline{x}) \neq 0$, par une remarque déjà utilisée de densité et de continuité .

On a le

Lemme 2. *Si il existe un voisinage ouvert de $\underline{x}$, où*

$$[(\tilde{b}_o)_1^2 + (\tilde{b}_o)_2^4]^2 + 4[(\tilde{b}_o)_3^2(\tilde{b}_o)_1^4 - (\tilde{b}_o)_1^2(\tilde{b}_o)_3^4] \not\equiv 0$$

alors le problème de Cauchy n'est pas bien posé dans ce voisinage.

Preuve. On se place en un point $\underline{\underline{x}}$ où:

$$[(\tilde{b}_o)_1^2 + (\tilde{b}_o)_3^4]^2 + 4[(\tilde{b}_o)_3^2(\tilde{b}_o)_1^4 - (\tilde{b}_o)_1^2(\tilde{b}_o)_3^4](\underline{\underline{x}}) \neq 0$$

et nous voulons démontrer que le problème de Cauchy n'est pas bien posé en $\underline{\underline{x}}$. On démontre le:

Lemme. *Pour tout voisinage ouvert G de $\underline{\underline{x}}$, assez petit, il existe une fonction vectorielle $Y_o(x) \in \mathcal{C}^\infty(G)$, telle que $Y_o(x) \neq 0$ dans G, il existe des fonctions $Y_j(x) \in \mathcal{C}^\infty(G)$, il existe une fonction scalaire $\psi \in \mathcal{C}^\infty(G)$ telle que: D_o Im $\psi(x) < 0$ dans G (où Im désigne la partie imaginaire), il existe $\delta = \pm 1$, tel que le développement formel:*

$$u = e^{i[\omega\delta x_1 + \omega^{1/2}\psi(x)]}\left[Y_o(x) + Y_1(x)\omega^{-1/2} + \right.$$
$$\left. + \ldots + Y_j(x)\omega^{-j/2} + \ldots\right],$$

vérifie formellement:

$$hu \equiv e^{i[\omega\delta x_1 + \omega^{1/2}\psi(x)]}\omega[Z_o(x) + Z_1(x)\omega^{-1/2} +$$
$$+ \ldots + Z_j(x)\omega^{-j/2} + \ldots] = 0$$

c'est-à -dire: $Z_o = Z_1 = \ldots = Z_j = \ldots = 0.$

Preuve. On cherche d'abord un développement formel:

$$\tilde{u} = e^{i[\omega\delta x_1 + \omega^{1/2}\psi(x)]}\left[\tilde{Y}_o(x) + \tilde{Y}_1(x)\omega^{-1/2} + \right.$$
$$\left. + \ldots + \tilde{Y}_j(x)\omega^{-j/2} + \ldots\right],$$

tel que:

$$\tilde{h}\tilde{u} = 0 \quad ; \quad \tilde{Y}_o \in \mathcal{C}^\infty(G) \quad \text{et} \quad \tilde{Y}_o \neq 0.$$

On trouve d'abord que $\tilde{Z}_o(x) = 0$, s'écrit:

$$\tilde{a}(x, D_o\varphi)\tilde{Y}_o(x) = 0,$$

soit (cf. 1.c):

$$\begin{pmatrix} J & 0 \\ 0 & J \end{pmatrix} \begin{pmatrix} \tilde{Y}_o^1 \\ \tilde{Y}_o^2 \\ \tilde{Y}_o^3 \\ \tilde{Y}_o^4 \end{pmatrix} = 0,$$

d'où:

$$\tilde{Y}_o^2 = \tilde{Y}_o^4 = 0;$$

ensuite $\tilde{Z}_1 = 0$ s'écrit:

$$\begin{pmatrix} J & 0 \\ 0 & J \end{pmatrix} \begin{pmatrix} \tilde{Y}_1^1 \\ \tilde{Y}_1^2 \\ \tilde{Y}_1^3 \\ \tilde{Y}_1^4 \end{pmatrix} + [D_o\psi I + \begin{pmatrix} J & 0 \\ 0 & J \end{pmatrix} \quad D_1\psi]\delta Y_o = 0;$$

la lère et la 3ème équation s'écrivent:

$$\begin{aligned} \tilde{Y}_1^2 + D_o\psi\delta\tilde{Y}_o^1 &= 0 \\ \tilde{Y}_1^4 + D_o\psi\delta\tilde{Y}_o^3 &= 0; \end{aligned}$$

ensuite, on a: $\tilde{Z}_2 = 0$:

$$\begin{pmatrix} J & 0 \\ 0 & J \end{pmatrix} \begin{pmatrix} \tilde{Y}_2^1 \\ \tilde{Y}_2^2 \\ \tilde{Y}_2^3 \\ \tilde{Y}_2^4 \end{pmatrix} + [D_o\psi I + \begin{pmatrix} J & 0 \\ 0 & J \end{pmatrix} \quad D_1\psi]\delta Y_1 +$$

$$- i\delta[D_o + \begin{pmatrix} J & 0 \\ 0 & J \end{pmatrix} \quad D_1 + \tilde{b}_o]\tilde{Y}_o = 0.$$

On obtient:

$$\tilde{Y}_2^2 + D_o\psi\delta\tilde{Y}_1^1 + D_1\psi\delta\tilde{Y}_1^2 - i\delta D_o\tilde{Y}_o^1 = 0$$
$$D_o\psi\delta\tilde{Y}_1^2 - i\delta((\tilde{b}_o)_1^2\tilde{Y}_o^1 + (\tilde{b}_o)_3^2\tilde{Y}_o^3) = 0$$
$$\tilde{Y}_2^4 + D_o\psi\delta\tilde{Y}_1^3 + D_o\psi\delta\tilde{Y}_1^4 - i\delta(D_o\tilde{Y}_o^3 + (b_o)_1^3\tilde{Y}_o^1) = 0$$
$$D_o\psi\delta\tilde{Y}_1^4 - i\delta((\tilde{b}_o)_1^4\tilde{Y}_o^1 + (\tilde{b}_o)_3^4\tilde{Y}_o^3) = 0.$$

La 2ème et la 4ème équation doivent avoir une solution non nulle en $\tilde{Y}_o^1$ et $\tilde{Y}_o^3$; on choisit ψ tel que:

$$\left[(D_o\psi)^2 + i\delta(\tilde{b}_o)_1^2\right]\left[(D_o\psi)^2 + i\delta(\tilde{b}_o)_3^4\right] + (\tilde{b}_o)_3^2(\tilde{b}_o)_1^4 \equiv 0,$$

soit:

$$(D_o\psi)^2 = \frac{1}{2}\Big(- i\delta[(\tilde{b}_o)_1^2 + (\tilde{b}_o)_3^4] \pm \pm i\sqrt{[(\tilde{b}_o)_1^2 + (\tilde{b}_o)_3^4]^2 + [(\tilde{b}_o)_3^2(\tilde{b}_o)_1^4 - (\tilde{b}_o)_1^2(\tilde{b}_o)_3^4]}\Big)$$

on choisit δ et le signe $\pm$ de sorte que:

$$(D_o\psi)^2 \quad \text{ne soit pas réel} \quad \geq 0,$$

et ensuite, on peut choisir $D_o\psi$ tel que $Im\ D_o\psi < 0$. De $\tilde{Z}_3 \equiv 0$, on tire

$$[(D_o\psi)^2 + i\delta(\tilde{b}_o)_1^2]Y_1^1 + i\delta(\tilde{b}_o)_3^2\tilde{Y}_3^1 + - 2iD_o\psi D_o\tilde{Y}_o^1 + C_1^1\tilde{Y}_1^1 + C_3^1\tilde{Y}_1^3 = 0$$
$$i\delta(\tilde{b}_o)_1^4 Y_1^1 + [(D_o\psi)^2 + i\delta(\tilde{b}_o)_3^4]\tilde{Y}_1^3 + - 2iD_o\psi D_o\tilde{Y}_o^3 + C_1^2\tilde{Y}_1^1 + C_3^2\tilde{Y}_1^3 = 0,$$

où les coefficients C sont analytiques et connus.

On écrit que ces équations en $\tilde{Y}_1^1$ et $\tilde{Y}_1^3$ sont compatibles et on exprime, à l'aide de la 2ème équation de $\tilde{Z}_2 = 0$, $\tilde{Y}_o^3$ en fonction de $\tilde{Y}_o^1$; on obtient:

$$(\tilde{b}_o)_1^4\big(2(D_o\psi)^2 + i\delta[(\tilde{b}_o)_1^2 + (\tilde{b}_o)_3^4]\big)D_o\tilde{Y}_o^1 + C'(x)\tilde{Y}_o^1 = 0,$$

où C' est analytique connu; compte tenu de l'hypothèse de l'énoncé du lemme précédent, le coefficient de $\tilde{D}_o\ \tilde{Y}_o^1$ est analytique et jamais nul dans un voisinage convenable de $\underline{\underline{x}}$; on choisit donc $\tilde{Y}_o^1 \in$

$\mathcal{C}^\infty$, $\tilde{Y}_o^1(\underline{\underline{x}}_o, x_1)$ à support compact, $\tilde{Y}_o^1 \neq 0$ dans G; ensuite $\tilde{Y}_o^3$ est déterminé , puis $\tilde{Y}_1^2$ et $\tilde{Y}_1^4$; on continue l'identification et on détermine tous les $\tilde{Y}_j$ par récurrence.

En revenant à h, on a, formellement:

$$h\Delta\tilde{u} = 0;$$

on pose:

$$u = \Delta\tilde{u};$$

u est de la forme:

$$u = e^{i[\omega\delta x_1 + \omega^{1/2}\psi(x)]}\left[\Delta_o(x)\tilde{Y}_o(x) + Y_1(x)\omega^{-1/2} + \ldots + Y_j(x)\omega^{-j/2} + \ldots\right],$$

avec: $Y_o(x) = \Delta_o(x)\tilde{Y}_o(x) \neq 0$, $Y_j(x) \in \mathcal{C}^\infty(G)$ tel que:

$$h\,u \;=\; 0.$$

Lemme. *On pose:*

$$u_J(x) = e^{i[\omega\delta x_1 + \omega^{1/2}\psi(x)]}[Y_o(x) + .. + Y_J(x)\omega^{-J/2}];$$

on désigne par $\| \quad \|_{k,K}$ *une semi-norme de* $\mathcal{C}^\infty(G)$ *quelconque,* k *entier* ≥ 0, K *compact de* G*; on a:*

$$\|hu_J\|_{k,K} = 0(\omega^{k-N_J}) \cdot \exp\left[\sup_K \big(-Im\ \psi(x)\big)\omega^{1/2}\right],$$

$$si \quad \omega \to +\infty$$

où on peut choisir J *assez grand pour que* $k - N_J < 0$*; de plus;*

$$\|u_J\|_{0,K} = \exp\left[\sup_K \big(-Im\ \psi(x)\big)\omega^{1/2}\right] \cdot [y + o(\omega)],$$

$$y \neq 0, \quad si \quad \omega \to +\infty$$

enfin, si: $\underline{\underline{x}} = (\underline{\underline{t}}, \underline{\underline{x}}_1)$

$$\|u_J(\underline{\underline{t}}, x_1)\|_{k,K\cap\{x_o = \underline{\underline{t}}\}} =$$

$$0(\omega^k)\exp\left[\sup_K \big(-Im\ \psi(\underline{\underline{t}}, x_1)\big)\omega^{1/2}\right] \cdot [y + o(\omega)]$$

$$si \quad \omega \to +\infty$$

Fin de la preuve du Lemme 2. Pour tout voisinage assez petit G de $\underline{\underline{x}}$, $\forall k$, $\forall K$, $\forall C > 0$, il existe ω, tel que:

$$\|u_J\|_{0,K} > C\{\|hu_J\|_{k,K} + \|u_J(\underline{\underline{t}}, x_1)\|_{k,K\cap\{x_o=\underline{\underline{t}}\}}\},$$

cela résulte du fait que:

$$-Im\ \psi(\underline{\underline{t}}, x_1) < -Im\ \psi(x_o, x_1) \quad \text{si} \quad \underline{\underline{t}} < x_o.$$

On déduit alors du théorème du graphe fermé de Banach que le problème de Cauchy n'est pas bien posé en $\underline{\underline{x}}$, d'où le Lemme 2.

Lemme 3. *Si au voisinage de* $\underline{\underline{x}}$*:*

$$[(\tilde{b}_o)_1^2 + (\tilde{b}_o)_3^4]^2 + 4[(\tilde{b}_o)_3^2(\tilde{b}_o)_1^4 - (\tilde{b}_o)_1^2(\tilde{b}_o)_3^4] \equiv 0,$$

alors le problème Cauchy n'est pas bien posé pour h*, dans ce voisinage.*

PREUVE. Le principe de la démonstration étant celui de la précédente, nous ne donnerons que les détails caractéristiques du lemme. On distingue deux cas, que nous expliquerons dans deux nouveaux lemmes.

Lemme 4. *Si il existe un voisinage ouvert de* $\underline{\underline{x}}$*, où:*

$$(\tilde{b}_o)_1^3 - 2D_o\Big(\frac{(\tilde{b}_o)_1^2}{(\tilde{b}_o)_3^2}\Big) - (\tilde{b}_o)_4^2\Big(\frac{(\tilde{b}_o)_1^2}{(\tilde{b}_o)_3^2}\Big)^2 \not\equiv 0,$$

alors le problème de Cauchy n'est pas bien posé dans ce voisinage.

PREUVE. On se place en un point $\underline{\underline{x}}$, où l'expression ci-dessus est $\neq 0$. On construit un développement

$$u = e^{i[\omega\delta x_1 + \omega^{1/4}\theta(x) + \omega^{1/2}\psi(x)]}[Y_o(x) + Y_1(x)\omega^{-1/4} +$$
$$+ \ldots + Y_j(x)\omega^{-j/4} + \ldots]$$

où $Y_o(x) \neq 0$ dans G, $Y_j \in \mathcal{C}^\infty(G)$, $\psi \in \mathcal{C}^\infty(G)$, $D_o\ Im\psi < 0$ dans G, $\theta \in \mathcal{C}^\infty(G)$, tel que $hu = 0$.

On construit d'abord $\tilde{u}$ analogue tel que:

$$\tilde{h}\tilde{u} = 0;$$

on écrit que $\tilde{Z}_o = \tilde{Z}_1 = \tilde{Z}_2 = \tilde{Z}_3 = \tilde{Z}_4 = 0$; à ce stade, on obtient encore:

$$[(D_o\psi)^2 + i\delta(\tilde{b}_o)_1^2]\tilde{Y}_o^1 + i\delta(\tilde{b}_o)_3^2\tilde{Y}_o^3 = 0$$
$$i\delta(\tilde{b}_o)_1^4\tilde{Y}_o^1 + [(D_o\psi)^2 + i\delta(\tilde{b}_o)_3^4]\tilde{Y}_o^3 = 0;$$

on choisit ψ tel que:

$$(D_o\psi)^2 = \frac{-i\delta[(\tilde{b}_o)_1^2 + (\tilde{b}_o)_3^4]}{2} \quad \text{, qui n'est jamais nul;}$$

on choisit δ et le signe $\pm$ de sorte que $D_o\ Im\psi < 0$.

On écrit: $\tilde{Z}_5 = \tilde{Z}_6 = 0$ et dans $\tilde{Z}_6 = 0$, on écrit que la 2ème et la 4ème équation en $\tilde{Y}_1^1$ et $\tilde{Y}_1^3$ sont compatibles; compte tenu de ce qui précède, on obtient une équation en $\tilde{Y}_0^1$ de la forme:

$$\{D_o\psi(x)[D_o\theta(x)]^2 + c \cdot C(x)\}\tilde{Y}_o^1(x) = 0,$$

c est numérique $\neq 0$ et $C(\underline{x})$ est $\neq 0$, d'après l'hypothèse du lemme 4; on choisit $\theta \in \mathcal{C}^\infty$ de sorte que le coefficient de $\tilde{Y}_o$ soit nul. On écrit $\tilde{Z}_7 = 0$ et on trouve que $\tilde{Y}_o^1$ satisfait une équation différentielle du 1° ordre en x_o, de la forme:

$$D_o\tilde{Y}_o^1 + c'(x)\tilde{Y}_o^1 = 0,$$

où c' est analytique; on choisit $\tilde{Y}_o^1 \in \mathcal{C}^\infty$ et on détermine de même les $\tilde{Y}_j$ suivants; ensuite on finit comme dans le Lemme 2.

Lemme 5. *Si au voisinage de $\underline{x}$:*

$$(\tilde{b}_o)_1^3 - 2D_o\left(\frac{(\tilde{b}_o)_1^2}{(\tilde{b}_o)_3^2}\right) - (\tilde{b}_o)_4^2\left(\frac{(\tilde{b}_o)_1^2}{(\tilde{b}_o)_3^2}\right)^2 \equiv 0$$

alors le problème de Cauchy n'est pas bien posé pour h, en $\underline{x}$.

PREUVE. On considère à nouveau le développement:

$$u = e^{i[\omega\delta x_1 + \omega^{1/2}\psi(x)]}[\tilde{Y}_o(x) + \ldots + \tilde{Y}_j(x)\omega^{-j/2} + \ldots].$$

On écrit $\tilde{Z}_o = \tilde{Z}_1 = \tilde{Z}_2 = 0$, comme précédemment et on obtient ψ, $Im\ D_o\psi < 0$; $\tilde{Z}_3 = 0$ est alors satisfaite; on considère $\tilde{Z}_4 = 0$; et on a ainsi une équation différentielle du 2ème ordre en $\tilde{Y}_o^1$:

$$D_o^2\tilde{Y}_o^1(x) + d(x)D_o\tilde{Y}_o^1(x) + d'(x)\tilde{Y}_o^1(x) = 0$$

où les coefficients sont analytiques. On finit comme précédemment.

Le Lemme 1 est donc démontré et par suite la Proposition L_1. *On suppose* maintenant *la condition* L_1 *satisfaite.*

3.c. Proposition L_2. *La condition L_2: "$\mathcal{L}_o(\Lambda_1)$ est divisible par ξ_o", est nécessaire pour que le problème de Cauchy soit localement bien posé.*

PREUVE. L_2 s'exprime par l'annulation d'une fonction analytique $\mathcal{M}_2(x)$. On va démontrer que, si en un point $\underline{x}$, $\mathcal{M}_2(\underline{x}) \neq 0$, le problème de Cauchy n'est pas bien posé en ce point. On utilise l'invariance de la condition L_2 par l'action de l'opérateur Δ; on obtient pour $\tilde{h}$, que: $\mathcal{M}_2(\underline{x}) \neq 0$ équivaut à la condition L_2 sur $\tilde{b}_o$ donnée à la fin du 2.b. On a donc à démontrer le:

Lemme.

i) Si $(\tilde{b}_o)_3^2(\tilde{b}_o)_1^4 \not\equiv 0$, *si*

$$\left[(\tilde{b}_o)_1^3 - 2D_o\left(\frac{(\tilde{b}_o)_1^2}{(\tilde{b}_o)_3^2}\right) - (\tilde{b}_o)_4^2\left(\frac{(\tilde{b}_o)_1^2}{(\tilde{b}_o)_3^2}\right)^2\right](\underline{x}) \neq 0,$$

alors le problème de Cauchy n'est pas bien posé pour h, en $\underline{x}$;

ii) Si $(\tilde{b}_o)_3^2 \not\equiv 0$ *et* $(\tilde{b}_o)_1^4 \equiv 0$, *si* $(\tilde{b}_o)_1^3(\underline{x}) \neq 0$, *alors le problème de Cauchy n'est pas bien posé en* $\underline{x}$;

iii) Si $(\tilde{b}_o)_1^4 \not\equiv 0$ *et* $(\tilde{b}_o)_3^2 \equiv 0$, *si* $(\tilde{b}_o)_4^2(\underline{x}) \neq 0$, *alors le problème de Cauchy n'est pas bien posé en* $\underline{x}$.

PREUVE. Nous considérons seulement le cas i) les autres cas étant analogues. On construit un développement de la forme:

$$\tilde{u} = e^{i[\omega\delta x_1 + \omega^{1/3}\psi(x)]}\big[\tilde{Y}_o(x) + \tilde{Y}_1(x)\omega^{-1/3} + \\ + \dots + \tilde{Y}_j(x)\omega^{-j/3} + \dots\big],$$

tel que $\tilde{h}\tilde{u} = 0$.

On écrit $\tilde{Z}_o = \tilde{Z}_1 = \tilde{Z}_2 = \tilde{Z}_3 = \tilde{Z}_4 = \tilde{Z}_5 = 0$; on obtient:

$$(D_o\psi)^3 = (\tilde{b}_o)_3^2(\tilde{b}_o)_1^3 + (\tilde{b}_o)_4^2(\tilde{b}_o)_1^4 - 2(\tilde{b}_o)_3^2 D_o\left[\frac{(\tilde{b}_o)_1^2}{(\tilde{b}_o)_3^2}\right],$$

le deuxième membre n'est pas nul, compte tenu de l'hypothèse de ce lemme; on peut choisir $\psi \in \mathcal{C}^\infty$ tel que: $Im\ D_o\psi < 0$; $\tilde{Z}_6 = 0$ détermine alors $\tilde{Y}_o^1$ par une équation différentielle du 1° ordre.

On finit comme précédemment:

On suppose maintenant L_1 *et* L_2 *satisfaites.*

3.d. Proposition L_3. *La condition L_3 : "$\mathcal{L}_o(\Lambda_2)$ est divisible par ξ_o", est nécessaire pour que le problème de Cauchy soit localement bien posé.*

PREUVE. L_3 s'éxprime par l'annulation d'une fonction analytique $\mathcal{M}_3(x)$. On va démontrer que, si en un point $\underline{x}$, $\mathcal{M}_3(\underline{x}) \neq 0$, le problème de Cauchy n'est pas bien posé en ce point. On utilise l'invariance de L_2 par l'action de Δ; on obtient pour $\tilde{h}$, que $\tilde{\mathcal{M}}_2(\underline{x}) \neq 0$ équivaut à la condition L_3 sur $(\tilde{b}_o)$ et $(\tilde{b}_1)$ donnée à la fin du 2b). On a donc à démontrer le:

Lemme. *Si* $(\tilde{b}_o)_3^2(\tilde{b}_o)_1^4 \neq 0$, *si:*

$$(\tilde{b}_o)_1^2\frac{(\tilde{b}_o)_4^2}{(\tilde{b}_o)_3^2}\left[(\tilde{b}_o)_1^3 - D_o\frac{(\tilde{b}_o)_1^2}{(\tilde{b}_o)_3^2}\right] + D_o\left[(\tilde{b}_o)_1^3 - D_o\frac{(\tilde{b}_o)_1^2}{(\tilde{b}_o)_3^2}\right] + \\ - \left[\frac{(\tilde{b}_o)_1^2}{(\tilde{b}_o)_3^2}(\tilde{b}_1)_1^2 + (\tilde{b}_1)_1^4\right] - (\tilde{b}_o)_1^4\left[\frac{(\tilde{b}_1)_3^2}{(\tilde{b}_o)_3^2} + \frac{(\tilde{b}_1)_3^4}{(b_o)_3^2}\right](\underline{x}) \neq 0,$$

alors le problème de Cauchy n'est pas bien posé pour h en $\underline{x}$.

PREUVE. On considère le cas i). On construit un développement de la forme:

$$\tilde{u} = e^{i[\omega\delta x_1 + \omega^{1/4}\psi(x)]}\left[\tilde{Y}_o(x) + \tilde{Y}_1(x)\omega^{-1/4} + \right.$$
$$\left. + \ldots + \tilde{Y}_j(x)\omega^{-j/4} + \ldots\right],$$

tel que:

$$\tilde{h}\tilde{u} = 0.$$

On écrit: $\tilde{Z}_o = \ldots = \tilde{Z}_8 = 0$; on obtient:

$$(D_o\psi)^4 = \delta C(x),$$

où $C(x)$ est $\neq 0$, compte tenu de l'hypothèse de ce lemme; on choisit ψ tel que $ImD_o\psi < 0$; en écrivant $\tilde{Z}_9 = 0$, on obtient que : $\tilde{Y}_o^1$ satisfait une équation différentielle d'ordre 1 en x_o. On finit comme précédemment.

Remarque. Avec des hypothèses supplémentaires, on peut obtenir des résultats analogues, lorsque les coefficients de h sont seulement C^∞, pour la condition nécessaire; on supposera ainsi en particulier que l'ensemble où le rang est constant est partout dense; on peut aussi se placer dans $\mathbf{R}^n$.

4. Condition suffisante pour que le problème de Cauchy soit localement bien posé.

4.a. Nous considérons l'opérateur du 2.b

$$h(x, D) = ID_o + \alpha(x)D_1 + b(x)$$

$$\alpha(x) = \begin{pmatrix} 0 & \mu(x) & 0 & 0 \\ 0 & 0 & 0 & 0 \\ 0 & 0 & 0 & \nu(x) \\ 0 & 0 & 0 & 0 \end{pmatrix}.$$

Nous ferons lorsque $b_3^2 b_1^4 \not\equiv 0$ *l'hypothèse* suivante:

μ et ν sont des germes irreductibles.

Comme il y a beaucoup de cas à envisager, nous ferons ensuite des *hypothèses* qui conduiront à l'étude d'un des cas; les autres se traitent de façon semblable:

i) *μ et ν sont premiers entre eux*

ii) *$\mu\nu$ ne divise pas b_1^4.*

On déduit de ces hypothèses et de la condition de Levi L_1 que

$$b_1^2 = -k\nu\rho\sigma, \quad b_3^4 = k\mu\rho\sigma, \quad b_3^2 = -k\mu\nu\sigma^2, \quad b_1^4 = k\rho^2,$$

où ρ et σ sont des germes premiers entre eux, k un germe quelconque.

On explicite ensuite L_2 et L_3, en posant:

$$U = \mu\sigma b_1^3 - D_o\rho, \quad V = \sigma\nu b_2^4 - D_o\rho$$
$$\mu S = \rho b_3^1 - D_o(\mu\sigma), \quad \nu T = \rho b_4^2 - D_o(\nu\sigma).$$

On obtient:

$$\rho(S+T) = \sigma(U+V)$$
$$\sigma D_o(\rho S - \sigma U) + (\rho S - \sigma U)(T+S) = 0.$$

Le système à résoudre s'écrit (avec $f = 0$; les cas général se traite de même):

$$\begin{aligned}
&1) \quad D_o u_1 + \mu D_1 u_2 + b_2^1 u_2 + b_3^1 u_3 + b_4^1 u_4 = 0\\
&2) \quad D_o u_2 - k\nu\sigma(\rho u_1 + \mu\sigma u_3) + b_4^2 u_4 = 0\\
&3) \quad D_o u_3 + \nu D_1 u_4 + b_1^3 u_1 + b_2^3 u_2 + b_4^3 u_4 = 0\\
&4) \quad D_o u_4 + k\rho(\rho u_1 + \mu\sigma u_3) + b_2^4 u_2 = 0,\\
&u_1(t,x_1) = g_1(x_1),\ u_2(t,x_1) = g_2(x_1),\\
&u_3(t,x_1) = g_3(x_1),\ u_4(t,x_1) = g_4(x_1),
\end{aligned}$$

où les g sont les données de Cauchy sur $x_o = t$.

On définit de nouvelles inconnues

$$\begin{aligned}
w &= \rho u_2 + \nu\sigma u_4\\
w_1 &= \rho u_1 + \mu\sigma u_3\\
w_2 &= -\mu D_1 w - U u_1 - \mu S u_3.
\end{aligned}$$

On obtient, compte tenu des conditions de Levi L_2, L_3:

5) $D_o w = -V u_2 - \nu T u_4$
6) $D_o w_1 = w_2 + C_1 u_2 + C_2 u_4$
7) $D_o w_2 = D_o \mu \mu^{-1} w_2 + C_3 w_1 + C_4 w + C_5 u_2 + C_6 u_4$.

On fait *l'hypothèse*: μ *divise* $D_o \mu$, c'est-à-dire que, à un facteur inversible près, μ ne dépend que de x_1; (cette hypothèse est inutile si on peut choisir $\rho = 1$);
C_1, C_2, C_3, C_4, C_5 et C_6 sont des germes analytiques dépendant des données,

$$\begin{aligned} w(t,x_1) &= \rho(t,x_1) g_2(x_1) + (\nu\sigma)(t,x_1) g_4(x_1) \\ w_1(t,x_1) &= \rho(t,x_1) g_1(x_1) + (\mu\sigma)(t,x_1) g_3(x_1) \\ w_2(t,x_1) &= -\mu(t,x_1) D_1 w(t,x_1) - U(t,x_1) g_1(x_1) + \\ &\quad - \mu(t,x_1) S(t,x_1) g_3(x_1). \end{aligned}$$

On résout le système différentiel ordinaire $\{(2), (4), (5), (6), (7)\}$ en u_2, u_4, w, w_1, w_2; on obtient ensuite u_1 et u_3 par (1) et (3). On vérifie que les u obtenus vérifient le système initial.

Les cas où $b_3^2 b_1^4 \equiv 0$ se traitent aisément.

On obtient donc le:

Théorème. *Sous les hypothèses indiquées, pour que le problème de Cauchy soit localement bien posé pour l'opérateur h, il suffit que les conditions de Levi L_1, L_2, L_3 soient satifsaites.*

4.b. Lorsque le rang généralisé est constant, on peut ramener l' étude de h à celle de $\tilde{h}$, comme au 1.d; les calculs précédents s'adaptent aisément avece $\mu = \nu = 1$ et on obtient le:

Théorème. *Si le rang généralisé est constant, pour que le problème de Cauchy soit localement bien posé, pour l'opérateur h, il faut et il suffit que les conditions de Levi L_1, L_2, L_3 soient satisfaites.*

Bibliographie

[1] R.Berzin, J.Vaillant, J. Math. Pures et Appliquées **58** (1979), 165-216.

[2] K.Kajitani, Publ. R.I.M.S. Kyoto **15** (1979), 519-550.

[3] W.Matsumoto, J. Math. Kyoto Univ. **21** (1981), 47-84 et 251-271.

[4] V.M.Petkov, Math. Nachr. **93** (1979), 117-131.

[5] H.Flaschka, G.Strang, Advances in Math. **6** (1971), 347-379.

[6] J.Vaillant, Ann. Inst. Fourier **151** (1965), 225-311.

[7] J.Vaillant, Hyperbolic equations, Pitman Res. Notes in Math. **158**, Longman, 1987.

[8] J.Vaillant, C.R. Acad. Sc. Paris **304**, Série I (1987), 379-384.

[9] J.Vaillant, C.R. Acad. Sci. Paris **305**, Série I (1987), 377-380.

[10] J.Vaillant, Recent developments in hyperbolic equations, Pitman Res. Notes in Math. **183**, Longman, 1988.

Mathématiques, Université de Paris VI
Tour 45-46, 5ème étage
4, Place Jussieu
F-75252 PARIS cedex 05

ON THE WEIERSTRASS INTEGRALS OF THE CALCULUS OF VARIATIONS OVER BV VARIETIES: RECENT RESULTS OF THE MATHEMATICAL SEMINAR IN PERUGIA

CALOGERO VINTI

Dedicated to Ennio De Giorgi on his sixtieth birthday

1. Introduction. The Weierstrass integrals of the Calculus of Variations are defined by means of an integration algorithm for set functions, i.e. in terms of a suitable limit process over finite sums. The primary idea is due to Weierstrass, about one century ago, but successively many Authors adopted this approach to variational problems and settled its formal definition; we mention among the others Tonelli, Burkill, Menger, Aronszajn, Kempisty, Radó, Ringemberg, Kober and Cesari.

Around the sixties I resumed the study of such functionals and in the following years a great deal of research has been done in the Mathematical Seminar of Perugia (on this subject we refer to [9b, c, d, e]). As it is well-known, the Weierstrass type approach to the Calculus of Variations has big advantages with respect to other classical integration processes as Lebesgue, Lebesgue-Stieltjes and Serrin ones.

In fact, besides a direct and easy definition, in any case, it represent the measure of a geometric quantity connected to the underlying variety, as Menger pointed out.

On the other hand it was this geometric and constructive nature itself constrained the application field of Weierstrass integrals to continuous varities anyhow.

The first extensions to wider classes, as $W^{1,1}$ or BV, dates the end of seventies and refer only to some special functionals like variation, lenght and area. While the question of the definition (of Weierstrass integrals) over BV varieties for more general integrands remained unsolved for a long time.

After few partial answers, only recently some results have been reached which can be considered satisfactory. The aim of this note is just to illustrate the latest developments of this research.

2. The non-parametric Weierstrass integral of the Calculus of Variations. Existence, integral representation, semi-continuity. An abstract formulation for the non-parametric Weierstrass integral of the Calculus of Variations is the following; it is inspired by the definition of Weierstrass parametric functional proposed by Cesari in [5a]. Let $(A, \mathcal{G})$ be a topological space and let $\{I\}$ denote a family of nonempty subsets of A that we will call intervals. A finite system $D = \{I_1, I_2, \ldots, I_N\}$ is a finite collection of non-overlapping intervals, i.e. $I_i^0 \neq \emptyset$ and $I_i^0 \cap \bar{I}_j = \emptyset$, $i \neq j$, $i,j = 1,2,\ldots,N$. Let as consider a family $\mathcal{D}$ of finite systems and let $\delta : \mathcal{D} \to \mathbf{R}^+$ be a "mesh" function which makes $\mathcal{D}$ a direct set, in a natural way. As it is well-known ([5a]) an interval function $\Phi : \{I\} \to \mathbf{R}^m$ is said to be *Burkill-Cesari integrable* if the below limit exists

$$\lim_{\delta(D)\to 0} \sum_{I\in D} \Phi(I) = BC \int_A \Phi.$$

Let $f : K \times \mathbf{R}^n \to \mathbf{R}^m$ be a given function with K compact metric space and let $p : \{I\} \to K$, $\varphi : \{I\} \to \mathbf{R}^n$, $\lambda : \{I\} \to \mathbf{R}^+$ be given interval functions.

We denote by $\Phi : \{I\} \to \mathbf{R}^m$ the interval function defined by

$$\Phi(I) = \lambda(I) f(p(I), \frac{\varphi(I)}{\lambda(I)}).$$

Following [9a] the Burkill-Cesari integral of Φ, when it exists, is called the *non-parametric Weierstrass integral* of the Calculus of Variations or *W-integral* briefly.

In the following we shall use the conditions:

(f_1) $f(k,\cdot)$ is convex, $k \in K$;

(f_2) the function $f(k,x)(1+\|x\|)^{-1}$ is continuous on K uniformly with respect to $x \in \mathbf{R}^n$;

(f_3) there exists $H \in \mathbf{R}^+$ such that $|f(k,x)| \leq H(1+\|x\|)$, $(k,x) \in K \times \mathbf{R}^n$;

(f_4) there exists $M \in \mathbf{R}^+$ such that $f(k,x) \geq -1 + M\|x\|$, $(k,x) \in K \times \mathbf{R}^n$.

Usually (see McShane [6], Vinti [9a]) a suitable parametric integrand $F : K \times \mathbf{R}^{n+1} \to \mathbf{R}^m$ is associated to the function f :

$$F(k;(s,x)) = \begin{cases} |s| f(k, \frac{x}{|s|}) & \text{if } s \neq 0 \\ \lim_{\xi \to 0} |\xi| f(k, \frac{x}{|\xi|}) & \text{if } s = 0. \end{cases}$$

By definition, F is positively homogeneous of degree one in (s,x) and moreover the following characterization, due to Boni [1a], holds.

Theorem 1. *Under assumption* (f_1), *the function* F *is well defined and globally continuous if and only if the integrand* f *satisfies* (f_2) *and* (f_3).

Now we recall some definitions that will be used in the following. The function φ is said to be of *bounded variation* (BV) if

$$V(\varphi) = \overline{\lim}_{\delta(D) \to 0} \sum_{I \in D} \|\varphi(I)\| < +\infty.$$

The function φ is said to be *quasi-additive* (q.a.) (see [5a]) if

(q.a.) for every $\epsilon > 0$ there exists a number $\eta = \eta(\epsilon) > 0$ such that, if $D_0 = [I] \in \mathcal{D}$ with $\delta(D_0) < \eta$ then there exists a number $\lambda = \lambda(\epsilon, D_0) > 0$ such that, if $D = [J] \in \mathcal{D}$ with $\delta(D) < \lambda$, then we have

$$\sum_I \|\varphi(I) - \sum_{J \subset I} \varphi(J)\| < \epsilon \quad \text{and} \quad \sum_{J \not\subset I} \|\varphi(J)\| < \epsilon.$$

The function p satisfies condition (γ) (see [2]) when

(γ) for every $\epsilon > 0$ there exists a number $\eta = \eta(\epsilon) > 0$ such that, if $D_0 = [I] \in \mathcal{D}$ with $\delta(D_0) < \eta$, then there exists a number $\lambda = \lambda(\epsilon, D_0) > 0$ such that, if $D = [J] \in \mathcal{D}$ with $\delta(D) < \lambda$, then we have

$$\max_I \max_{J \subset I} d(p(I), p(J)) < \epsilon.$$

As it is well-known, quasi-additivity condition plays a fundamental role in the theory of Burkill-Cesari integral; in fact the following result holds (see [5a], [4]):

Theorem 2. *If φ is* q.a. *then $BC \int_A \varphi$ exists. If φ is* q.a. *and BV, then $BC \int_M \varphi$ exists for every $M \subset A$.*

Note that the couple of conditions q.a. and BV is a "key couple". In fact not only it solves the problems of the existence and the hereditability for the Burkill-Cesari integral, but it has another important prerogative pointed out by Cesari in [5b]: it allows the extension of the integral to a measure.

In other words, under suitable assumptions enriching the base setting, if φ is q.a. and BV a regular measure of bounded variation $\mu : \mathcal{B} \to \mathbf{R}^n$ exists, over the Borel sets, such that $\mu(G) = BC \int_G \varphi$, $G \in \mathcal{G}$ and $\|\mu\|(G) = BC \int_G \|\varphi\|$, where $\|\mu\|$ denotes the total variation of μ.

In order to get this last result, a somewhat stronger assumption on φ is necessary, i.e. φ is supposed to be (o)-quasi additive ((o)-q.a.) (see [5b], [3b]).

By virtue of Theorem 1, the following existence result for the W-integral (see [3b]) is a consequence of the classical existence theorem proved by Cesari [5a] in the parametric case.

Theorem 3. *If f satisfies (f_1), (f_2) and (f_3), p satisfies condition (γ) and φ is* q.a. *and BV, then Φ is* q.a. *and BV.*

As a consequence of this result, many classical integrals of the Calculus of Variations, over continuous BV curves or surfaces, can be considered as particular W-integrals (see the survey [9b,c,d,e]). Moreover also the weighted generalized variation, lenght and area are special cases of this general formulation (see [1b],[7],[3a, c, e]).

For a long time many attemps have been made in order to include also the case of complete integrands over BV varieties (not necessarily continuous nor Sobolev's) in this setting; but they had not success because of the too restrictive assumption (γ) on the function p.

Recently Brandi-Salvadori [3f,g] improved Theorem 3 in view of this aim. More precisely they introduced a new global quasi-additivity condition on the system $(p;(\lambda,\varphi))$ which makes full use of the power of the quasi-additivity type properties.

The interval function $(p;(\lambda,\varphi))$ is said to be Γ-*quasi-additive* $(\Gamma - q.a.)$ if

$(\Gamma - q.a.)$ For every $\epsilon > 0$ there exist two numbers $0 \leq \sigma = \sigma(\epsilon) \leq \epsilon$ and $\eta = \eta(\epsilon) > 0$ such that, if $D_0 = [I] \in \mathcal{D}$ with $\delta(D_0) < \eta$, there exists a number $\lambda = \lambda(\epsilon, D_0) > 0$ such that, if $D = [J] \in \mathcal{D}$ with $\delta(D) < \lambda$ then we have

$$\sum_I \|(\lambda,\varphi)(I) - \sum_{J\in\Gamma_I}(\lambda,\varphi)(J)\| < \epsilon,$$

$$\sum_I \| \sum_{J\subset I, J\notin\Gamma_I} (\lambda,\varphi)(J)\| < \epsilon,$$

$$\sum_{J\not\subset I} \|(\lambda,\varphi)(J)\| < \epsilon,$$

where Γ_I is a subfamily (even empty) of the set

$$\{J \subset I : d(p(I), p(J)) < \sigma\}.$$

It is easy to see that if p satisfies condition (γ) and (λ,φ) is q.a., then $(p;(\lambda,\varphi))$ is $\Gamma - q.a.$. Furthermore if $(p;(\lambda,\varphi))$ is $\Gamma - q.a.$ then (λ,φ) is q.a. but p does not satisfy condition (γ) in general.

Moreover the following theorem holds ([3f,g]).

Theorem 4. *If f satisfies $(f_1),(f_2)$ and (f_3), $(p;(\lambda,\varphi))$ is $\Gamma-q.a.$ and (λ,φ) is BV, then Φ is $q.a.$ and BV.*

The extension of the W-integral to BV class, in the case of complete integrands, is consequence of the result above. On this purpose see the various applications given in [3f,g,h,i] both in the parametric and in the nonparametric case, and in [10] for the Fubini-Tonelli integral in the sense of Weierstrass.

Furthermore we point out the following proposition which extend the classical Tonelli's theorem on the lenght of a curve. It makes use of the two measures μ and ν relative to the Burkill-Cesari integral of φ and λ respectively

Theorem 5. *Suppose that f is non negative and satisfies (f_1), (f_2) and (f_3); $(p;(\lambda,\varphi))$ is $\Gamma-q.a.$ and (λ,φ) is $(o)-q.a.$ and BV. Moreover suppose that a map $\pi : A \to K$ exists such that*

$$\lim_{\delta(D)\to 0} \sum_{I\in D} p(I)\chi_{I^0}(a) = \pi(a)$$

ν-almost everywhere.

Then there exists a ν-measurable function $\frac{\delta\mu}{\delta\nu} : A \to \mathbf{R}^n$ such that

$$\lim_{\delta(D)\to 0} \sum_{I\in D} \frac{\mu(I)}{\nu(I)}\chi_{I^0}(a) = \frac{\delta\mu}{\delta\nu}(a)$$

ν-almost everywhere and we have

$$(*) \qquad BC\int_A \Phi \geq \int_A f(\pi(a), \frac{\delta\mu}{\delta\nu}(a))d\nu.$$

Furthermore, if f satisfies (f_4) then the equality sign holds in (*) if and only if φ is absolutely continuous with respect to λ. In this case $\delta\mu/\delta\nu$ coincides with the Radon-Nikodym derivative $\delta\mu/\delta\nu$.

Also the following integral representation holds, in terms of a Lebesgue-Stieltjes algorithm. It is a consequence of an analogous result proved in the parametric case, extending previous Cesari's theorem ([5b]) to BV varieties.

Theroem 6. *Suppose that f satisfies (f_1), $(f_2,)$ and (f_3), $(p;(\lambda,\varphi))$ is $\Gamma-q.a.$ and (λ,φ) is $(o)-q.a.$ and BV.*

Let $m : \mathcal{B} \to \mathbf{R}_0^+$ be a measure such that (ν, μ) is absolutely continuous with respect to m, and suppose that an m-measurable function $\xi : A \to K$ exists such that

$$\lim_{\delta(D)\to 0} \sum_{I\in D} p(I)\chi_{I^\circ}(a) = \xi(a)$$

m-almost everywhere.

Then we have

$$BC\int_A \Phi = \int_A F(\xi(a), \frac{d\nu}{dm}(a), \frac{d\mu}{dm}(a))dm.$$

Both these last theorems are proved in [3f,g] where the authors make use of suitable convergence results for martingales.

In Weierstrass formulation always the problem of semicontinuity appeared rather hard to be solved.

A first result on this subject is due to Warner [11b] in 1968.

However he took into consideration a kind of convergence which proved too restrictive. In fact it could be applied only in the case of continuous BV varieties, uniformly convergent.

Afterward Brandi-Salvadori [3d,h] proposed some variants to Warner's result in order to widen the field of applications. On the pattern of their recent existence result, they introduced a global convergence condition on the sequence $(p_n; (\lambda_n, \varphi_n))_n$.

The sequence $(p_n; (\lambda_n, \varphi_n))_n$ is said to be Δ-*convergent* to $(p_0; (\lambda_0, \varphi_0))$ if

(Δ) given any subsequence and for every $\epsilon > 0$ and $\mu > 0$, there is a finite system $D_0 = [I] \in \mathcal{D}$, with $\delta(D_0) < \mu$, and a furhter subsequence $(k_n)_n$ such that, for every $n \in \mathbf{N}$, a number $\xi_n = \xi_n(\epsilon, \eta, n)$ can be determined in such a way that, if $0 < \xi < \xi_n$, then $D = [J] \in \mathcal{D}$ exists with $\delta(D) < \xi$ and

$$\sum_I \|(\lambda_0, \varphi_0)(I) - \sum_{J\in\Delta_I} (\lambda_{k_n}, \varphi_{k_n})(J)\| < \epsilon$$

where Δ_I is a subfamily (even empty) of the set

$$\{J \subset I : d(p_0(I), p_{k_n}(J)) < \epsilon\}.$$

The following semicontinuity result holds.

Theorem 7. *Suppose that f is non-negative, seminormal and satisfies (f_1), (f_2) and (f_3). Let $(p_n;(\lambda_n,\varphi_n))_n$ be a sequence of Γ – q.a. interval functions which Δ-converges to $(p_0;(\lambda_0,\varphi_0))$ and such that $(\lambda_n,\varphi_n)_n$ are equi BV.*

Then we have

$$\liminf_{n\to\infty} BC\int_A \phi_n \geq BC\int_A \Phi.$$

See [3h] for many applications of this proposition.

Note that it contains as particular cases the well-known results already proved for the weighted generalized variation, lenght and area.

We point out that the equi BV assumption on $(\lambda_n,\varphi_n)_n$ can be omitted if f satisfies condition (f_4).

3. The W-integral over a BV curve. Comparison with the Lebesgue, Lebesgue-Stieltjes and Serrin integral functionals. In order to illustrate some recent results on the W-integral in BV class, we take into consideration the particular case of the W-integral over a BV curve which itself is very expressive (for the details see [3i]). Let $f: K\times\mathbf{R}\to\mathbf{R}$, with $K\subset\mathbf{R}^2$ compact and convex, be a function which satisfies the assumptions (f_1), (f_2) and (f_3). Let $x:[a,b]\to\mathbf{R}$ be a BV curve (neither necessarily continuous nor in $W^{1,1}$). We denote by E_x the set of the points where x is essentially continuous; as it is well-known $[a,b]\backslash E_x$ is at most countable.

Let $\{I\}$ be the family of the closed sub-intervals of $[a,b]$ whose endpoints belong to E_x and let $\mathcal{D}_x$ be the collection of all the finite systems of the type $D=[I_1,I_2,\ldots,I_N\}$ with $\bigcup_{i=1}^N I_i=[t_1,t_{N+1}]$. Let us consider the "mesh" function $\delta(D)=\max\{(t_1-a),(b-t_{N+1}),|I_i|,\ i=1,2,\ldots,N\}$. For every $I\in\{I\}$, $I=[\alpha,\beta]$, we denote by $t_I\in I\cap E_x$ an arbitrary point and we put $|I|=\beta-\alpha$, $\Delta x(I)=x(\beta)-x(\alpha)$, $\mu_I^-=\min(x(\alpha),x(\beta))$, $\mu_I^+=\max(x(\alpha),x(\beta))$.

Finally let $\Phi_x:\{I\}\to\mathbf{R}$ be the function defined by

$$\Phi_x(I)=\begin{cases}\dfrac{|I|}{|\Delta x(I)|}\displaystyle\int_{\mu_I^-}^{\mu_I^+} f(t_I,u,\frac{\Delta x(I)}{|I|})du, & \text{if } \Delta x(I)\neq 0\\ |I|f(t_I,x(t_I),0), & \text{if } \Delta x(I)=0.\end{cases}$$

Note that, by virtue of the assumptions on f, we have that

$$\Phi_x(I) = |I| f\left(t_I, \sigma_I, \frac{\Delta x(I)}{|I|}\right)$$

for a suitable choice $\sigma_I = \overline{co}\, x(I)$. Thus, in the case that x is continuous and BV, we reduce to the usual set function already adopted in this situation (see [9a]).

In the following the Burkill-Cesari integral of Φ_x will be briefly denoted by $W(x)$.

The following result can be proved as an application of Theorem 4.

Theorem 8 (Existence). *$W(x)$ exists over every BV curve such that* graph $x \subset K$.

Note that the integral $W(x, M)$ exists for every $M \subset [a, b]$ and it is a set function nonadditive in general. In fact for every $a \leq \alpha < \gamma < \beta \leq b$ we have

$$\begin{aligned} W(x, [\alpha, \beta]) &= W(x,]\alpha, \beta[), \\ W(x, [\alpha, \beta]) &= W(x, [\alpha, \gamma]) + W(x, [\gamma, \beta]) + S\gamma, \end{aligned}$$

with

$$S_\gamma = \begin{cases} \dfrac{1}{|\Delta x(\gamma)|} \displaystyle\int_{u_\gamma^-}^{u_\gamma^+} F(\gamma, u, 0, \Delta x(\gamma))du, & \text{if } \Delta x(\gamma) \neq 0 \\ 0, & \text{if } \Delta x(\gamma) = 0 \end{cases}$$

where

$$\Delta x(\gamma) = x(\gamma + 0) - x(\gamma - 0), \quad u_\gamma^- = \min\{x(\gamma + 0), x(\gamma - 0)\},$$

and

$$u_\gamma^+ = \max\{x(\gamma + 0), x(\gamma - 0)\}.$$

Moreover $W(x, \cdot)$ can be extended to a measure, by virtue of the results mentioned in the previous section. In other words a regular measure of bounded variation $m_W : \mathcal{B} \to \mathbf{R}$ exists such that

$m_W(G) = W(x, G)$ for every open set $G \subset [a, b]$; $m_W(\{t\}) = S_t$, for every $t \in [a, b]$; $m_W([\alpha, \beta]) = W(x, [\alpha, \beta]) + S_\alpha + S_\beta$.

As regards the integral representation of $W(x)$, put

$$I(x) = \int_a^b f(t, x(t), x'(t))dt$$

and denote by μ the variation-measure associated to x, the following results hold. They are an application of Theorem 6 and Theorem 5 respectively.

Theorem 9 (Representation). *For every measure m such that (dt, μ) is absolutely continuos with respect to m we have*

$$W(x) = \int_a^b F(t, \xi(t), \frac{dt}{dm}(t), \frac{d\mu}{dm}(t))dm.$$

In particular, if $x \in W^{1,1}$, then $W(x) = I(x)$.

Theorem 10 (Comparison between $W(x)$ and $I(x)$). *If f is non-negative and satisfies assumption (f_4), we have*

$$W(x) \geq I(x)$$

and the equality sign holds if and only if $x \in W^{1,1}$.

The following propositions deal with the approximation and the semicontinuity of the W-integral. Thus, as a consequence, they lead a natural comparison with the corrisponding Serrin functional (see [3h] for the details).

Theorem 11 (Approximation). *For every BV curve x with graph $x \subset K$, there exists a sequence $(p_n)_n$ of equi BV poligonals such that $p_n \xrightarrow{L_1} x$, graph $p_n \subset K$, $n \in \mathbf{N}$, and*

$$\lim_{n\to\infty} W(p_n) = \lim_{n\to\infty} I(p_n) = W(x).$$

Theorem 12 (Semicontinuity). *Suppose that f is non-negative, then for every sequence $(x_n)_n$ of equi BV curves such that $x_n \xrightarrow{L_1} x$ and graph $x_n \subset K$, $n \in \mathbf{N}$, we have*

$$\liminf_{n\to\infty} W(x_n) \geq W(x).$$

Let us consider the Serrin functional ([8])

$$S(x) = \inf_{(x_n)_n} \liminf_{n\to\infty} I(x_n),$$

where the greatest lower bound is taken with respect to all the sequences $(x_n)_n$ of equi BV curves in $W^{1,1}$ such that $x_n \xrightarrow{L_1} x$.

Then the following result holds.

Theorem 13 (Comparison between $W(x)$ and $S(x)$). *Suppose that f is non-negative, then for every BV curve x with graph $x \subset K$ we have*

$$W(x) = S(x).$$

Note that if f satisfies (f_4), then the equi BV assumption in Theorem 12 can be omitted. In this case $W(x)$ coincides with the Serrin functional defined with respect to L_1-topology.

References

[1.a] M.Boni, *Quasi additività e quasi subadditività nell'integrale ordinario del Calcolo delle Variazioni alla Weierstrass,* Rend. Ist. Mat. Univ. Trieste **6** (1974), 51-70.

[1.b] M.Boni, *Variazione generalizzata con peso e quasi additività,* Atti Sem. Mat. Fis. Univ. Modena **25** (1976), 195-210.

[2] M.Boni, P.Brandi, *Teoremi di esistenza per l'integrale del Calcolo delle Variazioni nel caso ordinario,* Atti Sem. Mat. Fis. Univ. Modena **23** (1974), 308-327.

[3.a] P.Brandi, A.Salvadori, *Sull'area generalizzata,* Atti Sem. Mat. Fis. Univ. Modena **24** (1979), 33-62.

[3.b] M.Boni, P.Brandi, *The non parametric integral of the Calculus of Variations as a Weierstrass integral. I - Existence and representation,* J. Math. Anal. Appl. **107** (1985), 67-95.

[3.c] M.Boni, P.Brandi, *The non parametric integral of the Calculus of Variaitions as a Weierstrass integral. II - Some applications,* J. Math. Anal. Appl. **112** (1985), 290-313.

[3.d] M.Boni, P.Brandi, *Existence, semicontinuity and representation for the integrals of the Calculus of Variations. The BV case*, Atti Conv. celebrativo I centenario Circolo Matematico di Palermo (1984), 447-462.

[3.e] M.Boni, P.Brandi, *L'integrale del Calcolo delle Variazioni alla Weierstrass lungo curve BV e confronto con i funzionali integrali di Lebesgue e Serrin,* Atti Sem. Mat. Fis. Univ. Modena **35** (1987).

[3.f] M.Boni, P.Brandi, *A quasi additivity type condition and the integral over a BV variety,* to appear.

[3.g] M.Boni, P.Brandi, *On the non-parametric integral over a BV surface,* Nonlin. Anal., to appear.

[3.h] M.Boni, P.Brandi, *On the lower-semicontinuity of certain integrals of the Calculus of Variations,* J. Math. Anal. Appl., to appear.

[3.i] M.Boni, P.Brandi, *On the definition and properties of a variational integral over a BV curve,* to appear.

[4] J.C.Brekenridge, *Burkill-Cesari integrals of quasi additive interval functions,* Pacific J. Math. **37** (1971), 635-654.

[5.a] L.Cesari, *Quasi additive set functions and concept of integral over a variety,* Trans. Amer. Math. Soc. **102** (1962), 94-113.

[5.b] L.Cesari, *Extension problem for quasi additive set functions and Radon-Nikodym derivatives,* Trans. Amer. Math. Soc. **102** (1962), 114-146.

[6] E.J.McShane, *Existence theorems for ordinary problems of the Calculus of Variations,* Ann. Scuola Norm. Sup. Pisa **3** (1934), 181-211.

[7] M.Ragni *Sull'approssimazione e rappresentazione dell'integrale di Burkill classico e generalizzato,* Atti Sem. Mat. Fis. Univ. Modena **28** (1980), 112-125.

[8] J.Serrin, *On the definition and properties of Certain Variational integrals,* Trans. Amer. Math. Soc. **101** (1961), 139-267.

[9.a] C.Vinti, *L'integrale di Weierstrass e l'integrale del Calcolo delle Variazioni in forma ordinaria,* Atti Accad. Scienze, Lett. Arti, Palermo **19** (1958-59), 51-62.

[9.b] C.Vinti, *Una breve panoramica dei risultati ottenuti da un gruppo di ricercatori dell'Università di Perugia su questioni connesse con il Calcolo delle Variazioni,* Rend. Circolo Mat. di Palermo **26** (1977), 131-155.

[9.c] C.Vinti, *Non linear integration and Weierstrass integral over a manifold, connection with theorems on martingales,* J. Opt. Theor. Appl. **41** (1983), 213-237.

[9.d] C.Vinti, *Teoremi di esistenza, rappresentazione e approssimazione per l'integrale del Calcolo delle Variazioni,* Atti del Convegno Celebrativo dell'80° anniversario della nascita di Renato Calapso (Messina-Taormina, 14 Aprile 1981).

[9.e] C.Vinti, *The integrals of the Calculus of Variations as Weierstrass-Burkill-Cesari integrals,* in "Contributions to Modern Calculus of Variations", Pitman Res. Notes in Math. **148** (1987).

[10] G.Vinti, *L'integrale non parametrico di Fubini-Tonelli alla Burkill-Cesari su coppie di curve* BV, to appear.

[11.a] G.Warner, *The Burkill-Cesari integral,* Duke Math. J. **35** (1968), 61-78.

[11.b] G.Warner, *The generalized Weierstrass-type integral* $\int f(\xi, \phi)$, Annal. Scuola Norm. Sup. Pisa **22** (1968), 163-192.

Dipartimento di Matematica
Università di Perugia
Via Pascoli
I-06100 PERUGIA

VARIABLE STRUCTURE CONTROL OF SEMILINEAR EVOLUTION EQUATIONS

TULLIO ZOLEZZI(*)

Dedicated to Ennio De Giorgi on his sixtieth birthday

Abstract. The variable structure control theory (see [1]) is extended to distributed systems described by semilinear evolution equations in Banach spaces. The finite-dimensional variable structure control theory is applied to Faedo-Galerkin approximations of the given control system. States fulfilling the (infinite-dimensional) sliding condition are obtained in the limit as dimensions increase. A definition of approximability (generalizing the finite-dimensional one firstly given in [4] and related to G-convergence) is introduced in order to avoid ambiguous systems and properly relate system's behavior near the sliding manifold to that on such a manifold. Some results of [8] are thereby extended.

Introduction. Aim of this work is the extension of variable structure control theory to distributed systems described by first order evolution equations in Banach spaces.

(*) Work partially supported by M.P.I. (fondi 40%, progetto "Teoria del controllo e ottimizzazione dei sistemi dinamici").

Variable structure control methods employed in (finite-dimensional) ordinary differential equations are particularly effective and flexible tools, see [2] and the reference therein. Significantly better performances may be obtained by keeping the (finite-dimensional) state vector on a suitably chosen sliding manifold during the required time interval. This is achieved by injecting a discontinuous feedback law in the system dynamics. Performances so obtained either cannot be achieved by smooth feedback control laws, or they can,but only by employing more complex control structures. By using such techniques we may obtain (for some classes of control systems) stable behavior, accurate tracking, robust performance, insensitivity with respect to disturbances or variations in the plant parameters.

For such reasons we believe that it is of interest to extend such a methodology to distributed control systems.

We refer to [1],[2] for an overview of finite dimensional variable structure control theory and applications.

The scope of the basic theory has been extended to more general systems in [3],[4],[5] by generalizing the notion of equivalent control and isolating the physically relevant property of approximability. Such a property characterizes those finite-dimensional control systems in which real states (verifying only approximately the control objective due to disturbances, imperfections etc.) converge to a well-defined ideal state (which satisfies exactly the sliding condition) as the imperfections disappear. Further results about variable structure observed systems may be found in [6]. For chattering reduction and application to robotics see e.g. [7] and the references therein.

We believe that this paper is the first attempt to consider, in a somewhat general mathematical setting, variable structure control theory for distributed systems. Results about variable structure control of the heat equation are obtained in [8]. Generalization of some results therein was a motivation for the present work.

We consider control systems described by semilinear evolution equations in separable reflexive Banach spaces, under monotonicity and weak continuity assumptions. The control variable enters linearly in the right-hand side. We assume that problem's data are compatible, i.e. any approximate finite-dimensional state vector can be sent (say at time $t = 0$) on the given approximate finite-dimensional sliding manifold. Such a problem has well-known solutions (see [1]). In this paper we focus on the problem, how to keep the state variable over the sliding manifold during the given interval.

By using a Faedo-Galerkin method (for any fixed basis) we apply variable structure control results to the finite-dimensional approximations thereby obtained, then we take limits as the dimensions diverge. This procedure is constructive in principle and avoids using a direct extension of the equivalent control notion (see [1]), which requires stringent invertibility assumptions. The convergence is then shown to an ideal state of the infinite-dimensional control system. A natural definition of approximability is introduced, which is related to the physical meaning of the convergence behaviour for real states in the given control systems. The approximability is then shown to hold for some classes of semilinear infinite-dimensional control systems.

We point out that a mere existence theory (giving infinite-dimensional controls keeping the state vector on the sliding manifold) may be meaningless from the control-theoretic point of view, since ambiguous behavior of the system *near* the sliding manifold should be properly related to that *on* this manifold (see [1] for a discussion in the finite-dimensional setting from an engineering point of view, and [4], [5] for the relevant mathematical description).

In order not to obscure the main ideas, no attempt is made here to obtain the most general results. As far as the evolution equations (with fixed open-loop control) are concerned, we adopt the framework of [9]. This approach is particularly well adapted to get weak convergence of Faedo-Galerkin approximations. On the other hand a direct solution (which we shall briefly discuss later) based on such an approach needs working with time-derivatives in a possibly very large space. This requires a possibly strong assumption about the differentiability of the sliding manifold, and a restrictive invertibility assumption. An approach based on semigroup theory will be presented elsewhere.

Some interesting related problems (which we hope to answer in future work) are the following: weakening boundedness assumptions about control laws: stronger convergences and error estimates; the meaning of solutions to infinite-dimensional state equations when discontinuous feedback is used; relations with approximability and convergence for the corresponding finite-dimensional approximations; variable structure boundary control problems; output feedback.

Part of the results of this paper was presented at the IFIP-TC7 workshop of control problems for systems described by partial differential equations, Gainesville, February 1986.

Problem statement, notations and common assumptions. We are given positive numbers $T, C, p > 1$, two real Hilbert spaces H, Z, two real reflexive Banach spaces V, W, a closed convex subset $U \subset W$, a fixed $y_o \in H$, and mappings

$$A : V \longrightarrow V^*, \qquad B : W \longrightarrow V^*, \qquad s : D \longrightarrow Z$$

where V^* is the dual space of V, and D is either V or an open (sufficiently large) subset of H. All spaces are assumed separable.

We assume that

$$V \subset H \subset V^*$$

with continuous imbeddings, V dense in H and $H = H^*$ in the usual way.

The control system we shall consider is given by

$$\begin{cases} \dot{y} + A(y) = Bu & \text{in} \quad [0, T], \\ y(0) = y_o \end{cases} \tag{1}$$

$$s[y(t)] = 0, \qquad 0 \leq t \leq T. \tag{2}$$

Admissible controls are all functions

$$u : [0, T] \longrightarrow U$$

such that $u \in L^{p'}(0, T, W)$ and

$$\int_o^T \|u\|^{p'} \, dt \leq C^{p'} \tag{3}$$

where $\frac{1}{p} + \frac{1}{p'} = 1$. Here the state variable y takes values in V. A solution y to (1) is meant in the sense of [9], Theorem 1.2 p.162.

The problem we shall consider is to find an admissible control u such that (1), (2) are verified (so that (2) defines the *sliding manifold*, i.e. the control objective).

The following hypotheses (4), (5), (6) are assumed throughout the paper (see Remark 1 below for a weaker coercivity condition).

(4) A is a monotone, hemicontinuous mapping such that
$< Av, v > \geq \alpha\|v\|^p, \quad \|A(v)\| \leq c_1\|v\|^{p-1} + c_2$
for some constants $\alpha > 0, c_1$ and c_2, and all $v \in V$.

Here $\|\cdot\|$ denotes the norm in any Banach space different from H. The norm in H is denoted by $|\cdot|$, the scalar product either in H or in Z by $(\cdot,\cdot)$, and $< \cdot,\cdot >$ is the duality pairing between V^* and V.

(5) B is a bounded linear operator.

(6) s is continuously Fréchet differentiable on D.

The Fréchet differential of s at x will be denoted by $s'(x)$. The following definition will be used in the sequel. The mapping $A : V \mapsto V^*$ is *p-weakly continuous* iff $y_n \rightharpoonup y$ in $W^p(0,T)$ imply $A(y_n) \rightharpoonup A(y)$ in $L^{p'}(0,T,V^*)$, where (see [9] p.57) $y \in W^p(0,T)$ iff $y \in L^p(0,T,V)$ and $\dot{y} \in L^{p'}(0,T,V^*)$ and the norm there is defined as $\|y\|_{L^p} + \|\dot{y}\|_{L^{p'}}$, $A(y_n)(t) = A[y_n(t)]$ and $\rightharpoonup$ denotes weak convergence. We shall denote strong convergence by $\rightarrow$ and weak-star convergence by $\rightharpoondown$.

Example 1 (2-weakly continuous mappings A).

(i) A is a bounded linear operator from V to V^*.

(ii) Suppose that Ω is an open bounded subset of $\mathbf{R}^n$. Consider $H = L^2(\Omega)$, $V = H_0^1(\Omega)$,

$$A(y) = -\sum_{i,j=1}^{N} \frac{\partial}{\partial x_i}\left[a_{ij}(x,y)\frac{\partial y}{\partial x_j} + b_i(x,y)\right],$$

where a_{ij}, b_i are real-valued Carathéodory's functions on $\Omega \times \mathbf{R}$ such that

$$\alpha|\lambda|^2 \leq \sum_{i,j=1}^{N} a_{ij}(x,y)\lambda_i\lambda_j$$

for all $y \in \mathbf{R}$, a.e. $x \in \Omega$, every $\lambda \in \mathbf{R}^N$ and some constant $\alpha > 0$;

$$\left| \sum_{j=1}^{N} a_i j(x, y)\lambda_j \right| \leq h(x) + \omega(|y| + |\lambda|)$$

for every $i = 1, \ldots, N$, $\lambda \in \mathbf{R}^N$, a.e $x \in \Omega$, some constant ω, and $h \in L^2(\Omega)$;

$$|b_j(x, y)| \leq k(x) + \omega|y|$$

for every $i = 1, \ldots, N$ and y, a.e. $x \in \Omega$, some $k \in L^2(\Omega)$. Then if $y_n \rightharpoonup y$ in $W^2(0, T)$, by theorem 5.1 p. 58 of [9] we have $y_n \rightarrow y$ in $L^2(0, T, L^2(\Omega))$. So it is easily seen that the non-linear mapping A is 2-weakly continuous. (The corresponding stationary case is studied in [10], section 3, where weak continuity from V to V^* is characterized for Leray-Lions type operators).

Variable structure control for Faedo-Galerkin approximations. In this section we solve an approximate version of the control problem (1), (2) by using the relevant finite-dimensional variable structure control theory.

A *basis* $v_1, \ldots, v_n, \ldots$ of V is any countable subset thereof such that each $v_1, \ldots, v_n$ is linearly independent, and setting

$$V_n = \text{sp}\{v_1, \ldots, v_n\},$$

we have

$$\text{cl} \bigcup_{n=1}^{\infty} V_n = V,$$

where cl denotes closure.

We fix three bases, $\{v_n\}$ in V, $\{w_n\}$ in W and $\{z_n\}$ in Z. Given n and $x = (x_1, \ldots, x_n)' \in \mathbf{R}^n$ we denote by

$$A_n(x) \in \mathbf{R}^n$$

the vector of components

$$< A\left(\sum_{i=1}^{n} x_i v_i\right), v_j >, \qquad j = 1, \ldots, n.$$

If $u = \sum_{i=1}^{m} u_{im} w_i \in \mathrm{sp}(w_1, \dots, w_m)$, then $u_m \in \mathbf{R}^m$ will denote the vector of components u_{im}, $i = 1, \dots, m$. The vector of components

$$(s(x), z_j), \qquad j = 1, \dots, m$$

will be denoted by $s_m(x)$. Finally for any m we consider a sequence $U_m \subset \mathrm{sp}(w_1, \dots, w_m)$ of closed convex subsets of W such that $u_m(t) \in U_m$ a.e., $m = 1, 2, \dots$, and $u_m \rightharpoonup u_o$ in $L^{p'}(0, T, W)$ for some subsequence imply $u_o(t) \in U$ a.e.

Given $\overline{x}_{in}$ such that

$$\sum_{i=1}^{n} \overline{x}_{in} v_i \to y_o \quad \text{in } H \quad \text{as } n \to +\infty$$

we consider the following approximate version of (1), (2) for fixed n, m with $n \geq m$:

$$(7) \qquad \begin{cases} H_n \dot{x}(t) + A_n[x(t)] = B_{nm} u_m, \\ x(t) \in \mathbf{R}^n, \qquad x(0) = (\overline{x}_{in}, \dots, \overline{x}_{nn})', \\ \sum_{i=1}^{m} u_{im} w_i \in U_m; \quad s_m\left(\sum_{i=1}^{n} x_i(t) v_i\right) = 0, \\ 0 \leq t \leq T. \end{cases}$$

Here $x_1, \dots, x_n$ are the components of x, u_{im} those of u_m; H_n is the $n \times n$ matrix whose (i, j)-th element is given by $< v_j, v_i >$; B_{nm} is the $n \times m$ matrix whose (i, j)-th element is $< B w_j, v_i >$.

A solution to (7) is any pair u, x such that x is absolutely continuous in $[0, T]$, u is measurable there and (7) holds a.e. in $[0, T]$. Of course $u(t) = u^*[x(t)]$ in feedback form is allowed.

Remark. Under suitable non singularity conditions (see (8), (9) below), (a.e.) solution to (7) just defined are Filippov solutions (see [11] p. 101-102) corresponding to discontinuous feedback, and conversely, if the equivalent feedback control u^* (to be defined later, see (10)) is used in (7). Then, if such conditions are met, we may ignore both discontinuous feedback in (7) and the Filippov solution concept. The reason behind it is that the control variable u appears affinely in (7), so that the (finite-dimensional) equivalent control of [1] may be successfully used. In the non linear (with respect to u)

setting, ambiguous behaviour may arise (see [1]) and the concept of approximability (see [4] or [5]) becomes relevant.

We see that, given n and m, (7) defines a finite-dimensional control problem of the variable structure type since a sliding condition appears there. Existence of solutions to (7) will be proved below.

Let us denote by s_{mn} the restriction of s_m to V_n. We remark that $s_{mn} \in C^1(D \cap V_n)$, and we denote by $(\partial/\partial x)s_{mn}(\overline{x})$ its $m \times n$ jacobian matrix evaluated at $\overline{x}$.The following condition will be required in the next existence theorem.

(8) The rank of $\dfrac{\partial s_{mn}}{\partial x}(\overline{x})$ is m for all $\overline{x} \in D \cap V_n$ such that $s_m(\overline{x}) = 0$.

Remark. Following [1], [3], if we assume that

$$\begin{aligned} &\frac{\partial s_{mn}}{\partial x}(\overline{x}) H_n^{-1} B_{nm} \quad \text{is non singular for every } \overline{x} \in V_n \\ &\text{near } \{x \in V_n : s_m(x) = 0\}, \end{aligned} \tag{9}$$

then the *equivalent control* for (7) is the unique solution $u^*_{nm} = u^*_{nm}(x)$ to

$$\frac{\partial s_{mn}}{\partial x}(x) H_n^{-1} \left(B_{nm} u - A_n(x)\right) = 0. \tag{10}$$

Therefore

$$u^*_{nm}(x) = \left(\frac{\partial s_{mn}}{\partial x}(x) H_n^{-1} B_{nm}\right)^{-1} \frac{\partial s_{mn}}{\partial x}(x) H_n^{-1} A_n(x).$$

From known results in [9], [1], [3], [4], [5] we get the following fundamental facts. Assume (8), (9) and the compatibility condition

$$\overline{x} \text{ as in (9) implies } u^*_{nm}(x) \in U_m. \tag{11}$$

Then if in (7) $s_m(x(0)) = 0$,

(a) the feedback control system (7) with $u = u^*$ satisfies the control objective;

(b) states of (7) corresponding (a.e) to the equivalent control are Filippov solutions to (7) corresponding to piecewise continuous feedback control laws and conversely;

(c) following the terminology of [1], real states converge to uniquely defined ideal states of (7) as disturbances disappear (which cause approximate satisfaction of the control objective).

Fact (a) is an existence theorem which will be generalized next. For a more precise statement of fact (b) see Theorem 1 p.545 of [3] (or Theorems 1 and 2 in [5]). Fact (c) comes from affinity of (7) in the control variable: see [1] (especially Theorem p.45), and [4] or [5] for the relevant property (i.e. approximability) behind this fact.

In the next theorems we prove existence of solutions to (7) (for fixed n, m) under less stringent conditions than (9) (i.e. without requiring the existence of the corresponding equivalent control), by using the basic tangential condition of viability theory (see [11]).

Lemma 1. *(a) Given n, there exists an unique solution y_n in $[0,T]$ to*

$$\begin{cases} < \dot{y}, v_j > + < A(y), v_j > = < Bu, v_j >, \ j = 1, \dots, n, \\ y(t) \in V_n, y(0) = \sum_{i=1}^{n} \overline{x}_{in} v_i, \end{cases}$$

for any admissible control u.

(b) There exists $k > 0$ such that $k \geq \sup\{\int_0^T \|y_n(t)\|^p dt + |y_n(t)| \ : 0 \leq t \leq T;\ u$ admissible control, $n = 1, 2, 3, \dots\}$.

Proof. From uniform boundedness in $L^{p'}(0,T,V^*)$ of Bu, u admissible control, and the estimate (1.17) p.159 of [9] we get

$$\frac{1}{2}|y_n(t)|^2 + \alpha \int_o^t \|y_n\|^p dt \leq \|B\|C \left(\int_o^t \|y_n\|^p dx \right)^{1/p} + \frac{1}{2}|y_n(0)|^2,$$

which gives existence and uniqueness in the large of y_n. Since $y_n(0)$ is bounded in H, the estimate follows. □

Throughout the rest of the paper we shall denote by E a fixed open set in H containing the ball

$$\{x \in H : |x| \leq k\}$$

where k is given by Lemma 1. By known results ([9], chapter 2, p.156) and Lemma 1 we see that $y(t) \in E, 0 \leq t \leq T$, for all admissible states y of (1), (2).

The next existence theorem covers the case when the control objective is defined on H.

Theorem 1. *Suppose $D = E$. Given $m \leq n$, assume that $s_m(x) = 0$ for some $x \in V_n \cap E$, condition (8) and the following hypothesis:*

$$\text{(12)} \quad \begin{array}{l} U_m \textit{ is compact and convex, and for all } x \in V_n \cap E \\ \textit{with } s_m(x) = 0 \textit{ there exists some solution } u \in U_m \textit{ to (10).} \end{array}$$

Then (7) has a solution if

$$x_o = \sum_{i=1}^{n} \overline{x}_{in} v_i \in E \quad \textit{and} \quad s_m(x_o) = 0.$$

PROOF. Set

$$F(x) = H_n^{-1} B_{nm} U_m - H_n^{-1} A_n(x), x \in V_n \cap E, K = s_m^{-1}(0) \cap E \cap V_n.$$

Then we apply theorem 1 p.180 of [11] to the differential inclusion problem

$$x(t) \in K, \dot{x}(t) \in F[x(t)], 0 \leq t \leq T.$$

All required assumptions are readily checked. In particular (8) implies that the tangential condition (required in that theorem)

$$F(x) \cap T_K(x) \text{ non empty if } x \in K$$

(notation of [11]) comes from (12), since by (8) the contingent cone $T_K(x)$ is but the tangent space to the sliding manifold K at x. The conclusion is then obtained by Filippov implicit function lemma. □

The next theorem deals with the case when the control objective acts on V only.

Theorem 2. *Let the assumptions of Theorem 1 hold and $D = V$. Assume moreover that either*

$$\{A_n(x) : x \in \mathbf{R}^n, s_m\left(\sum_{i=1}^{n} x_i v_i\right) = 0\} \quad \textit{is bounded,}$$
$$\textit{or} \quad \{x \in V_n : s_m(x) = 0\} \quad \textit{is compact.}$$

Then (7) has a solution if

$$s_m\left(\sum_{i=1}^{n} \overline{x}_{in} v_i\right) = 0.$$

PROOF. We apply Theorem 1 p.180 of [11] with F and K as in the proof of Theorem 1. In particular we have boundedness of

$$F(K) = \cup\{F(x) : x \in V_n \quad \text{and} \quad s_m(x) = 0\},$$

as required in that theorem. The conclusion follows by Filippov's lemma. □

Remark. Further existence criteria for solutions to (7) may be obtained by using growth conditions which relax boundedness assumptions required in Theorems 1 and 2. Application of results in [15] gives further existence theorems for (7).

Convergence. The next step is to exploit the behaviour of solutions to (7) as $(n, m) \to +\infty$. (This means $n \to +\infty$ if Z is finite-dimensional, $n \to +\infty, m \to +\infty$ and $n \geq m$ otherwise). In the limit we shall obtain admissible states of (1) which satisfy the control objective (2). Given a solution $\overline{u}_{nm}, x_{nm}$ to (7) we shall consider their components $\overline{u}_{jnm}, x_{inm}$ and set

$$y_{nm} = \sum_{i=1}^{n} x_{inm} v_i, \quad u_{nm} = \sum_{j=1}^{m} \overline{u}_{jnm} w_j.$$

Subsequences will be denoted as the original sequences.

Theorem. *Assume A is p-weakly continuous and s is affine on $D = V$. Then for any sequence $\overline{u}_{nm}, x_{nm}$ of solutions to (7) such that some subsequence of u_{nm} fulfils (3) and*

$$u_{nm} \rightharpoonup u_o \text{ in } L^{p'}(0,T,W),$$

there exists $y \in W^p(0,T)$ such that u_o, y solves (1), (2) and (for the same subsequence)

$$\begin{aligned} y_{nm} &\rightharpoonup y \quad \text{in} \quad W^p(0,T), \\ y_{nm} &\rightarrow y \quad \text{in} \quad L^\infty(0,T,H). \end{aligned}$$

PROOF. For any $n \geq m$ we have from (7)

$$\begin{gathered} < \dot{y}_{nm}, v_j > + < A(y_{nm}), v_j > = < Bu_{nm}, v_j >, \; j = 1, \ldots, n, \\ y_{nm}(0) \rightarrow y_o \quad \text{in } H, \\ s_m\left(y_{nm}(t)\right) = 0, \quad 0 \leq t \leq T. \end{gathered}$$

Since (by assumption) u_{nm} is bounded in $L^{p'}(0,T,W)$, there exist subsequences weakly converging to some admissible u_o. Take any subsequence of the corresponding one of y_{nm}. By the estimate (1.17) p.159 of [9] some further subsequence satisfies

$$y_{nm} \rightharpoonup y \quad \text{in } L^p(0,T,V), \quad y_{nm} \rightarrow y \quad \text{in } L^\infty(0,T,H),$$

moreover $y_{nm} \in W^p(0,T)$. By (4) there exists x such that (for some subsequence)

$$A(y_{nm}) \rightharpoonup x \quad \text{in} \quad L^{p'}(0,T,V^*).$$

Remembering that

$$Bu_{nm} \rightharpoonup Bu_o \text{in} L^{p'}(0,T,V^*)$$

we follow the proof given in [9] p.159, obtaining in the distributional sense on $[0,T]$

$$\frac{dy}{dt} + x = Bu_o, \quad y(0) = y_o.$$

Then $y \in W^p(0,T)$. Now let $\varphi \in C_o^1(0,T,V)$. Since both y_{nm} and y are in $W^p(0,T)$,

$$\int_o^T < \dot{y}_{nm}, \varphi > dt = -\int_o^T < \dot{\varphi}, y_{nm} > dt \to$$

$$\to -\int_o^T < \dot{\varphi}, y > \ dt = \int_o^T < \dot{y}, \varphi > \ dt.$$

By density we get

$$\dot{y}_{nm} \rightharpoonup \dot{y} \quad \text{in} \quad L^{p'}(0,T,V^*).$$

Therefore $y_{nm} \rightharpoonup y$ in $W^p(0,T)$, and by p-weak continuity of A we see that

$$\dot{y} + A(y) = Bu_o,$$

thus (u_o, y) solves (1). Moreover the original subsequence y_{nm} converges as required since (1) has uniqueness (by Theorem 1.2 p.162 of [9]). Now we verify that $y(t)$ belongs a.e. in $(0,T)$ to the sliding manifold (2). Since $s = a - b$, where a is linear bounded from V to Z and $b \in Z$, we get

$$(a(y_{nm}(t)), z_j) = (b, z_j), \quad j = 1, \ldots, m,$$

so that for any $t_o \in (0,t)$ and $\epsilon > 0$ sufficiently small

$$\int_{t_o-\epsilon}^{t_o+\epsilon} < a^*(z_j), y_{nm}(t) > dt = 2\epsilon(b, z_j).$$

Taking firstly limits as $n, m \to +\infty$ and then as $\epsilon \to 0$ we get for a.e. t_o and all j

$$(a(y(t_o)), z_j) = (b, z_j). \qquad \square$$

Remark. Assume A bounded linear, $p = 2$. Then a simpler proof of Theorem 3 may be given, as follows. Fix j and $\varphi \in C^1(0,T)$ with $\varphi(T) = 0$. For any $n \geq m > j$

$$< \dot{y}_{nm}, \varphi(t)v_j > + < A(y_{nm}), \varphi(t)v_j > = < Bu_{nm}, \varphi(t)v_j > .$$

Integrating by parts we get

$$-\int_0^T < y_{nm}, \dot{\varphi}(t) v_j > dt + \int_0^T < y_{nm}, \varphi(t) A^* v_j > dt =$$

$$= (y_{nm}(0), \varphi(0) v_j) + \int_0^T < B u_{nm}, \varphi(t) v_j > dt.$$

Taking limits as $(n, m) \to +\infty$

$$\begin{aligned} & -\int_0^T < y, \dot{\varphi} v_j > dt + \int_0^T < A(y), \varphi v_j > dt = \\ & \quad = (y_0, \varphi(0) v_j) + \int_0^T < B u_0, \varphi v_j > dt. \end{aligned} \tag{13}$$

If $\varphi \in C_0^1(0, T)$,then for all j

$$\frac{d}{dt} < y, v_j > + < A(y), v_j > = < B u_0, v_j >$$

in the distributional sense. This implies

$$\dot{y} + A(y) = B u_0$$

and $\dot{y} \in L^2(0, T, V^*)$, so that we can integrate by parts in (13) obtaining $y(0) = y_0$. From now on the proof is the same as before.

We remark that, by using special bases v_j as the eigenfunctions basis in $V = H_0^1(\Omega)$ given by

$$(v_j, x)_{H_0^1(\Omega)} = \lambda_j (v_j, x)_{L^2(\Omega)}, \quad x \in H_0^1(\Omega),$$

a priori boundedness of y_{nm} in $W^p(0, T)$ is known ([9], chapter 1, section 6.3).

Theorem 4. *Assume that $D = E$ and the imbedding of V in H is compact. Then the conclusions of theorem 3 are obtained under the same assumptions thereof except affinity of s.*

PROOF. From the proof of Theorem 3 we know that

$$y_{nm} \rightharpoonup y \text{ in } W^p(0, T).$$

The imbedding of $W^p(0,T)$ in $L^p(0,T,H)$ is compact ([9], Theorem 5.1 p.58), therefore for a subsequence and a.e. $t \in (0,T)$

$$y_{nm}(t) \to y(t) \text{ in } H.$$

Taking limits (for any fixed j) as $(n,m) \to +\infty$ in

$$(s[y_{nm}(t)], z_j) = 0$$

we get the required conclusion, i.e.

$$s[y(t)] = 0, \text{ all } t \in [0,T]$$

since y is an H-valued continuous function. □

Theorem 5. *Assume $D = V$ and B completely continuous. Then the conclusions of Theorem 3 hold.*

PROOF. By complete continuity, $Bu_{nm} \to Bu_0$ in V^*. As in the proof of Theorem 3, by estimate (1.17) p.159 of [9],some subsequence of $y_{nm} \in W^p(0,T)$ satisfies

$$y_{nm} \rightharpoonup y \text{ in } L^p(0,T,V), \quad y_{nm} \rightharpoonup y \text{ in } L^\infty(0,T,H).$$

Then we may follow the proof of Theorem 1.1 in [9] (p.159-161), taking advantage from the strong convergence of Bu_{nm} ($= f$ in the notation of that proof). □

We remark explicitly that p-weak continuity of A is not required in the above theorem. The last convergence theorem deals with a non linear control objective of special form defined on V.

Theorem 6. *Given r,q let $h_1, \ldots, h_r$ any fixed elements of H, $s_1 \in C^0(R^n, R^q)$, and*

$$s(v) = s_1[(v,h_1), \ldots, (v,h_r)], v \in V.$$

Under the assumptions of Theorem 3 (except affinity of course) and uniform boundedness of v_j in V, the conclusions of Theorem 3 hold.

PROOF. We use notations as in the proof of Theorem 3. By Lemma 1, $|y_{nm}(t)|$ is uniformly bounded in $[0,T]$. Therefore $(y_{nm}(\cdot), v_j)$ is uniformly bounded. If $0 \leq t' < t'' \leq T$ then (for any j and n,m sufficiently large)

$$\int_{t'}^{t''} \frac{d}{dt}(y_{nm}, v_j)dt = (y_{nm}(t'') - y_{nm}(t'), v_j) = \\ = \int_{t'}^{t''} < Bu_{nm} - A(y_{nm}), v_j > dt$$

thus giving

$$| y_{nm}(t'') - y_{nm}(t') | \leq k_1(t'' - t')^{1/p}\left[\left(\int_{t'}^{t''} \|u_{nm}\|^{p'} dt\right)^{1/p'} + \right. \\ \left. + \left(\int_{t'}^{t''} \|y_{nm}\|^{p} dt\right)^{1/p'}\right] + k_2(t'' - t')$$

for suitable constants k_1,k_2 (remembering (4) and Lemma 1). Therefore a diagonal selection gives us,for a subsequence and any j,

$$(y_{nm}(\cdot), v_j) \to (y(\cdot), v_j)$$

uniformly in $[0,T]$.

As a matter of fact this holds for the original subsequence y_{nm}. By density of $sp(v_1, \ldots, v_n, \ldots)$ in H and continuity of s_1 we get the conclusions. □

Remark. Further convergence results may be obtained by exploiting boundedness of Filippov solutions to (7) corresponding to discontinuous feedback control laws,and growth estimates for the equivalent control of (7),under conditions implying equivalence between Filippov and (a.e) solutions to (7) (as given in Theorems 1,2 of [5]).

Remark. An existence theorem for solutions to a viability problem when A is linear is proved in [14] by a convergence procedure somewhat similar to that used in the present paper (without reference to control problems).

A direct solution and some related difficulties. A formal calculation, analogous to that used in finite-dimensional setting (see [1]) gives us the following direct "solution" to (1),(2). If u, y solve the above control problem, then at least formally

$$s'[y(t)][Bu(t) - A(y(t))] = 0, 0 \leq t \leq T. \tag{14}$$

If moreover $s'(y)B$ is an isomorphism for all y near the sliding manifold, then an "equivalent control" could be defined as in [1], i.e.

$$u^*(y) = [s'(y)B]^{-1}s'(y)A(y).$$

We did not follow such an approach for the following reasons. Condition (14) requires that for all y with $s(y) = 0$, $s'(y)$ may be extended to the whole of V^* as a linear bounded mapping.This is sometimes a restrictive condition. More important,the isomorphic character of $s'(y)B$, required for a direct definition of the equivalent control in the infinite-dimensional setting, may be restrictive assumption (even false,for example when Z is finite dimensional). The existence of $(s'(y)B)^{-1}$ is (probably) related to stronger convergence of approximate finite-dimensional states y_{nm} obtained by solving (7), and to the approximability property to be introduced in the next section.

It would be interesting to relate the closed-loop system

$$\begin{cases} \dot{y} + A(y) = B(s'(y)B)^{-1}s'(y)A(y) \\ y(0) = y_0 \end{cases}$$

to discontinuous feedback injected in (1),(2) (through an appropriate definition of solution: it is not known whether the definition introduced in [18] fits this framework).

Approximability. Following the (finite - dimensional) terminology of [1],the ideal sliding modes of the control system (1), (2) should be closely approximated by real states of the system (i.e. states fulfilling only approximately the sliding condition (2)). This physically relevant property of a variable structure control system has been discussed in depth in [1], and precisely formalized in [4], [5] in the finite-dimensional setting under the name of approximability. As shown there, it is related to G-convergence (see [16], [17]) of differential equations (a theory started by De Giorgi).

We extend the definition of such a property to infinite-dimensional variable structure control systems (1), (2) as follows (avoiding any notion of equivalent control).

Definition. *The control system (1),(2) fulfils the* approximability property *iff*

$$\dot{y}_n + A(y_n) = Bu_n, \quad n = 1, 2, 3, \ldots,$$
$$u_n \textit{ admissible controls}, \; s(y_n) \to 0$$

uniformly in $[0,T]$ *as* $n \to +\infty, y_n(0) \to y_0$ *in* H*, imply the existence of* $y \in W^p(0,T)$ *and of an admissible control* u *such that* u,y *solve (1),(2) and for the original sequence*

$$y_n \to y \textit{ in } L^p(0,T,H).$$

Remark. In the finite-dimensional setting this definition reduces to that given in [4] or [5] if $s'(y)B$ is non singular for all y near the sliding manifold,

$$\left| \frac{d}{dt} s(y_n(t)) \right| \le M(t) \text{ for some } M \in L^q(0,T), \; q > 1,$$

$$|A(x)| \le c' + c''|x| \text{ for all } x.$$

It is easily seen that the approximability property is equivalent to each of the following:

$$\begin{array}{ll} \text{(16)} & u_n, y_n \text{ as in the above definition, } u, \, y \text{ a solution to (1),(2),} \\ & y_n(0) \to y(0) \text{ in } H \text{ imply } y_n \to y \text{ in } L^p(0,T,H) \end{array}$$

$$u_i, y_i \text{ a solution to (1),(2), } i = 1,2, y_1(0) = y_2(0) \text{ imply } y_1 = y_2, \tag{17}$$

under the following assumption

(18) A is p-weakly continuous, $s \in C^0(E, Z)$ and the imbedding of V into H is compact (without requiring (6)).

Therefore when A is linear and s is affine with linear part s_0, then approximability holds if V is compactly imbedded in H and

$$\dot{z} + A(z) = Bu, \; z(0) = 0, s_0(z) = 0, \; u \in L^2(0,T,W) \text{ imlpy } z = 0.$$

By (17) we see that approximability is a form of (essential) uniqueness property for the given control system, as far as the sliding behavior is concerned. Then the underlying compactness for solutions to (1),(2) gives the desired convergent behavior of real states to ideal ones.

It is known (see [1], p.45) that approximability holds in the finite - dimensional setting if the control variable enters affinely in the dynamics (and more generally for many non linear control systems, see [4],[5], but not for all, see a counter-example in [1], p.64).

The next theorem generalizes this fact to infinite-dimensional control systems (1),(2) (under somewhat stringent assumptions). Let us set

$$L(y) = B[s'(y)B]^{-1}s'(y).$$

Theorem 7. *The control system (1),(2) fulfils the approximability property if $D = V$, (18) holds, $s'(y)$ can be extended as a bounded linear operator from the whole V^* into Z for all $y \in V$, and moreover*

$$s'(y)B \text{ is an isomorphism for all } y \in V; \tag{19}$$

$$< A(y) - A(z), y - z > \geq \alpha \|y - z\|^p, \; \alpha > 0, \; p > 1; \tag{20}$$

$$\|L(y)A(y) - L(z)A(z)\| \leq c\|y - z\|^{p-1}, \; 0 < c < \alpha, \text{ with } L(0)A(0) = 0. \tag{21}$$

PROOF. Let us verify (17). From (19) we see that y_1, y_2 (perhaps modified on a set of Lebesgue measure 0, see [12] chapter 1, sections 3 and 4) are both solutions to (15), which may be written as

$$\dot{y} + f(y) = 0, \quad y(0) = y_0$$

where

$$f(y) = A(y) - L(y)A(y)$$

satisfies the assumptions of Theorem 1.2 p.162 of [9], which entails uniqueness in [0,T]. □

Remark. Assuming (19), Theorem 7 applies if A is bounded, linear, strongly monotone map and s is continuous affine on V^* (taking of course $p = 2$). Different assumptions may be placed on AL in order to obtain uniqueness for (15) (see e.g. [13]).

Remark. The above theory applies if A is as in Example 1. In particular, linear control systems (1),(2) are covered. More general control systems as

$$\dot{y} + A(t, y) = B(t)u + f(t)$$

may be treated under analogous assumptions.

Remark. Weaker coercivity conditions on A than

$$< Av, v > \geq \alpha ||v||^p, v \in V, \alpha > 0 \tag{22}$$

may be required to obtain the above results.

If $p = 2$ and in Theorem 3 we replace (22) by

$$< Av, v > \geq \alpha ||v||^p - \alpha_1 ||v||; \alpha, \alpha_1 > 0; v \in V$$

then the conclusions hold (with the same proof, since (1.17) p.159 of [9] is now replaced by a similar estimate which still implies boundedness (and weak convergence) of y_{nm} in both $L^2(0, T, V)$ and $L^\infty(0, T, H)$, as easily verified).

Example. The following control system

$$\begin{cases} y_t = y_{xx} + u, \\ y(0,x) = y_0(x), \\ y_x(t,0) = y_x(t,1) = 0; \ 0 \le t \le T, \ 0 \le x \le 1; \\ s(y) = \int_0^1 y(\cdot, x)dx - b \end{cases}$$

(where b is a given constant) lacks approximability (since (17) fails).

Remark. Null controllability results using the variable structure control theory developed in this paper, together with a more general convergence theory and appropriate examples, will presented elsewhere.

References

[1] V.I.Utkin, *Sliding modes and their application in variable structure systems,* English translation, MIR Publishers, Moscow, 1978.

[2] V.I.Utkin, *Variable structure systems. Present and future,* Automat. Remote Control. **44** (1984), 1105-1120.

[3] G.Bartolini, T.Zolezzi, *Non linear control of variable structure systems,* Lecture Notes in Control and Information Sci. **62**, part.1 (1984), 542-549.

[4] G.Bartolini, T.Zolezzi, *Variable structure systems non linear in the control law,* IEEE Trans. Automat. Control. AC-**30** (1985), 681-684.

[5] G.Bartolini, T.Zolezzi, *Control of nonlinear variable structure systems,* J. Math. Anal. Appl. **118** (1986), 42-46.

[6] G.Bartolini, T.Zolezzi, *Dynamic output feedback for observed variable structure control systems,* Systems Control Letters **7** (1986), 189-193.

[7] J.J.E.Slotine, *The robust control of robot manipulators,* Internat. J. Robotics Research **4** (1985).

[8] V.J.Orlov, V.I.Utkin, *Use of sliding modes in distributed system control problems,* Automat. Remote Control **43** (1982), 1127-1135.

[9] J.L.Lions, *Quelques méthodes de résolution des problèmes aux limites non linéaires,* Dunod and Gauthier-Villars (1969).

[10] D.Giachetti, *Controllo ottimale in problemi vincolati,* Boll. Un. Mat. Ital. **2B** (1983), 445-468.

[11] J.P.Aubin, A.Cellina, *Differential inclusions,* Springer, 1984.

[12] V.Barbu, Th.Precupanu, *Convexity and optimization in Banach spaces,* Ed. Academiei-Sijthoff and Noordhoff, 1978.

[13] H.Brézis, *Operateurs maximaux monotones,* North-Holland, 1973.

[14] Shi Shuzhong, *Viability theory for partial differential inclusions,* preprint. See also *Théorèmes de viabilité pour les inclusions aux dérivèes partielles,* C. R. Acad. Sci. Paris **303** (1986), 11-14.

[15] N.H.Pavel, *Differential equations,flow invariance and applications,* Pitman Res. Notes in Math. **113** (1984).

[16] E.De Giorgi, G.Buttazzo, *Limiti generalizzati e loro applicazione alle equazioni differenziali,* Le Matematiche **36** (1981), 53-64.

[17] E.De Giorgi, *Convergence problems for functional and operators,* Proceedings International Meeting on Recent Methods in Nonlinear Analysis, Rome 1978, edited by De Giorgi-Magenes-Mosco, 131-188, Pitagora, 1979.

[18] E.De Giorgi, M.Degiovanni, A.Marino, M.Tosques, *Evolution equations for a class of nonlinear operators,* Atti Acc. Naz. Lincei, Rend. Cl. Sci. Fis. Mat. Nat. **75** (1983).

Dipartimento di Matematica
Università di Genova
Via L.B.Alberti,4
I-16132 GENOVA

Progress in Nonlinear Differential Equations and Their Applications

Editor
Haim Brezis
Department of Mathematics
Rutgers University
New Brunswick, NJ 08903
U.S.A.
and
Département de Mathématiques
Université P. et M. Curie
4, Place Jussieu
75252 Paris Cedex 05
France

Progress in Nonlinear Differential Equations and Their Applications is a book series that lies at the interface of pure and applied mathematics. Many differential equations are motivated by problems arising in diversified fields such as Mechanics, Physics, Differential Geometry, Engineering, Control Theory, Biology, and Economics. This series is open to both the theoretical and applied aspects, hopefully stimulating a fruitful interaction between the two sides. It will publish monographs, polished notes arising from lectures and seminars, graduate level texts, and proceedings of focused and refereed conferences.

We encourage preparation of manuscripts in some such form as LaTex or AMS TEX for delivery in camera ready copy, which leads to rapid publication, or in electronic form for interfacing with laser printers or typesetters.

Proposals should be sent directly to the editor or to: Birkhäuser Boston, 675 Massachusetts Avenue, Suite 601, Cambridge, MA 02139.